Studies in Logic
Volume 84

Witness Theory
Notes on λ-calculus and Logic

Volume 74
Dictionary of Argumentation. An Introduction to Argumentation Studies
Christian Plantin. With a Foreword by J. Anthony Blair

Volume 75
Theory of Effective Propositional Paraconsistent Logics
Arnon Avron, Ofer Arieli and Anna Zamansky

Volume 76
Argumentation and Inference. Proceedings of the 2nd European Conference on Argumentation. Volume I
Steve Oswald and Didier Maillat, eds.

Volume 77
Argumentation and Inference. Proceedings of the 2nd European Conference on Argumentation. Volume II
Steve Oswald and Didier Maillat, eds.

Volume 68
Logic and Philosophy of Logic. Recent Trends in Latin America and Spain
Max A. Freund, Max Fernández de Castro and Marco Ruffino, eds.

Volume 79
Games Iteration Numbers. A Philosophical Introduction to Computability Theory
Luca M. Possati

Volume 80
Logics of Proofs and Justifications
Roman Kuznets and Thomas Studer

Volume 81
Factual and Plausible Reasoning
David Billington

Volume 82
Formal Logic: Classical Problems and Proofs
Luis M. Augusto

Volume 83
Reasoning: Games, Cognition, Logic
Mariusz Urbański, Tomasz Skura and Paweł Łupkowski, eds.

Volume 84
Witness Theory. Notes on λ-calculus and Logic
Adrian Rezuş

Studies in Logic Series Editor
Dov Gabbay   dov.gabbay@kcl.ac.uk

Witness Theory
Notes on λ-calculus and Logic

Adrian Rezuş

© Individual author and College Publications, 2020
All rights reserved.

ISBN 978-1-84890-326-5

College Publications
Scientific Director: Dov Gabbay
Managing Director: Jane Spurr

http://www.collegepublications.co.uk

All rights reserved. No part of this publication may be reproduced, stored in a retrieval system or transmitted in any form, or by any means, electronic, mechanical, photocopying, recording or otherwise without prior permission, in writing, from the publisher.

Foreword

§ 1 *The 'grammar' of proving.* This book is concerned with the *main concept of the science of logic*, the concept of (a) *proof.* Putting things colloquially, it is about proof theory handled *exclusively* by λ-calculus methods.

The objects of study in this kind of enterprise are what I am calling, here – following colloquialisms going back to L. E. J. Brouwer (1881–1966) –, *witnesses.*

Roughly speaking, a witness is just a *logical proof* of a tautology (a valid first-order formula, for instance) in a given (proof-) context.

In order to be able to write down witnesses, we need a 'witness-language', i.e., an appropriate *proof-formalism.* We have at hand specific extensions of the 'pure' λ-calculus for the purpose.

In its simplest variant, the calculus has a single 'syntactic category', *the λ-terms.* The terms of the calculus – called *witness-* or *proof-terms*, here – make up a recursive set supposed to be generated from *atomic data*, including variables – called, here, *witness-* or *proof-variables* –, with the help of appropriate *operators*, called *witness* or *proof operators.* This is, more or less, as in traditional algebra, except for the fact that one has also at hand *abstraction-operators* (i.e., syntactical *variable-binding* mechanisms) for the purpose, thereby inducing a distinction between *free* and *bound variables.*

Furthermore, the witness-terms are supposed to be *decorated.* The association *(witness-) term ≃ decoration* is also recursive, and 'rigid', so to speak, in the sense that each witness-term has a *unique decoration*, given by appropriate inductive clauses ('rules', also called 'typing rules' or 'type-assignment rules'). Formally, the decorations – called also 'types' sometimes – are just the formulas of a given logic (first-order classical logic, for instance), so that the objects of concern are actually pairs (a,A), where a is a λ-term and A is a logic formula. In the end, we have a bi-dimensional (recursive) grammar, generating separately decorations (i.e., 'types' = formulas), on the one hand, and witness-terms, on the other.

The *intended interpretation* of the resulting formalism consists of saying that a pair (a:A) represents the fact that the term a is a witness of its (rigidly) associated decoration A (i.e., that it is a proof of A). If the witness-term a is 'correctly constructed' – according to the said 'rules' – we use a turnstile for emphasis and write ⊢ a : A to specify this (reading 'a is a witness for A').

Since a witness-term a may contain free variables, witnessing is relative to a set of (decorated) witness-variables, namely the free variables contained in a. In this setting, a (decorated) witness-variable is supposed to represent a *proof assumption*, while a finite sequence $\hat{\Gamma}$ of (decorated) witness-variables is ('represents') a *witness-* or a *proof-context.* In general, where $\hat{\Gamma} \equiv [x_1{:}A_1],...,[x_n{:}A_n]$, $a \equiv a[x_1{:}A_1,...,x_n{:}A_n]$, for $n \geq 1$, if a is a ('correctly constructed') witness for A, in the (proof-) context $\hat{\Gamma}$, we can write this explicitly as $\hat{\Gamma} \vdash a : A$.

At this point, the reader may want to notice the fact that the explicit notation is rather redundant, because the (proof-) context can be retrieved from the (syntactic) structure of the witness a. As a matter of fact, if $\hat{\Gamma} \vdash a : A$, for some $\hat{\Gamma}$, possibly empty, i.e., if a is 'correctly constructed', then the formula A, thereby 'witnessed', can be also 'derived' from the (syntactic) structure of the witness-term a.

In other words, *proving* amounts to *witness-term construction*, while *proof-checking* consists, essentially, of *establishing the 'wellformedness' of a witness-term*, according to the 'rules of term-construction', a purely 'grammatical' affair, so to speak.

Typically, the witnesses of the tautologies (logically valid formulas) are going to be represented by closed λ-terms, also called *combinators*, in jargon, and – as the witness-variables are supposed to represent *assumptions* (in a concrete proof) – the witness operators are representing, *in abstracto*, so-called *logical rules of inference*.

The 'witness grammar' described generically in the above allows also defining *substitution in witness-terms*. This concept is to be defined recursively, again, 'by induction on the structure of a witness-term', and amounts to the (global) CUT-rule (*Schnitt*), a so-called 'structural' rule of inference in Gerhard Gentzen's *Inauguraldissertation* (1934–1935), or else to the specific 'transition' (*Übergang*) called *Schluss* in Frege's **Grundgesetze der Arithmetik** (1893).

If *non-logical* (e.g., arithmetical, or else, in general, 'mathematical') symbols / constants are used explicitly in the (proof-) context $\hat{\Gamma}$, then the witness is ('represents') a proof of a *mathematical theorem*.

In fact, this is the usual concern of the logician, except, perhaps, for the fact that one is confronted, here, with a *different level of abstraction*.

Moreover, unlike in the usual presentations of (a) logic, the witness operators are supposed to be *defined explicitly* by 'characteristic' *equational conditions*, as in algebra.

Alternatively, in this setting the logical rules of inference are represented by *open* λ-terms, i.e., by λ-terms containing free (decorated / 'typed') variables.

A *witness theory* is just a consistent (i.e., non-trivial) *equational theory* on witnesses. In colloquial, informal parlance (like, e.g., in the title of this book), the *witness theory* is the resulting theoretical 'discipline'. Practically, the latter amounts to a specific set of exercises in λ-calculus [*sic*].

As expected, the equations involved in this kind of (formal) game are representing *proof-isomorphisms*.

> With a simile from category theory, the 'art déco' part of a given witness theory (understood as an equational theory) is very much similar to the object part in a given category (to its 'graph'), while the proof-isomorphisms would match the 'characterisation' of the morphisms, given usually by so-called 'commuting diagrams' in categories.

In view of the equational ('algebraic') component of the approach, one can, even, ignore the decorations (i.e., the logic formulas themselves) and study such equational theories in a 'type-free' régime. In fact, we *define* the witness operators by ensuring first the [Post-] *consistency* – i.e., the non-triviality – of the corresponding 'type-free' λ-theories (cf. the Part II of the book).

In the case where the underlying logic contains (propositional, first-, second- or higher-order) quantifiers, the corresponding witness-syntax is slightly more involved: one has an additional 'syntactic category' of *scalars* (such that the scalars and the witness-terms are disjoint) and additional *scalar witness / proof operators*.

For the first-order case, for instance, the scalars are meant to represent 'individual terms' **t**, whereas the corresponding decorations / 'types' are first-order formulas, to be generated inductively from atomic formulas P[**t**], where P is a first-order predicate and **t** := [t_1,...,t_n], n > 0, is a finite sequence of individual terms. In the case of the (first-order) classical logic, the universal quantifier [∀] is sufficient, and its associated witness / proof operators – Λ and ▶, correspoding to Universal Generalisation and Instantiation, resp. — are analogous to the witness / proof operators of the purely implicational fragment of the propositional, quantifier-free minimal calculus of Johansson, λ and ▷, correspoding to the 'deduction theorem' – viewed as a rule of inference – and to the rule of detachment (*modus ponens*) resp. Equivalently, the abstraction operators λ and Λ can be replaced by a finite number of primitive combinators (i.e., closed λ-terms in the extended syntax), so that we end up with a purely *algebraic setting*, in the traditional sense (i.e., we have *equational theories* or *algebraic varieties* à la Birkhoff-Mal'cev).

This being said, the reader might eventually realise that the abstract witness-theoretical setting described here corresponds to what (s)he already knew, long ago, from her / his logic textbooks, for the various logic-presentations (via axiomatic, 'natural deduction', sequent and / or tableaux, resp. refutation systems, etc.) are, in the end, nothing but *elliptic, underspecified* ways of dealing with witnesses and witness operators: they correspond to witness theory done *by hand-waving*, so to speak.

Otherwise, the approach described schematically here is relatively well-known – mainly since the early nineteen seventies, from the literature on the proof theory of intuitionistic logic – under the ad hoc label 'formulas-as-types', resp. as *Curry-Howard Correspondence* (or even *Isomorphism*) [CHC, for short].

The fact that there is hardly anything *intuitionistically specific* in this kind of approach has been realised only in recent times, however, during the nineteen eighties or so.

To put things in slightly different terms, witness theory is the (*abstract*) *study of witness operators* or else of *proof-isomorphisms*, for that matter.

A specific derivative of this enterprise concerns the study of *proof complexity*, in terms of appropriate *notions of reduction*, as usually understood in λ-calculus, a subject touched incidentally upon in this book (see, e.g., the section on *proof-détours*, in Part II).

On the *technical* side, the *departure* from the usual (theoretical) concerns with λ-calculus consists of the fact that the basic formalism to start with – in this book – is *not* the 'pure' λ-calculus (vastly studied nowadays, mainly in view of its alleged 'applications' in computing, via the somewhat superficial

viii

connection (λ-*definability*; cf. Barendregt 1981, 1984², Rezuş 1981) with the theory of *general recursive functions* and the concept of *Turing computability*), but, rather, a *proper equational extension* of it, known as *'extended λ-calculus'* or as *λ-calculus with 'surjective pairing'* (**λπ**, in jargon).¹

The basic idea of the book relies on the observation (Rezuş, cca 1985–1986) that, under an appropriate decoration ('typing'), the λπ-calculus is already a *witness theory* – i.e., a 'proof-system' – for *classical logic*.

Algebraically, the extended calculus is equationally equivalent to a *Lambek C-monoid* (Lambek & Scott 1986), a fall-out from the study of *cartesian closed categories* (CCC's, for short).² In view of this, the *abstract proof theory of classical logic* turns out to be a proper *alegbraic subject*, and the slogan to be retained consists of saying that – qua logicians – we are, essentially, *dealing in monoids*, a rather weird and specific variety thereof, indeed.³

§ 2 *The structure of the book.* The notes contained in this volume have been written during the last ten years (2009–2019) and thus, save for the inevitable backward and/or cross-references, one can read them as separate essays. Personally, I'd advise the *chronological order*, as appearing in the references, and as mentioned explicitly in the titles of the chapters listed in the Table of Contents, but the reader can choose her own way into the book, according to casual moods, interests or preferences.

The book is organised in a *topical* (i.e., more or less, *logical*) order.

Part I covers the *epistemic background* to witness theory, in historical terms.

Part II consists of a *technical introduction* to the subject, and contains a survey of the λ-calculus prerequisites, in a general setting.

Part III is rehearsing, in a way, nearly the same matter in *historical order*, tracing back – in witness-theoretical terms – the *origins of modern proof theory* to the (great-) grandfathers of the discipline: Chrysippus, Frege, the Polish pioneers (Łukasiewicz, and Jaśkowski), and Gentzen (mainly the *Nachlass*, pub-

---

¹ Actually, the 'mathematical models' of the ('pure') λ-calculus (Dana S. Scott 1969; cf., e.g., Scott 1973) are already models of **λπ**.

² Scott's D_∞-construction (Scott 1973) ensures the fact that *there are* such things like Lambek C-monoids (so, one should probably speak about *Lambek-Scott monoids*, instead). Surprisingly, one can also establish this (i.e., the consistency of **λπ**) by 'constructive means', i.e., 'syntactically', using an appropriate conservativity argument (Støvring 2006).

³ Honestly speaking, **λπ** – or else the theory of *cartesian folds* epitomised in the Part II of this book – accounts for the *quantifier-free* segment of classical logic alone. As mentioned in the above, however, the accommodation of the quantifiers, in this setting – via *scalars* – is straightforward, and can be done in so-called 'type-free' terms, as well, i.e., by ignoring the decorations. This in contrast with various 'constructive type-theories' advertised currently on the (logical) market. Otherwise, the 'constructive' label – attached emphatically to this kind of endeavours – is, rather, a mere propagandistic *quid pro quo*, inviting *conceptual confusions*, because – no need to quote Hilbert on this – the (meta-) mathematics of *classical proof theory* has been always meant to be *constructive*.

lished recently [2017]), and on the specifics of the *Curry-Howard Correspondence* [CHC], as applying to classical logic.

Finally, Part IV is focused on a typical application of witness theory to a practical logic problem (axiomatisability via singleton bases), a subject derived, essentially, from Leśniewski and Tarski. It consists of a longer gloss on a couple of (meta-) theorems in pure λ-calculus proved by Alfred Tarski around 1925 (i.e., before the [pure] λ-calculus was officially born), as well as on my first published paper (Rezuş 1982).

The last Part (V) is meant to provide some *Guidelines* to the reader. Chapter 18 – to be used in conjunction with the *Index of subjects* – contains a *synopsis* of the *concepts* and *notations* used most frequently in the book. The Bibliography includes also a guide – a kind of friendly *Vademecum* (Chapter 19) – to the relevant literature, updating – selectively though – the previous BHK-bibliography of Rezuş 1991, 1993^R (last revised as of 2000).

Given the character of the book, some of my friends and colleagues advised to include as many *cross-references* as possible, in order to facilitate the reader's access to individual topics. This explains why I have casually provided comments in the cumulative list of references. As there was no point in mentioning the author's name in the global *Index of names*, I prepared a separate index containing (inner) cross-references to the various Chapters of the book, as well as to a part of my previous publications and/or prepublications not included here but still relevant to the subject.

§ 3 *Acknowledgements.* In the long run, in matters of institutional, financial, and other kind of help, my debts go to several Dutch academic institutions, mainly to the Universities of Utrecht, Eindhoven, and Nijmegen, and to the Dutch National Science Research Foundation [NWO, The Hague], that have supported initial segments of my research on the subject.

On logic and λ-calculus matters, I have been indebted, along the years, in various ways to several persons, including Henk Barendregt, Nuel Belnap Jr., Corrado Böhm [†], Nicolaas G. de Bruijn [†], Alonzo Church [†], Dirk van Dalen, Branden Fitelson, John Halleck, J. Roger Hindley, David Meredith [†], Robert K. Meyer [†], Willem L. van der Poel, V. Frederick Rickey, Dana S. Scott, Jonathan P. Seldin, Anne S. Troelstra [†], *et al.*

Stylistical improvements are due to J. Roger Hindley, and to Lloyd Humberstone, whereas Mark van Atten, Afrodita Iorgulescu, and Tyler Burge have gently supplied last minute bibliographical updates for G. Kreisel, G. C. Moisil, and A. Church, resp.

On the editorial side – including the LATEX-typsetting – I have relied on the expert advice of Jane Spurr (College Publications, London).

Nijmegen, 15 November 2019 – 17 February 2020

Contents

Foreword v

Part I The epistemic background

1 *...en avançant dans le brouillard...* (2019d) 3
  § 1 The Formularian Age 3
  § 2 'The unreliability of the logical principles' 5
  § 3 The formularian reshaping of Brouwer's 'logic' 8
  § 4 A shift of paradigm: witness theory 9
  § 5 The Kolmogorov interpretation of witnessing 10
  § 6 'Traditional' versus 'mathematical' logic in historical prospect ... 12
  § 7 On seeing out of the fog 13
  § 8 Polish new formularian standards and anomalies 16
  § 9 *Vergessenheit* 17

2 What is a 'challenging problem'? (2010a) 19
  § 1 'Service to humanity' 19
  § 2 Wonder and challenge 21
  § 3 Epistemic *vs* technical challenge 25

Part II Background to Witness Theory

3 Cartesian folds and Witness Theory (2017b) 33

4 Witness theory for inferential systems (2017a) 55

5 On the Kolmogorov $\lambda\gamma$-calculus (2017c) 65

6 Proof-détours in classical logic (2017e) 69

7   Appendix: Does (η) really matter? (2017f) 73

Part III The origins of modern proof theory

8   **An ancient logic (2016)** 79
    Foreword 80
    § 1 Chrysippean logical grammar 81
        Propositions and entailments............................. 81
        Rejections / refutations / (logical) conflict 83
        Atomic propositions and the Chrysippean negation 83
        Proper atoms 84
        Atomic constants 84
        Complex propositions 84
    § 2 Chrysippean proof semantics 88
        Analysis and polar projections 88
        The Chrysippean logic as a rejection / refutation system 88
        Dilution and the fallacies of relevance 90
        A 'dialectical theorem' 92
        Proper logical rules 93
        Chrysippus' logic is classical logic........................ 94
        Derivable rules 95
        A historical aside 96
    § 3 Redundancy, fibrations and formal systems 97
        Syntactic fibrations 98
        An Ancient Logic, Formulation 1 99
        An Ancient Logic, Formulation 2 99
    § 4 Alternative Formulations and witness theory 100
        Inferential Formulations 100
        Double-negation enrichments............................100
        Witness theory 101
    § 5 Bibliographical notes 101

9   **Im Buchstabenparadies (2009)** 103
    § 1 The Many Names of *The Truth*..........................104
        The Originary Symbolic Dichotomy (*Zeichenarten*) 104
        Ontology first 106
        Names, Fathers, and Sons 107
        Willkommen im Paradies................................110
    § 2 The Logic of *Grundgesetze*..............................115
        The New Rules of Logic 115
    § 3 Appendix. *The Name of a Rose*..........................124

10 'Begriffsschrift' as *Beweisschrift* (2019a) 127
   Introduction 127
   § 1 Frege's 'Begriffsschrift'-logic 130
   § 2 Restoring the logical constants: *verum* and *falsum* 144
   § 3 Frege and the Law of Non-Contradiction 153
   § 4 The equational proof-normalisation of Double Negation 157
   § 5 Extending the witness-theoretical picture 162
   § 6 Frege's quantifiers and Hilbert's *epsilons* 172

11 Łukasiewicz, Jaśkowski and natural deduction (2017) 181

12 The Curry-Howard Correspondence revisited (2019b) 209

Part IV Normal singleton bases

13 Tarski's Claim, thirty years later (2010) 217

14 Tarski singleton bases: 1925–1932 (2019) 227
   Foreword 227
   § 1 Prerequisites 229
   § 2 Non-organic singleton bases (1925–1932) 231
   § 3 Tarski's first singleton basis (1925) 234
   § 4 A Łukasiewicz singleton basis (1932) 237

15 Old folklore on λ-calculus generators (2015b) 241

16 Implications and generators (2019c) 245

17 Appendix: Bolesław Sobociński 1932, § 1 257

Part V Guidelines

18 Catalogue of concepts and notations 271

19 *Vademecum*: a guide to the bibliography 287

References 295

Internal cross-references 351

Index of subjects 353

Index of names 363

Part I

The epistemic background

1
...en avançant dans le brouillard... **(2019d)**

§ 1 The Formularian Age

To begin with the beginning, I shall be concerned, first, with an attempt to understand and assess critically a nearly omni-present *conceptual disease* of our times, originating, essentially, with Frege's **Begriffsschrift**-booklet (henceforth **BS**), published in 1879, i.e., about 140 years ago. This amounts to the fact we are living, in logic, in a *Formularian Age*.

The qualifier ('formularian') does not refer to the more or less programmatic 'formalism', in foundational logico-mathematical studies, or to the alleged 'formalists' – Hilbert and his fellow *Göttinger*, for instance – but, rather, to a side-effect of the pioneering approaches in modern logic.

Roughly speaking, in logic, a *formularian* is a person – involved in theorising (about) logic – who thinks, exclusively, in terms of formulas – as expressing propositions and/or propositional schemes – or in terms of sets and/or visualisable configurations of formulas and their properties (as, e.g., *provability*, *decidability*, and the like), not in terms of proofs, proof-operations and proof-properties. In plain terms, the formularian theorist is addressing proofs by proxy.

As a consequence, for a formularian thinker, a 'logic' is just a collection of provable (resp. valid or true) formulas, a set of tautologies, more or less (a wrong idea, in the end, because one can have distinct logics sharing exactly the same stock of provable formulas, but with different derivable rules of inference), while a rule of inference – in a given logic – counts as a kind of cookbook-like recipe, which is best understood if written down vertically, on a piece of paper – appealing thus, in 2D, to the eye – so that the corresponding meta-conditional, as well as the 'transition' from premisses to conclusion – i.e., the rule as such – are left *un-formalised*.[1] The qualifier has been suggested by Peano's **Formulaire**

---

[1] In traditional, philosophical – here: Kantian – terms, the formularian theorist is taking *Vorstellungen* for *Begriffe*. The appeal to *visualisation* and to *visual*

de mathématiques, a copious collection of formalised mathematical propositions, issued in several editions, in Italy and in France, during 1894–1908 (cf. Peano 1894, 1895–1903, 1908). Giuseppe Peano (1858–1932) was, next to the young – third Earl – Russell, one of the few competent readers of Frege's logical writings, and, among other things, an industrious propagandist for what he used to call 'mathematical logic', in Italian, in French and, even, in artificial, awfully un-grammatical Latin (*latino sine flexione* or *interlingua*)², later on. The term 'mathematical logic' made fortune in Britain and in continental Europe, due mainly to Bertrand Russell, and took over the alternative, rather confusing French term *logistique* and the colloquial appellation 'symbolic logic', popular mainly overseas, giving, ultimately, an official name to the new (formularian) enterprise for the mathematical community at large.³

Like Frege, Peano was *pasigraphically inclined* and fond of persuading people worldwide – his fellow Italian mathematicians included – that mathematical propositions can be transcribed in a convenient, formal notation, based, essentially, on Frege's 1897-booklet, although deprived of its, rather weird, 'Japanese' features.⁴

---

ingredients is best summarised by the *dictum* – a popular saying oft attributed to Georg Kreisel – '*I know what is a proof when I see one*'. *Prima facie*, this sounds deep and philosophical – mainly because Kreisel did also notice the fact that the so-called *Beweistheorie* of David Hilbert is, rather, a talk about *provability*, a property of formulas, i.e., neither a (mathematical) theory nor about proofs – but, on reflection – unless Kreisel had in mind his Mind's Eyes – the slogan is superficial and somewhat à la Wittgenstein (1956, 1976), who used to think a proof has to be *visually perspicuous* and *convincing* in order to be a proof.

² Cf., e.g. Peano 1903–1904, for a programmatic statement. A relevant sample (from Peano 1908): [*O*]*mne progessu de Mathematica responde ad introductione de signos ideographico vel symbolos* [...] *inter duo systema symbolico, illo que contine minore numero de symbolos es, in generale, plus perfecto*, etc. For the classically trained reader, this might sound like rephrasing *Divina Commedia* in substandard American English – or else in macaronic Latin ['dog Latin'], for that matter – but the mathematical pasigraphers of the time took this kind of enterprise seriously, in the long shadow of Leibniz, more or less, who pondered, occasionally, on Chinese ideograms in his search for a *characteristica universalis*. Notably, Giovanni Vacca (1872–1953), one of the assistants and close collaborators of Peano, 'a highly cultured mathematician from Genoa [...], [a] polyglot with solid knowledge of both classical and modern languages, [...] [a] refined bibliophile, [...] [and] an esteemed historian of the sciences in his days' was effective in the (re-) discovery of Leibniz's logic manuscripts (Luciano 2011), and taught, for a long period of time (1922–1947), Chinese Language and Literature in Florence (O'Connor & Robertson 1997).

³ Incidentally, Russel used to speak about 'symbolic logic', in his 1903, as a 'synonym for formal logic'.

⁴ Peano even devoted a mathematical journal – he managed to edit together with a few, otherwise respectable mathematical colleagues – to this kind of silly and quite aggressive advertising. For details, see, e.g., Grattan-Guinness 2000, **5**. For Frege's pasigraphical *penchants*, see Kreiser 2001, **§§ 3.1–3.3**. The reaction of the contemporary *mathematical* establishment to the pasigraphical endeavours of the

Subsequently, Bertrand Russell and David Hilbert, as well as Gerhard Gentzen, Kurt Gödel, and some Polish philosophers and mathematicians working in the area – as, e.g., Stanisław Leśniewski, Jan Łukasiewicz, Alfred Tarski, Mordchaj Wajsberg, Bolesław Sobociński, etc. – were *dedicated formularians*, and the trend went into a veritable, world-wide industry, mainly in continental Europe (Germany, Poland, Italy, and Romania), in Soviet Union, and, last but not least, in the United States.

> In this respect, Gödel is the paradigm: the proof of his famous (first) Incompleteness Theorem – of Peano Arithmetic [**PA**] – for instance, is a piece of pure formularian thinking.[5] Further, in spite of his explicit concern with inference rules and proof-complexity matters, Gentzen's Main Theorem (*Hauptsatz*) in his Göttingen *Inauguraldissertation* (1933, in print: 1934–1935) served, essentially, to yield a 'sub-formula property' – entailing casually consistency for first-order (intuitionistic and classical) logic – while Tarski even managed – about the same time, circa 1930, – to evacuate entirely the concept of (a) proof from logic, by using the (formularian) idea of (a) *consequence relation* – adapted from set-theoretic topology – still *en vogue*, with minor revisions, in our times.

As a matter of fact, the direct – and, in a way, more successful – inheritor of the formularian standpoint that has emerged from the pioneering work of Frege – as well as of that of Boole, Peirce and their followers, viz. the so-called 'algebraic logic' – is the modern (Tarskian) *model theory*.

On the other hand, a closely related – and rather benefic – side-effect of this kind of approach in logic consisted of the fact it caused mathematicians to pay attention to the more general idea of (an) *algebraic structure*.[6]

§ 2 'The unreliability of the logical principles'

A *conceptual break* with this kind of somewhat misleading trend in logic has been prompted by the Dutch mathematician Luitzen Egbertus Jan Brouwer (1881–1966), about a century ago (PhD Dissertation, Amsterdam 1907), for

---

early formularians – Peano and Russell included – is best illustrated by the sarcastic remarks of Henri Poincaré, in his 1905–1906, and 1906. (Honestly, however – and in spite of his indebtness to Frege and Peano – Russell himself was rather immune to ideographical and Latin para-linguistic excesses and preferred plain Middle English for formularian propagandistic purposes.) See also Detlefsen 1992, 2011–2102, and Grattan-Guinness 2000, *loc. cit.*, etc. On Peano, see Kennedy 1974, 1980, 2002. The relation Peano-Russell has been documented, e.g., in Kennedy 1973, 1975.

[5] One can prove the said meta-theorem by *proper* mathematical means, without making appeal to selfreferential liars and gödelisation.

[6] Notably, lattices were first called 'algebraic stuctures', a way of speaking that is still surviving in the modern Russian mathematical terminology. As an aside, for Grigore Moisil (1935) and Valery Glivenko (1938), for instance, (formularian) 'logic' and 'abstract' algebra were, more or less, the same thing.

whom logic was a piece of mathematics, and had to be justified mathematically – and explicitly so – in proper *proof-theoretic terms*, and, more specifically, in terms of *mathematical evidence* or else *mathematical witnesses*.

Like Henri Poincaré (1854–1912) in France, Brouwer himself had little esteem for the contemporary 'logistic' developments of the formularians, despised formalism and formalisation and used to think that logic has *empirical origins*, being rooted into our everyday, limited experience with *finite* data. This is why, he claimed, some of the traditional 'principles of logic' are 'unreliable' (Dutch: *onbetrouwbaar*), and thus unsuited in mathematics proper, mainly in our reasoning about the continuum.

Among the Brouwerian logical *unreliables* – in mathematics – one may count, for instance, the genuinely classical *tertium non datur*, the traditional reasoning by *reductio ad absurdum*, Cardano's *consequentia mirabilis*, half of the Double Negation 'Law', as well as half of the traditional Contraposition 'Laws', half of the so-called 'De Morgan Laws', etc. On the other hand, most of the Aristotelian syllogistic *modi* – although despicable, because absolutely trivial – were perfectly reliable for Brouwer. In particular, Brouwer trusted both the 'Law' of Identity and the 'Law' on Non-Contradiction [LNC], although he did not insist on formulating the latter one in exact terms.

> Brouwer's explicit interest in logic was rather negligible. He managed to prove – informally – a specific, proper substitution instance of the Double Negation 'Law', viz. the fact that triple negations are intuitionistically equivalent (as understood by his standards) to simple negations: *triplex negatio negat*. So, unlike $\neg\neg p \to p$, the tautology $\neg\neg\neg p \to \neg p$ was reliable for Brouwer, whence also the equivalence, $\neg\neg\neg p \leftrightarrow \neg p$, because the other half of the Double Negation 'Law' is, anyway, intuitionistically valid. In other words, Brouwer proved an instance of a more general meta-theorem found by Glivenko in 1928.

In fact, the reflection on the status of mathematical reasoning – mainly in (classical) analysis – was a concern for other mathematicians active by the turn of the previous century, Brouwer was not an isolated voice. Putting aside the negative reaction of Poincaré to *logistique*, already alluded to before, Brouwer had illustrious predecessors as Leopold Kronecker (1845–1918), in Germany, and Émile Borel (1871–1956), in France, to say the least.

In 1904, for instance, Jules Molk (1857–1914), a former PhD student of Kronecker and the editor-in-chief of the French Encyclopédie des sciences mathématiques pures et appliquées – based upon Felix Klein's encyclopaedia [Enzyklopädie der mathematischen Wissenschaften mit Einschluss ihrer Anwendungen] – summarised Kronecker's 'standpoint' on *reductio ad absurdum* in (classical) anaysis in quasi-Brouwerian terms, *avant la lettre*:

> [L]'évidence logique d'un raisonnement ne suffit pas pour légitimer l'emploi de ce raisonnement en Analyse. Pour avoir donné une démonstration mathématique d'une proposition, il ne suffit pas, par exemple, d'avoir établi que la proposition contraire implique contradiction. Il faut donner un procédé permettant d'obtenir, au moyen d'un nombre fini d'opérations arithmétiques

au sens ancien du mot, effectuées sur les éléments que l'on envisage, le résultat qu'énonce la proposition à démontrer. C'est ce procédé qui constitue l'essence de la démonstration; il ne vient pas s'y ajouter.[7]

Obviously, Brouwer did not see Molk's gloss (1904) on Pringsheim, while writing up his 1907, 1908a, otherwise he would have been happy to reproduce it in full, in order to support his own, more specific claims: Molk was a mathematical authority, not only in France, by then.

Although less inclined to go into logical details, Henri Poincaré was, by far, more vocal a defendor of the 'constructivistic' mathematical tradition against the new logico-formularian invasion than the young Brouwer.

Putting things in ad hoc ideological terms, the 'progress' was represented, in the epoch, by the formularian trend(s) in logic, while the mathematical 'constructivists' were, in a way, siding with the 'reactionary' *ancien régime*. It is only about two decades later that Brouwer and Weyl would count as revolutionary Bolsheviks[8], opposing explicitly the (old) 'constructivists' and the formularian reshaped mathematical establishement.[9]

---

[7] 'The logical evidence of a mathematical argument [*raisonnement*] is not sufficient in order to justify [*légitimer*] the use of this argument in Mathematical Analysis. While giving a mathematical proof of a [specific] proposition it is not enough to establish, for instance, the fact that its negation [*la proposition contraire*] implies a contradiction, we must give an explicit procedure [*procédé*], allowing to establish the result announced by the proposition one has to prove by means of finitely many arithmetical operations performed on the elements of concern, [where 'arithmetical operation' is to be] taken in the traditional [*ancien*] sense of the word. It is [exactly] this procedure that makes up the essence of the proof, this is not a mere additional ingredient.' (My translation, AR.) Cf. Molk & Pringsheim 1904, under *Nombres irrationnels*, [§] *10. Point de vue de L. Kronecker*, pp. 159–160, expanding copiously on Pringsheim 1898 [§ *8 Die volkommene Arithmetisierung im Sinne Kronecker's*, p. 58]. (The long, colloquial elaboration – credited to Kronecker, here – is, obviously, due to the Frenchman, otherwise a *connoisseur* in matters concerning his teacher's views.) See also Molk's PhD Dissertation, Molk 1885, *passim* (mainly pp. 3–5, 8), the comments of Mark van Atten *ad loc.*, in van Atten & Sundholm 2014, p. 16, and the introduction to the new English translation of Brouwer 1908a, by the latter two authors, in van Atten & Sundholm 2017, § *11 Precursors*, pp. 12–15. In retrospect, the Molk-quote above would nearly make Brouwer into a kind of plagiarist, if we ignore the fact that such ideas were rather common, by the end of the xix-th century, among mathematicians, both in Germany and in France.

[8] The term 'Bolshevik menace', in this context, comes from Frank Plumpton Ramsey (Ramsey 1925, p. 380, resp. Ramsey 1931, p. 56), a 'bourgeois thinker', according to Ludwig Wittgenstein.

[9] For the larger historical background, see Reid 1986, van Dalen 1999, 2000, 2001, 2005, and, possibly, Grattan-Guinness 2000, concerning the so-called 'logicists'. (The *batrachomyomachia* – frogs and mice battle, in Einstein's ironical terms – opposing personally Brouwer and Hilbert has been documented in van Dalen 1990.) For the history of mathematical 'constructivism' (in the xx-th century), see, e.g., the succinct survey Troelstra 2011.

§ 3 The formularian reshaping of Brouwer's 'logic'

The debate around the subject made rage amongst philosophers and mathematicians, mainly during the mid- and the late nineteen twenties[10], and ended up short with the *constat* 'Brouwer's logic' was nothing but a *proper fragment of classical logic* that can be, moreover, *formalised* by exactly the same standards as those advocated by Brouwer's – and Poincaré's – arch-ennemies, the formularians.

The principal contributors to the *formularian reshaping* of Brouwer's ideas were two Russian – by then already Soviet – mathematicians (Andreĭ Nikolaevič Kolmogorov 1925, 1932, and Valeriĭ Ivanovič Glivenko 1928, 1929) and a Dutchman, Arend Heyting (around 1928, in print: 1930), actually a former PhD student of Brouwer in Amsterdam. By default, and *faute de mieux*, the outcome was called 'intutionistic logic'.

> There were minor disagreements, as regards the way of understanding negation in Brouwer, so – in retrospect – one can distinguish among the official *Glivenko-Heyting formalisation* of 'logical intuitionism', and a proper formal subsystem of it, due mainly to Kolmogorov, and discussed at length, later on, by the Norwegian mathematician Ingebrigt Johansson (1936–1937), under the label *Minimalkalkül*. The alternative idea of abandoning altogether negation in intuitionism was not very appealing to the formularians, and has been dismissed subsequently by Brouwer himself (cf. Brouwer 1948).

This was, more or less, a Pyrrhic victory of the formularian faction, because things turned out considerably *more complex*, and because, moreover, both Kolmogorov and Heyting went on *justifying conceptually* the intuitionistic logic primitive connectives, in mathematical terms, as actually meant by Brouwer.

Summing up, on the one hand, Kolmogorov and Glivenko – as well as, slightly later, Gödel and Gentzen – noticed the fact that classical logic is not really *more* than the (formalised) 'intuitionistic' logic, because one can find *provability preserving translations* of the former into the latter, and, on the other hand, the mathematical attempt to justify (ideologically, more or less) the Brouwerian ideas on proof and proving in mathematics – the so-called *BHK-interpretation* (short for 'Brouwer-Heyting-Kolmogorov') of the intuitionistic formalism – disclosed a *very different way of approaching theoretical logic* as such. Putting things in plain terms, the logical rules of inference, vastly neglected in the formularian approach, had to be considered as *mathematical operators* (as proof or witness operators).

Ultimately, Brouwer's ideas were relevant far beyond the *parochial* logic subject he was actually advocating – i.e., the 'intuitionistic' logic – although they took more than a half of a century to reach – backwards – the formularian logico-mathematical establishment.

Practically, the explicit mathematical concern with proofs – and proof-properties – in logic goes back to the late nineteen sixties.

---

[10] Cf. e.g., Surdu 1976, 1977 and, more recently, the copious documentation collected in Hesseling's Utrecht PhD Dissertation 1999 (now Hesseling 2003).

The *formalisation* of the mathematical *proofs themselves* emerged into the picture in a more *pragmatic* context, with no explicit reference to Brouwer, and is also due to a Dutch mathematician, Nicolaas Govert de Bruijn [within the AUTOMATH project – short for 'automated mathematics' – developed at the Eindhoven University of Technology, in the Netherlands, since around 1966–1967], the actual founding father of *witness theory*. N. G. de Bruijn was a defendor of *classical mathematics*, his interest in Brouwer's doctrines, in intuitionism or in 'constructivist' mathematical trends was rather limited – if not even non-existent – while, ultimately, his ideas regarding proof-formalisation should be understood as a reaction to the formularian – Peano-Russell – tradition (cf. de Bruijn 1978, 1980, 1981, Rezuş 1983, Barendregt & Rezuş 1983). Characteristically, the first extensive example of a fully formalised mathematical proof-text ('book') in AUTOMATH [AUT-QUE] – due to de Bruijn's student L. S. van Benthem Jutting , in his Eindhoven PhD Dissertation (1977) – was Edmund Landau's **Grundlagen der Analysis** (Landau 1930) For details, see the Introduction to Rezuş 1983, mainly § *0.2 Tracing back motivations*.

§ 4 A shift of paradigm: witness theory

Witness theory is a relatively recent theoretical subject consisting of the study of the *formal constituents of proofs*, as they occur in everyday mathematics. The objects of concern are themselves mathematical objects, called here *witness operators*, an abstraction of the idea of (a) *proof operation* in intuitionism, as emerging from work done during the nineteen 60's and the early 70's by Stephen C. Kleene, Georg Kreisel, William Howard, Anne S. Troelstra *et al*. In a way, the witness operators are the 'fundamental particles' of mathematics, while any particular witness theory – ultimately, a kind of algebra – is supposed to record their properties, or else, specifically, their *equational behaviour*.[11]

The witness operators do not manifest themselves *in isolation*, and cannot be 'observed' or 'detected' as such in (mathematical) 'nature', so to speak, i.e., in the mathematician's practice. First because (i) in actual mathematical proofs, the 'working mathematician' is mainly paying attention to (*mathematical*) *contents*, and only incidentally to *logical form* – a very different phenomenological *Einstellung*, in Husserlian jargon[12] – and, next, because (ii) the 'formal' proofs, as studied in 'mathematical' or 'symbolic' logic, are ultimately, composite and rather complex objects. In fact, the witness operators 'occur' in the 'phenomenal' (mathematical) world only as *rules of inference*, and that's how – and why – the logicians have noticed them, in the first place, as 'atoms of meaning'. Yet,

---

[11] Cf. Kreisel's way of explaining to outsiders (a British Lord and his wife) the early work of one of his PhD students, Henk Barendregt, on combinatory logic: *When we reason, we make steps, deductive steps. These can be smaller or larger steps. Barendregt studies the smallest possible such steps, the so-called atomic steps, and the way they can be combined to form larger ones.* (Barendregt 1996, p. 6.)

[12] On 'mathematical proofs' *vs* the 'logical point of view' on proofs, see, e.g., Kreisel 1968, 1968a, 1971, 1973b, 1976, 1977, 1979, 1979a, Feferman 1979, Rota 1997, etc.

such 'atoms' are complex 'natural (mathematical) objects', and a 'phenomenally' detectable rule of inference would rarely, if ever, 'manifest' the proper mathematical 'nature' of the witness operator it is supposed to represent: most of the time, the finitistic ('phenomenal') appearance of a rule of inference hides *infinitistic* ingredients, covering a variety of unrelated equational behaviours (different 'operator algebras', so to speak). Roughly speaking, the 'real' objects dealt with in logic – the witnesses of provable formulas, or else the 'proofs' of mathematical theorems, in current (mathematical) parlance – are complex 'constructions' made up of 'primary', intuitive data ('mathematical evidence', as they say) and witness operators.

In this sense, the intuitionistic logic is a laboratory for the study of conceptual phaenomena occurring endemically in logic as such (i.e., in the so-called classical logic, the 'logic of classical mathematics').

§ 5 The Kolmogorov interpretation of witnessing

A proper *mathematical* way of understanding the idea of a witness in logic (proof theory) consists of making appeal to a simile from traditional algebra.

As in algebra, more or less, a witness *operator* is supposed to be defined by *characteristic equations*.

With an analogy, in a monoid, for instance, we have two 'operations', a binary ∘, and a null-ary 'unit' **1**, where ∘ is supposed to be 'associative' – viz. one has a disguised universally quantified statement

(∘) ⊢ (f ∘ g) ∘ h = f ∘ (g ∘ h) –

and, moreover, a statement to the effect that the 'constant' **1** is a 'bilateral unit' w.r.t. ∘ – i.e.,

(**1**) ⊢ **1** ∘ f = f = f ∘ **1** –

(the latter being, in fact, a disguised existential statement), so that the 'definition' of the monoid is given by specific equations, 'characterizing' simultaneously both ∘ and **1**.

Now, in traditional mathematical parlance, one takes the term 'equation' in two, slightly different senses, namely, an 'equation' can be *either*

(1) an equality, F = G (quantified universally outermost, say), as the case of the associativity 'law' (∘) above *or*
(2) an (algebraic) expression F(x) = 0, containing one or more 'unknowns' x, as, e.g., in the case of the quadratic

$$(Q^2) \qquad F(x) := ax^2 + bx + c = 0,$$

where the said 'equation' (Q^2) is supposed to have 'solutions' (for x) in the complex plane (for a, b, c rational).

If we pay attention to the underlying grammar, in the first case we have a (grammatically) indicative statement expressing an *asserted proposition*, while, in the second case, we have a (grammatically) interrogative sentence, i.e. a *question* – expressing a particular (mathematical) problem – or else, equivalently, a command-sentence, expressing an *instruction*, the *command* being: 'find an x such that (Q^2) holds'.

So a 'mathematical problem' of this kind is of the form:

(Q(**x**)) $?x_1 \ldots ?x_n.A(x_1, \ldots, x_n)$

where $A(x_1, \ldots, x_n)$ is an 'open' statement in x_i, $0 < i < n+1$, $n > 0$, and the task ('solve the problem!') consists of finding 'values' c_i for the x_i's (in a certain set), such that the 'equality' (taken in the first sense), as expressed by $A(c_1, \ldots, c_n)$, holds.

In general, mathematical problems are not necessarily 'numerical'[13], however, so one can replace the n-tuple **x** := $< x_1, \ldots, x_n >$ occurring in (Q(**x**)), by a specific (mathematical) *construction* **c**, whence the 'problem' is of the form:

(Q(**c**)) $?\mathbf{c}.A(\mathbf{c})$.

Recall now that we usually 'define' – in traditional, classical mathematics – the *algebraic numbers* as being those numbers that are 'solutions' to algebraic equations (with 'solution' taken in the sense above), in contrast, say, with those called 'trancendental' (like π or Euler's e-constant) that are not so.

With this kind of simile, the witnesses can be best compared with the algebraic numbers: they are 'solutions to mathematical problems', taken in the general sense above. More definitely, Q(**c**) – a famous open mathematical problem, say – can be answered, classically, by either providing a witness a such that $\vdash a : A$ or else by providing a witness ā such that $\vdash \bar{a} : \neg A$, where A is the obvious statement-transform of Q(**c**) and $\neg A$ is its (classical) negation.[14] The situation where we don't have an answer – either a witness a or a witness ā – is pointless and by the way, because we did not solve the problem Q(**c**).

This takes care of the so-called 'constructive' cases, and corresponds, more or less, to the 'problem-interpretation' of the intuitionistic logical formalism, due to A. N. Kolmogorov (1932).

What about the 'transcendentals' mentioned in the analogy above? Our 'algebraic' simile still works in this case, although in a derived sense, namely at 'higher-order' level: we just quantify over propositions expressed by statements of a certain form.

---

[13] In the light of Gödel 1931, we can, of course, *translate* non-numerical ('syntactic', say) data into arithmetical – i.e. 'numerical' – data, typically, into 'gödelised' **PA**-statements, but this is a different story.

[14] Note that, in the latter case, the formula-'answer' – if any – to Q(**c**) must be (classically equivalent to) an *existential* statement.

§ 6 'Traditional' versus 'mathematical' logic in historical prospect

The historians of science (mathematics included, and, in particular, the academic involvement in a proper recovery of the history of logic, *lato sensu*) have become only recently aware of the fact their main concerns are *epistemically sensitive*, and thus, possibly, *prone to error*.

The new incentive has been mainly – though not only – prompted by *philosophers*, to wit, by 'philosophers of science'. This kind of *awareness* is more or less inscribed in the nature of things, so to speak, the need for an inter-disciplinary approach in history-writing is derived, essentially, from the fact that *reflection on science* is always involved in *doing science*: from Plato and Aristotle – via Galileo, Descartes, Newton, Leibniz, Kant, Hegel – to Husserl, Einstein, and Gödel – and Heidegger, perhaps – we have been always bound to include a *critical historical dimension* in our *purely scientific* endeavours.

The well-known and oft quoted, pious metaphor *nanos gigantum humeris insidentes* ['dwarfs standing on the shoulders of giants', implying discovering 'truth' by building upon previous discoveries] – oft credited to Sir Isaac Newton [15] – makes up only half of the 'historical truth'; most of the time, the 'giants' have been rather 'criticised' by all those modest 'dwarfs'[16] who *reflected upon* their heritage: historical reception is rarely, if ever, a pure, lavish repetition of previous (historical) data.

The fact is that, until recently, science – and, in particular, mathematics – history-writing has been awfully *descriptive* and, more relevantly perhaps, a pure piece of *whig history*, history written from the point of view of the alleged 'conqueror', *du point de vue du vainqueur*, as they say: from the so-called 'modern point of view'.

Yet, what if the 'moderns' were wrong, now and then? A typical case in point – in the history of logic, our main area of interest, here – is Jan Łukasiewicz (qua historian of logic): Łukasiewicz meant to examine ('from a modern point of view', i.e., from the point of view of the newly emerging *formularian* logic) both Aristotle and the Stoic logic tradition. While illuminating – in part – on Chrysippean and, in general, on Stoic logic matters, his alleged 'reconstruction'

---

[15] *If I have seen further it is by standing on ye sholders of Giants* [sic]. (Letter to Robert Hooke, dated February 5, 1675, in: Newton 1675.) The last half of the phrase [*standing ... Giants*] is also inscribed on the edge of the British two pound coin issued in 1997, alluding to Newton. The original simile is, actually, about five centuries older and comes from a xii-th century monk, Bernard de Chartres, a French Neo-Platonist philosopher, and a former chancellor of the Cathedral School of Chartres until 1124. Bernard's *dictum* was recorded by John of Salisbury in his *Metalogicon* (completed in 1159). Cf. John of Salisbury 1955, Book III, Chapter 4, p. 167. For other references, see, e.g., Klibansky 1936 (Raymond Klibansky also edited the works of Thierry and Bernard de Chartres), and the learned comments of Robert K. Merton 1965, 1985^R, and Umberto Eco, in Merton 1993^R, Eco 2019.

[16] Cf. the title of Hesseling 2003 for an appropriate diversion.

of Aristotle's 'logic' was biased by 'modern' preconceptions: we can, of course, retrieve Aristotle's syllogistic as a purely formal 'deductive system' – by 'modern' (read: formularian) standards – but this is too generous a 'reconstruction': unlike Chrysippus, a hundred years afterwards, Aristotle and his Peripatetic followers did not have the slightest idea as to what we mean by a 'propositional connective' nowadays, for instance, let alone the modern theory of (first-order) quantification. By the same token, Łukasiewicz missed the main *historical* fact as regards the Stoics, namely that the father of (what we call) classical logic was, actually, Chrysippus (cf. Rezuş 2016 and the references mentioned there). In retrospect, however, there is no point in blaming Łukasiewicz and his peers (Frege, Peano, Russell, Tarski, Hilbert, and Gödel included) for their way of thinking (i.e., for being *formularians*): it would be, more or less, like blaming Robert Boyle – and the Ancient 'atomists' (Leucippus, Democritus & co), for that matter – for ignoring the basics of the NMR-spectroscopy.

In this guise, the fomularian way of thinking in logic is, by our *current* standards, not only obsolete, but also (epistemically) misleading.

§ 7 On seeing out of the fog

The motto of Hesseling's book on L. E. J. Brouwer[17] comes from Milan Kundera. The quote sounds as follows:

> '*L'homme est celui qui avance dans le brouillard. Mais quand il regarde en arrière pour juger les gens du passé il ne voit aucun brouillard sur leur chemin. De son present, qui fut leur avenir lointain, leur chemin lui paraît entièrement clair, visible dans toute son etendu. Regardant en arrière, l'homme voit le chemin, il voit les gens qui s'avancent, il voit leurs erreurs, mais le brouillard n'est pas là. Et pourtant, tous, Heidegger, Maïakovski, Aragon, Ezra Pound, Gorki, Gottfried Benn, Saint-John Perse, Giono, tous ils marchaient dans le brouillard, et on peut se demander: qui est le plus aveugle? Maïakovski qui en écrivant son poème sur Lénine ne savait pas ou menerait le léninisme? Ou nous qui le jugeons avec le recul des decennies et ne voyons pas le brouillard qui l'enveloppait ?*'[18]

---

[17] See Hesseling 2003 and the reviews by Mark van Atten (CNRS, Paris), in van Atten 2004, and by Jeremy Avigad (CMU, Pittsburgh PA), in Avigad 2006.

[18] 'Man proceeds in the fog. But when he looks back to judge people of the past, he sees no fog on their path. From his present, which was their faraway future, their path looks perfectly clear to him, good visibility all the way. Looking back, he sees the path, he sees the people proceeding, he sees their mistakes, but not the fog. And yet all of them – Heidegger, Maïakovski, Aragon, Ezra Pound, Gorki, Gottfried Benn, St-John Perse, Giono – all were walking in fog, and one might wonder: who is more blind? Majakovski, who as he wrote his poem on Lenin did not know where Leninism would lead? Or we, who judge him decades later and do not see the fog

Putting aside Kundera's political (anti-communist) *penchants*, as well as Hesseling's reference to the 'gnomes' (in Dutch: *kaboutermannetjes*, as opposed to a – Dutch – Giant!) of the time – i.e., during the twenties – in his title, I don't think Kundera's (essentially political) metaphor is a very accurate description of the situation under focus in the book. It might work, however, as a mere *façon de parler*, and only for a while, since, in retrospect – 'avec le recul des decennies', as Kundera would have said – it became more and more clear to most of us[19] that modern logic – also called 'mathematical', 'symbolic' or even 'formal'[20] – was born in the midst of a terrible fog, indeed. Let alone the more or less traditional psychologistic mishandling of logic, this very act of birth happened to the point of *obliterating* – and even *letting to fall into oblivion* – the main concern of logic itself, a concern which consists of a proper, *theoretical study of the* (*logical*) *concept of* (*a*) *proof*. As if, with every step in advance – *en avançant dans le brouillard !* – we are bound to make two or three steps aside / side-wise, doomed to err within the same overwhelming fog.

As a matter of fact, the so-called 'modern' logic – a rather modest collection of para- (and peri-) mathematical notational endeavours, modelled on the algebraic habits of the time, not significantly older than my own grand-parents (cf. Frege's **Begriffsschrift** [**BS**], 1879, and his **Grundgesetze der Arithmetic** [**GGA**], Bd. 1, 1893, or Peirce's famous paper of 1885) – was born *formularian*[21], and focused almost exclusively on the formal [read: *formulaic*, or *by formulas*] way of expressing propositions and/or would-be propositional structures, as well as properties thereof.

Specifically, the so-called 'inference rules' played a minor role in both Frege's **BS**, and in Peirce's pioneering paper of 1885 (the corresponding axiomatic presentations [Peirce had also propositional quantifiers] – paradigmatic pieces of *Satzlogik*, so to speak – used just *modus ponens* for 'material' implication,

---

that enveloped him?' From: Milan Kundera **Les Testaments trahis**, Gallimard, Paris 1993. English version in: **Testaments Betrayed** (translation by Linda Asher), Faber and Faber, London and Boston 1995, p. 240. (DEH's note)

[19] Well, at least to some of us, and not necessarily 'intuitionists'.

[20] A rather unfortunate label, because logic is either 'formal' or it is no logic at all. Hegel's **Wissenschaft der Logik** (1812–1816), for instance, a popular example of 'contentistic' – as opposed to merely 'formal' – logic is neither a piece of 'science' nor about 'logic' (it is, at best, the title of an amusing book, written by a muddy-thinker [well, foggy-thinker, to emulate Hesseling's way of speaking] in matters logical proper). And Kant didn't fare any better with his 'transcendental' considerations, either.

[21] As mentioned before, the proper allusion, here, is to Peano's many editions of his **Formulaire** (Bocca *et co*, Turin, 1895–1908, etc.) – *ad hoc* collections of logico-mathematical formulas he managed to publish and to talk obsessively about, in Italian, in French, and in pseudo-Latin (*latino sine flexione*) by the turn of the previous century – not to formalisation as such. By (grammatical) derivation, a *formularian* (person) is, thus, going to be a logician who is thinking in terms of formulas alone, *id est* in the fog.

and the simplest quantifier-rules for instantiation and generalisation), while Frege's *Regellogik* of **GGA** contained only 'flat' rules of the form

$$\vdash A_1, \ldots, \vdash A_n \Rightarrow \vdash B,$$

with no conditional (entailment-like) premisses, as well as 'transitions' [*Übergänge*], allowing validity-preserving transformations of such specific (rule-) schemes. In other words, Frege did not distinguish, not even in his *Regellogik*, between a bare (true) conditional formula

$$\vdash A_1 \to \ldots \to .A_n \to B,$$

and the corresponding (valid) entailment

$$A_1, \ldots, A_n \vdash B,$$

(so that, in particular, he won't have understood the point behind the Deduction Theorem [Tarski 1921], i.e., the 'implication-introduction' rule of the Natural Deduction [ND] style presentations of [classical and intuitionistic] logic – Rezuş 2017, 2017e, 2019a – that have emerged about forty years later).

Even the rather late-to-come 'rule-logics' of Stanisław Jaśkowski (Warsaw 1926–1927, in print: 1934) and Gerhard Gentzen (Göttingen *Inauguraldissertation*, defended in May 1933, in print: 1934–1935), who managed to identify *the most general concept of a logical rule of inference*, remained in the darkness of the foggy formularian (mis-) understanding of logic, since the logical rules of inference were thought, at best, as being mere *sequences* and/or *configurations of formulas*, whereas the main (logic) concepts to be taken care of – theoretically – were just properties of formulas (as, e.g., *provability*) and/or sequences thereof (e.g., *derivability*, for rules).[22]

Whereupon, a (logic) *proof* became, at best, even in the improved Jaśkowski-Gentzen setting, a bare (visual) arrangement of formulas: they were not 'true mathematical' objects, worth deserving a proper (theoretical) study, as such. Whence, ultimately, a logic (the 'classical' logic, for instance) amounts, in this view, either to

[A] a set of (true) formulas ('tautologies'), at worst, or to
[B] a set of (valid) inference rules, at best.

A very wrong idea, in fact, since one can easily manufacture counter-examples, in both cases. A counter-example to [A] has been given by Henry Hiż, in Hiż 1959, while, as regards [B], Gentzen and his followers have never understood, even to-date, the rather elementary fact that a so-called 'cut-free' logic is *not the same thing* as the corresponding logic with derivable 'cut' (here, 'cut' – a substitution rule [sic] – is to be taken in Gentzen's terms).

---

[22] The alternative, tree – or nested-block – formula-arrangements were just meant to please the (pseudo-) mathematical audience, I suspect. For, in fact, in the most general case, a ND-derivation is *not* a [labelled] tree, but a rather sophisticated (term-) graph (as Nicolaas G. de Bruijn has explicitly noticed a while ago). On the other hand, a so-called Gentzen 'L-derivation' is, essentially, a special case of a ND-style derivation [standing for a *refutation*], if understood properly.

§ 8 Polish new formularian standards and anomalies

During the early thirties, Tarski managed to standardise – with some assistance from Łukasiewicz and a few other members of the Lvov-Warsaw logic 'School' – the subject in *model-theoretic* terms (essentially, in terms of *compact closure operators* – as in set-theoretic topology, more or less – and dubbed *consequence relations*, for the purpose), based on the very same formularian tradition.[23] The Tarskian reformulation made proof theory dispensable [!] and is still *en vogue*, nowadays, with some modifications and refinements (due, mainly, to Dana S. Scott): most people keep 'defining' logics by consequence relations, i.e., by hand-waving, without paying attention to proper proof-theoretic details, doomed as (too) 'syntactically sensitive'.[24]

The conceptual trouble is that a logic – actually, just a fragment of classical logic, by the time Tarski, Łukasiewicz *et al.* were pondering on the subject[25] – came to be identified, thereby, with a bare *set of tautologies* (valid formulas). This turned out to be a very wrong idea, because parochial, again: the rules of inference – actually: the proof operators – were thereby neglected.

As mentioned earlier, the first (*technical*) counter-example to the initial, inadvertent formularian doctrine claiming that *a logic is a set of formulas* is due to a Polish logician, as well, viz. to Henry Hiż (1957, 1958, 1959), also known as a theoretical linguist.

This concerns a more subtle distinguo, viz. that between *rule-derivability* – a proof-theoretic concept (actually: *explicit definability* for proof operators) and mere *rule-admissibility* (actually: closure under a given rule of inference). Roughly speaking, a rule is *admissible* for / in a given logic $\mathcal{L}$ if its addition does not increase the set of provable formulas in $\mathcal{L}$.[26]

In essence, Hiż invented a three-valued logic – let us call it $\mathcal{H}_3[\rightarrow, \neg]$, say – on the signature $[\rightarrow, \neg]$ (implication, negation), whose tautologies (valid

---

[23] Cf., e.g., Tarski 1930, and the historical cross-references recorded in the collections Tarski 1956, 1986.

[24] Somewhat surprisingly, Tarski's *radical* approach – to proof theory, in essence – has been attacked, in recent times, from a different ideological corner, by formularian thinkers, although in *model-teoretical* terms, this time. Cf. Etchemendy 1990, etc. There is a vast literature on this (see, e.g., the editoral references in Tarski 2002), of no real concern here, since the debate looks, ultimately, like a family quarrel amongst differently minded formularian factions, rather than an attempt to a proper theoretical reflection on the main concepts of logic, as actually meant by Tarski, a while ago. (Cf. Tarski 1936, 1936a, resp. 2002, and the recent comments appearing in Tarski 1969, 1986a.) Incidentally, in this debate – around the concept of (a) *logical consequence* ['following logically from'] – the newest jihad of Jean-Yves Girard (2011, 2016 etc.) against Alfred Tarski – and the Greek particle *meta* – is *by the way* and *of no consequence*.

[25] Initially, with, as only example to talk about, the Łukasiewicz *multi-valued* logics.

[26] The distinction *admissibility / derivability* for rules of inference appers rather late in (formularian) logic – around the early nineteen fifties – and is due to Paul Lorenzen and Haskell B. Curry, independently. Needless to say, (rule-) admissibility is best understood model-theoretically.

formulas) are exactly the classical $[\to,\neg]$-tautologies, although $\mathcal{H}_3[\to,\neg]$ does not contain all classically valid rules of inference as derivable rules. Moreover – and somewhat ironically – the Hiż-logic $\mathcal{H}_3[\to,\neg]$ is also 'extendible', i.e., unlike classical logic, it is not Post-complete, and, in particular, it admits of (infinitely many) extensions $\mathcal{H}_3[\to,\neg] + C$, where $C \equiv \neg A$ is to be accounted for as a tautology, for an arbitrary $[\to,\neg]$-formula A. Hiż's $\mathcal{H}_3[\to,\neg]$ has a finite basis relative to a finite set of rules *not* including ($\triangleright$), however, i.e., *modus ponens* for ($\to$), is not derivable (although admissible) in $\mathcal{H}_3[\to,\neg]$. Incidentally, Hiż's idea was likely that of damaging the famous *modus ponens*, although he was not very radical on this matter, because (i) his logic has a derivable special case (a substitution instance) of the said rule, and because (ii) he managed to damage other classically valid rules during the process, in particular the Double Negation rules and, more signficantly perhaps, the rule corresponding to the traditional 'Law' of Non-Contradiction. Model-theoretically (see also Nowak 1992), Hiż's (three-valued) implication is closely related to Frege's own ways of handling truth-values in **BS** (mainly in connection with his 'horizontal' operator; cf. Simons 1996, Pelham 1999, etc.), although the Hiż (three-valued) 'negation' is weird enough to deserve a separate talk. In particular, $\mathcal{H}_3[\to,\neg]$ admits of a witness-theoretical description, as well. There are, actually, many variations on Hiż's idea. Notably, slightly afterwards, during the early nineteen sixties, a French logician – Jean Porte – and – in Poland – Witold A. Pogorzelski (cf. Pogorzelski 1994, etc.) came out with similar *modus ponens*-related examples. While confronted with such – otherwise obvious – *conceptual inconveniences*, some model-theorists (mainly Polish, experts in logical 'matrices' and the like, i.e., essentially, formularian-based model theory) have invented the ad hoc concept of a *structural consequence relation*, supposed to take care of substitutions. (Cf. Łoś & Suszko 1958 etc.) In algebraic / formularian terms, substitutions – as understood in quantifier-free logics – are just *homomorphisms of formulas*, of course. In fact, the ('traditional') algebraists can cope quite elegantly – in the same long shadow of Leibniz – with the ideas of *replacement* and *substitution*, at least as long as abstraction operators are not present in their equations. This kind of *conceptual patching* does not answer the problem and shares the same parochial way of thinking about logic: the basic concept of logic – that of (a) proof – being *badly mishandled*, if not, even, *programatically ignored*.

§ 9 *Vergessenheit*

The slightly – un-intended, though – Heideggerian overtones of my talk, here, about 'oblivion' (*Vergessenheit*) might sound, at a first look, misplaced and one might want, perhaps, to use a milder term, like 'oversight', instead. Yet oblivion – *Vergessenheit* – it was, since the Ancients (Chrysippus and his followers, for instance) managed to handle correctly, some twenty-two centuries ago already, both *entailments*[27] and *general 'transitions'* (in Frege's terms), i.e., validity-preserving entailment transformations. We – this includes our teachers and their

---

[27] An *entailment* is a finite sequence of propositions [or just formulas] where exactly one element of the sequence is tagged (as 'conclusion'). A *refutation* or a *rejection*

teachers, as well – have inherited of this kind of oblivion, too big a [conceptual] mishandling to be accounted for as a mere 'oversight'. And it is very likely that without the insistence of the early intuitionists[28], who aimed at *understanding the proofs themselves* in *proper mathematical terms*, we would have remained, even nowadays, in the proto-formularian fog! Doing (Tarskian) model theory – *id est* a kind of (formularian) algebra – instead.

24 October 2015 – 23 October 2019

(*elenchos*, in Greek) is a special kind of entailment, where the tagged proposition (formula) is just *falsum* (f, say, i.e., an arbitrary, paradigmatic, false proposition). Chrysippus noticed the fact that general entailments can be defined in terms of rejections ['conflict', in his terms], provided one can tell, for every proposition [formula] A, what is *the polar opposite* [or 'the contradictory'], A*, of A. Whence, a Gentzen *Sequenz* $A_1, \ldots, A_m \vdash B_1, \ldots, B_n$, ('multiple on the right') is just a refutation, i.e., a specific entailment of the form $A_1, \ldots, A_m, B_1^*, \ldots, B_n^* \vdash f$. In other words, there is no conceptual distinction between a 'natural deduction' [ND] formulation and a Gentzen 'logistic' [L-] formulation of classical logic. More precisely, as far as classical logic is concerned (Gentzen's *bête noire*, in a way, and the manifest reason he [thought he] had to invent L-sequents, 'cut-elimination' [his *Hauptsatz*] and the like), a Gentzen L-formulation is just *a special case* of a ND-formulation.

[28] And of a few other people – not necessarily intuitionist mathematicians, like Andreĭ Nikolaevič Kolmogorov (1903–1987) and Valeriĭ Ivanovič Glivenko (1900–1940), for instance – that managed to *see out of the fog* at a rather early stage (1925, resp. 1928), at least in as far the so-called 'intuitionistic' logic was concerned. Even Kurt Gödel (1906–1978) – a formularian-trained thinker, a mathematical 'Platonist', and a promotor of 'classical' logic, as well – was able to see, now and then, 'out of the fog': he clarified actually – this happened during the early thirties – a few typical formularian mis-understandings regarding 'Brouwer's [new] logic' (as a matter of fact, Heyting's formularian, axiomatic presentation, including first-order intuitionistic arithmetic). The full story has been told, with copious details, in Hesseling's Utrecht PhD Diss. (1999, cf. Hesseling 2003), taken here as a pretext.

2
What is a 'challenging problem'? (2010a)

§ 1 'Service to humanity'

We are, most of us, involved in *solving problems*, most of our time. Some of them are 'real' or 'really important' (as they say), other ones are merely 'academic' (they – the other ones – would say).

A few years ago – sometime during the previous millennium – one of my 'polytechnical' friends (an electrical engineer, expert in high power technologies and so on) managed to circumscribe *exactly* – to my mind – the first kind: she claimed she was, simply, 'in the service of humanity'[1]. That's 'real' and even 'really important' (think, e.g., of the poor state of the electrical networks in the States these days). The other kind is equally easy to exemplify: take, for instance, the problem Professor Andrew Wiles (well, along with plenty of others) was confronted with, while still a teenager: the Fermat Conjecture, now a Theorem. This one is, definitely 'real' – it's a 'real problem', one might say – and probably even 'important' in some sense; yet significantly less 'serviceable to humanity', I suspect. We can certainly survive without Fermat (and, even, Professor Wiles[2]) for a long while – a millennium or more, from now on. It is, nonetheless, less obvious that we can survive that long – by our current standards and habits – without people like my power-planting friend.

One might object to the previous example to the effect that I'm confusing things like 'applied' and 'pure' science (mathematics[3], in particular). Because even though both my IEEE friend as well as Professor Wiles are actually 'using' highly sophisticated mathematics, the maths involved are of a different 'kind', in each case: the latter is just 'pure science', while the IEEE people are dealing in 'applied science', as they use to call it.

---

[1] This because she was mainly involved in designing / fixing power plants... and/or doing 'service' for the IEEE community, in her free time.

[2] A sad fact of life, alas!

[3] Sic: the plural is British (as in Ancient Greek)! Other people use to think that *mathematic* (sic) is one!

Wrong idea! How do we know Fermat-Wiles won't eventually yield plenty of so-called 'applications', within a century or, even, less? Thereby becoming equally 'serviceable' to the humanity next to come, as, e.g., to our grand-sons and grand-daughters, their sons and daughters, and so on.

Look at the recent whereabouts of the Galois theory: from a purely academic business, this piece of 'pure' algebraic wisdom has become nearly unavoidable in cryptography, nowadays. And don't tell me cryptography is no good 'for humanity', or else that it's only good for spies and secretive persons or governments. After all, cryptographers – highly theoretical folk, Alan Turing included – contributed significantly to the end of the Second World War. And even wars are human affairs, you know. Unfortunately.

"Wait!", my would-be opponent would likely exclaim. "You're awfully rhetorical, Sir! This kind of argument destroys already your previous *distinguo* between IEEE science – as you've put it, in the footsteps of your electrical friend – *versus* Fermat-Wiles."

Not at all! Certainly, my *distinguo* is 'timed' and, above all, 'empirical', so to speak. After all, humanity has survived even without IEEE engineers, for a long while. Past a certain point in time (called also 'human history'), it won't be the case any more, I suspect. But this does not mean the IEEE people are dealing in 'applied' matters alone, while Professor Wiles – and, possibly, a couple of his friends and fans – are dealing in (mere) 'pure' matters. Both 'kinds' of science (ultimately: mathematics) are based on theoretical endeavours, on 'contemplation', so to speak. And, they're also both based on an 'appetite of knowledge', as old Aristotle would have had it (cf. **Metaph.**, at the very beginning). Usually, I / we just wonder 'why' [such and such is the case]! Sometimes my / our stubborn questioning – '*mirare*' in Romanian: it's easy to find out the Latin therein – would end up in something 'serviceable' for the rest of us, and sometimes not.

Roughly speaking, there is no real point in distinguishing between 'pure' and 'applied' science (mathematics, in particular). It's a matter of time, not one of... essence.

Ultimately, there is no 'applied science' (and no 'applied mathematics', for that matter).

Alternatively, in order to avoid unintended – otherwise boring – professional conflicts ('my guild in better than yours!', and the like), I might be able to refine the difference suggested above in terms of a single discipline, the calculus, in mathematics (also called 'mathematical analysis', in old Europe).

Suppose we are confronted – for some reason – with a huge amount of calculations, involving a high mathematics background. A talented – perhaps even an average – mathematician (expert in 'mathematical analysis') would eventually solve our problem(s), pending time and/or manpower, by using extant / well-known computing methods and recipes. Confronted with the same set of problems, her neighbour next-door – an expert in TCS (one can even forget about the T [of 'theoretical'], here) – would do something else: she will first re-visit the 'theory' thoroughly, rephrase the 'task' in her own (guild's) terms, extract an algorithm for the task and then 'implement' it, as a piece of software. Ultimately, she'd end up with similar solutions to our original problem(s), except perhaps for the fact the latter are going to be more efficient: in the latter case it might takes minutes / hours / weeks to get a solution, instead of days / months / years. Now we are speaking about two 'applied' scientists (and both of them mathematicians, in the end). Yet, if 'time' is at premium, as it is usual, nowadays, only the latter would be really 'serviceable to humanity' (at least to a part of it). For the other one(s) would be always too late, in this respect.

§ 2 Wonder and challenge

"Somewhat too Baconian, perhaps (a reference to the famous Lord High Chancellor of Great Britain), your power-planting friend!", my would-be opponent would, once more, object. "For not every kind of science is ultimately 'power', even if 'serviceable to humanity'. Even though there is no 'applied science', ultimately, as you *timely* claimed, you are still confusing the 'science' itself with its specific uses, now and then. As a matter of fact, the recent (2000) Knight Commander of the Order of the British Empire – as well as plenty of string theorists, say, to change the running examples for a while – won't really care about being or becoming 'serviceable', anyway. They're just... wondering, like those venerable Greeks you already mentioned would have said. There is always a kind of *challenge*, involved. But there is still no 'service' intended in actual problem-solving. And, for sure, no 'humanity welfare', or anything like that, among the reasons a scientist may have to 'do science'. I am usually becoming a 'scientist' just because I'm curious – or only *neugierig*, in German – there is an *appetite for knowledge*, to begin with, indeed."

Right! This is, is fact, my next point. – I'd rather dissent, however, on reasons and motivations.

There are at least two distinct ways of 'wondering about'. 'I wonder why', meaning, more or less, 'how' for 'why'. This is a typical starting point for a becoming / future scientist. If you 'wonder' about the right kind of things, you may end up eventually with a Nobel prize (or a Fields medal, for that matter), because you're ultimately 'serviceable' to the rest of us. On the other hand, you'd never get such a high and explicit ('human'!) recognition if you'd wonder about 'being and beings', for instance (like Leibniz, long ago, or, more recently, Martin Heidegger). And only the latter kind of questioners won't really care about 'serviceability', to humanity and/or whatever else. The former kind would, at least implicitly.

In other words, while questioning *en philosophe*, we don't 'really' care about the world itself, humanity included (you can even write it with a capital H, if you like: it would amount to the same thing). So the philosopher can – and will – ask his / her kind of question(s) forever. A ('real') scientist won't even recognise such 'questions' – the 'philosophical' ones – as ('real') questions.

Because a ('real') scientific question is one for which we can – sooner or later – get (converging, and thus... 'serviceable'!) answers.

There is no (converging) point in wondering – again and once more, after ages – *à la* Leibniz: 'Why is there Being rather than Nothing?'. (*Pace* Herr Professor Heidegger, scientists won't bother about Nothing, anyway.)

This does not mean all 'philosophical' (or 'philosophically looking') questions are 'un-scientific' (or just 'pseudo-problems', as one of our former friends once claimed). We might usually ask 'wrong' – even 'stupid' – questions now and then ('awfully philosophical', at a first look). The next generation(s) – one or more – would repeat such questions, in slightly different terms, by refining them – by 'getting to the point' – so that would-be answers would come, eventually, in sight.

Whence a kind of (meta-) question I intended to ask from the very beginning (sorry for the long détour: I was unable to put things in a simpler way):

'What about the *motive ground* of a (*real*) question? What's *behind* a question that makes me try, again and again, to get an answer (to it)?'

It's an offending state of affairs, I'd say. The French would likely call it a *défi*, something to be taken, more or less, *personally*. This comes about to the English 'challenge', more or less. (Britons and most – English speaking – descendants thereof would rather have it as a kind of club – sport, say – affair.)

Apparently, there is no generic, no 'metaphysical' (meta-) answer to this. Because, simply, why is not every question 'challenging' to me? Or else 'equally challenging', indeed? Subjective conditions, knowledge / background, all this is immaterial, here: I can modify my state of mind, I can also learn, etc. etc.

Well, if you'd ask me about a specific something, there might be no 'challenge' in your question / problem. Because – simply – I already know the answer: so – while answering – I'm eventually acting as a mere tutor, as a teacher or so, no funny business. You can also use a book, instead, with the same effect, if not even with more profit.

If, however, I don't know the answer to your question / problem in advance (and, moreover, if I'm aware of the fact nobody else does), your question / problem would certainly look 'challenging' to me.

Still, there are different kinds of 'challenge'. Planning to go into this next.

Yet, before addressing the *difference*, here is a good place – as any other, perhaps – to add a parenthesis.

§ 2.1 A bad (meta-) dream and the drunk little monads

We can first think of 'science' as of a vast repository of data, a kind of encyclopedia, let's say.

If the answer to the question / problem is not *out there*, in the Universal Data Base, then the question / problem looks 'challenging' to me / to [all of] us: worth taking the trouble to look into it.

This is a rather superficial view on 'science', though.

First, 'real science' is highly 'regionalised': this would amount to many, apparently unrelated 'encyclopedias' instead.

There are 'sciences' not a single 'science', indeed: the 'science itself' looks rather like a maze of un-communicating containers, not like a unique Holy Grail.

On the book-metaphor, we are only confronted with disparate chapters in a novel written by an absent-minded author – at different times – inspired by very different, unrelated 'real-life' episodes.

On a more 'philosophical' metaphor, the true picture of 'real science' ressembles to a world *à la* Leibniz, one without a pre-established harmony, though: a collection of un-communicating monads, with no Over-All Monad to integrate – and to harmonise – little poor monads – the folk below – from above.

The older neo-positivists' (meta-) dream of a would-be 'Unity of Science' is, at best, a reductive idea, one based – more or less – on the model of a single God-blessed 'discipline': the Theoretical Physics. This was a piece of wishful (meta-) thinking, in the end. Because we are actually dreamers of many Holy Grails, as many as the historically attested 'scientific disciplines' (if not even more, since – as we go – we are always tempted to invent new 'disciplines' from scratch, from trifles, or from something like that).

Second, even if granted the (multi-) disciplinary perspective, the would-be – intended – coherence of a single scientific 'discipline' is not always automatically granted: the 'real' little poor monads are often schizophrenic entities, or else the monads actually populating our 'scientific' world would look like drunk people, *derilium tremendes*, rather than lucid, sober, and thus respectable folk. And this picture reflects – unfortunately – the rule of the game, not mere exceptions.

In spite of all appearances, even the Paradigmatic Queen so beloved by the earlier neo-positivists – Theoretical Physics – is nothing but a fat, highly delirious, old lady, nowadays (look at those many, highly imaginative 'string theorists', and to their [highly] 'theoretical' *zizanies*!).

With a popular joke (stolen from the world of mathematics, more or less), the only 'unifying principle' in contemporary physics seems to be the label 'Theoretical Physics' displayed on the wall of the physics departments and/or institutes, 'round the world. Just like that![4]

On the drunkard metaphor, the 'real', immediate / urgent challenge would consist of making people sober, first.

Looking for (a lost) coherence: a very different kind of problem-solving, indeed! Rather – if compared with usual: punctual / local problem-solving – a would-be meta-business.[5]

---

[4] The original mathematical joke amounts to the observation that the fictious French mathematician Nicolas Bourbaki has invented a new French singular noun, from a French noun defective of singular, *mathématiques*. As a matter of fact, the French plural comes from an Old Greek [plural] phrase: there was no *mathématique* in Old Greek, either; *tà* mathēmatiká were simply 'the mathematical disciplines', even for Plato and Aristotle.

[5] This has nothing to do with 'metaphysics', 'meta-physics', and the like, by the way.

§ 2.2 'System', the Kantian dream, the actual theory-building, and the perpetual *zizanie*

The ideal situation for a specific 'scientific discipline' is best described by an old Kantian dream: the *Normative Idea*.

In 'real life', this points out to something like the 'systematic character' of a (particular piece of) science: in a given 'discipline', things must 'hang together' first: there is no 'science' otherwise.

Whence, the meta-business alluded to above, a normative idea ('look for coherence first!') should come *before* the actual (punctual / local) business of problem-solving. If our 'science' ('scientific discipline') consists a mere *bric-à-brac* collection of conflicting data and statements, there is little chance we can ever solve local / punctual problems in this area; the said 'science' is, in fact, useless.

Aside. There is a famous general theorem on computation, in Lambda-Calculus (due to the father of the business, Alonzo Church, and to one of his students, J. Barkley Rosser), the so-called *Confluence Theorem* (also known as the *Church-Rosser Theorem*; CR, for short). Roughly speaking, the CR Theorem says that the result of any particular computation [$3+(2\times 3)+(5\times 2)$, say], in a reasonably 'coherent' computation system, must not depend of the order of performing smaller computations steps. So, it would be reasonable to ask from a specific computation system to satisfy this kind of (minimal) requirement [here: *confluence*, in technical terms]. Indeed, imagine the disaster, otherwise: a kind of algebra / arithmetic where $3+2+1$ would lead to different results, according to the path of computing you might choose: 'do first $3+2$' or 'do first $2+1$'. – This is mainly to stress the fact that a 'coherence' requirement might not be a mere ['traditional'] logic requirement, where coherence is taken to mean logical non-contradiction or consistency.

The meta-business above – characterized by the (more or less Kantian 'normative') recommendation / requirement: 'look for coherence first!' – has been always among the main concerns of 'ordinary' scientists; so obvious that it was not even worth mentioning as a separate 'scientific' task. Whatever the philosophical ancestry (Aristotelian, Galilean, Cartesian or so), it is only in recent times that 'professional' philosophers managed to 'take over', and grab the subject for their own. The modern catchword for this kind of [meta-] concern is 'epistemology'.[6]

In ordinary scientific practice, the 'normative' requirement, mentioned above, would amount to the fact that any 'scientific discipline' should have a 'systematic character'; specifically, to the fact that any 'particular science' must be presented / organized as a 'deductive theory'.

Initially, the implied – more or less managerial – task looked like a paradigmatic business for ('professional') logicians. In a longer run, in spite of an

---

[6] In plain English: 'theory of science'. To be fair, there is no 'theory of science' so far, there are just some – otherwise rather smart – people around, thinking 'theoretically' about such things.

endemic – oft much too loud – agitation (around the early thirties till the late forties), in both Europe and overseas, it quickly turned out the logicians 'proper' have actually little – or nothing – to say about the subject itself. In retrospect, the extant contributors to the so-called 'logic of science' have done more damage than otherwise.[7]

In the meantime (during the last half a century, say), we managed to learn / acknowledge a few more things, nevertheless.

If the main (meta-) business in 'science' – specific 'disciplines', I mean – is the so-called Theory-Building, then we must take for granted the actual state of affairs (and, possibly, remember that this was *always* the 'current' state of affairs in 'science', along the ages, at any specific point in time), namely, the fact that there is no 'normal science' (to use a recent catchword), except for a while or two (!): the (historical) rule of the game is, rather, the 'theoretical conflict': a nasty thing to cope with, indeed!

In other words: 'true / real science' is, in fact, a Perpetual Quarrel among Dis-Agreeing Factions – it's actual locus is a Battle-Field, down-here, rather than a Serene Privileged Room, high-there, in the Ivory Tour.

§ 3 Epistemic *vs* technical challenge

Coming back to 'challenge(s)', my concern in the title. –

There are plenty of things to say about.

Let me first display a few examples of 'challenging' questions / problems in my sense (in logic, my preferred spare-time, for a long while).

Rather elementary, I'd say (the required background can be reviewed / explained in a few minutes – and /or a few lines in print).

[*A conceptual*] *Aside.* If I remember well, Henk Barendregt pointed out, in conversation, already thirty years ago or so, to the fact that my colloquial uses of the (epistemic) qualifier 'elementary' (as applied to mathematics problems, mainly) might [have] look[ed] oft [those times] rather loose and even confused. The point concerned standard – yet careless – ways of of speaking in mathematics: Peano Arithmetic [**PA**] was / is still usually referred to as 'elementary arithmetic'. A very wrong idea, if you really care about complexity matters and things like that. So, I'd rather think twice or more – since – before referring to a specific **PA**-question as being 'elementary'. Slightly later, I had the opportunity

---

[7] Honestly speaking, the logicians are *not* and *should not* be concerned with such things. 'From a logical point of view', if 'physics' – or 'mathematics', for that matter: set-theoretically re-shaped as **ZFC**, say – would turn out to be 'inconsistent', then much worse for the 'physics' – and/or the **ZFC**-based 'mathematics'! Why bother? For – remember, please! – these are *worldly* affairs logicians are not and should not be concerned with. – The fact is that recent, serious logicians would typically avoid the subject, nowadays, half a century – or so – later, the associated message being, roughly said: 'mind your own business, folks; we don't really care about yours!', etc.

to dissert at length about colloquial proof-qualifiers like 'obvious', 'easy', 'as ever' etc. [in the introduction to **Abstract** AUTOMATH, 1983, kind of a digest, recording endless conversations with Nicolas G. de Bruijn, in Eindhoven, on *proving in mathematics* and so on], and even managed, next, to put together a little private theory of the would-be 'degrees of elementarity' in mathematical practice. (Whence also my disappointment as regards the current talk about complexity in logic, mathematics, TCS, and the like.) Here, I'd rather avoid such subtleties, however. In particular, I'm going to use the qualifier 'elementary' only as regards the way of formulating / expressing a particular problem / statement, *not* as a qualifier of the would-be methods / techniques etc. that might be needed in order to solve a problem and/or to prove a particular statement.

§ 3.1 A problem from Nuel Belnap Jr.

Suppose you would have asked me, some thirty five or forty-so years ago (around 1975, thus), to prove the following little puzzle (in propositional logic):

Let [B], [CB], [I] be formulas as below, and $\mathcal{BB}'$ be a 'propositional logic' with implication [→] only, axiomatised by [B], [CB] and *modus ponens* for → (the substitution rule being tacitly assumed).

[B] $(p \to q) \to ((r \to p) \to (r \to q))$ – [prefixing]
[CB] $(p \to q) \to ((q \to r) \to (p \to r))$ – [suffixing]
[I] $p \to p$ – [identity]
[mp] $A \to B, A \Rightarrow B$ – *modus ponens*, detachment for →.

Show that no substitution instance of identity [I] is derivable in $\mathcal{BB}'$ (*idest*: there is no proof of [I] from the [B] + [CB] axioms, with *modus ponens* and substitution).

This is a rather 'elementary' problem in (propositional) logic. Easy to state for everybody, no 'advanced' knowledge required to grasp its meaning, etc.

Yet, 'challenging', indeed! Thirty five or forty years ago, I won't have had the slightest idea on how to prove the claim.[8]

I'd classify this query as an *epistemic* challenge (it is, more or less, clear why).

The specific point, here, is in the fact that the problem can be solved in different ways, and at different 'levels of abstraction'.[9]

---

[8] Historically, the Belnap conjecture – now a theorem – stated above remained, stubbornly, an open problem for more than fifteen years. At the time of writing, there are several distinct ways of answering it. In each case, the techniques therein involved are surprisingly 'advanced', in mathematical terms. No connection whatsoever with the popular wisdom on propositional logics you may eventually learn about, in a first year logic-course, in mathematics and/or in philosophy departments, etc.

[9] Just try the problem *by hand*, without cheating (looking on the web etc.): if you end up with an answer in less than a year, say (!), then you are a smart mathematician, likely (although – to be fair – you won't thereby contribute a bit to the welfare of humanity, alas).

§ 3.2 A problem from Alfred Tarski

Suppose somebody [else] would have asked some time ago to prove the following equally puzzling (meta-) statement:

> Let [K], [D] be (purely implicational) formulas, in a propositional language as above ($\to$ stands for implication).
>
> [K] $p\to(q\to p)$ – irrelevance,
> [D] $p\to(q\to((p\to(q\to r))\to r))$ – pairing.
>
> Let $\mathcal{L}$ be a finitely-axiomatisable propositional logic, i.e., a logic with finitely many axioms and *modus ponens* for $\to$ (with, as earlier, the substitution rule assumed tacitly). Show that $\mathcal{L}$ is axiomatisable with a single axiom (*modus ponens*, and substitution), if both [K] and [D] are derivable in $\mathcal{L}$.

Historically, this statement was claimed in 1925, without a proof in print, by Alfred Tarski. The Pole is, among other things, the founder of modern Model Theory, a branch of [contemporary] Mathematical Logic, by the way. Parenthetically, the first proof in print of Tarski's claim is due to me [1978–1979, in print 1982], using a technique involving concepts *not available* in 1925 [i.e., λ-calculus, invented around 1927–1928, in the United States, by Alonzo Church; cf. Church 1931–1932].

Aside. A meta-puzzle: how did Alfred Tarski manage to obtain such a result, in 1925? The meta-puzzle points out to a famous – and similar, in fact – epistemic situation, namely, to the case of Fermat's claim. If Fermat did really have a proof of the 'Fermat theorem', his proof was, definitely, not like the one Andrew Wiles was able to produce a few centuries later. Wiles' proof won't fit the margin of that famous Greek book, either; but the real point is in the fact that it won't fit the overall epistemic context, anyway. Because Professor Wiles was inheriting of a very different kind of mathematical wisdom. Pierre Fermat could not have ever dreamed about elliptic curves. At least not in this context. In the end, if we are to be 'really fair', Andrew Wiles has *not* actually solved the *original* – 'historical' – problem; namely: 'prove the Fermat Theorem with the conceptual means of his time'!

On the other hand, my own *meta-conjecture* might consist of saying that Tarski anticipated – somehow – λ-calculus, combinatory logic and the like. Because there is no other – reasonable – explanation: we cannot prove Tarski's claim by *essentially other* means![10]

This is also an 'epistemic challenge', like the one under [1] above.

Specific point, here: very likely, this problem can be solved in only one way and Alfred Tarski was apparently aware of it, *avant la lettre*. (How?)[11]

---

[10] All this would leave Fermat as mysterious and as cryptic as he has been so far, I'm afraid.

[11] If you manage to do it *by hand*, without ever mentioning λ-calculus, combinators and the like, then you are certainly at least as smart as Alfred Tarski!

§ 3.3 A problem from Larry Wos

Suppose now you are a member of the relatively new [scientific] AAR-community (the *Association of Automated Reasoning*) – so you are already familiar with various AR (Automated Reasoning) software packages (OTTER, etc.) and/or techniques of [soft-] proof – and the current President of the AAR (actually, it's Larry Wos, these days[12]) would have displayed – at a recent AAR-meeting – a rather long formula – [R] say – containing circa hundred symbols or so, claiming [R] is axiomatizing classical propositional logic CL, with *modus ponens* (for 'material' implication) alone. The Wos Question / Problem is: show this using AR software (*idest*: prove [R] from a particular axiom system for CL, using OTTER, for instance).

Rather tricky, I would say, if you are not already familiar with the Tarski-story, already mentioned above, under **§ 3.2**.

Here, formula [R] (they call it a 'Rezuş formula' nowadays, although it should have been 'Tarski formula', as a matter of fact) may remain anonymous: one can easily obtain explicit examples thereof from the solution to the previous problem, anyway.

The Wos Question / Problem is, however, *not* of the same type as under **§ 3.1** and **§ 3.2**.

This question / problem is, rather, an example of a *technical* challenge. Because associated with a specific *technical* constraint (here: 'use AR software').[13]

Acknowledgements. I owe the last *distinguo* above to Branden Fitelson (now [2016], at Northeastern University, Boston MA), who pointed out to the fact that – generalizing slightly – actual conditions / constraints occurring in our current problem-solving endeavours may not be always 'purely epistemic', but rather – and typically so, in 'real life' – of the kind claimed by my power-planting friend, the IEEE engineer, a while ago. Moreover, such ('real-life') constraints would usually make things less obvious than they might appear at a first look, *in abstracto* ('laboratory conditions' and so on). The latter *distinguo* – epistemic vs technical challenge – deserves further reflection. Putting things in different terms: 'does *common science* need translation in order to become *serviceable*?' For, if so, this would involve a different kind of knowledge, apparently.

Thanks are also due to John Halleck (University of Utah, Salt Lake City, UT) for making comments, similar to those of Branden Fitelson – including a huge amount of technicalities on so-called 'Rezuş formulas' and the like.

> "I think what we have here is a difference of cultures, and cultural context. The context in which Dr. Wos is publishing is people working on automated therorem proving programs, trying to make them solve a wider class of theorems. Dr. Wos is himself famous for 'pushing the

---

[12] These notes were written in September-October 2010.
[13] Actually, at the meeting of the AAR, in June 2010, Larry Wos gave a different, more complex, example, based on the same Tarski-story as above. I confess I did not first see the *challenge*, whereupon I certainly owe him an apology.

envelope' and developing techniques to do just that. [...] At the current 'state of the art', there are huge areas where there are human proofs, and automated theorem proving does very badly and either can not solve them at all, or only some programs can. His [Wos'] challenges are not intended to be particularly challenging to people, and certainly not to any half-way educated person with access to the literature. What they ARE intended to be is something that is very difficult for PROGRAMS to do, and are areas where research in methods is sorely needed. And the Rezuş axiom for BCIW, while never intended for human display, is perfect as a test of an area that (as he pointed out) computer programs are very weak." [JH: 11 August 2010]

Nicely said! Although I'd have rather mentioned a 'diference of *mathematical* culture(s)', instead!

Postscript [3 October 2010]. Apparently based on a previous approach due to Dolph Ulrich (2004–2005), John Halleck has provided, a few days ago [23 September 2010], a genuine solution to the 'challenging puzzles' § **3.2** and § **3.3** above. Actually, Halleck did not prove the claim of § **3.2** *by hand*, the old-fashioned way; he used specific software instead, enabling him to produce derivations very close to the Łukasiewicz-style derivations (*modus ponens cum substitution*) of the late twenties and early thirties. For a short λ-calculus version of Halleck's proof, see the *Addendum* to my note *Tarski's Claim, thirty years later* (Rezuş 2010).

September–October 2010, revised 12 May 2016

Part II

Background to Witness Theory

3

Cartesian folds and Witness Theory (2017b)

In this survey, we discuss specific *equational theories* where the operators might be *abstraction operators*. In applications, such theories – essentially λ-calculi (called, for convenience, *folds*, here) – are meant to codify *the proof theory of classical logic*, understood as *a mathematical theory of proving*.

§ 1 Subsystems of extended λ-calculus and scalar extensions

§ 1.1 *Syntactic categories: witnesss-terms and scalars.* It is assumed that we have at hand at least one syntactic category of *terms* – called *witness terms* (*w-terms* for short); metavariables: a, b, c, ... – which is both *atomic* and *substitutive* in the sense of Haskell B. Curry. In other words, [1] the terms are generated from *witness atoms* (*w-atoms* or *w-variables*) x, y, z, ... and appropriate operators – called *witness operators* – such that [2] *the substitution operation* for w-atoms in w-terms makes sense.

Notation. Where a and c are w-terms and x is a w-atom occurring possibly 'free' in c – shown as in c[x] – the notation c[x:=a] (reading 'x becomes a in c') is supposed to be defined in the expected way, by induction on the structure of the w-term c. This can be iterated such as to obtain the *simultaneous substitution operation* c[**x**:=**a**], where **x** is a sequence of n pairwise distinct *w-atoms* and **a** a sequence of w-terms (of the same *length*: lh(**a**) = lh(**x**) = n > 0).

§ 1.1.1 *Scalars.* We also allow a second atomic syntactic category of terms called *scalars* – metavariables: s,t – generated from *atomic scalars* u,v and *scalar operators.* The latter is supposed to be substitutive, as well, and may contribute to the grammar of the w-terms. We have a corresponding notion of substitution for atomic scalars that may occur, possibly 'free', in w-terms; notation c[u:=t] (reading 'u becomes t in c'); *mutatis mutandis*, this can be also iterated in the expected way. Usually, scalars are to be taken modulo a scalar equality ($\doteq$).[1]

---

[1] Atomic scalars and w-atoms are supposed to be distinct, of course. In some cases, we can also take scalar equality ($\doteq$) to be syntactic identity ($\equiv$). In the first place, scalars serve to accommodate quantifiers in a witness-theoretic setting. Specifically,

§ 1.1.2 *Subterms, subterm replacement.* Subterms occurring in a given term are determined, inductively, in the expected way. In particular, scalars may occur in w-terms, not as (sub-) w-terms, though. We usually write c[a] := c[x:=a] for the fact that a is a subterm of c[a] and, similarly, c[t] := c[u:=t] for the fact that a scalar t occurs in c[t]. Inductive subterm replacement is denoted analogously; e.g., c[a::=b] stands for the result of replacing each occurrence of a in c[a] by b.

Intuitively, since (appropriately decorated) witness-terms are ultimately meant to represent *proofs* in logic, *proving* amounts – formally – to *w-term construction* [sic].

§ 1.1.3 *Abstraction operators.* An abstractor Ⓢ has associated term-form

$$\text{Ⓢ}(\mathbf{u},\mathbf{x}).a[\mathbf{u},\mathbf{x}],$$

where $\mathbf{u}$ is a sequence of m pairwise distinct *scalar atoms* [lh($\mathbf{u}$) = m] and $\mathbf{x}$ is a sequence of n pairwise distinct *w-atoms* [lh($\mathbf{x}$) = n], with m + n > 0. If the sum-length lh($\mathbf{u}$) + lh($\mathbf{x}$) = p > 0, the abstractor is said to be *p-adic*. If lh($\mathbf{u}$) = 0, the abstractor is said to be *pure*, if lh($\mathbf{u}$) > 0, the abstractor is said to be *scalar*, while, if, moreover, lh($\mathbf{x}$) > 0, the abstractor is said to be *mixed*.

The régime of *free* and *bound* variables is determined in the usual way. (If c ≡ Ⓢ($\mathbf{u},\mathbf{x}$).a, then the atomic scalars of $\mathbf{u}$ and the w-atoms of $\mathbf{x}$ are said to be free in a and bound in c.) We assume systematic alphabetical relettering (α-conversion) for bound variables, i.e., where $\mathbf{u}, \mathbf{v}$, resp. $\mathbf{x}, \mathbf{y}$ are sequences of pairwise distinct atomic scalars resp. w-atoms of the same length, and a[$\mathbf{v},\mathbf{y}$] ≡ a[$\mathbf{u}:=\mathbf{v},\mathbf{x}:=\mathbf{y}$], one has

(α) ⊢ Ⓢ($\mathbf{u},\mathbf{x}$).a[$\mathbf{u},\mathbf{x}$] ≡ Ⓢ($\mathbf{v},\mathbf{y}$).a[$\mathbf{v},\mathbf{y}$].

A w-term without free variables is said to be *closed* – or a *combinator* – otherwise it is *open*. A monadic abstractor ♮ – term-form ♮x.a[x] – is said to be *strict* if x occurs actually free in a[x], *linear* if it occurs free at most once in a[x], and *strictly linear* if x occurs free exactly once in a[x]. *Strict* (resp. *strictly linear*) w-terms are the terms containing no other abstractors than the strict (resp. strictly linear) ones.

In what follows, all operations involved in w-term construction – including the abstractors – are supposed to be 'algebraic', i.e., they satisfy due monotony conditions (in jargon: the equality is said to be 'compatible with the operations'). This means that we have:

(ξκ) ⊢ $a_i = b_i \Rightarrow\; \vdash \kappa(a_1,..., a_n) = \kappa(b_1,..., b_n)$, 0 < i < n + 1,
(ξⓈ) ⊢ a = b ⇒ ⊢ Ⓢ($\mathbf{u},\mathbf{x}$).a = Ⓢ($\mathbf{u},\mathbf{x}$).b,

---

in the case of *first-order quantifiers*, the scalars are *individual terms* and the scalar operators can be left unspecified. In the case of *propositional quantifiers*, the scalars are *propositionally quantified formulas* [sic] and ≐ is just syntactic identity. *Second-order quantifiers* are to be handled analogously. For another, though rather *ad hoc*, use of scalars [actually: pseudo-scalars], see the digression on λγ̄-calculi, in **§ 1.5.1**.

for n-ary operations κ (n > 0) and abstraction operators Ⓢ, with abstraction terms of the form Ⓢ(**u**,x).c.

Examples. In particular, whe shall encounter monadic abstractors ♮, with term-forms ♮x.a, pure dyadic abstraction operators ∫, with term-forms ∫(x,y).c (where x, y are distinct), as well as scalar abstractors $\bar{\gamma}$, Λ (monadic, with term-forms $\bar{\gamma}$u.a, Λu.a, resp.) and Σ (dyadic, with term-forms Σ(u,z).e) supposed to satisfy:

(ξ♮) ⊢ a = b ⇒ ⊢ ♮x.a = ♮x.b (for ♮ := λ, ρ, ∂, ϵ, γ, ε),
(ξ♮) ⊢ a = b ⇒ ⊢ ♮u.a = ♮u.b (for ♮ := $\bar{\gamma}$, Λ),
($\xi\int$) ⊢ a = b ⇒ ⊢ ∫(x,y).a = ∫(x,y).b (x ≠ y),
($\xi\Sigma$) ⊢ a = b ⇒ ⊢ Σ(u,x).a = Σ(u,x).b.

§ 1.2 *Folds.* An equational theory containing a monadic abstractor ♮ and a binary operation 'cut' ⸰ (infix notation) – term-forms: ♮z.a, c⸰a – satisfying the condition

(β♮) ⊢ (♮x.b[x]) ⸰ a = b[x:=a],

is said to be *a fold*. A fold is *extensional* if, moreover, the [♮,⸰]-pair satisfies also

(η♮) ⊢ ♮x.(c⸰x) = c (x not free in c).

For extensional folds one has, indeed, the Leibniz-condition:

(ext) ∀x: ⊢ f⸰x = g⸰x ⇒ ⊢ f = g, provided x is not free in f, g.

In other words, a fold is just a λ-calculus. As we are mainly concerned with extensional folds, we shall use the term 'fold' for the extensional variety, unless otherwise specified.[2]

---

[2] In spite of the fact a so-called 'weak combinatory logic' – an equational theory based on atomic constants ('basic combinators') and a binary cut operator (▷, 'application') – makes up an *algebraic variety* [an equational class], à la Birkhoff and Mal'cev, it is not a fold by present standards, since the (definable) abstractor that can be associated to the primitive cut ▷ may not be monotone, viz., even though an abstractor ♮ satisfying both (β♮) and (η♮) is 'algorithmically' – i.e., constructively – definable in such theories, the monotony condition (ξ♮) fails. One might want to reserve the label *combinatory folds* for such equational theories (also known as 'applicative structures' or 'combinatory algebras'), by requiring the generic substitution condition: (lsv) ⊢ a = b ⇒ ⊢ c[x:=a] = c[x:=b], in place of monotony for the corresponding abstractor. Obviously, (lsv) yields (rsv) ⊢ a = b ⇒ ⊢ a[x:=c] = b[x:=c], as well as monotony for the cut-operator. So defined, combinatory folds would correspond exactly to the *Howard-Hindley λ-calculi* of Rezuş 1981, II.1 (but see Scott 1963), while the 'algebraic' combinatory folds – where the definable abstractor associated to the primitive cut operator is monotone – would be just folds, in the present sense (and, in the end, a combinatory fold with (ext) would be the same thing as an extensional fold). Although purely 'algebraic' (we get thereby 'applicative structures' – *algebraic varieties*, thus – in the background), the latter

For n > 1, a *n-fold* is the result of joining together n distinct folds, with no further additional equational constraints. For n = 1, one has *mono-folds*, for n = 2, one has *di-folds*, etc.³

For extensions, one may want to specify, colloquially, the underlying fold(s), so we say, for instance, that *the extended λ-calculus* (cf. Lambek & Scott 1986) – denoted here by $\lambda\pi$ – is *based on* its underlying mono-fold λ – i.e., on *pure λ-calculus* – or, explicitly, on $[\lambda,\triangleright]$, where λ is its *basic abstractor* and $\triangleright$ *its associated cut*. The order in which we write the 'cut' operator ≀ of a mono-fold based on [♮-≀] is arbitrary, although fixed, usually, by the characteristic condition (β♮).⁴

Incidentally, a mono-fold ♮, based on [♮,≀], is said to be *β-only* if it does not satisfy (η♮).

§ 1.2.1 *Mono-folds: basic properties.* Most theoretical studies in λ-calculus concern exclusively mono-folds, and mainly the pure λ-calculus λ, based on $[\lambda,\triangleright]$, and/or subsystems thereof, i.e., ultimately equational theories with a rather limited import in logic (proof theory) as such⁵. We review first some (λ-calculus) folklore on mono-folds.

---

way of speaking might lead to confusions in the case of 'multi-cut' (resp. 'multi-applicative') theories / structures, as, e.g., in many-folds (see below). Things tend to be even more complex and confusing, since one encounters 'combinatory' theories / structures that are apparently 'cut-free', where the 'cut' (i.e., the 'application') operator) is not a primitive, although it is explicitly definable in terms of 'basic combinators' and, possibly, other operations (cf. **§ 1.2.3** and the brief discussion of *C-monoids* in **§ 1.6.2**). On *historical* reasons, the traditional algebraist won't admit of primitive abstraction operators in algebraic structures, thereby generating a somewhat artificial separation between *syntax* (here: equational theories, viewed as formal systems) and *models* ('algebraic structures'), in contexts where abstractions operators are present. In the latter case, we have, in fact, algebraic structures with *infinitely many operations*. [In general, the abstraction operators can be 'eliminated' inductively, so we don't have a real conceptual conflict with the traditional way of coping with algebraic structures.] Here, we intend to overcome formally this inconvenience, by handling folds (λ-calculi) as *equational classes*, resp. as *algebraic varieties*. [Added in proof: 12 September 2019. As the professional algebraist would, usually, tolerate the use of *inductive definitions* in algebraic contexts, this way of speaking is, is fact, not much of a novelty. See, e.g., the discussion of Selinger 2002, for a typical approach to the 'algebraic character' of a (mono-) fold. Even though conceptually illuminating, Selinger's tutorial is of limited interest for our purposes, since it ultimately concerns only special cases, viz, the mono folds.]

³ Informally, one might speak about *many-folds*, as well (no pun intended with the various *manifolds*, occurring in current mathematical parlance).

⁴ The notational choice for cuts is guided by readability considerations. Historically, however, it is worth noting the fact that writing 'arguments' first – as for the pure ∂-fold below, where the β-redex reads, officially (c ⋆ ∂z.e) – has precedence (Frege **GGA**, 1:1893).

⁵ Characteristically, see, e.g., Barendregt 1981, the 'Bible' of pure λ, where specific logic applications are discussed in its Appendixes A and B, while cartesian folds are mentioned obliquely, including Barendregt 1974 – just three times, in the whole

3 Cartesian folds and Witness Theory (2017b)

§ 1.2.1.1 *Monoids* (Curry 1930). Every extensional mono-fold ♮ – based on [♮,≀] – is a monoid (under *composition* ∘, given by a∘b := ♮z.(a≀(b≀z)), with z not free in a, b, and *identity*, defined by I := ♮z.z).

Notational conventions. Further, we assume association to the left for the cut operator ≀. For convenience, let also $F(X_1)...(X_n) := F≀X_1≀ ... ≀X_n$ (n > 0).

§ 1.2.1.2 *Frege-Church sequences* (Frege 1893, Church 1932). Define, for n > 0,

(df p) $<a_1, ..., a_n> := ♮z.(z≀a_1≀ ... ≀a_n)$, z not free in a_i (0 < i < n + 1),
(df K_i^n) $K_i^n := ♮x_1. ... .♮x_n.x_i$ (0 < i < n + 1),
(df P_i^n) $P_i^n := <K_i^n>$ (0 < i < n + 1).

We have

⊢ $P_i^n(<a_1, ..., a_n>) = a_i$ (0 < i < n + 1),
⊢ $<a_n> ∘ ... ∘ <a_1> = <a_1, ..., a_n>$.

§ 1.2.1.3 *Strictly linear combinators.* Set

(df I) I := ♮x.x,
(df B) B := ♮x.♮y.(x∘y),
(df CB) CB := ♮x.♮y.(y∘x),
(df C) C := ♮x.♮y.♮z.(x≀z≀y),
(df CC) CC := ♮x.♮y.♮z.(y≀z≀x),
(df CI) CI := ♮x.♮y.(y≀x),
(df D) D := ♮x.♮y.(<x,y>).

Straightforward calculations yield:

⊢ C ∘ C = I,
⊢ C(C(a)) = a,
⊢ CI = C(I) = ♮x.<x>,
⊢ D = C ∘ CI = ♮x.♮y.(<y>∘<x>),
⊢ C = CC(CC)(CC) = ♮x.♮y.(<y>∘x),
⊢ C(a) = CB(a) ∘ CI = ♮x.(<x>∘a),
⊢ C(a)(b) = <b>∘a,
⊢ CB = C(B), ⊢ B = C(CB),
⊢ CB(a) = <a> ∘ B,
⊢ B(a) = <a> ∘ CB,
⊢ CC = C(C) = B ∘ CI = ♮x.♮y.(<x>∘y),
⊢ CC(a)(b) = <a>∘b,

---

treatise, and, mainly, in Exercises [that's why I used to say that modern proof theory is, essentially, a matter of doing the exercises in his book] – whereas Rezuș 1981 – containing also typical formularian applications to proof theory (Church's 'ordinal logic') – is devoted entirely to *strict mono-folds*, i.e., to λ-theories containing strict [monadic] abstractors alone. The logic applications appearing in Barendregt et al. 2013 are discussed briefly in Rezuș 2015a.

$\vdash \mathsf{CC}(a) = \natural x.\mathord{<}a\mathord{>}\circ x.$

§ 1.2.2 *Bases.* The strictly linear terms of a pure mono-fold $\natural$ can be generated from the strictly linear basis BCI := $\{\mathsf{B, C, I}\}$ and w-variables. With $\mathsf{K} := \mathsf{K}_1^2$, $a\Box b := \natural z.((a\wr z)\wr(b\wr z))$, z not free in a, b, and $\mathsf{S} := \natural x.\natural y.(x\Box y)$, the linear terms of $\natural$ can be generated from the linear basis BCK := $\{\mathsf{B, C, K}\}$ (and w-variables), the strict terms of $\natural$ can be generated from the strict basis SBCI := BCI $\cup$ $\{\mathsf{S}\}$ (and w-variables), whereas, analogously, the basis SK := $\{\mathsf{S, K}\}$ generates all terms of $\natural$. There are many alternative choices for the above, even singleton bases ('generators'). For instance, with $\mathsf{W} := \natural x.\natural y.(x\wr y\wr y)$, one can use the basis BCKW := BCK $\cup$ $\{\mathsf{W}\}$ in place of SK. In general, any set of terms closed under equality, and containing K and D [$\equiv \mathsf{D}_0$] – resp. K and $\mathsf{D}_1 := \natural x.\natural y.\natural z.(\mathord{<}x,y,z\mathord{>})$ – admits of a singleton basis.[6]

§ 1.2.3 *Strictly linear cut definability* (Church 1937). Finally, set: $f \ni a := \mathord{<}f,a\mathord{>} \circ \mathsf{CC}$. We have $\vdash f \ni a = \mathord{<}a\mathord{>} \circ \mathsf{C}(f) = f(a)$ [$\equiv f\wr a$]. In other words, in extensional mono-folds, the cut operation ('application') is explicitly definable in strictly linear terms[7].

§ 1.2.4 *Linear decomposition.* Any extensional mono-fold $\natural$ – based on [$\natural,\wr$] – admits of a *linear decomposition* by setting

(df $\sharp$) $\sharp z.a[z] := \natural z.(a[z]\wr z)$, the *iterator* of the decomposition, and
(df $\flat$) $\flat(c) := \natural z.c$ (z not free in c), its *cancellator*,

whereupon we have:

($\natural$) $\vdash \natural z.c[z] = \sharp z.\flat(c[z])$, and, moreover,
($\beta\sharp$) $\vdash (\sharp z.a[z]) \wr c = a[z:=c] \wr c$,
($\eta\sharp$) $\vdash \sharp z.a = a$, if z is not free in a,
($\sharp\sharp$) $\vdash \sharp x.\sharp y.a[x,y] = \sharp z.a[x,y:=z]$, and
($\beta\flat$) $\vdash \flat(e)\wr c = e$,
($\sharp\flat$) $\vdash \sharp z.(\flat(a[z])\wr z) = \sharp z.a[z]$,

as well as monotony for the defined operators ($\sharp$ and $\flat$), so that, upon turning ($\natural$) into a definition, the conditions ($\beta\sharp$), ($\eta\sharp$), ($\beta\flat$), and ($\sharp\flat$) suffice in order to retrieve the initial [$\natural,\wr$]-fold, i.e., ($\beta\natural$) and ($\eta\natural$).

In particular, if $\bar{z}_j$, $0 < j < n+1$, $n > 0$, are all the occurences of the free variable z in a[z], then, in view of ($\sharp\sharp$),

---

[6] The latter result, due to Alfred Tarski (1925) – actually without a proof in print, one of the first 'practical' applications of witness theory, somewhat *avant la lettre*, indeed – is relevant in logic (axiomatisability problems), because it allows an effective construction of single-axiom systems for various logics. Cf. Rezuş 1982, 2010, 2019, for a proof of the original claim of Tarski and for several extensions thereof to systems without K, resp. to strict and strictly linear witness terms. The extension of Tarski's result appearing in Rezuş 1982 is based on the idea of *restricted solvability*, going back to Kleene (1935), Barendregt (1977), and Rezuş (1978). See Rezuş 1980, 1981, 1982, for exact references and further details.

[7] Note that extensionality is required here.

$\vdash \natural z.a[z] = \natural \bar{z}_1. \ldots .\natural \bar{z}_n.a[\bar{z}_1, \ldots, \bar{z}_n]$,

i.e., $\natural$ can be replaced by *stricty linear* uses of it, $\natural x.a[x]$, where x occurs in a[x] *exactly once*.

If the initial mono-fold is the pure extensional λ-calculus **λ**, based on [λ,▷], then the iterator is $\bar{W}x.a[x] := W{\triangleright}(\lambda x.a[x])$ and its cancellator is $\bar{K}(c) := K{\triangleright}c$, where W and K are the usual combinators.

Analogously, for the mono-fold ∂, based on [∂,⋆], satisfying

($\beta\partial$) $\vdash$ c ⋆ (∂z.e[z]) = e[z:=c],
($\eta\partial$) $\vdash$ ∂z.(z⋆a) = a (z not free in a),

the iterator is $\epsilon z.a[z] := \partial z.(z{\star}a[z])$ and its cancellator is $\varpi(e) := \partial z.e$, with z not free in e, and we have, as an exercise in notation:

(∂) $\vdash$ ∂z.c[z] = ϵz.ϖ(c[z]), and
($\beta\epsilon$) $\vdash$ c ⋆ (ϵz.a[z]) = c ⋆ a[z:=c],
($\eta\epsilon$) $\vdash$ ϵz.a = a, if z is not free in a,
($\epsilon\epsilon$) $\vdash$ ϵx.ϵy.a[x,y] = ϵz.a[x,y:=z], and
($\beta\varpi$) $\vdash$ c ⋆ ϖ(e) = e,
($\epsilon\varpi$) $\vdash$ ϵz.(z⋆ϖ(a[z])) = ϵz.a[z],

so that ∂ can be formulated equivalently (Rezuş 2017, 2017a) as an equational theory $\epsilon\varpi$ with ($\beta\epsilon$), ($\eta\epsilon$), ($\beta\varpi$), ($\epsilon\varpi$) and the expected monotony conditions.

§ 1.2.5 *Kolmogorov β-only mono-folds.* Any extensional mono-fold $\natural$ can simulate a genuine β-only fold, its underlying *Kolmogorov fold* λ_β^K, say, based on $[\lambda^K, \triangleright^K]$, defined by $\lambda^K x.b[x] := \langle\natural x.b[x]\rangle$ and $c{\triangleright}^K a := \natural y.(c{\triangleright}\langle a,y\rangle)$, y not free in a, c.[8]

§ 1.3 *Pairings and cartesian folds.* A *pairing* is a triple [π,π$_1$,π$_2$] – where π is a binary operation and π$_1$, π$_2$ are singulary [single-place] operations, with term-forms $\prec$a,b$\succ$:= π(a,b), **j**(c) := π$_j$(c), j :=1, 2, resp.– satisfying:

($\beta\pi_1$) $\vdash$ **1**($\prec$a,b$\succ$) = a,
($\beta\pi_2$) $\vdash$ **2**($\prec$a,b$\succ$) = b,

together with the expected monotony conditions.

A pairing is *extensional* (or *surjective*) if it satisfies also:

($\eta\pi$) $\vdash$ $\prec$**1**(c),**2**(c)$\succ$ = c.

An equational theory is said to be *sub-cartesian*, if it is equipped with a pairing, and *cartesian*, if it is equipped with an extensional pairing.

---

[8] For a proof-theoretical application, see Rezuş 2017a, **§ 2.3**, and Rezuş 2017b. Cf. also with the simulation of [λ,▷] in the (extensional) cartesian fold $\partial\pi$, discussed below (**§ 1.3**).

In particular, the extended λ-calculus **λπ** is an (extensional) cartesian fold.[9]

As shown above (under **§ 1.2.1.2 Frege-Church sequences**), every fold is sub-cartesian. Folds are not necessarily cartesian, however[10].

Less known is the fact that any cartesian (extensional) fold contains a di-fold, and, in fact, a n-fold, for each n > 1. Indeed, let $\partial\pi$ be a (extensional) cartesian mono-fold based on $[\partial,\star]$, with grammar[11]:

w-terms:: a, b, c, d, := x | ∂z.e | c ⋆ a | ≺a,b≻ | **1**(c) | **2**(c) | ...

satisfying, thus, at least $(\beta\partial)$, $(\eta\partial)$ and $(\beta\pi_1)$ $(\beta\pi_2)$, $(\eta\pi)$ and define:

(df λ) λx.b[x] := ∂z.(**2**(z) ⋆ b[x:=**1**(z)]), z fresh for b[x],
(df ▷) c ▷ a := ∂y.(≺a,y≻⋆c), y not free in a and c.

One has, immediately:

(β**p**) ⊢ ≺a,c≻ ⋆ f = c ⋆ (f ▷ a), and

($\beta\lambda$) ⊢ (λx.b[x]) ▷ a = b[x:=a],
($\eta\lambda$) ⊢ λx.(c▷x) = c (x not free in c),

as well as the expected monotony conditions for the defined operations.

On a slightly different plan, one can define, in any cartesian mono-fold $\partial\pi$, a (pure) dyadic *split* abstraction operator $\int$:

(df $\int$) $\int$[x,y].c[x,y] := ∂z.c[x:=**1**(z),y:=**2**(z)] (z fresh for c[x,y], x≠y).

The latter yields, once more:

(λ) ⊢ λx.b[x] = $\int$(x,y).(y⋆b[x]), y not free in b[x], and

($\int \lambda\partial$) ⊢ $\int$(x,y).e[x,y] = λx.∂y.e[x,y],

($\beta \int$) ⊢ ≺a,b≻ ⋆ ($\int$(x,y).c[x,y]) = c[x:=a,y:=b],
($\eta \int$) ⊢ $\int$(x,y).(≺x,y≻ ⋆ c) = c (x,y not free in c),
($\zeta \int$) ⊢ ∂z.e[z] = $\int$(x,y).e[z:=≺x,y≻] (x,y fresh for e[z]),

as well as:

($\beta\lambda^*$) ⊢ ≺a,c≻ ⋆ (λx.b[x]) = c⋆(b[x:=a]),

---

[9] The fact that there are β-only cartesian folds (not necessarily extensional, thus) has been first shown model-theoretically, by Dana S. Scott (the Plotkin-Scott Graph Model, cf. Scott 1976).

[10] In view of Barendregt 1974, extensional pairing is not definable in pure λ-calculus. As a matter of fact, this important negative result can be extended to cyclic mono-folds, to n-folds, in general, as well as to some proper extensions thereof. Practically, except for $\partial\pi(\Pi)$, all equational theories of concern here are sub-cartesian: as regards logic applications, the world of sub-cartesianness is rather vast.

[11] This is just a typographically (!) isomorphic copy of the extended λ-calculus, with ∂ for λ and cuts written in the reversed order (as in Frege **GGA** 1:1893).

$(*) \equiv (\eta\mathbf{p}) \vdash \lambda x.\partial y.(\prec x,y\succ \star c) = c$ (x, y not free in c),
$(*) \equiv (\zeta\mathbf{p}) \vdash \partial z.e[z] = \lambda x.\partial y.e[z:=\prec x,y\succ]$ (x,y fresh for e[z]).

In general, one can define

(λ_0) $\lambda_0 x.b := \partial x.b[x]$,
$(\triangleright_0)$ $c \triangleright_0 a := a \star c$, and for $n \geq 0$,
(λ_{n+1}) $\lambda_{n+1}x.b[x] := \lambda_n z.(b[x:=\mathbf{1}(z)] \triangleright_n \mathbf{2}(z))$ (z fresh for b[x]),
$(\triangleright_{n+1})$ $c \triangleright_{n+1} a := \lambda_n y.(c \triangleright_n \prec a,y\succ)$ (y not free in a, c),

whence, for any $n \geq 0$,

$(\beta\lambda_n) \vdash (\lambda_n x.b[x]) \triangleright_n a = b[x:=a]$,
$(\eta\lambda_n) \vdash \lambda_n x.(c \triangleright_n x) = c$ (x not free in c),

by induction on n. For an alternative iteration, see Rezuş 2017, 2017a.

§ **1.4** *Scalar theories.* For theories based on a scalar syntax, a *scalar* (or a mixed) *pairing* is a triple $[\Pi,\Pi_1,\Pi_2]$ – where Π is a binary operation and Π_1, Π_2 are singulary [single-place] operations, with term-forms [w-terms] $\downarrow_t(c) := \Pi(t,c)$, $\theta(c) := \Pi_2(c)$, and [scalars] $\iota(c) := \Pi_1(c)$ – satisfying:

$(\beta\Pi_1) \vdash \iota(\downarrow_t(c)) \doteq t$,
$(\beta\Pi_2) \vdash \theta(\downarrow_t(c)) = c$,
$(\eta\Pi) \vdash \downarrow_t(\theta(c)) = c$, if $\iota(c) \doteq t$,

together with the expected monotony conditions for the primitive operations. An equational theory is *scalar* if it is equipped with a scalar pairing.[12]

§ **1.4.1** *Scalar mono-folds.* Let $\partial\Pi$ be a scalar (extensional) mono-fold based on $[\partial,\star]$, i.e., with grammar including:

scalars :: s, t := u | ι(a) | ...
w-terms:: a, b, c, d, e := x | ∂z.e | c $\star$ a | $\downarrow_t$(c) | θ(c) | ...

and satisfying, thus, at least:

$(\beta\partial) \vdash c \star (\partial z.e[z]) = e[z:=c]$,
$(\eta\partial) \vdash \partial z.(z \star a) = a$ (z not free in a),

$(\beta\Pi_1) \vdash \iota(\downarrow_t(c)) \doteq t$,
$(\beta\Pi_2) \vdash \theta(\downarrow_t(c)) = c$,
$(\eta\Pi) \vdash \downarrow_t(\theta(c)) = c$, if $\iota(c) \doteq t$,

together with the expected monotony conditions. Define, in $\partial\Pi$,

---

[12] A scalar pairing is, essentially, what is usually meant by *Hilbert's epsilon symbol*. Cf. Hilbert's 1922, 1923, 1928. Roughly speaking, the construction corresponds to the so-called *no-counterexample interpretation* of the universal quantifier (Georg Kreisel 1949–1950), already popular in the Hilbert school. See also § **2.2**, below.

(df Σ) Σ(u,x).e[u,x] := ∂z.e[u:=\(z),x:=θ(z)], z fresh for a[u,x],
(df Λ) Λu.a[u] := Σ(u,z).(z⋆a[u]), z fresh for a[u],
(df ▶) c ▶ t := ∂z.($\downarrow_t$(z)⋆c), z not free in c.

This yields:

($\beta\Lambda$) ⊢ (Λu.a[u]) ▶ t = a[u:=t],
($\eta\Lambda$) ⊢ Λu.(c▶u) = c, (u not free in c),

($\Sigma\Lambda\partial$) ⊢ Σ(u,x).e[z] = Λu.∂x.e[u,x],

($\beta\Lambda^*$) ⊢ $\downarrow_t$(c) ⋆ (Λu.a[u]) = c⋆(a[u:=t]),
($\eta\Lambda^*$) ⊢ Λu.∂z.($\downarrow_u$(z) ⋆ c) = c (u, z not free in c),
($\zeta\Lambda^*$) ⊢ ∂z.e[z] = Λu.∂x.e[z:=$\downarrow_u$(x)] (u, x fresh for e[z]),

($\beta\Sigma$) ⊢ $\downarrow_t$(a) ⋆ (Σ(u,x).c[u,x]) = c[u:=t,x:=a]),
($\eta\Sigma$) ⊢ Σ(u,x).($\downarrow_u$(x) ⋆ c) = c (u, x not free in c),
($\zeta\Sigma$) ⊢ ∂z.e[z] = Σ(u,x).e[z:=$\downarrow_u$(x)] (u, x fresh for e[z]),

and monotony conditions for the defined operations.

§ 1.5 *Derived witness operators:* $\lambda\gamma$- *and* $\lambda\varepsilon\varpi$-*calculi* (Rezuş 1986). Consider the di-fold $\boldsymbol{\lambda\partial}\mathbf{p}_\zeta$, based on [$\lambda$,▷] and [$\partial$,⋆], resp., satisfying thus

($\beta\lambda$) ⊢ (λx.b[x]) ▷ a = b[x:=a],
($\eta\lambda$) ⊢ λx.(c▷x) = c (x not free in c),

($\beta\partial$) ⊢ c ⋆ (∂z.e[z]) = e[z:=c],
($\eta\partial$) ⊢ ∂z.(z⋆a) = a (z not free in a),

and, additionally,

(β**p**) ⊢ ≺a,c≻ ⋆ f = c ⋆ (f ▷ a), and
(ζ**p**) ⊢ ∂z.e[z] = λx.∂y.e[z:=≺x,y≻] (x,y fresh for e[z]).

Formally, $\boldsymbol{\lambda\partial}\mathbf{p}_\zeta$ contains the bare di-fold $\boldsymbol{\lambda\partial} := \partial + \lambda$ and is a subsystem of $\boldsymbol{\partial\pi}$. Note also that, from ($\eta\partial$) and (ζ**p**), we can obtain:

(η**p**) ⊢ λx.∂y.(≺x,y≻ ⋆ c) = c (x, y not free in c),

as well, whence, in particular, one has also subsystems $\boldsymbol{\lambda\partial}\mathbf{p}_\beta := \boldsymbol{\lambda\partial} + (\beta\mathbf{p})$ and $\boldsymbol{\lambda\partial}\mathbf{p}_{\beta\eta} := \boldsymbol{\lambda\partial} + (\beta\mathbf{p}) + (\eta\mathbf{p})$. (See below for the nomenclature.)
Define now (in $\boldsymbol{\lambda\partial}\boldsymbol{\mu}_\zeta$)

(df μ) μ(a,b) := λz.(b ▷ (z ▷ a)) (z not free in a, b).
(df $\dot{\tau}$) $\dot{\tau}$(c) := λx.(c⋆x) (x not free in c),
(df γ) γz.e[z] := ∂z.e[z:=$\dot{\tau}$(z)].

We have, first

($\beta\dot{\tau}$) ⊢ $\dot{\tau}$(c) ▷ a = c ⋆ a,

3 Cartesian folds and Witness Theory (2017b)

$(\mu\dot{\tau})\ \vdash \mu(a,\dot{\tau}(b)) = \dot{\tau}(\prec a,b\succ)$,

whence also

$(\bar{\beta}\gamma)\ \vdash \gamma x.(x\triangleright(\gamma y.e[x,y])) = \gamma z.e[x,y:=z]$,
$(\eta\gamma)\ \vdash \gamma z.(z\triangleright a) = a$ (z not free in a),
$(\zeta\gamma)\ \vdash \gamma z.e[z] = \lambda x.\gamma y.e[z:=\mu(x,y)]$,

$(\dot{\tau}\gamma)\ \vdash \dot{\tau}(c) \triangleright \gamma z.e[z] = e[z:=\dot{\tau}(c)]$,
$(\xi\gamma)\ \vdash \gamma z.e_1 = \gamma z.e_2 \Rightarrow\ \vdash \gamma z.(f\triangleright e_1) = \gamma z.(f\triangleright e_2)$,

and, where $c[a] \equiv c[x:=a]$,

$(\hat{\beta}\gamma)\ \vdash \gamma x.c[x\triangleright(\gamma y.e[x,y])] = \gamma z.c[e[x,y:=z]]$,

i.e., we have obtained the calculi $\boldsymbol{\lambda\gamma}$ and $\boldsymbol{\lambda\gamma\dot{\tau}}$ of Rezuş 2017a.[13]
Furthermore, with

(df ε) $\varepsilon z.a[z] := \gamma z.(z\triangleright a[z])$,
(df ϖ) $\varpi(e) := \gamma z.e\ [\equiv \partial z.e]$ (z not free in e),

one has

$(\gamma\varepsilon\varpi)\ \vdash \gamma z.e[z] = \varepsilon z.\varpi(e[z])$,

and, from the latter group, we get

$(\hat{\beta}\varepsilon\varepsilon)\ \vdash \varepsilon x.c[x\triangleright\varepsilon y.a[x,y]] = \varepsilon z.c[z\triangleright a[x,y:=z]]$,
$(\eta\varepsilon)\ \vdash \varepsilon z.a = a$ (z not free in a),
$(\varepsilon\varepsilon)\ \vdash \varepsilon x.\varepsilon y.a[x,y] = \varepsilon z.a[x,y:=z]$
$(\zeta\varepsilon)\ \vdash \varepsilon z.a[z] = \lambda x.\varepsilon y.(a[z:=\mu(x,y)]\triangleright x)$, and

$(\hat{\beta}\varepsilon\varpi)\ \vdash \varepsilon z.c[z\triangleright\varpi(e[z])] = \varepsilon z.c[e[z]]$,
$(\zeta\varpi)\ \vdash \varpi(e) = \lambda x.\varpi(e)$ (x not free in e),
$(\varepsilon\varpi)\ \vdash \varepsilon z.\varpi(z\triangleright a[z]) = \varepsilon z.a[z]$,

giving rise to the calculi $\boldsymbol{\lambda\varepsilon(\varpi)}$ of Rezuş 2017a.

A digression: $\lambda\bar{\gamma}$-calculus. Consider the following modified grammar:

scalar atoms :: u, v, ...
w-atoms :: x, y, z, ...
w-terms :: a, b, c, d, e, f := x | λx.b | c $\triangleright$ a | $\bar{\gamma}$u.e | u $\ltimes$ a

---

[13] Here, $\boldsymbol{\lambda\gamma} := \boldsymbol{\lambda} + (\hat{\beta}\gamma) + (\eta\gamma) + (\zeta\gamma)$, and $\boldsymbol{\lambda\gamma\dot{\tau}} := \boldsymbol{\lambda\gamma} + (\dot{\tau}\gamma)$, including monotony conditions.

Here, the only scalar terms are the scalar atoms[14] [sic], and we have an implicit restriction on w-terms, viz. the $\bar{\gamma}$-abstractions $\bar{\gamma}u.e$ are such that, if the scalar u occurs free in e, then it occurs only in subterms of the form (u⋉a). We denote by $e[(u⋉a)::=(v⋉b)]$ the operation of replacing subterms of the form (u⋉a) by subterms of the form (v⋉b) in e.

Let $\boldsymbol{\lambda\bar{\gamma}}$ be the mono-fold based on $[\lambda,\triangleright]$, satisfying thus $(\beta\lambda)$ and $(\eta\lambda)$, with, moreover:

$(\beta\bar{\gamma})$ ⊢ v ⋉ ($\bar{\gamma}$u.e[u]) = e[u:=v],
$(\eta\bar{\gamma})$ ⊢ $\bar{\gamma}$u.(u⋉a) = a (u not free in a),
$(\zeta\bar{\gamma})$ ⊢ $\bar{\gamma}$u.e[u⋉f] = λx.$\bar{\gamma}$v.e[u⋉f::=v⋉(f⊳x)] (x, v fresh for e[u]),

where e[u:=v] stands for substitution of scalar atoms in w-terms (renaming).

Obviously, $\boldsymbol{\lambda\bar{\gamma}}$ can be interpreted in $\boldsymbol{\lambda\partial p_\zeta}$ (even in $\boldsymbol{\lambda\gamma\dot{\tau}}$); it can be thus viewed as a subsystem of $\boldsymbol{\partial\pi}$, as well.[15]

§ 1.6 *The world beyond.* In view of the *explicit definability* results mentioned in the above, the largest fold we need being concerned with is $\boldsymbol{\partial\pi\Pi}$, the scalar cartesian (extensional) mono-fold based on $[\partial,\star]$.[16] The context allows, however, some improvements.

§ 1.6.1 *Inversions.* An *inversion* is a pair $[\uparrow,\downarrow]$ of singular (single-place) operations with

$(\beta\uparrow)$ ⊢ $\downarrow(\uparrow(a))$ = a, and
$(\eta\uparrow)$ ⊢ $\uparrow(\downarrow(c))$ = c.

An equational theory is said to be *cyclic* if it contains an inversion.

One can establish easily the fact that any di-fold – whence also any cartesian mono-fold – contains at least one inversion.

Let $\boldsymbol{\rho\partial}$ be a di-fold, whith mono-folds $\boldsymbol{\rho}$ and $\boldsymbol{\partial}$ based on $[\rho,\cdot]$ and $[\partial,\star]$, resp., i.e., we have:

$(\beta\rho)$ ⊢ $(\rho x.e[x]) \cdot a$ = e[x:=a],
$(\eta\rho)$ ⊢ $\rho x.(c \cdot x)$ = c (x not free in c),

---

[14] Recall that scalars are not w-terms.
[15] The equational theory $\boldsymbol{\lambda\bar{\gamma}}$ is a *typographic variant* [sic] of the so-called $\lambda\mu$-calculus of Michel Parigot (cca 1991–1992). As a matter of fact, Parigot's term-syntax is more restrictive. Cf., e.g., Sørensen & Urzyczyn 2006 for a discussion and exact references.
[16] Technically, this is a *unessential extension* of the usual (extended) λ-calculus $\boldsymbol{\lambda\pi}$ (sic). So, we actually evolve between $\boldsymbol{\partial}$ and $\boldsymbol{\partial\pi\Pi}$. – Somewhat metaphorically, the difference between the 'pure' [extensional] mono-fold $\boldsymbol{\partial}$ and the scalar cartesian [extensional] mono-fold $\boldsymbol{\partial\pi\Pi}$ – both based on $[\partial,\star]$ – is that between the real unit segment [0,1] and an infinitely-dimensional vector space, or else, in proof-theoretical terms, that between the tiny 'logic' of minimal / intuitionistic implication and second-order classical logic, i.e., ultimately, classical analysis. – For consistency matters, see **§ 1.7**.

($\beta\partial$) ⊢ c ⋆ (∂z.e[z]) = e[z:=c],
($\eta\partial$) ⊢ ∂z.(z ⋆ a) = a (z not free in a),

and the expected monotony conditions. Define

(df Δ) Δ(c) := ∂z.(c · z) [z fresh for c] and
(df ∇) ∇(a) := ρx.(x ⋆ a) [x fresh for a];

this yields immediately the inversion [Δ,∇], with

($\beta\Delta$) ⊢ ∇(Δ(c)) = c, and
($\eta\Delta$) ⊢ Δ(∇(a)) = a.

Conversely, any cyclic mono-fold contains a di-fold. Indeed, let $\partial\Delta$, for instance, be a mono-fold based on [∂,⋆], satisfying ($\beta\partial$) and ($\eta\partial$), equipped with an inversion [Δ,∇], satisfying ($\beta\Delta$) and ($\eta\Delta$). Defining

(df ρ) ⊢ ρx.e[x] := ∂z.e[x:=Δ(z)],
(df ·) ⊢ c · a := ∇(a) ⋆ c,

yields the second mono-fold, ρ, based on [ρ,·], satisfying ($\beta\rho$) and ($\eta\rho$).
Alternatively, let $\rho\Delta$ be a mono-fold based on [ρ,·], satisfying ($\beta\rho$) and ($\eta\rho$), equipped with an inversion [Δ,∇] satifying ($\beta\Delta$) and ($\eta\Delta$). Defining now

(df ∂) ∂z.e[z] := Δ(ρz.e[z]), [sic]
(df ⋆) c ⋆ a := ∇(a) · c,

yields the mono-fold ∂, based on [∂,⋆], with

($\beta\partial$) ⊢ c ⋆ (∂z.e[z]) = e[z:=c],
($\eta\partial$) ⊢ ∂z.(z⋆a) = a (z not free in a).

The reader can easily establish the fact that a cartesian mono-fold contains infinitely many distinct inversions (and so infinitely many distinct cyclic mono-folds), a detail that might be of interest for group-theorists perhaps.

§ 1.6.2 *C-monoids.* From the above, we know that every (extensional) cartesian mono-fold ♮π contains a di-fold and thus also a [definable] *inversion* [↑,↓], with (β↑) and (η↑). An alternative way of establishing this consists of defining directly (cf., *mutatis mutandis*, the usual 'curry-ing' / 'un-curry-ing' operators):

(df ↑) ↑(c) := ♮x.♮y.(c≀≺x,y≻) [x, y fresh for c],
(df ↓) ↓(a) := ♮z.((a ≀ **1**(z)) ≀ **2**(z)) [z fresh for a].

Altogether, this means that every (extensional) cartesian mono-fold is a *C-monoid*, in the sense of Lambek (cf. Lambek & Scott 1986), i.e., that it is a cartesian monoid (in the category-theoretic sense) containing also an inversion. Explicitly, if ♮π in an extensional cartesian fold – based on [♮,≀] – where composition ∘ and identity I are as above, one can define *cartesian pairs*, by (df P) [a,b] := ♮z.≺a≀z,b≀z≻ [z fresh for a,b],

whence, by lifting the projections **j** (j := 1, 2) to ('categorical') combinators $\bar{1}$:= ♮z.**1**(z), $\bar{2}$:= ♮z.**2**(z), one checks easily that ♮π is a *cartesian monoid* (in the category-theoretic sense), i. e.,

(βP$_1$) ⊢ $\bar{1}$ ∘ [a,b] = a,
(βP$_2$) ⊢ $\bar{2}$ ∘ [a,b] = b,
(ηP) ⊢ [$\bar{1}$ ∘ c, $\bar{2}$ ∘ c] = c,

and thus a (Lambek) *C-monoid* (because of the inversion).

§ 1.6.3 *Derived projections.* Worth mentioning perhaps is also the fact that, in extensions of the di-fold **λ∂p**$_\beta$:= **λ∂** + (β**p**) – already mentioned, in passing, earlier; cf. also below – one can also simulate projections **p**$_j$ (j := 1,2) for the primitive pair-construct π. For convenience, let us turn (∫ λ∂) into a definition

(df $\bar{\int}$) ⊢ $\bar{\int}$(x,y).e[x,y] := λx.∂y.e[x,y],

setting, next [with z fresh for c],

(df **p**$_1$) **p**$_1$(c) := ∂z.(c ⋆ $\bar{\int}$(x,y).(z⋆x)),
(df **p**$_2$) **p**$_2$(c) := ∂z.(c ⋆ $\bar{\int}$(x,y).(z⋆y)).

Then (π,**p**$_1$,**p**$_2$) makes up a pairing, indeed, i.e., one has

(β**p**$_1$) ⊢ **p**$_1$(≺a,b≻) = a,
(β**p**$_2$) ⊢ **p**$_1$(≺a,b≻) = b,

although, unlike (π,π$_1$,π$_2$), it is *not* extensional.

§ 1.7 *Consistency matters.* The extended λ-calculus is known to be (Post-) consistent by a model-theoretic construction ('Scott-models', due to Dana S. Scott 1969; cf. Scott 1973). So, (extensional) cartesian mono-folds are consistent λ-theories.[17] The result extends easily to the corresponding scalar theories [sic]. This ensures consistency for each formal system (equational theory) taken into account here.

§ 1.8 *An inventory of sub-cartesian folds and scalar extensions.* The preceding discussion yields about a dozen of equationally distinct λ-calculi (most of them surveyed in Rezuş 2017, 2017a, i.e., folds in the present sense.

For further reference, we identify the main subsystems worth taking into account (monotony conditions are tacitly included). In general, we shall pay attention to *extensional* theories.

∂ := (β∂) + (η∂)

---

[17] Surprisingly, one can also show consistency – for ∂π, say – by syntactic means alone, 'constructively', so to speak. Cf., for instance, Støvring 2006. In the end, we don't really models in order to show consistency.

$$\boldsymbol{\lambda} := (\beta\lambda) + (\eta\lambda)$$
$$\boldsymbol{\lambda\partial} := \boldsymbol{\partial} + \boldsymbol{\lambda}$$

$$\boldsymbol{\lambda\partial p_\beta} := \boldsymbol{\lambda\partial} + (\beta p)$$
$$\boldsymbol{\lambda\partial p_{\beta\eta}} [\equiv \boldsymbol{\lambda\partial p}] := \boldsymbol{\lambda\partial} + (\beta p) + (\eta p)$$
$$\boldsymbol{\lambda\partial p_\zeta} := \boldsymbol{\lambda\partial} + (\beta p) + (\eta p) + (\zeta p)$$

$$\boldsymbol{\partial\lambda^*} := \boldsymbol{\partial} + (\beta\lambda^*) + (\eta\lambda^*)$$
$$\boldsymbol{\partial\lambda^*_\zeta} := \boldsymbol{\partial} + (\beta\lambda^*) + (\eta\lambda^*) + (\zeta\lambda^*)$$

$$\boldsymbol{\partial\!\int} := \boldsymbol{\partial} + (\beta\!\int) + (\eta\!\int)$$
$$\boldsymbol{\partial\!\int_\zeta} := \boldsymbol{\partial} + (\beta\!\int) + (\eta\!\int) + (\zeta\!\int)$$

As formulated in the above, the ζ-systems are slightly redundant.[18]

Not all systems are equationally distinct. Where $T_1 \preceq T_2$ means that T_1 is a subsystem of T_2, and $T_1 \simeq T_2$ stands for equational equivalence, one has

$$\boldsymbol{\lambda}, \boldsymbol{\partial} \preceq \boldsymbol{\lambda\partial} \preceq \boldsymbol{\lambda\partial p_\beta} \preceq \boldsymbol{\lambda\partial p_{\beta\eta}} \simeq \boldsymbol{\partial\lambda^*} \simeq \boldsymbol{\partial\!\int} \preceq \boldsymbol{\lambda\partial p_\zeta} \simeq \boldsymbol{\partial\lambda^*_\zeta} \simeq \boldsymbol{\partial\!\int_\zeta} \preceq \boldsymbol{\partial\pi}.$$

If, moreover, T is any one of the systems mentioned in the above, containing at least $\boldsymbol{\lambda\partial}$, one can define scalar extensions:

$$T\boldsymbol{\Lambda} := T + (\beta\Lambda) + (\eta\Lambda)$$
$$T\boldsymbol{\Lambda^*} := T + (\beta\Lambda^*) + (\eta\Lambda^*)$$
$$T\boldsymbol{\Lambda^*_\zeta} := T + (\beta\Lambda^*) + (\eta\Lambda^*) + (\zeta\Lambda^*)$$

$$T\boldsymbol{\Sigma} := T + (\beta\Sigma) + (\eta\Sigma)$$
$$T\boldsymbol{\Sigma_\zeta} := T + (\beta\Sigma) + (\eta\Sigma) + (\zeta\Sigma),$$

and we have also

$$T \preceq T\boldsymbol{\Lambda} \preceq T\boldsymbol{\Lambda^*} \simeq T\boldsymbol{\Sigma} \preceq T\boldsymbol{\Lambda^*_\zeta} \simeq T\boldsymbol{\Sigma_\zeta} \preceq \boldsymbol{\partial\pi\Pi}.$$

Everywhere in the above, the inclusions are *strict*.

To the lists above, one may also add systems containing an inversion $[\Delta, \nabla]$, taken as a primitive. The latter are contained in $\boldsymbol{\partial\pi(\Pi)}$, as well.

As mentioned before, the $\lambda\gamma(\dot{\tau})$-calculi are subsystems of $\boldsymbol{\lambda\partial p_\zeta}$. In particular, they would admit of scalar $(\Lambda(^*)$- and/or Σ-) extensions, too.

§ 2 Art déco ('typing')

§ 2.1 *Witness theories.* Formally, a *witness theory* is just a decorated fold, where a decoration (also called 'typing' sometimes[19] or 'stratification') is a

---

[18] Note that the equational theory called $\boldsymbol{\lambda\partial p}$ in Rezuş 2017, 2017a corresponds to $\boldsymbol{\lambda\partial p_\zeta}$, here. (In the above, we used, for convenience, the label $\boldsymbol{\lambda\partial p}$ as a mere abbreviation for the $\beta\eta$-theory $\boldsymbol{\lambda\partial p_{\beta\eta}}$.)

[19] As we do not know what is a 'type' in general and as all our 'types' are going to be logic formulas expressing propositions and/or propositional schemes, anyway – we should perhaps refrain from speaking about 'types', 'typing', 'type-assignment' and the like, in proof theory.

partial mapping ('assignment') from w-terms a[**u**,**x**], containing, possibly – even fictiously so – (pairwise distinct) atomic scalars from **u** := [u_1, ..., u_m] and (pairwise distinct) w-atoms from **x** := [x_1, ..., x_n], to triples

$$(\hat{\Gamma}[\mathbf{u},\mathbf{x}] \vdash a[\mathbf{u},\mathbf{x}] : A[\mathbf{u}]) := (\hat{\Gamma}[\mathbf{u},\mathbf{x}], a[\mathbf{u},\mathbf{x}], A[\mathbf{u}]),$$

reading: 'a[**u**,**x**] is a witness of A[**u**] relative to $\hat{\Gamma}$[**u**,**x**]', where

1) A[**u**] is a formula of classical logic in a given signature, and
(2) $\hat{\Gamma}$[**u**,**x**] – called *assumption context* – is a finite graph associating to each w-atom (i.e., w-variable) in **x** a formula possibly containing atomic scalar parameters from **u**.

In detail, for **u** := [u_1, ..., u_m] and **x** := [x_1, ..., x_n] as above, one has an assumption context $\hat{\Gamma}$[**u**,**x**] ≡ **u**,$\hat{\mathbf{x}}$, where $\hat{\mathbf{x}}$:= [x_1:A_1, ..., x_n:A_n] is a finite list of formula-decorated w-atoms, and the formulas A_i, ($0 < i < n+1$), may contain scalars from **u**. In particular, if the assumption context is empty, the (decorated) w-term a is a *combinator* (closed w-term) and A is a closed formula (expressing a *proposition*).

We exemplify, next, the case where the scalars are instantiated to so-called *individual terms* in (classical) first-order logic.[20]

For convenience, the preferred *primitive* signature for first-order classical logic will be [**f**,→,¬,∀], where **f** stands for *falsum*, denoting an arbitrary false proposition, → stands for material implication, ¬ stands for classical negation, and ∀ stands for the (first-order) universal quantifier. It is also convenient to set A↛B := ¬(A→B), as well as ∃u.¬A[u] := ¬(∀u.A[u]) [sic], perhaps.

Alternatively, the primitive [→,¬]-pair can be replaced by a primitive △ (i.e., the **nand**-connective), whereupon one can set ¬A := A△A, A∧B := ¬(A△B), A→B := A△¬B, etc., as usual.[21]

In presence of **f** (*falsum*), we have redundancy (since classical negation can be also defined 'inferentially'; we ignore the defined notion in such contexts).

The main tenet of witness theory consists of saying that

a (*logical*) *rule of inference is a witness operator* (and conversely).[22]

---

[20] The case where the scalars are meant to represent *propositionally quantified formulas* is completely analogous. See Rezuş 2015, 2017 for details. Although it requires some additional formal provisions, the extension to second-order classic logic is, more or less, straightforward. Cf., e.g., Church 1956, 1972.

[21] Analogousy, while putting **f** (*falsum*) back, for classical logic one can also use the slightly redundant primitive signature(s) [**f**,△,¬,(∀)], for instance (ignoring thus the fact that [classical] negation is also definable as self-incompatibility, etc. in this setting).

[22] In the sequel, we ignore the so-called 'structural rules'. Except for the so-called (CUT)-rule – Gentzen's *Schnitt*, a *substitution* provision used freely in our explicit definitions appearing in the above – they state, witness-theoretically, (*entailment*) *equivalences* [sic], and concern, essentially, formal proof-context / assumption manipulations. Cf., e.g., Rezuş 2009 or Rezuş 2019a for an account. Recall that, in

This yields a correspondence *witness operators* ≅ *(logical) rules of inference*, also known, for the intuitionistic case, as *Curry-Howard Correspondence* or *Isomorphism*.[23]

§ 2.2 *Basic rules of inference.* The *basic witness operators* – corresponding to *primitive rules of inference* in $\partial \pi \Pi$ – discussed in the above are:

(∂) $\hat{\Gamma} \vdash \partial x{:}\neg A.e[x] : A$, if $\hat{\Gamma}[x{:}\neg A] \vdash e[x] : \mathbf{f}$
  [*reductio ad absurdum*],
($\star$) $\hat{\Gamma} \vdash c \star a : \mathbf{f}$, if $\hat{\Gamma} \vdash c : \neg A$, and $\hat{\Gamma} \vdash a : A$
  [the 'law of non-contradiction'],

(π) $\hat{\Gamma} \vdash \prec a,c \succ \; : A \twoheadrightarrow B$, if $\hat{\Gamma} \vdash a : A$, and $\hat{\Gamma} \vdash c : \neg B$,
(π_1) $\hat{\Gamma} \vdash \mathbf{1}(c) : A$, if $\hat{\Gamma} \vdash c : A \twoheadrightarrow B$,
(π_2) $\hat{\Gamma} \vdash \mathbf{2}(c) : \neg B$, if $\hat{\Gamma} \vdash c : A \twoheadrightarrow B$,

(Π) $\hat{\Gamma} \vdash \downarrow_t(a) : \exists u.\neg A[u] \; [\equiv \neg \forall u.A[u]]$, if $\hat{\Gamma} \vdash a : \neg A[u{:=}t]$
  [*counterexample*],
(Π_1) $\hat{\Gamma} \vdash \backslash(c) \equiv t$, if $\hat{\Gamma} \vdash c : \exists u.\neg A[u]$
  [*counterexample-scalar*: 'some u such that ¬A[u]' (Hilbert's epsilon)],
(Π_2) $\hat{\Gamma}[v] \vdash \theta(c) : \neg A[v]$, if $\hat{\Gamma} \vdash c : \exists u.\neg A[u]$ (v fresh for c and $\hat{\Gamma}$)
  [*counterexample-witness*: a witness for ¬A[v], for some v = \(c)][24].

§ 2.3 *Derivable rules of inference.* All other witness operators defined in terms of the above correspond to *derivable rules of inference* and are *inheriting* of

---

witness theory, there is *no conceptual distinction* between a Natural Deduction and a Gentzen L-style presentation of (classical) logic. The latter refer actually to specific – misleading and delusional, both elliptic and redundant – visual, graphical and/or pictorial *representations of witness-terms*.

[23] See, e.g., Sørensen & Urzyczyn 2006. [Added in proof: 12 September 2019.] Notably, for the intuitionistic logic, the correspondence is one-one, while, for the classical logic – and logics sharing classical features – the correspondence is many-one, so that a witness operator can be also seen as a way of *disambiguating* the intuitive idea of a logical rule of inference. Cf. Rezuş 2019a, 2019b for a discussion. [In what follows, the turnstile ($\vdash$) states (also) the fact that a given w-term is *correctly decorated* ('typed'), so we can use, informally, both $\vdash$ and the definition sign (:=) in the same context.]

[24] With full formal details, one should have indexed the decorated pairings on formulas, writing, e.g., $\mathbf{j}_{A,B}(c)$ (j :=1,2), as well as $\backslash_A(c)$ and $\theta_A(c)$, for the appropriate 'projections', etc. In particular, 'Hilbert's epsilon' – here: $\backslash_A(c)$ – should correspond, colloquially, to the *indefinite article* (reading, in German: *ein v so daß c ist ein Beweis für ¬A[v]*), more or less as in the would-be original Greek reading of Chrysippus [sic]. Hilbert – cf. his 1923, etc. – had no explicit θ-operator because – like everybody else, in those times: the early twenties – he had no proof/witness-notation at all. The latter is, however, implicit in his informal explanations of the idea. In the end, two millennia away, Hilbert's views can be seen to fit the 'ancient' conceptual setting recovered in Rezuş 2016.

decorations from the basic ones. The reader can easily check by herself the following examples (as derived from the explicit definitions given previously):

($\int$) $\hat{\Gamma} \vdash \int(x{:}A, y{:}\neg B).e[x,y] : A{\rightarrow}B$, if $\hat{\Gamma}[x{:}A][y{:}\neg B] \vdash e[x,y] : \mathbf{f}$,

(λ) $\hat{\Gamma} \vdash \lambda x{:}A.b[x] : A{\rightarrow}B$, if $\hat{\Gamma}[x{:}A] \vdash b[x] : B$
[the 'deduction theorem'],

($\triangleright$) $\hat{\Gamma} \vdash (c \triangleright a) : B$, if $\hat{\Gamma} \vdash c : A{\rightarrow}B$, and $\hat{\Gamma} \vdash a : A$
[modus ponens],

(ρ) $\hat{\Gamma} \vdash \rho x.e[x] : \neg A$, if $\hat{\Gamma}[x{:}A] \vdash e[x] : \mathbf{f}$
[the intuitionistic $\neg$-introduction],

($\cdot$) $\hat{\Gamma} \vdash c \cdot a : \mathbf{f}$, if $\hat{\Gamma} \vdash c : \neg A$, and $\hat{\Gamma} \vdash a : A$ [sic]
[the intuitionistic 'law of non-contradiction'][25],

(Σ) $\hat{\Gamma} \vdash \Sigma(u, x{:}\neg A[u]).e[u,x] : \forall u.A[u]$, if $\hat{\Gamma}[u][x{:}\neg A] \vdash e[u,x] : \mathbf{f}$
[indirect generalisation from no counterexamples],

(Λ) $\hat{\Gamma} \vdash \Lambda u.a[a] : \forall u.A[u]$, if $\hat{\Gamma}[u] : a[u] : A[u]$
[direct generalisation],

($\blacktriangleright$) $\hat{\Gamma} \vdash c \blacktriangleright t : A[u{:=}t]$, if $\hat{\Gamma} \vdash c : \forall u.A[u]$
[direct instantiation].

For the derived witness operators $\gamma, \varepsilon, \varpi, \bar{\gamma}, \ltimes$, mentioned above, we have (with $\sim A := A{\rightarrow}\mathbf{f}$):

(γ) $\hat{\Gamma} \vdash \gamma x{:}{\sim}A.e[x] : A$, if $\hat{\Gamma}[x{:}{\sim}A] \vdash e[x] : \mathbf{f}$
[inferential *reductio ad absurdum*], resp.

---

[25] Note that the pairs $[\partial, \star]$ and $[\rho, \cdot]$ are *distinct*, and, in particular, so are the corresponding cuts. If we leave out the witness information, we cannot make the due distinction between the corresponding rules of inference, of course. (Classically, they are distinct [equationally], in witness-theoretical terms. In intuitionism, the problem does not arise, since one of them [$\star$] is just missing, mainly because Brouwer claimed its (fold) cognate [i.e., ∂, here] is *onbetrouwbaar* ['unreliable', in Dutch].) On the other hand, confusing ('inferentially') ρ and λ, as well as $\triangleright$ (*modus ponens*) and $\cdot$ ('the law of non-contradiction', as understood intuitionistically) is, at best, a *practical expedient hiding a notational oversight*, if not even a *logical blunder*. — The real problem goes, however, beyond a mere ideological debate regarding classical vs intuitionistic logic, since both logics – and mainly the intuitionistic one, as justified by the BHK interpretation of the intuitionistic implication – would eventually make $\rightarrow$ into a *paradigmatic carrier of inference*, via the Official Deduction Theorem (λ). Ultimately, classical logic can avoid this (ideological) inconvenience, by explaining the classical $A{\rightarrow}B$ as $A \triangle \neg B$, i.e., by taking logical incompatibility – or logical conflict ($\triangle$, **nand**) – as a starting point, à la Chrysippus, in a way (see the discussion of *fibrations* in Rezuş 2016). There is no such a way out in intuitionism, however.

(ε) $\hat{\Gamma} \vdash \varepsilon$x:~A.a[x] : A, if $\hat{\Gamma}$[x:~A] $\vdash$ A
[inferential *consequentia mirabilis*],
(ϖ) $\hat{\Gamma} \vdash \varpi_A(e)$ [$\equiv \bar{\gamma}$x:~A.e, x not free in e] : A, if $\hat{\Gamma} \vdash$ e : **f**
[the *ex falso* rule], and (with primitive negation ¬, if necessary)
($\bar{\gamma}$) $\hat{\Gamma} \vdash \bar{\gamma}$u:~A.e[u] : A, if $\hat{\Gamma}$[u:~A] $\vdash$ e[u] : **f**
[scalar (restricted) *reductio ad absurdum*],
($\ltimes$) $\hat{\Gamma}$[u:~A] $\vdash$ u $\ltimes$ a : **f**, if $\hat{\Gamma} \vdash$ a : A
[scalar (restricted) cut].

On the alternative primitive signature [**f**,△,∀], we have:

(π) $\hat{\Gamma} \vdash \prec$a,b$\succ$: A∧B, if $\hat{\Gamma} \vdash$ a : A, and $\hat{\Gamma} \vdash$ b : B,
(π_1) $\hat{\Gamma} \vdash$ **1**(c) : A, if $\hat{\Gamma} \vdash$ c : A∧B,
(π_2) $\hat{\Gamma} \vdash$ **2**(c) : B, if $\hat{\Gamma} \vdash$ c : A∧B,

whence also

($\int$) $\hat{\Gamma} \vdash \int$(x:A,y:B).e[x,y] : A△B, if $\hat{\Gamma}$[x:A][y:B] $\vdash$ e[x,y] : **f**, etc.

§ 2.4 *More inversions*. Witness-theoretically, the inversions correspond to *transforms* from explicit (primitive) to 'inferential' negation and backwards, as well as to various 'double negation laws'.

(1) If the inversion [△,∇] is taken as a primitive – as, e.g., in $\partial\Delta$ or in $\rho\Delta$, based on [**f**,→,¬] – then we decorate it in the *intended* way, i.e.

(Δ) $\hat{\Gamma} \vdash \Delta$(c) : A, if $\hat{\Gamma} \vdash$ c : ¬¬A,
(∇) $\hat{\Gamma} \vdash \nabla$(a) : ¬¬A, if $\hat{\Gamma} \vdash$ a : A.

(2) One checks easily the fact that the decorated di-fold $\rho\partial$[**f**,→,¬] yields (from the definitions of **§ 1.6.1**) the *intended decoration* for the inversion [△,∇].

(3) Consider now the 3-fold $\lambda\rho\partial$, obtained by adding the mono-fold $\rho := [\rho,\cdot]$, satisfying ($\beta\rho$) and ($\eta\rho$), to the di-fold $\lambda\partial$. Obviously, $\lambda\rho\partial$ is consistent (it is a subsystem of $\lambda\pi$), and so is its decorated version [witness theory] $\lambda\rho\partial$[**f**,→,¬]. In particular, $\lambda\rho\partial$ has $\lambda\rho$ and $\lambda\partial$ as 'natural' fragments, so to speak, and similarly for their corresponding decorated cognates, on the same signature. Let, as ever, ~A := A→**f** ['inferential classical negation'].

(3.1) Define first, in the $\lambda\rho$[**f**,→,¬]-segment, ignoring context-parametrisations:

(df τ) $\vdash \tau$(c) := λx:A.(c·x) : ~A, if $\vdash$ c : ¬A (x not free in c),
(df $\bar{\tau}$) $\vdash \bar{\tau}$(c) := ρx:A.(c⊳x) : ¬A, if $\vdash$ c : ~A (x not free in c).

One can check easily the fact that we obtain an inversion [$\tau,\bar{\tau}$], with

($\beta\tau$) $\vdash \bar{\tau}(\tau$(c)) = c, if c : ¬A,
($\eta\tau$) $\vdash \tau(\bar{\tau}$(c)) = c, if c : ~A.

(3.2) Next, set, in the $\lambda\partial$[**f**,→,¬]-segment, this time,

(df ↑) ⊢ ↑(c) := ∂z:¬A.(c ▷ z) : A, if c : ∼¬A (z not free in c),
(df ↓) ⊢ ↓(a) := λz:¬A.(z ⋆ a) : ∼¬A, if a : A (z not free in a).

The pair [↑,↓)] does not make up an inversion, but one can set, in the full system **λρ∂[f,→,¬]**,

(df Δ) ⊢ Δ(c) := ↑(τ(c)) : A, if c : ¬¬A,
(df ∇) ⊢ ∇(a) := τ̄(↓(a)) : ¬¬A, if a : A.

This yields (cf. **§ 1.6.1**)

(Δ) ⊢ Δ(c) = ∂z:¬A.(c · z) : A, if c : ¬¬A (z fresh for c) and
(∇) ⊢ ∇(a) = ρx:¬A.(x ⋆ a) : ¬¬A, if a : A (x fresh for a),

whence also the (classically) intended equational behaviours [(βΔ),(ηΔ)], turning the defined notions into an inversion [Δ,∇], i.e.,

(βΔ) ⊢ ∇(Δ(c)) = c, for ⊢ c : ¬¬A, and
(ηΔ) ⊢ Δ(∇(a)) = a, for ⊢ a : A.

§ 2.5 Note that **λρ∂[f,→,¬]** – a witness theory / proof system for the full (quantifier-free) classical logic – is just a *genuinely classical* extension of the [f,→,¬]-fragment of the *Minimalkalkül* (Johansson 1937). The only uncommon feature of this equational theory [an extension of the pure λ-calculus] consists of the fact one has *two primitive witness operators* corresponding to 'the law of non-contradiction', viewed as a rule of inference. From the above, it is obvious that one can obtain exactly the same effect by assuming that the double negation 'laws' – taken, both, as primitive notions (here: witness operators), more or less as in Frege's **Begriffsschrift** (Frege 1879) – make up an inversion. [Added in proof (29 October 2019). Actually, if we assume a *double negation isomorphism* – via equational inversion-conditions – on the top of the minimal system **λρ[f,→,¬]**, we end up with *infinitely many equationally distinct witness operators* corresponding to the 'law of non-contradiction' (viewed, intuitively, as an inference rule). On the other hand, in the 'ρ-free', still genuinely classical, subsystem **λ∂[f,→,¬]** we do not encounter this kind of proliferation, but we loose (equationally) the minimal – intuitionistically valid – segment **λρ[f,→,¬]**. (Incidentally, in **λ∂[f,→,¬]**, a conceptually simple, 'constructive' rule of inference, like the so-called 'negation introduction' (ρ) is going to be simulated in a roundabout way, in genuinely classical terms, as, e.g., via specific uses of the ∂-operator – representing the classical *reductio ad absurdum* – and the Δ-operator, representing the classical 'double negation elimination' rule.) In 'inferential' λγ-calculi we do not encounter such proliferations either, but this is because the 'law of non-contradiction' is therein simulated by a specific instance of *modus ponens*. See Rezuş 2019a for details.]

§ 3 Acknowledgements

Most of the results surveyed in **§ 1** go back to a research project bearing the title *Subsystems of type-free λ-calculus and combinatory logic*, submitted, in

1978, to the Chair of Dirk van Dalen [Logic and Foundations of Mathematics], University of Utrecht, subsequently supervised by Henk Barendregt. The title was conveniently found by Dirk van Dalen, and was meant to cover some specific work on proof theory I have done previously, during the mid- and late seventies. The initial project concerned, mainly, the proof theory of *relevance*, *modal* and *intuitionistic logics*.[26]

The interest in the proof theory of *classical* logic emerged explicitly a bit later, during the early and mid-eighties, when I was involved in research on various 'typed' λ-calculi, directly related to the formalisation of mathematical proofs in λ-calculus terms, as, e.g., the AUTOMATH systems of Nicolaas G. de Bruijn (Eindhoven University of Technology) – cf. de Bruijn 1978, 1980, 1981, Rezuş 1983, Barendregt & Rezuş 1983 – the Girard-Reynolds 'impredicative' type-theory, Rezuş 1986, and the constructive type theory of Per Martin-Löf, Rezuş 1986a.

The earliest *public* reference, advertising the subject – labelled *witness theory* here – for a specific λ-calculus audience was likely the Nijmegen talk of 1988, repeated in Karlsruhe (for GMD, the German National Research Center for Mathematics and Computing Science [*Datenverarbeitung*]), in Groningen (for the *Informatica Colloquium* of the Computer Science Department), in Rome (for the Mathematical Institute 'Guido Castenuovo', University *La Sapienza*), and in several other places, during 1989–1993. A formal exposition of the first-order ['typed'] $\lambda\gamma$-calculi is contained in the lectures Rezuş 1990 and in Rezuş 1991 (see also the extended abstract Rezuş 1993 contributed to the *Dirk van Dalen Festschrift*, Utrecht 1993).

The specific applications of witness theory to *relevance logics* have been recorded in several [unpublished] reports, dated 1987–1988, meant for (American and Australian) colleagues working in the area. For relevance logics – with or without modalities – vs subsystems called 'linear' later on, see Rezuş 1987.[27]

---

[26] Although the logic described in Rezuş 1981 – Church's 'ordinal logic' – was, at bottom, *classical* in nature, the *transfinite ordinal hierarchy*, including *Peano arithmetic* and a fragment of *classical analysis*, retained a *relevance logic* flavour, in view of the fact that the construction was based on Church's |! strict $\lambda\delta$-calculus.

[27] In theoretical computer science, the term 'linear', as applied to logic, emerged – more or less as an advertising label – around 1987, in connection with Girard's paper, Girard 1987 [otherwise *not properly refereed* before going to print], although the 'technical' terminology – referring to *strictness* and *linearity* in λ-calculus and in witness theory – goes back to colloquialisms I used publicly about a decade before. To wit: "The name *linear λ-term* originated with Adrian Rezuş around 1979; Komori 1987 calls the same class of terms BCK λ-*terms*." (Hindley 1989). Taking things modulo Curry-Howard, Jean-Yves Girard used, apparently, the term 'linear' (first in lectures, Paris 7, 1986–1987) in the same sense I used the label 'strictly linear', while reserving the term 'affine' for what I called 'linear'. In the end, this is not too important, since, anyway, in logic and in λ-calculus, the contrast BCK- vs BCI- [linear vs strictly linear] goes back to Carew A. Meredith (cca 1950), a former student of Jan Łukasiewicz in Dublin, while 'linearity' – in my sense – goes back to the PhD Dissertation of Frederic B. Fitch (Yale University, New Haven CT 1934).

The earlier software-side of the story has been documented mainly for the Nijmegen Department of Computer Architecture and Operating Systems, around 1983–1984, and, later on, for the Dutch National Science Research Foundation (NWO, previously ZWO), 1984–1987.

The most general theoretical setting ('witness theory'), as described in the above – as well as in Rezuş 2017, 2017a – has been disseminated lately in private and/or in informal public (resp. online) talks.

Historical issues have been addressed in Rezuş 2009, 2010, 2017, 2019, 2019a, 2019b.

Older and/or historical references alluded to – but not mentioned explicitly – in the main text can be retrived from Rezuş 1981b, 1982a (λ-calculus) and the bibliography of Rezuş 1991 (proof theory, BHK, and Curry-Howard), updated in 2000.

On the personal side, I am indebted, in various ways to many λ- and logic-colleagues and friends, including the late Robert K. Meyer (ANU, Canberra ACT), Nicolaas G. de Bruijn (Eindhoven) – the actual founder of witness theory – Corrado Böhm (Rome), and Anne S. Troelstra (Amsterdam), as well as to Henk Barendregt (Utrecht and Nijmegen), Nuel Belnap Jr. (Pittsburgh PA), Dirk van Dalen (Utrecht), and, last but not least, J. Roger Hindley (Swansea, Wales UK).

9 June 2017 – 29 October 2019

4

Witness theory for inferential systems (2017a)

This note is a detailed *technical* comment on Rezuş 2017, § 7 meant to establish the relation between the equational theory $\lambda\partial p$ – *loc. cit.*, § 7.4 – and *the $\lambda\gamma$-calculi* of Rezuş 1990, 1991, 1993. The text is, essentially, self-contained.

§ 1 *The calculus $\lambda\partial p$ and its subsystems.* The calculus $\lambda\partial p$ – a proper extension of the pure (extensional) λ-calculus λ – is the 'type-free' counterpart of a *witness theory for classical propositional logic* – based on *falsum* f, material implication [C] and negation [N], as primitives – suggested by and based on work done by Jan Łukasiewicz and his student Stanisław Jaśkowski (on *natural deduction*) during the early nineteen twenties. Cf. Rezuş 2017. On the other hand, $\lambda\gamma$ – also a proper extension of pure λ – is the 'type-free' counterpart of a *witness theory for classical propositional logic*, based on *falsum* and material implication alone, with negation defined *inferentially*, à la Peirce 1885. The latter (due to the present author: work done during the late seventies and the early eighties) goes back to much older findings due to Andreĭ Nikolaevič Kolmogorov (1925) and Valeriĭ Ivanovič Glivenko (1928, 1929) in connection with *intuitionistic logic*. See also Prawitz 1965 and Rezuş 1990, 1991, 1993.[1]

For convenience, we revisit, in § 1.1, Rezuş 2017, § 7.4, with full proofs. An (infinite) iteration similar to one presented in Rezuş 2017, § 7.5 is discussed in § 1.2. In § 1.3, we show that $\lambda\gamma$ is a subsystem of $\lambda\partial p$ [notation: $\lambda\gamma \preceq \lambda\partial p$]. Some variants and neighbours of $\lambda\gamma$ are taken into account in § 2. Finally, § 3

---

[1] Everywhere here, 'witness theory' is to be taken, more or less, in the sense of the *Curry-Howard Correspondence* (Haskell B. Curry, circa 1937, William A. Howard 1969). On the *historical* side, the label is rather a misnomer, since the 'correspondence' has been vastly anticipated by Alfred Tarski (1925) – somewhat *avant la lettre*, indeed, cf. Rezuş 1982, 2010 – Saunders MacLane (PhD Dissertation, Göttingen 1934), Gerhard Gentzen (circa 1938) – cf. von Plato 2017, p. 55 – Carew A. Meredith (during the early 1950's), Dag Prawitz (1965), Hans Läuchli (1965), Nicolaas G. de Bruijn (1967), etc. See also Sørensen & Urzyczyn 2006 and Rezuş 2017.

is devoted to the corresponding 'typed' systems, while **§ 4** mentions also some work *related to* $\lambda\gamma$ done by some other people during the 1990's.[2]

Recall first that, in Rezuş 2017, we had:

$$(\preceq_\partial) \quad \boldsymbol\lambda \preceq \boldsymbol{\lambda\partial} \preceq \boldsymbol{\lambda\partial p}_\beta \preceq \boldsymbol{\lambda\partial p} \preceq \boldsymbol{\lambda\pi},$$

where $T_1 \preceq T_2$ means that T_1 is a subsystem of T_2, $\boldsymbol\lambda$ is the pure (extensional) λ-calculus and $\boldsymbol{\lambda\pi}$ is the (extensional) λ-calculus with 'surjective pairing'. To make things self-contained, $\boldsymbol{\lambda\partial p}$ extends the pure λ-calculus $\boldsymbol\lambda$ – based on the usual [$(\beta\lambda)$-$(\eta\lambda)$]-conditions – with a monadic abstractor ∂, a binary 'cut'-operator $\star$, and pairs π (viz. with terms of the form: ∂z.e, c$\star$a and $\prec$a,b$\succ$, resp.), satisfying $(\beta\partial)$, $(\eta\partial)$ and $(\beta\mathbf{p})$, $(\eta\mathbf{p})$ together with the expected monotony conditions for the new operations. Formally, $\boldsymbol{\lambda\partial p}$ consists of the following (equational) 'postulates':

$(\beta\lambda)$ $\vdash$ $(\lambda$x.b[x]) $\triangleright$ a $=$ b[x:=a],
$(\eta\lambda)$ $\vdash$ λx.(c$\triangleright$x) $=$ c (x not free in c),

$(\beta\partial)$ $\vdash$ c $\star$ (∂z.e[z]) $=$ e[z:=c],
$(\eta\partial)$ $\vdash$ ∂z.(z$\star$a) $=$ a (z not free in a),

$(\beta\mathbf{p})$ $\vdash$ $\prec$a,c$\succ$ $\star$ f $=$ c $\star$ (f $\triangleright$ a).
$(\eta\mathbf{p})$ $\vdash$ ∂z.e[z] $=$ λx.∂y.e[z:=$\prec$x,y$\succ$].

The subsystem $\boldsymbol{\lambda\partial}$ is just the fragment without primitive pairs (π) and thus without $(\beta\mathbf{p})$ and $(\eta\mathbf{p})$, while $\boldsymbol{\lambda\partial p}_\beta$ is obtained from $\boldsymbol{\lambda\partial p}$ by leaving out $(\eta\mathbf{p})$. Putting $\boldsymbol{\lambda\partial}$ and $\boldsymbol{\lambda\partial p}_\beta$ aside, we are going to insert, in what follows, two more subsystems between the pure $\boldsymbol\lambda$ and $\boldsymbol{\lambda\pi}$.[3]

For further reference, $\boldsymbol{\lambda\pi}$ extends the pure λ-calculus $\boldsymbol\lambda$, with a 'surjective pairing', i.e., with pairs (π) and projections (π_j) – term-forms $\prec$a,b$\succ := \pi$(a,b), and $\mathbf{j}$(c) $:= \pi_j$(c), j $:= 1, 2$, supposed to satisfy:

$(\beta\pi_1)$ $\vdash$ $\mathbf{1}(\prec$a,b$\succ)$ $=$ a,
$(\beta\pi_2)$ $\vdash$ $\mathbf{2}(\prec$a,b$\succ)$ $=$ b,
$(\eta\pi)$ $\vdash$ $\prec\mathbf{1}$(c),$\mathbf{2}$(c)$\succ$ $=$ c.

§1.1 ($\preceq_\partial$) We show first $\boldsymbol{\lambda\partial p} \preceq \boldsymbol{\lambda\pi}$, in some detail. For convenience, we use a notational shift for the pure part ($\boldsymbol\lambda$) of $\boldsymbol{\lambda\pi}$, by renaming $\flat \equiv \lambda$ and $\top \equiv \triangleright$, so that the pair [($\flat$),($\top$)] will be said to satisfy ($\beta\flat$) and ($\eta\flat$), and so on, whereas the corresponding calculus becomes $\boldsymbol{\flat\pi}$ with 'pure' part $\boldsymbol\flat$.[4]

Define, in $\boldsymbol{\flat\pi}$ [$= \boldsymbol{\lambda\pi}$]:

---

[2] In retrospect, none of them were, apparently, aware of [1] the (historical) connections with early antecedents in logic nor with [2] the (technical) – otherwise rather obvious – relation of their subject to $\boldsymbol{\lambda\pi}$ and to the closely related category theoretic lore dating from the early (nineteen) seventies.
[3] In general, everywhere here, all inclusions are *strict*.
[4] This notational artifice is just to save typography.

(df ↑) ↑(c) := ♮z.((c ⊤ **1**(z)) ⊤ **2**(z)),
(df ↓) ↓(f) := ♮x.♮y.(f⊤≺x,y≻),
(df ∂) ∂z.e[z] := ↑(♮z.e[z]),
(df ⋆) c ⋆ a := ↓(a) ⊤ c,
(df ∫) ∫[x,y].c[x,y] := ∂z.c[x:=**1**(z),y:=**2**(z)], z fresh for c[x,y],
(df λ) λx.b[x] := ∂z.(**2**(z)⋆b[x:=**1**(z)]), z fresh for b[x],
(df ▷) c ▷ a := ∂y.(≺a,y≻⋆c), y not free in a and c.

Remark. Note that

(λ∫) ⊢ λx.b[x] ≡ ∫(x,y).(y⋆b[x]), y not free in b[x].

1.1.1 Theorem. In ♮π [≡ λπ], one has:

(β↑) ⊢ ↓(↑(a)) = a,
(η↑) ⊢ ↑(↓(c)) = c,
(β∂) ⊢ c ⋆ (∂z.e[z]) = e[z:=c],
(η∂) ⊢ ∂z.(z⋆a) = a (z not free in a),
(βλ) ⊢ (λx.b[x]) ▷ a = b[x:=a],
(ηλ) ⊢ λx.(c▷x) = c (x not free in c),
(β**p**) ⊢ ≺a,c≻ ⋆ f = c ⋆ (f ▷ a),
(η**p**) ⊢ ∂z.e[z] = λx.∂y.e[z:=≺x,y≻],
(βλ*) ⊢ ≺a,c≻ ⋆ (λx.b[x]) = c⋆(b[x:=a]),
(ηλ*) ⊢ λx.∂y.(≺x,y≻ ⋆ c) = c (x, y not free in c),
(β∫) ⊢ ≺a,b≻ ⋆ (∫(x,y).c[x,y]) = c[x:=a,y:=b],
(η∫) ⊢ ∫(x,y).(≺x,y≻ ⋆ c) = c (x,y not free in c),
(∫λ∂) ⊢ ∫(x,y).e[x,y] = λx.∂y.e[x,y],
(∂∫) ⊢ ∂z.e[z] = ∫(x,y).e[z:=≺x,y≻].

Proof. One shows, successively: (β↑): by (β♮), (η♮) and (βπⱼ) [j := 1, 2]; (η↑): by (β♮), (η♮) and (ηπⱼ); (β∂): by (β♮) and (β↑); (η∂): by (η♮) and (η↑); (βλ): by (β∂), (η∂) and (βπⱼ) [j := 1, 2]; (ηλ): (β∂), (η∂) and (ηπⱼ) (β**p**): by (β∂) and (df ▷); (η**p**): by (β∂), (ηπ) and (df λ); (βλ*): by (βλ) and (β**p**); (ηλ*): by (η∂) and (η**p**); (β∫): by (β∂) and (βπⱼ) [j := 1, 2]; (η∫): by (η∂) and (ηπ); (∫λ∂): by (β∂); (∂∫): from (η**p**) and (∫λ∂). □

Remark. The Theorem proves (⪯∂). In fact, we have established, by the same token, slightly more, viz.

(⪯∂⁺) T ⪯ ♮π [≡ λπ], with T := λ∂**p**, ∂λ***p**, ∂∫**p**,

where $\partial\lambda^*\mathbf{p}$ and $\partial\int\mathbf{p}$ are as defined in Rezuş 2017, but we had equational equivalence ($\simeq$) for the three systems (*loc. cit.*, § 7.4), i.e. $\lambda\partial\mathbf{p} \simeq \partial\lambda^*\mathbf{p} \simeq \partial\int\mathbf{p}$.

§ 1.2 *Iterations.* In general, one can define

$\lambda_0 \text{x.b} := \partial \text{x.b}[x]$,
$c \triangleright_0 a := a \star c$, and for $n \geq 0$,
$\lambda_{n+1}\text{x.b}[x] := \lambda_n z.(b[x:=\mathbf{1}(z)] \triangleright_n \mathbf{2}(z))$, z fresh for b[z],
$c \triangleright_{n+1} a := \lambda_n y.(c \triangleright_n \prec a, y \succ)$,

whereupon one obtains the

1.2.1 Theorem. For any natural number n,

$(\beta\lambda_n) \vdash (\lambda_n \text{x.b}[x]) \triangleright_n a = b[x:=a]$,
$(\eta\lambda_n) \vdash \lambda_n \text{x.}(c \triangleright_n x) = c$ (x not free in c),

Proof. One shows both conditions simultaneously, by induction on n. The basis of the induction is the pair $[(\beta\partial),(\eta\partial)]$. □

Remark. The $[(\lambda),(\triangleright)]$-pair defined as above yields the usual pure 'eastern' λ-calculus, $\boldsymbol{\lambda} \equiv \boldsymbol{\lambda}^{\triangleright}$, as shown. The reader may try to go west, in order to obtain the 'western' variant $\boldsymbol{\lambda}^{\triangleleft}$, say, by defining, instead, a $[(\lambda^{\triangleleft}),(\triangleleft)]$-pair:

(df $\lambda^{\triangleleft}$) $\lambda^{\triangleleft}\text{y.a}[y] := \partial z.(\mathbf{1}(z)\star a[y:=\mathbf{2}(z)])$, z fresh in a[y],
(df $\triangleleft$) $c \triangleleft b := \partial x.(\prec x, b \succ \star c)$, y not free in b and c.

An analogous 'western' iteration of the construction is also available.

Remark (*Monoids and inversions, C-monoids*). As is well-known, the extensional pure λ-calculus (here: $\natural \equiv \boldsymbol{\lambda}$) is a monoid (under *composition* ∘, given by a∘b := $\natural z.(a_\top(b_\top z))$, with z not free in a,b, and *identity*, defined by I := $\natural z.z$), while $\natural\pi$ [≡ $\boldsymbol{\lambda}\pi$] – known as 'extended λ-calculus' in category theory – contains also a [definable] *inversion*, i.e., a pair $[(\uparrow),(\downarrow)]$ of singuary operations satisfying $(\beta\uparrow)$ and $(\eta\uparrow)$ above. Altogether, this means that $\natural\pi$ [≡ $\boldsymbol{\lambda}\pi$] is a C-monoid, in the sense of Lambek & Scott 1986. Explicitly, if $\natural\pi$ in as above, one can define *cartesian pairs*, by

(df P) $[a,b] := \natural z.\prec a_\top z, b_\top z \succ$,

whence, by lifting the projections **j** (j := 1, 2) to combinators $\bar{\mathbf{1}} := \natural z.\mathbf{1}(z)$, $\bar{\mathbf{2}} := \natural z.\mathbf{2}(z)$, one checks easily that $\natural\pi$ [≡ $\boldsymbol{\lambda}\pi$] is a cartesian monoid.

(βP_1) $\bar{\mathbf{1}} \circ [a,b] := a$,
(βP_2) $\bar{\mathbf{2}} \circ [a,b] := b$,
(ηP) $[\bar{\mathbf{1}} \circ c, \bar{\mathbf{2}} \circ c] := c$,

and thus a *C-monoid* (because of the inversion). The equational *equivalence* ('extended λ-calculi' ≃ C-monoids) is established in Lambek & Scott 1986.[5]

In view of *the infinite iteration effect* mentioned in the above – as well as in Rezuş 2017, § 7.5, on a slightly different route – C-monoids share an infinitistic feature with the 'extended λ-calculi', otherwise absent from CCC's [cartesian closed categories]: a C-monoid is an object that 'contains itself' in infinitely many distinct ways, so to speak, a rather weird algebraic structure, indeed.

§ 1.3 We consider, next, subsystems of **λ∂p**. Define, in **λ∂p**,

(df $\dot{\tau}$) $\dot{\tau}(c) := \lambda z.(c \star z)$, where z iz not free in c,
(df μ) $\mu(a,b) := \lambda z.(b \triangleright (z \triangleright a))$, where z is not free in a, b,
(df γ) $\gamma z.e[z] := \partial z.e[z := \dot{\tau}(z)]$.

With the usual (pure λ-calculus) shorthand notation $a \circ b := \lambda z.(a \triangleright (b \triangleright z))$ and $<c> := \lambda z.(z \triangleright c)$, where z is not free in a,b,c, we have, immediately:

($\dot{\tau}$) $\vdash \dot{\tau}(c) \triangleright a = c \star a$,
(μ) $\vdash \mu(a,b) = b \circ <a>$.

1.3.1 Lemma. ($\mu\dot{\tau}$) $\vdash \mu(a, \dot{\tau}(b)) = \dot{\tau}(\prec a, b \succ)$, for all a, b, in **λ∂p**.
Proof. Straightforward, using (βp). □

1.3.2 Theorem (**λγ** and **λγτ̇**). Where $c[a] \equiv c[x := a]$, one has, in **λ∂p**,

($\hat{\beta}\gamma$) $\vdash \gamma x.c[x \triangleright (\gamma y.e[x,y])] = \gamma z.c[e[x,y := z]]$,
($\eta\gamma$) $\vdash \gamma z.(z \triangleright a) = a$, if z is not free in a,
($\zeta\gamma$) $\vdash \gamma z.e[z] = \lambda x.\gamma y.e[z := \mu(x,y)]$,
($\dot{\tau}\gamma$) $\vdash \dot{\tau}(c) \triangleright \gamma z.e[z] = e[z := \dot{\tau}(c)]$.

Proof. The first two conditions are obtained by easy calculations. For ($\zeta\gamma$), use (ηp) and the ($\mu\dot{\tau}$)-lemma, while ($\dot{\tau}\gamma$) follows from ($\dot{\tau}$) and ($\beta\partial$). □

Let now **λγ** be the extension of pure λ-calculus **λ** with a new primitive (monadic) γ-abstractor – and terms γz.e – satisfying the conditions ($\hat{\beta}\gamma$), ($\eta\gamma$), ($\zeta\gamma$) above, as well as monotony for γ, and **λγτ̇** be the extension of **λγ** with a new primitive $\dot{\tau}$ – and terms $\dot{\tau}(c)$ – satisfying also the condition ($\dot{\tau}\gamma$) above, as well as monotony for $\dot{\tau}$.

We have, this time:

($\preceq_\gamma$)   **λ** $\preceq$ **λγ** $\preceq$ **λγτ̇** $\preceq$ **λ∂p** $\preceq$ **λπ**.

---

[5] Recall that a C-monoid (alternatively: a CCM, short for 'cartesian closed monoid') is what we obtain by leaving out the 'object'-part from a cartesian closed category (including the terminal object thus), so a C-monoid is, essentially, an algebra of 'arrows' [hom's]. Since a category can be viewed as a decorated ['typed'] monoid, a CCC is a just decorated CCM with an additional arrow – !, say – supposed to satisfy the condition $\vdash ! \circ a = !$ (in terms of λ-calculus, one has $\vdash !_f := \downarrow(f \circ \bar{2}) = K \triangleright f = \lambda x.f$ – x not free in f – where K is the usual combinator, so that $\vdash !_f \circ a = !_f$). C-monoids and weaker derivatives thereof, without ($\eta \uparrow$) and (ηP), occur in connection with work on *categorical models* of the ('type-free') λ-calculus.

Here, $\lambda\gamma$ is the decoration-free ('type-free') counterpart of the *basic* $\lambda\gamma$-*calculus* of Rezuş 1990, 1991, 1993.[6]

§ 2 *Variants and subsystems of* $\lambda\gamma$. A few remarks on variants, subsystems and neighbours might be useful.

§ 2.1 Note first that the 'full diagonalisation' condition ($\hat{\beta}\gamma$) – holding 'at any depth', so to speak – admits of an analysis into a 'surface diagonalisation' condition ($\bar{\beta}\gamma$) and a quasi-monotony rule ([ξ]γ), where

($\bar{\beta}\gamma$) ⊢ γx.(x▷(γy.e[x,y])) = γz.e[x,y:=z],
($\xi\gamma$) ⊢ γz.e_1 = γz.e_2 ⇒ ⊢ γz.(f▷e_1) = γz.(f▷e_2).

§ 2.2 If we leave out the 'infinitistic' condition ($\zeta\gamma$), the corresponding subsystem – $\lambda\gamma_{\mathbf{G}}$, say [with **G** short for 'Glivenko'] – is already contained in the pure λ-calculus, $\boldsymbol{\lambda}$. To see this, set γz.e[x] := λx.e[x:=<x>] (sic).[7]

§ 2.3 On the other hand, from ($\zeta\gamma$), we get, by ($\beta\lambda$),

($\zeta_\beta\gamma$) ⊢ γz.e[z] ▷ a = γz.e[z:=μ(a,z)].

It is an easy exercise in pure λ-calculus – more or less λ-folklore of the early sixties, if not older – to establish the fact that the system with ($\beta\lambda$) only – no ($\eta\lambda$) thus – and ($\hat{\beta}\gamma$), ($\eta\gamma$), ($\zeta_\beta\gamma$) – $\boldsymbol{\lambda_\beta\gamma_K}$, say [with **K** short for 'Kolmogorov'] – is already contained in pure $\boldsymbol{\lambda}$.

Indeed, where <a,b> := λz.(z▷a▷b) [with z not free in a, b] are the usual Frege-Church pairs, and <c> is as above, define, in $\boldsymbol{\lambda}$,

(df λ^K) λ^Kx.b[x] := <λx.b[x]>,
(df ▷^{K})   c ▷^{K} a := λy.(c ▷ <a,y>),
(df τ^K) τ^K(c) := <<a>> [sic], with *mutatis mutandis*,
(df μ^K) μ^K(a,b) := λ^Kz.(b▷^{K}(z▷^{K}a)), z not free in a,b.

Then the [(λ^K),(▷^{K})]-pair satisfies, *mutatis mutandis*, ($\beta\lambda^K$) – no ($\eta\lambda^K$), however – and we have an analogous ($\mu\dot{\tau}$)-lemma, reading, this time:

2.3.1 Lemma. ($\mu^K\tau^K$) ⊢μ^K(a,τ^K(b)) = τ^K(<a,b>), for all a, b, in pure $\boldsymbol{\lambda}$.
Proof. We have, first, an analogue [sic] of (β**p**), viz.
  (**p**) ⊢ (c ▷^{K} a) ▷ b = c ▷ <a,b>, by ($\beta\lambda$).
Next, after unpacking the definitions, we get
  (1) ⊢ τ^K(c) ▷ <a,b> = (a ▷ c) ▷ b, by ($\beta\lambda$), while
  (2) ⊢ τ^K(c) ▷^{K} a = a ▷ c,
follows from (1) and ($\eta\lambda$). The Lemma follows from (2) and (**p**) above. □

Finally, with γ^Kz.e[z] := λz.e[z:=τ^K(z)] (sic), we have the expected

---

[6] The full system of Rezuş 1990 amounts to a 'typed' variant of $\boldsymbol{\lambda\gamma\Lambda}$, with the additional primitive [(Λ)-(▶)]-pair, while Rezuş 1991, 1993 considered also an 'typed' extension $\boldsymbol{\lambda\gamma\pi\Lambda}$, with primitive pairs, projections and 'surjectivity of pairing'. For the [(Λ)-(▶)]-pair, see Rezuş 1990, 1991, 1993, 2017.

[7] This is, actually, what we get while reading Glivenko 1928, 1929 *in terms of proofs*.

2.3.2 Theorem (*The $\lambda_{\beta}\gamma_K$-calculus*). In pure $\boldsymbol\lambda$, the conditions $(\lambda\beta)$, $(\hat\beta\gamma)$, $(\eta\gamma)$, and $(\zeta_\beta\gamma)$ hold for λ, $\triangleright$, γ and μ, resp. replaced by λ^K, $\triangleright^K$, γ^K and μ^K, resp.
Proof. Straightforward calculations. For $(\zeta_\beta\gamma)$, use the $(\mu^K\tau^K)$-lemma. □

In $\boldsymbol\lambda$, we have the corresponding τ^K-extension of $\boldsymbol\lambda_{\beta}\boldsymbol\gamma_K$, as well. As already mentioned earlier, $(\beta\eta)$ fails for the 'Kolmogorov'-cognates. The reader may want to notice the fact that the simulation above reflects, ultimately, the essence of the so-called 'Kolmogorov translation' of classical propositional logic into a fragment of intuitionistic logic. See Rezuş 1991 [*Appendix 3*], for a discussion of the original translation.

§ 2.4 On a different route, the γ-segment of $\boldsymbol\lambda\boldsymbol\gamma$ admits of a 'linear decomposition', so to speak.[8] Define, in $\boldsymbol\lambda\boldsymbol\gamma$,

(df ε) $\varepsilon z.a[z] := \gamma z.(z\triangleright a[z])$,
(df ϖ) $\varpi(e) := \gamma z.e$, where z is not free in e.

We have, immediately:

$(\gamma\varepsilon\varpi)$ $\vdash \gamma z.e[z] = \varepsilon z.\varpi(e[z])$, and a

2.4.1 Lemma ($\boldsymbol\lambda\boldsymbol\varepsilon\boldsymbol\varpi$). In $\boldsymbol\lambda\boldsymbol\gamma$,

$(\hat\beta\varepsilon\varepsilon)$ $\vdash \varepsilon z.c[x\triangleright\varepsilon y.a[x,y]] = \varepsilon z.c[z\triangleright a[x,y:=z]]$,
$(\eta\varepsilon)$ $\vdash \varepsilon z.a = a$, if z is not free in a,
$(\varepsilon\varepsilon)$ $\vdash \varepsilon x.\varepsilon y.a[x,y] = \varepsilon z.a[x,y:=z]$
$(\zeta\varepsilon)$ $\vdash \varepsilon z.a[z] = \lambda x.\varepsilon y.(a[z:=\mu(x,y)]\triangleright x)$, and

$(\hat\beta\varepsilon\varpi)$ $\vdash \varepsilon z.c[z\triangleright\varpi(e[z])] = \varepsilon z.c[e[z]]$,
$(\zeta\varpi)$ $\vdash \varpi(e) = \lambda x.\varpi(e)$, if x is not free in e,
$(\varepsilon\varpi)$ $\vdash \varepsilon z.\varpi(z\triangleright a[z]) = \varepsilon z.a[z]$.

Proof. Easy calculations. [Note that $(\varepsilon\varepsilon)$ follows from $(\varepsilon\text{-}\varpi)$ conditions.] □

We can safely leave to the reader the task of formulating the alternative version, $\boldsymbol\lambda\boldsymbol\varepsilon\boldsymbol\varpi$ say, (equationally) equivalent to $\boldsymbol\lambda\boldsymbol\gamma$, based on $\{\lambda, \triangleright, \varepsilon, \varpi\}$ as primitives.[9]

---

[8] *Mutatis mutandis*, this holds for the pairs $[(\lambda),(\triangleright)]$ and $[(\partial),(\star)]$, as well.
[9] From $\boldsymbol\lambda\boldsymbol\varepsilon\boldsymbol\varpi$ one can extract its ϖ-free subsystem $\boldsymbol\lambda\boldsymbol\varepsilon$, say, based on $\{\lambda, \triangleright, \varepsilon\}$ alone. In particular, if $\bar z_j$, $0 < j < n+1$, $n > 0$, are all the occurrences of the free variable z in a[z], then, in view of $(\varepsilon\varepsilon)$, $\vdash \varepsilon z.a[z] = \varepsilon\bar z_1....\varepsilon\bar z_n.a[\bar z_1,...,\bar z_n]$, i.e., ε can be replaced by *stricty linear* uses of it, $\bar \varepsilon z.a[z]$, where z occurs in a[z] *exactly once*. The ε-operator – representing, after typing', the inferential version of Cardano's *consequentia mirabilis* (resp. the 'Rule of Clavius') – goes back to Haskell B. Curry, circa 1950. Notably, after 'typing' $\boldsymbol\lambda\boldsymbol\varepsilon$ is a witnes theory for the 'pure' part of his logic of 'strict negation'. (The first note, appearing in Curry's card index, mentioning the 'Law of Clavius' is a reference to / from Łukasiewicz, dated 29 September 1953 [information verified, in the Curry *Nachlass*, by Jonathan P. Seldin: correspondence 1992].) See Curry 1952, 1963, Seldin 1989, and Rezuş 1991 for details.

§ 2.5 At this point, the reader may want to notice that the analogous monadic abstractor $\epsilon z.a[z]$, say, defined, in $\boldsymbol{\lambda\partial p}$, by $\epsilon z.a[z] := \partial z.(z\star a[z])$, has a more transparent equational behaviour, viz., in $\boldsymbol{\lambda\partial p}$,

$(\beta\epsilon)$ $\vdash$ $c \star \epsilon z.a[z] = c \star a[z:=c]$,
$(\eta\epsilon)$ $\vdash$ $\epsilon z.a = a$, if z is not free in a,
$(\epsilon\epsilon)$ $\vdash$ $\epsilon x.\epsilon y.a[x,y] = \epsilon z.a[x,y:=z]$,
$(\zeta\epsilon)$ $\vdash$ $\epsilon z.a[z] = \lambda x.\epsilon y.(a[z:=\prec x,y\succ]\triangleright x)$,

while, with $\varpi(e) := \partial z.e$ [$\equiv \gamma z.e$], z not free in e, one has also:

$(\beta\varpi)$ $\vdash$ $c \star \varpi(e) = e$,
$(\epsilon\varpi)$ $\vdash$ $\epsilon z.\varpi(z\star a[z]) = \epsilon z.a[z]$, and
$(\partial\epsilon\varpi)$ $\vdash$ $\partial z.e[z] = \epsilon z.\varpi(e[z])$,

so that the $[(\partial),(\star)]$-segment of $\boldsymbol{\lambda\partial}$ admits of analogous 'linear decomposition' in terms of $\{\epsilon, \varpi, \star\}$. Indeed, in view of $(\partial\epsilon\varpi)$, one can show that $\boldsymbol{\lambda\partial}$ is equivalent to an equational theory – $\boldsymbol{\lambda\epsilon\varpi}$, say – based on the $[(\lambda),(\triangleright)]$-pair – satisfying the usual conditions $(\beta\lambda)$ and $(\beta\eta)$ – and a primitive triple $[(\epsilon),(\varpi),(\star)]$, satisfying $(\beta\epsilon)$, $(\eta\epsilon)$, $(\beta\varpi)$ and $(\epsilon\varpi)$. [Here, like for $(\epsilon\epsilon)$ above, $(\epsilon\epsilon)$ follows from the remaining ϵ-ϖ-conditions.]

§ 3 *'Typed' systems* ['*inferential witness theories*']. For the decorated ['typed'] systems $\boldsymbol{\lambda\gamma}[f,C]$ and $\boldsymbol{\lambda\gamma\dot\tau}[f,N,C]$, the *art déco* ['typing'] is as expected, viz., ignoring context-parametrisations:

[γ] $\vdash$ $\gamma z:C\alpha f.e[z] : \alpha$, if $[z : C\alpha f] \vdash e[z] : f$,
[$\dot\tau$] $\vdash$ $\dot\tau(c) : C\alpha f$, if $\vdash c : N\alpha$,

so $\dot\tau$ is, here, a kind of translation-device from a primitive negation $N\alpha$ to an 'inferential' negation, defined à la Peirce 1885 in terms of (material) implication [C] and *falsum* **f**, while the decorated γ is just the *inferential* version of *reductio ad absurdum*, so to speak.

Recall that, in $\boldsymbol{\lambda\partial p}$, ϵ is ['corresponds to'] *consequentia mirabilis*, also known as the 'Law of Clavius'. As noted earlier, ε, defined as above, is ['corresponds to'] the *inferential* version of *consequentia mirabilis*, while ϖ is ['corresponds to'] the usual *ex falso* rule (since we have $\vdash \varpi(e) = \partial z.e$, where z is not free in e, as well), i.e.

[ε] $\vdash$ $\varepsilon z:C\alpha f.a[z] : \alpha$, if $[z : C\alpha f] \vdash a[z] : \alpha$,
[ϖ] $\vdash$ $\varpi_\alpha(e) : \alpha$, if $\vdash e \cdot f$

Incidentally, the 'witness operator' corresponding to the *Rule of Peirce*, i.e., $C\alpha\beta \vdash \alpha \Rightarrow \vdash \alpha$, can be defined, in $\boldsymbol{\lambda\gamma}$, by:

(df $\wp$) $\vdash$ $\wp z:C\alpha\beta.a[z] := \varepsilon z:C\alpha f.a[z:=\lambda x:\alpha.\varpi_\beta(z\triangleright x)]$,

provided $[z:C\alpha\beta] \vdash a[z] : \alpha$, whence a 'witness' for the *Law of Peirce* would amount to a combinator ('typed' closed $\lambda\gamma$-term):

$\vdash \mathsf{P}_{\alpha,\beta} := \lambda\mathrm{y}{:}CC\alpha\beta\alpha.\varepsilon\mathrm{z}{:}C\alpha\mathbf{f}.(\mathrm{y}{\triangleright}(\lambda\mathrm{x}{:}\alpha.\varpi_\beta(\mathrm{z}{\triangleright}\mathrm{x}))) : CCC\alpha\beta\alpha\alpha.$

The extensions of the above to (propositional or first- resp. second-order) quantifiers are straightforward. See, *mutatis mutandis*, Rezuş 1990, 1991, 1993, 2017, for details.

§ 4 *Related work.* Niels Jakob Rehof and Morten Heine Sørensen 1994 studied several variants of **λγ** – called λ_{Δ}-*calculi* [10] – as well as 'typed' versions thereof, in connection with the the *the calculus of* control (*and* abort) of Matthias Felleisen (PhD Dissertation, Indiana University, Bloomington IN 1987); cf. also Rezuş 1991. Yet another variant of **λγ** – called *λμ-calculus* – with a more involved syntax (two sorts of variables, two sorts of terms, additional term-replacement operations etc.), has been proposed and studied by Michel Parigot (Paris 7) since about 1991–1992; cf., e.g., Parigot 1997 and Sørensen & Urzyczyn 2006. In view of computer science applications, the subject has been also addressed, on several occasions, during the 1990's, by Chetan R. Murthy (PhD Dissertation, Cornell University, Ithaca NY 1990) and by many others since. For details on *C-monoids* and *categorical models of the λ-calculus* – a subject originated with Joachim Lambek during the early 1970's – see, e.g., the PhD Dissertations of Karst Koymans (Utrecht 1984) and Pierre-Louis Curien (Paris 7, 1985), as well as the work of Adam Obtułowicz (Warsaw 1979 and later), Takanori Adachi (Tokyo 1982, 1983) and Hirofumi Yokouchi (Tokyo 1983), cited in the monograph Lambek & Scott 1986 and in the theses of Koymans (1984) and Curien (1985).

April 10 – June 26, 2017

---

[10] Their $\Delta\mathrm{z.e}$ is just $\gamma\mathrm{z.e}$, in the present syntax.

5

On the Kolmogorov $\lambda\gamma$-calculus (2017c)

The 'intensional', β-only fragment of the $\lambda\gamma$-calculus (Rezuş 1986–1987) is contained, modulo translation, in the 'pure' extensional $\lambda\beta\eta$-calculus, $\boldsymbol{\lambda}$. The idea behind this simulation goes back to Kolmogorov's [*O principe 'tertium non datur'*, Matematičeskiĭ Sbornik 32 (4), [30 XI] 1925, pp. 646–667] translation of classical propositional logic into Johansson's *Minimalkalkül* [*Der Minimalkalkül, ein reduzierter intutionischer Formalismus*, Compositio Mathematica 4 (1), 1937, pp. 119–136], a proper subsystem of intuitionistic propositional logic. We discuss an alternative, equivalent formulation (a 'linear decomposition') of the corresponding 'intensional' $\lambda\gamma$-calculus, based on proof operators corresponding to the intuitionistic *ex falso* rule [the Medieval's *ex falso quodlibet*] and to Cardano's genuinely classical rule *consequentia mirabilis*, also known as the 'Rule of Clavius'.

The following piece of – more or less trivial – λ-calculus folklore might explain, among other things, my apparently cryptical references to A. N. Kolmogorov (and V. I. Glivenko) while talking about the origins of the $\lambda\gamma$-calculus[1].

§ 1 *The Kolmogorov λ-calculi.* From Rezuş 2017b, **§1.2.5** and Rezuş 2017a, **§2.3**, it follows that the pure extensional λ-calculus $\boldsymbol{\lambda}$ ($\equiv \boldsymbol{\lambda}_{\beta\eta}$), based on [$\lambda$,$\triangleright$], can simulate the *Kolmogorov β-only mono-fold* $\boldsymbol{\lambda}_\beta^K$, say, based on [$\lambda^K$,$\triangleright^K$], i.e., a $\lambda\beta$-calculus, where, with the usual notation for Frege-Church tuples:

λ^Kx.b[x] := <λx.b[x]> $\equiv \lambda$z.λx.(z$\triangleright$b[x]),
z fresh for b[x], and
c$\triangleright^K$a := λy.(c$\triangleright$<a,y>) $\equiv \lambda$y.(c $\triangleright$ (λz.(z$\triangleright$a$\triangleright$y))),
y not free in a, c, and z $\neq$ y fresh for a.

---

[1] Cf. Rezuş 1990, 1991, 1993. Otherwise, the 'type-free' $\lambda\gamma$ and its decorated ('typed') version $\lambda\gamma$[**f**,$\rightarrow$] goes back (Nijmegen lectures, cca 1986–1987), explicitly, to Dag Prawitz's PhD Dissertation (Stockholm 1965). The historical background can be recovered from Rezuş 2017, and, *lato sensu*, from Rezuş 2016. For terminology and other technical details, see Rezuş 2017b.

This relies on an easy exercise, actually on a piece of λ-calculus folklore.[2]

Why 'Kolmogorov'? Because the basic idea behind this simulation is already implicit in the so-called 'Kolmogorov translation' (1925) of the classical propositional logic into Johansson's *Minimalkalkül* (1937), a proper subsystem of the intuitionistic propositional logic.[3] (Cf. Rezuş 2017b, **§ 1.2.5.**)

As a bonus (cf. Rezuş 2017a), the pure extensional λ-calculus $\boldsymbol{\lambda}$ can also simulate a 'Kolmogorov' γ-abstractor, $\gamma^K \equiv \hat{\gamma}$. Indeed, upon defining further:

$\hat{\tau}(c) := <<a>>$ [sic],
$\hat{\gamma}z.e[z] := \lambda z.e[z:=\hat{\tau}(z)]$ [sic], and
$\hat{\mu}(a,b) := \lambda^K z.(b \triangleright^K (z \triangleright^K a))$, z not free in a, b,

straightforward calculations yield, in pure $\boldsymbol{\lambda}$,

$(\hat{\mu}\hat{\tau})$ $\vdash \hat{\mu}(a,\hat{\tau}(b)) = \hat{\tau}(<a,b>)$,

i.e., the $(\mu^K \tau^K)$-Lemma of Rezuş 2017a, **§ 2.3**), whence also (*loc. cit.*, **§ 2.4**),

$(\hat{\beta}\hat{\gamma}^K_\beta)$ $\vdash \hat{\gamma}x.c[x \triangleright^K (\hat{\gamma}y.e[x,y])] = \hat{\gamma}z.c[e[x,y:=z]]$, with $c[a] \equiv c[x:=a]$,
$(\eta\hat{\gamma}^K)$ $\vdash \hat{\gamma}z.(z \triangleright^K a) = a$ (z not free in a), and
$(\zeta\hat{\gamma}^K_\beta)$ $\vdash (\hat{\gamma}z.e[z]) \triangleright^K a = \hat{\gamma}y.e[z:=\hat{\mu}(a,y)]$ (y fresh for a),

thereby obtaining a 'Kolmogorov' β-only ('type-free') $\lambda_\beta\gamma$-calculus $\boldsymbol{\lambda}^K_\beta\hat{\gamma} :=$ $\boldsymbol{\lambda}^K_\beta + (\hat{\beta}\hat{\gamma}^K) + (\eta\gamma^K) + (\zeta\gamma^K_\beta)$, say (expected monotony conditions included), a (β-only) mono-fold already contained in the pure extensional mono-fold $\boldsymbol{\lambda}$.

Recall that the extensional mono-fold $\boldsymbol{\lambda}\gamma$ – based on $[\lambda, \triangleright]$, too, and still a mono-fold, since $\hat{\gamma}$ satisfies only a [CBV-like] restricted β-condition – is

---

[2] The fact that, in view of the simulation above, the 'Kolmogorov' β-only λ-calculus is a fragment of the pure (extensional) λ-calculus should be also familiar to the reader from the so-called CBV $\Rightarrow$ CBN translation of the nineteen seventies (Gordon Plotkin 1975, relating 'call-by-value' and 'call-by-name' λ-evaluations).

[3] See Rezuş 1991 [*Appendix 3*], for a discussion of the original translation – in a decorated ['typed'] setting – and of a few variants thereof, including the so-called 'continuation-passing-style' [CPS] translation, based on suggestions from Timothy Griffin (1990) and Chetan R. Murthy (PhD Dissertation, Cornell University 1990, and later), in the footsteps of Matthias Felleisen (PhD Dissertation, Indiana University 1987), concerning his 'type-free' [CBV] λ-calculus with non-local control, etc. Specifically, Griffin noticed the fact that the original Kolmogorov translation can be connected to the CBV $\Rightarrow$ CBN translation of Gordon Plotkin (1975) and, more interestingly, to the semantics of the so-called *continuations* (Michael J. Fischer 1972 et al.). A similar observation has been made by Bruce Duba, somewhat earlier. The latter remarks are relevant, in the present context, because the corresponding simulation yields a β-only $\lambda(\gamma)$-calculus. In a 'type-free' setting, the simulation of the γ-abstraction operator is simpler (cf. Rezuş 2017a, **§ 2.3–2.4** or the summary following below). A more recent comment, connecting explicitly this kind of interpretation – mainly Griffin's – to Kolmogorov (1925), can be found in Coquand 2007, **§ 2.6**.

(equationally) the same thing as $\boldsymbol{\lambda}^K\hat{\gamma} + (\eta\lambda^K)$, so $\boldsymbol{\lambda}^K\hat{\gamma}$ would admit of the same 'classical' decoration as $\boldsymbol{\lambda}\gamma$. In other words, where – like γ – the abstractor $\hat{\gamma}$ stands for the inferential *reductio ad absurdum*, the decorated ('typed') $\boldsymbol{\lambda}^K\hat{\gamma}[\mathbf{f},\to]$ is, ultimately, a witness theory for classical (propositional) logic, as well. Full details can be retrieved from Rezuş 2017a, 2017b.

§ 2 *The Kolmogorov $\boldsymbol{\lambda}\varepsilon\varpi$-calculus.* In view of Rezuş 2017a, 2017b, one can also formulate the equivalent 'linear decomposition' of $\boldsymbol{\lambda}^K\hat{\gamma}$, resp. $\boldsymbol{\lambda}^K\hat{\gamma}[\mathbf{f},\to]$, by setting (Rezuş lectures 1986–1987, etc.):

$\hat{\varpi}(e) := \hat{\gamma}x.e \equiv \gamma x.e$, x not free in e, and
$\hat{\varepsilon}x.a[x] := \hat{\gamma}x.(x\triangleright^K a[x])$,

obtaining thus 'Kolmogorov' β-only versions $\boldsymbol{\lambda}^K\hat{\varepsilon}\hat{\varpi}$, resp. $\boldsymbol{\lambda}^K\hat{\varepsilon}\hat{\varpi}[\mathbf{f},\to]$, of $\boldsymbol{\lambda}\varepsilon\varpi$, resp. $\boldsymbol{\lambda}\varepsilon\varpi[\mathbf{f},\to]$, where $\boldsymbol{\lambda}\varepsilon\varpi$, resp. $\boldsymbol{\lambda}\varepsilon\varpi[\mathbf{f},\to]$ are as in Rezuş 2017a, 2017b.[4] Specifically, we get equations of the form:

$(\beta\hat{\varepsilon}\hat{\varepsilon}^K) \vdash \hat{\varepsilon}x.c[x\triangleright^K \hat{\varepsilon}y.a[x,y]] = \hat{\varepsilon}z.c[z\triangleright^K a[x,y:=z]]$,
$(\eta\hat{\varepsilon}^K) \vdash \hat{\varepsilon}z.a = a$, with z not free in a,
$(\hat{\varepsilon}\hat{\varepsilon}^K) \vdash \hat{\varepsilon}x.\hat{\varepsilon}y.a[x,y] = \hat{\varepsilon}z.a[x,y:=z]$,
$(\zeta_\beta\hat{\varepsilon}^K) \vdash \hat{\varepsilon}z.c[z] \triangleright^K a = \hat{\varepsilon}y.(c[z:=\hat{\mu}(a,y)]\triangleright^K a)$,
   where $\hat{\mu}(a,y) := \lambda z.(y\triangleright^K(z\triangleright^K a))$, $z \neq y$, and z is not free in a,
$(\hat{\beta}\hat{\varepsilon}\hat{\varpi}^K) \vdash \hat{\varepsilon}z.c[z\triangleright^K \hat{\varpi}(e[z])] = \hat{\varepsilon}z.c[e[z]]$,
$(\zeta_\beta\hat{\varpi}^K) \vdash \hat{\varpi}(e) \triangleright^K a = \hat{\varpi}(e)$,
$(\hat{\varepsilon}\hat{\varpi}^K) \vdash \hat{\varepsilon}z.\hat{\varpi}(z\triangleright^K a[z]) = \hat{\varepsilon}z.a[z]$.

Note that – like, *mutatis mutandis*, in the case of (the extensional) $\boldsymbol{\lambda}\varepsilon\varpi$ – the condition $(\hat{\varepsilon}\hat{\varepsilon}^K)$ is derivable from the remaining conditions. [We need $(\hat{\varepsilon}\hat{\varpi}^K)$, for this.] In this context, however – as $(\eta\lambda^K)$ is missing – one cannot derive 'extensionally lifted' η-versions of $(\zeta_\beta\hat{\varepsilon}^K)$, $(\zeta_\beta\hat{\varpi}^K)$, viz.

$(\zeta\varepsilon) \vdash \varepsilon z.c[z] = \lambda x.\varepsilon y.(c[z:=\mu(x,y)]\triangleright x)$,
   where $\mu(a,b) := \lambda z.(b \triangleright (z \triangleright a))$ (z not free in a, b), and
$(\zeta\varpi) \vdash \varpi(e) = \lambda x.\varpi(e)$ (x not free in e).

---

[4] In logic, ϖ and $\hat{\varpi}$ correspond to the intuitionistic *ex falso* rule – i.e., the *ex falso quodlibet* of the medievals – while ε and $\hat{\varepsilon}$ stand for the genuinely classical rule *consequentia mirabilis* of Girolamo Cardano (1501–1576), Bellissima & Pagli 1996, traditionally known also as the 'Rule of Clavius', after the German polymath Cristoph Klau SJ [*Pater Christophorus Clavius* (1538–1612)], 'the Euclid of the xvi-th century', a famous mathematics teacher at the Jesuit's *Collegium Romanum*, in Rome, who commented in passing upon and attributed it – by confusing use and mention – to (the 'true') Euclid, the alleged author of the **Elementa**. Incidentally, the $\hat{\varpi}$-free fragment of $\boldsymbol{\lambda}^K\hat{\varepsilon}\hat{\varpi}[\mathbf{f},\to]$ can be viewed as a witness theory for the $[\mathbf{f},\to]$-fragment of Curry's logic of 'strict negation' Curry 1952 (Notre Dame University lectures of 1948). Cf., e.g., Rezuş 1991, mainly Chapter VI, for details. Equationally, the $\widehat{\ldots}$-versions are just 'intensional' variants, so to speak.

As expected, the (extensional) mono-fold $\lambda\varepsilon\varpi$ can be obtained from $\lambda^K\hat\varepsilon\hat\varpi$ by adding 'extensionality' to the underlying pure mono-fold λ_β.

One can justify, thus, the 'characteristic' (equational) proof behaviour of classical (propositional) logic [based on implication and *falsum*, with negation defined inferentially], by using the (extensional) mono-fold λ. If, however, (classical) negation is taken as a primitive, we can use di-folds, instead, so that the whole construction becomes radically simpler (cf. Rezuş 2017a, 2017b).[5]

§ 3 On the *historical* side, the λ-calculus experts could have discovered – or formulated – the Curry-Howard Correspondence for the *classical* (propositional) logic, long before H. B. Curry (cca 1934) and W. A. Howard (1969), by reading carefully Kolmogorov (the famous *tertium non datur* paper of 1925), for instance[6], and/or, possibly, Glivenko's short notes of 1928, resp. 1929).

Ultimately, the 'intensional' proof theory of classical logic is already 'contained equationally', so to speak, in the 'pure' λ-calculus, λ.

13–26 June 2017

---

[5] For instance, a witness for the genuinely classical *Law of Peirce* (A→B→A→A) would amount, in the present setting, to a (genuine) $\lambda^K\hat\varepsilon\hat\varpi$-combinator $\hat P :=$ $\lambda^K f.\hat\varepsilon z.(f\triangleright^K(\lambda^K x.\hat\varpi(z\triangleright^K x)))$, decorated as appropriate. (Cf., *mutatis mutandis*, Rezuş 2017a, **§ 3**, for the analogous $\lambda\gamma$-version, simulated in $\lambda\pi$.) — In fact, the (extensional) fragment of the $\lambda\gamma$-calculus without the ζ-condition(s) above can be already simulated in the pure (extensional) λ-calculus, by setting $\gamma z.e[z] :=$ $\lambda z.e[z:=<z>]$ (cf. Rezuş 2017a, **§ 2.2**) – the basic idea in Kolmogorov 1925 and Glivenko 1928, while thinking in terms of proofs – so that, ultimately, Prawitz's construction of Prawitz 1965 – on the signature [**f**,→], say – which consists of eliminating recursively 'complex' applications of γz:C.e[z] (*reductio ad absurdum*), with C := A→B, in favour of 'atomic' applications thereoof, with A atomic, could have been justified in pure λ-calculus terms, as well. Actually, in his PhD Dissertation (1965), Prawitz ignored both λ- and γ-extensionality, and had a single additional condition (in terms of reduction), similar to $(\hat\beta\hat\gamma^K)$ above. Cf. Rezuş 1990 for minute details, applying to the first-order case.

[6] The apparent reason is in that they could not read Russian. (Kolmogorov's 1925 has been translated into English and published only in 1967. Glivenko's notes 1928, 1929 were, however, written and printed in French, in a Western European academic publication, a few years before Curry's first publications on the 'functionality theory'.) The fact is that they did not read (carefully) Prawitz 1965 either. (His monograph, in print since 1965, can be transcribed entirely in 'typed' λ-calculus terms.) Otherwise, Prawitz referred explicitly, in footnotes to papers in print, to both Kolmogorov and Glivenko, as well as to Curry, during the early 1970's.

6

Proof-détours in classical logic (2017e)

In what follows, if not already standard – Barendregt 1981, 1984[2], Baader & Nipkow 1998, TeReSe 2003 – the terminology is as in Rezuş 2017b. The proofs of some statements can be found in Rezuş 2017, 2017a.

§ 1 An 'extended' λ-calculus [Rezuş 2017b]. In the background, we have an abstract grammar with two syntactic categories – *scalars* and *witness terms* – that are both atomic and substitutive (in the sense of Haskell B. Curry). The inner structure of the scalars (the scalar 'constructors') can be ignored. The régime of metavariables is fixed by:

  atomic scalars :: u, v, w
  scalar terms :: s, t := u | ...
  atomic witness terms :: x, y, z
  witness terms :: a, b, c, d, e, f

while the witness terms (w-terms, for short) are given explicitly by:

  a, b, c, d, e, f := x | c $\star$ a | ∂z.e | $\prec$a,c$\succ$ | $\int$(x,z).e | $\downarrow_t$(c) | Σ(u,z).e.

Let $\partial\!\int\!\Sigma$ be the following scalar mono-fold based on $[\partial,\star]$, with 'characteristic' equations ('postulates'):

($\beta\partial$) $\vdash$ c $\star$ (∂z.e[z]) = e[z:=c],
($\eta\partial$) $\vdash$ ∂z.(z$\star$a) = a (z not free in a),

($\beta\!\int$) $\vdash$ $\prec$a,b$\succ$ $\star$ ($\int$(x,y).c[x,y]) = c[x:=a,y:=b],
($\eta\!\int$) $\vdash$ $\int$(x,y).($\prec$x,y$\succ$ $\star$ c) = c (x,y not free in c),

($\beta\Sigma$) $\vdash$ $\downarrow_t$(a) $\star$ (Σ(u,x).c[u,x]) = c[u:=t,x:=a]),
($\eta\Sigma$) $\vdash$ Σ(u,x).($\downarrow_u$(x) $\star$ c) = c (u, x not free in c),

and $\partial\!\int\!\Sigma_\zeta := \partial\!\int\!\Sigma + (\zeta\!\int) + (\zeta\Sigma)$, where

($\zeta\!\int$) $\vdash$ ∂z.e[z] = $\int$(x,y).e[z:=$\prec$x,y$\succ$] (x, y fresh for e[z]),

$(\zeta\Sigma) \vdash \partial z.e[z] = \Sigma(u,x).e[z:=\downarrow_u(x)]$ (u, x fresh for e[z]).

So, beyond the underlying mono-fold ∂ based on $[\partial,\star]$, characterised by $(\beta\partial)$ and $(\eta\partial)$ (a copy of the pure extensional λ-calculus), we have two kinds of pairs (term-forms: $\prec a,c \succ$ and $\downarrow_t(c)$ resp.), and two dyadic abstractors (term-forms: $\int(x,z).e$ and $\Sigma(u,z).e$, resp.).[1] The compatibility conditions are as expected.

From Rezuș 2017b we know that $\partial\!\int\!\Sigma_\zeta$ (resp. its ζ-free subsystem $\partial\!\int\!\Sigma$) is consistent, *qua* equational theory – formally: $\mathrm{Cons}(\partial\!\int\!\Sigma_{(\zeta)})$ – since it is (properly) contained in $\partial\pi\Pi$, the scalar extension of the cartesian fold $\partial\pi$.[2]

§ 2 Reduction. Let us focus on the ζ-free system $\partial\!\int\!\Sigma$ first. As expected, the conversion (equality) of $\partial\!\int\!\Sigma$ can be generated by a 'notion of reduction' – terminology as in Barendregt 1981, 1984[2] – by orienting the equalities above from left to right.

Let $\mapsto$ be the associated notion of reduction, $\longmapsto$ be its compatible closure, and $\twoheadrightarrow$ be the reflexive-transitive closure of $\longmapsto$. (Usually, $\twoheadrightarrow$ is called 'reduction'.) Then the equality of $\partial\!\int\!\Sigma$ is the symmetric closure of $\twoheadrightarrow$ and we have also the expected

2.1 Theorem. *The notion of reduction of $\partial\!\int\!\Sigma$ (the relation $\mapsto$) is confluent.*
Proof. As ever. [*Hint.* Examine critical pairs – Baader & Nipkow 1998, TeReSe 2003, etc. – or else use the 'residuation-method' of 'parallel' reductions, as in Takahashi 1989, 1995.] □

From this, we have, once more, a

2.2 Corollary. $\mathrm{Cons}(\partial\!\int\!\Sigma)$.

§ 3 Art déco (*'typing'*). We revert to 'typing' next. Our 'types' are formulas of *classical* first-order logic, otherwise just decorations on witness terms. (The 'typing' is rigid, à la Church, thus.) The scalars are supposed to represent 'individual' terms (as mentioned before, the inner structure of the scalars – to be specified explicitly for any particular first-order theory – can be ignored).

The underlying first-order logic signature is based on $[\mathbf{f},\neg,\rightarrow,\forall]$, i.e., falsum, (classical) negation, material implication and the (classical) universal quantifier, i.e., we have an abstract grammar

atoms :: P[**t**] (where **t** is a finite sequence of scalar terms),
formulas :: A, B, C, ... := $\mathbf{f}$ | P[**t**] | $\neg$A | A$\rightarrow$C | $\forall$u.C[u].

---

[1] Note that the scalar syntax replicates a segment of the 'pure' syntax. Syntactically, one can unify the notation for the pairs and the dyadic 'split'-abstractors resp., by writing $\langle a,c\rangle \equiv \prec a,c \succ$, and $\langle t,c\rangle \equiv \downarrow_t(c)$, resp., as well as $\mathbf{S}(x,z).e \equiv \int(x,z).e$, and $\mathbf{S}(u,z).e \equiv \Sigma(u,z).e$, resp., since [1] the declarations of the metavariables make things notationally unambiguous, and [2] the equational behaviour of the (pair-split) teams is 'isomorphic', so to speak.

[2] Note, however, that although $\partial\!\int\!\Sigma_{(\zeta)}$ is a sub-cartesian fold, it in *not* cartesian (we can simulate projections for the primitive pair $\prec...,...\succ$, yet the resulting pairing is not 'surjective').

We write, for convenience A↛C := ¬(A→C) and ∀̄u.A[u] := ¬(∀u.A[u]). The resulting witness theory is called $\partial\!\!\int\!\Sigma[\mathbf{f},\neg,\to,\forall]$.

§ 3.1 The primitive rules of inference.
Where $\hat{\Gamma}$ stands for an arbitrary proof-context, the *witness operators* (the *primitive rules of inference*) of $\partial\!\!\int\!\Sigma[\mathbf{f},\neg,\to,\forall]$ are:

(∂) $\hat{\Gamma} \vdash \partial x{:}\neg C.e[x] : C$, if $\hat{\Gamma}[x{:}\neg C] \vdash e[x] : \mathbf{f}$ [*reductio ad absurdum*],
($\star$) $\hat{\Gamma} \vdash c \star a : \mathbf{f}$, if $\hat{\Gamma} \vdash c : \neg C$, $\hat{\Gamma} \vdash a : C$ ['law of non-contradiction'],
($\int$) $\hat{\Gamma} \vdash \int(x{:}A,y{:}\neg C).e[x,y] : A{\not\to}C$, if $\hat{\Gamma}[x{:}A][y{:}\neg C] \vdash e[x,y] : \mathbf{f}$,
(π) $\hat{\Gamma} \vdash \prec a,c\succ : A{\to}C$, if $\hat{\Gamma} \vdash a : A$, and $\hat{\Gamma} \vdash c : \neg C$,

(Σ) $\hat{\Gamma} \vdash \Sigma(u,x{:}\neg C[u]).e[u,x] : \bar{\forall}u.C[u]$, if $\hat{\Gamma}[u][x{:}\neg C] \vdash e[u,x] : \mathbf{f}$,
(Π) $\hat{\Gamma} \vdash \downarrow_t(c) : \bar{\forall}u.C[u]$, if $\hat{\Gamma} \vdash c : \neg C[u{:=}t]$.

§ 3.2 The classical 'détour eliminations'.
Under the decoration above, the primitive $\beta\eta$-conditions – meant to define the witness operators of $\partial\!\!\int\!\Sigma[\mathbf{f},\neg,\to,\forall]$ – state *proper détour eliminations*, if viewed as reduction 'rules'. In the spirit of Chrysippus, the founder of classical logic – cf. Rezuş 2016 – a proof-détour (formally: a 'redex') points out to a way of *coping with a contradiction*.

§ 3.3 The ζ-rules.
What about the ζ-conditions of $\partial\!\!\int\!\Sigma_\zeta[\mathbf{f},\neg,\to,\forall]$? Consider the decorated witness terms (∂-abstractions) of the form $a := \partial z{:}\neg C.e[z]$ (where $e[z] : \mathbf{f}$, for $[z{:}\neg C]$), and call them *reductio-terms*. If C is an atom or $\mathbf{f}$ then a is said to be *atomic*, otherwise it is *complex*. A witness term containing no complex *reductio*-terms is said to be *déco-normal* [d-normal, for short]. Finally, let $\partial\!\!\int\!\Sigma^\star[\mathbf{f},\neg,\to,\forall]$ be *the d-normal fragment* of $\partial\!\!\int\!\Sigma_\zeta[\mathbf{f},\neg,\to,\forall]$.

If we are reading the ζ-conditions as '(improper) reductions' – oriented from left to right – they can serve to eliminate (recursively) complex *reductio*-terms in favour of atomic ones:

($\zeta\!\int$) $\vdash \partial z{:}(A{\not\to}C).e[z] \mapsto_\zeta \int(x{:}A,y{:}\neg C).e[z{:=}\prec x,z\succ]$, x,y fresh for e,
($\zeta\Sigma$) $\vdash \partial z{:}(\bar{\forall}u.C[u]).e[z] \mapsto_\zeta \Sigma(u,x{:}\neg C).e[z{:=}\downarrow_u(x)]$, u,x fresh for e.

It is then obvious that we can always apply the ζ-rules first, before attempting to apply any proper reduction 'rule', i.e., we have the rather trivial

3.3.1 Theorem. *The witness terms of $\partial\!\!\int\!\Sigma_\zeta[\mathbf{f},\neg,\to,\forall]$ are d-normalisable.*
Proof. Straightforward. □

Finally, the reader can establish by herself the expected

3.3.2 Theorem. *In $\partial\!\!\int\!\Sigma^\star[\mathbf{f},\neg,\to,\forall]$, all witness terms are bounded.*
Proof. Mutatis mutandis, as, e.g., in Rezuş 1990. Cf. López-Escobar 1990.[3] □

In other words, the (decorated) witness terms of $\partial\!\!\int\!\Sigma^\star[\mathbf{f},\neg,\to,\forall]$ are strongly normalisable.

---
[3] As a matter of fact, in view of the confluence result mentioned earlier, it is enough to establish weak normalisability for the corresponding terms.

§ 4 Remarks. A few remarks are in order.

(1) Our use of ζ-conditions parallels closely the use of the corresponding $(\zeta\gamma)$-conditions in $\lambda\gamma$-calculi – Rezuş 1990 – as suggested by Dag Prawitz in his PhD Dissertation (Stockholm 1965). Cf. also with the Kolmogorov β-only calculi of Rezuş 2017c.

(2) Incidentally, in the above one could have also used an alternative 'classical' decoration based on the primitive signature $[\mathbf{f},\triangle,\forall]$, where $\triangle$ stands for the 'Sheffer functor' **nand** (incompatibility), and $\neg A := (A \triangle A)$.

(3) Since the methods of proof hinted at here are rather standard, the reader can easily establish analogous results for the corresponding witness theories based on di-folds, equationally equivalent to $\partial \int \Sigma_{(\zeta)}[\mathbf{f},\neg,\rightarrow,\forall]$, described in Rezuş 2017b.

(4) For a recent update on the earliest history of normalisation proofs for *classical* logic, see Rezuş 2017d.

Acknowledgement. The author is indebted to J. Roger Hindley (Swansea, Wales, UK) for comments on a previous draft of this note.

16 July – 31 October 2017

7

Appendix: Does (η) really matter? (2017f)

The λ-calculus workers – including the pioneers, as, e.g., Church, Kleene and Rosser – have spent a significant amount of time on λ_β, i.e., on the pure λ-calculus without 'extensionality', resp. without the (η)-condition. Most applications of the λ-calculus – in computing, for instance – rely essentially on λ_β, indeed, and even on weaker theories.[1] I have shown elsewhere[2] that extensional λ-theories are oft useful in logic applications, although, in a way, *only as regards consistency-matters*. Whence a *question*: Does (η) have *specific* virtues, beyond the latter concern? Or else: Isn't, somehow, 'the consistency of (η)' a trivial affair? And, if so, in what sense? Let us make such questions more definite.

Notation. Where **T** is an equational theory, Cons(**T**) means '**T** is consistent', i.e., **T** does not prove $\vdash$ a = b for all a, b. Let **T**$_\beta$ be a λ-theory – extending λ_β, thus, i.e., the pure λ-calculus based on the characteristic condition (β), possibly in a *specific* additional vocabulary – and **T**$_{\beta\eta}$ = **T**$_\beta$ + (η), where, as usual,

(β) $\vdash$ (λx.b)(a) = b[x:=a], and
(η) $\vdash$ λx.f(x) = f, provided x not free in f.

Terminology. Here, the terminology is as in Barendregt 1981, 1984², resp. Rezuș 1981, viz. **T**$_\beta$:= λ_β + **T** is an equational extension of the pure λ-calculus λ_β in a vocabulary $\mathcal{L}(\mathbf{T})$ containing, as primitives, λ-abstraction, application and, possibly, additional algebraic constants subjected to first-order equational conditions stated explicitly in **T**.[3]

The 'first-order algebraic' stipulation refers to the fact that any additional constant κ in $\mathcal{L}(\mathbf{T})$ must be compatible with the equality (here: conversion), i.e., that we must have also 'monotony' in **T**: if κ is k-ary (k > 0), then

---
[1] See the note on Scott 1963, below.
[2] Cf., e.g., Rezuș 2017, where the cartesian folds [$\lambda_{\beta\eta}\pi$, say] are shown to be *essentially stronger* than the sub-cartesian folds (like, e.g., λ_β or $\lambda_{\beta\eta}$). In particular, $\lambda_{\beta\eta}\pi$ contains infinitely many distinct copies of $\lambda_{\beta\eta}$ and thus of itself, so to speak.
[3] As regards the terminology, Barendregt 1981, 1984² allows λ-theories to be inconsistent.

7 Appendix: Does (η) really matter? (2017f)

$(\mu\kappa) \vdash a_1 = b_1, ..., \vdash a_k = b_k \Rightarrow \vdash \kappa(a_1,...,a_k) = \kappa(b_1,...,b_k).$

In particular, κ might be an abstraction-operator, as well. If, for instance, κ is a monadic abstractor (like λ), then we must also have an associated ($\xi\kappa$)-condition:

$(\xi\kappa) \vdash a = b \Rightarrow \vdash \kappa x.a = \kappa x.b.$

So $\mathbf{T}_{\beta\eta}$ contains the pure extensional λ-calculus $\boldsymbol{\lambda}_{\beta\eta}$.

Further, a *Howard-Hindley* λ-*calculus* $\boldsymbol{\lambda_o}$ – equationally equivalent to a weak combinatory theory (the Schönfinkel-Curry or the Rosser combinatory theory, say), i.e., a λ-theory with 'extensionality type' **o**; cf., e.g., Rezuş 1981 – would lack ($\xi\lambda$) ['weak λ-extensionality'], but would admit of a substitutivity condition (lsv), instead, whence also (rsv), where

(lsv) $\vdash a = b \Rightarrow \vdash c[x:=a] = c[x:=b]$,
(rsv) $\vdash a = b \Rightarrow \vdash a[x:=c] = b[x:=c]$,

while $\boldsymbol{\lambda_o}$ + (ext) = $\boldsymbol{\lambda}_{\beta\eta}$, equationally speaking, where

(ext) $\vdash fx = gx \Rightarrow \vdash f = g$, provided x is not free in f, g.[4]

The (η)-conjecture. If Cons($\mathbf{T}_\beta$) then Cons($\mathbf{T}_{\beta\eta}$).

Remark [*the* (ext)-*conjecture*]. A *stronger* statement would consist of replacing $\mathbf{T}_\beta$ and $\mathbf{T}_{\beta\eta}$, resp., in the above, by $\mathbf{T_o}$ and $\mathbf{T_{ext}}$, resp., where $\mathbf{T_o}$ is a weak combinatory logic, or else a *Howard-Hindley* λ-*calculus*, and $\mathbf{T_{ext}} := \mathbf{T_o}$ + (ext). Obviously – as $\mathbf{T_o}$ is a sub-theory of $\mathbf{T}_\beta$, Cons($\mathbf{T}_\beta$) implies Cons($\mathbf{T_o}$), while $\mathbf{T_{ext}}$ and $\mathbf{T}_{\beta\eta}$ are equationally equivalent – the stronger statement implies the (η)-conjecture.

Question. Can we prove the (η)-conjecture above 'constructively', i.e., for instance, by syntactic means alone?

Examples. (1) $\mathcal{L}(\mathbf{T})$, the 'language' of $\mathbf{T}$, contains primitive pairs $\pi(a,b)$ and projections $\pi_1(c), \pi_2(c)$, and $\mathbf{T}$ consists of a 'surjective pairing', i.e., in $\mathbf{T}$ one has:

$(\beta\pi_j) \vdash \pi_j(\pi(c_1,c_2)) = c_j$ (for j := 1,2),
$(\eta\pi) \vdash \pi(\pi_1(c),\pi_2(c)) = c$,

i.e., in this case, $\mathbf{T}_{\beta\eta}$ is just the extensional $\lambda\pi$-calculus (with surjective pairing), $\boldsymbol{\lambda}_{\beta\eta}\pi$, while $\mathbf{T}_\beta$ is the corresponding equational sub-theory, $\boldsymbol{\lambda}_\beta\pi$, without ($\eta$).[5]

---

[4] In retrospect, and in view of Scott 1963, I should have, perhaps, better called the 'weak' λ-calculus (with 'extensionality type' **o**) a *Scott* λ-*calculus*, because both William Howard and J. Roger Hindley pondered on the weaker case later, and mainly in connection with the associated *notion of reduction*.

[5] The former λ-theory is interpreted in the extensional (Scott) D_∞-model, while the latter is also interpreted in the Plotkin-Scott *Graph Model*. Notably, de Vrijer 1987 has established Cons($\boldsymbol{\lambda}_\beta\pi$) syntactically, by a conservativity argument. See also Støvring 2006.

(2) Alternatively, **T** may also contain equations including symbols from the pure λ-calculus, as, e.g.,

$(\beta\pi_j\lambda) \vdash \lambda x.(\pi_j(c)) = \pi_j(\lambda x.c)$ (for j := 1,2),
$(\beta\pi) \vdash \pi(f,g)(c) = \pi(fc,gc)$,

thereby yielding together a *commuting $\lambda\pi$-calculus*.[6]

Remarks.

(1) If my *conjectures* are true, and if they can be established syntactically, then we don't really need *explicit* extensional λ-model constructions in order to prove the 'consistency of (η)' – resp. (ext) – separately. If so – with the first *example* given earlier – we can rely on Plotkin-Scott-Engeler models alone in order to ensure Cons($\lambda_{\beta\eta}\pi$), for instance.[7]

(2) Otherwise, a counterexample falsifying either one of the conjectures above would imply the existence of a consistent and *essentially intensional* λ-theory $\mathbf{T}_\beta$, i.e., one whose (λ-) extensional version is inconsistent.[8]

(3) Incidentally, I have also noticed, along the years, the fact that, after properly translating the issues in plain (and clean) λ-calculus terms, most of the current work in proof theory relying on Gentzen (Dag Prawitz, Jonathan P. Seldin, Anne S. Troelstra, Helmut Schwichtenberg, Sara Negri, Jan von Plato *et al.*) does *not* make appeal to *explicit extensionality assumptions*. This includes the usual way of handling intuitionistic logic along the Curry-Howard Correspondence, of

---

[6] Note that, in $\lambda_{\beta\eta}\pi$, the first two conditions $(\beta\pi_j\lambda)$, (j := 1,2), imply $(\beta\pi)$. The cumulative λ-theory, $\lambda_{\beta\eta}\pi^c$, say – extending $\lambda_{\beta\eta}\pi$ properly – can be interpreted in an extensional (Scott) λ-model. See, e.g., Durfee 1997, for details. Støvring 2006 has also established Cons($\lambda_{\beta\eta}\pi^c$) syntactically, whence also Cons($\lambda_{\beta\eta}\pi$), improving on de Vrijer 1987. Although rather weird at a first look (the resulting calculus identifies pairs and λ-lifted 'cartesian' pairs, for instance), $\lambda_{\beta\eta}\pi^c$ and some sub-systems thereof have interesting applications in computer science (György E. Révész, during the nineteen eighties and later).

[7] Ignoring thus the fact that the Graph Model can also simulate the extensional D_∞-construction (cf. Scott 1976 or Koymans 1984, Chapter 3). As mentioned earlier, Cons($\lambda_{\beta\eta}\pi$) has been established syntactically, by a conservativity argument, in Støvring 2006. The result relies, ultimately, on Cons($\lambda_{\beta\eta}$), while Cons($\lambda_{\beta\eta}$) can be already obtained by a confluence argument ('syntactically', thus), the 'limit' case [**T** = ∅] of the (η)-*conjecture*.

[8] For a closely related, although rather tricky, example, see, e.g., the strict (type-free) 'intensional' λ-calculus $\lambda\epsilon$, exemplified in Rezuş 1981, II.3.3 (a proper sub-system of Church's strict [= λ_βI-] calculus, corresponding, after 'typing', to the *pure theory of entailment* of Alan Ross Anderson and Nuel Belnap Jr.; cf. also the PhD Dissertation (1977) of Glen Helman, cited there, and/or Anderson, Belnap, Dunn *et al.* 1992, **§71**, for typed variants), where the 'field' of well-formed $\lambda\epsilon$-terms is not closed under η-reduction resp. η-equality. In this case, however, one can rather say that there is no such a thing like $\lambda\epsilon + (\eta)$, *by definition*, not that the would-be outcome [equational theory] is inconsistent.

course.[9] For the classical realm, one can also extract plenty of β-only examples from Rezuş 2017.[10]

15 August 2017

---

[9] *Aside.* The extensionality conditions for the intuitionistic disjunction and existence proof-operations correspond, exactly, to the so-called *disjunction-* and *existence properties*. As exceptions to this remark, one can mention the category-theoretic accounts of intuitionistic proof theory, and, possibly, the very last *extensional* variants of Martin-Löf's *Constructive Type Theory*, as handled within the recent HoTT-project.

[10] Note, however, that I have heavily used extensionality assumptions (as available in, e.g., $\lambda_{\beta\eta}$ resp. $\lambda_{\beta\eta}\pi$) – otherwise apparently unavoidable – in order to ensure consistency (i.e., the fact that the corresponding proof operations are properly defined) in nearly all cases of concern. Even the β-only 'Kolmogorov $\lambda\gamma$-calculus' discussed in Rezuş 2017a requires Cons($\lambda_{\beta\eta}$), as a justification. Yet, we can usually trade, in such contexts, syntactic consistency proofs – relying, e.g., on confluence and normalisation – for model-theoretical arguments; we don't really need extensional ('type-free') λ-models, for most logic applications of the λ-calculus.

Part III

The origins of modern proof theory

8
An ancient logic (2016)

> *Het bekend zijn met de geschiedenies van een bepaalde wetenschappelijke discipline is een noodzakelijke stap voor het juiste begrip van de huidige ontwikkelingen (indien deze er zijn), terwijl het verleden alleen juist begrepen kan worden door de juiste plaats te vouden binnen de huidige kennis.* (*Stelling* [Proposition] 8, in Rezuş 1981a.)

This 'theorem' ['proposition', *Stelling*, in Dutch] was part of the dissertation Rezuş 1981. The Dutch quote is Henk Barendregt's version of my English original, which I lost. Here is a backwards translation, with some approximations, for the benefit of my Dutchless readers:

> 'The fact of being familiar with the history of a specific scientific discipline is required in view of a correct understanding of the current developments – in as far they exist – while the past can be understood correctly only by finding its right place within the contemporary knowledge.'

The self-quote illustrates best my way of understanding Gentzen's **LK** (Gentzen 1934–1935) via Chrysippus. And conversely, perhaps, although the latter step is not necessarily yet another piece of *whig historiography*.

Foreword

It is a pity that the pioneers of modern logic – also called 'mathematical' or 'symbolic' – as, e.g. Boole, Frege, Peirce, Peano, Russell, etc., did not spend some time on the Stoic, mainly Chrysippean, fragments on logic that have survived: the effort could have been rewarding.

They might have had a historical excuse: the comprehensive edition of Hans von Arnim, **Stoicorum Veterum Fragmenta** [SVF], was published by the turn of the previous century [three volumes, 1903–1905, the fourth one, containing Adler's index, is dated 1924], von Arnim's edition was incomplete in as far logic was concerned, the received views (Carl [von] Prantl, Eduard Zeller *et al.*) on Stoic logic were rather inaccurate – to say the least – while the first competent person to realise the logic significance of the Stoic corpus, Jan Łukasiewicz (circa 1923), managed to publish an account of his findings only in 1934.

Relevant studies of Stoic logic re-emerged, in the footsteps of Łukasiewicz, only after the WWII, during the late forties and the fifties (due to Benson Mates and Oskar Becker; cf. Mates' UCB PhD Dissertation 1948, published in 1953, and Becker's notes *Über die vier Themata der stoischen Logik*, 1957).

Thus far, we have about a dozen – or so – of technical studies in print, on the subject, worth mentioning. On this line of research, most authors have been involved in 'reconstructing' a would-be 'Stoic logic' in modern terms. The main trouble is in the fact that there is no general agreement as to what is to 'reconstruct', technically speaking.

The present attempt to a *conceptual* reconstruction consists of a condensed summary of work done previously (on **Proof Structures in Traditional Logic** [1994–2007]). The full discussion, bearing the working title **Chrysippus and His Modern Readers** – half-stolen, *mutatis mutandis*, from Charles Lutwidge Dodgson (1832–1898), also known as Lewis Carroll (cf. **Euclid and His Modern Rivals**, 1879, 1885[2]) – contains a critical examination of the previous 'reconstructions' against the extant ancient sources and has been deferred for a later publication.

Specifically, what follows – in this first installment of the projected monograph – is a set of remarks on *the conceptual structure* of the logic of Chrysippus of Sol[o]i (cca 279 – cca 206 BCE), written from the point of view of modern proof theory. The reader is supposed to be familiar with the post-Fregean logic in general, and, mainly, with the recent work on *natural deduction* (Jaśkowski, Gentzen, Prawitz, etc.), including the *Curry-Howard Correspondence*, as applied to *classical logic*.

There is a good reason to put 'technical' comments before historical and/or philological minutiae. Because – first of all – if we agree on the fact that Chrysippus was a reputed logician, as otherwise claimed by a longstanding tradition, then there are not too many distinct ways of saying, once more, in our terms, what he meant to say. On the other hand, the reconstruction of a *logical theory* – of the kind Chrysippus is confronting us with – is, essentially, a mathematical – perhaps also a philosophical – endeavour, not a mere philologico-historical concern. In this respect, I am, *prima facie*, addressing a *logico-mathematical audience*, not the classical philologist, the historian of ideas or the (generalist) philosopher, in particular. Putting things is a slightly different way: like Chrysippus, I am a logician – also deeply concerned with the history of his subject (as most of my colleagues and/or co-workers, in fact) – and, as such, I think I understand *what he said*, as well as *what he could* and *what he could not say*, about twenty-two centuries ago. In this sense, the present comments should make up a mere piece of *historical data recovery*, rather than an attempt to a radical re-interpretation of known historical facts.

Acknowledgements. I am indebted to Susanne Bobzien (All Souls College, University of Oxford) for reviving my interest in one of my preferred spare time subjects – viz. the logic of Chrysippus – and, indirectly, to my friend and teacher of Ancient Greek, Petru Creţia (1927–1997), the editor of Plato in Romanian.[0]

§ 1 Chrysippean logical grammar

The main claim of the present notes consists of saying that Chrysippus' logic was what we call, nowadays, 'classical' logic. I shall first focus on the quantifier-free fragment of what I take to be 'Chrysippus' logic'.[1]

A preliminary observation: as a logician, Chrysippus payed attention to *logical form*, as opposed to mere *grammatical expression*. He was also concerned with the study of grammar, as well as with the relation between logical forms and their expression in natural language. On the other hand, he did not propose a formal, symbolic notation for logic constructs (the idea occurred to other people about twenty centuries later). Whence a good deal of Chrysippean – and, in general, Stoic – considerations on ambiguity and the like[2].

The basic Stoic (actually Chrysippean) logic concepts are: *proposition* (axiōma), *polar opposition* (or [logical] conflict), *entailment* (argument or even 'syllogism', as a special case), and *rule(s) of inference* (thema(ta), more or less).

Propositions and entailments

For Chrysippus, the propositions (axiōmata) are abstract entities[3]; they can be either *simple* (atomic) or *complex*.

---

[0] These notes have first appeared in print as **An Ancient Logic** (*Chrysippus and His Modern Readers I*), LAP – Lambert Academic Publishing, Saarbrücken 2016 [ISBN 978-3-330-01661-3].

[1] As it appears, the claim is neither new nor very original, but some recent authors have claimed otherwise, in the meantime. Rather unconvincingly, *on technical grounds alone*, to my mind. Cautiously, I shall, however, avoid, *prima facie*, polemic remarks on current research attempting to show something else. For convenience, the discussion of the *Stoic quantifiers* and a detailed scrutiny of the sources – in the guise of supporting *textual evidence* for my remarks – are deferred and will appear as separate notes.

[2] Cf., e.g., Atherton 1993, on this.

[3] In proof theory, we need not be concerned with their metaphysical status.

According to their meaning – 'semantically' thus – they fall into *opposita*, contradictory (better: 'polar') pairs.[4]

In modern terms, a *Stoic argument* (logos, and oft also sullogismos, as a special case) is a finite sequence of propositions, where exactly one is tagged (as a *conclusion*). I shall use next *entailment* as a technical, neutral term, instead[5]. With this terminology, Stoic logic is an 'entailment logic', not a 'propositional' logic (*Satzlogik*), à la Frege (**Begriffsschrift**, 1879 [**BS**]), Peirce, Russell or Łukasiewicz.[6]

Formally, with $\Gamma := (A_1, ..., A_n)$, $n > 0$, an entailment can be written down as $\Gamma \vdash C$ (the elements of Γ are called assumptions, lēmmata, in technical Chrysippean terminology), where the tag is the turnstile $\vdash$ itself, in the guise of punctuation (read 'therefore' or 'yields'), and the conclusion C (sumperasma, or epiphora, in Stoic jargon) occurs last. Alternative reading: 'C is a consequence of Γ'.[7]

---

[4] Contradiction is at the root of Ancient (Greek) logic. Roughly, for the Ancient Greeks to prove something is to prove a contradiction. Worth mentioning here is the pre-Stoic tradition on this subject: the Pythagoreans, with their – somewhat empirical – tables of *opposita*, Heraclitus' metaphysics of conflict (polemos), the Eleats' obsessive interest in contradictory arguments and in *reductio ad absurdum*, the Sophist's specious, somewhat defective and argutious use of contradictions, Socrates' 'maieutic art' of refuting an opponent, en tō agora, Aristotle's 'square of oppositions' [Peri herm.], his *very* specific concern with defective refutations [Soph. el.], etc.)

[5] No relation to Anderson & Belnap 1975, 1992, where the technical term 'entailment' is reserved for specific conditional propositions, as expressed by formulas. The closest Anderson-Belnap approximation would be, likely, 'first degree entailment', in this context.

[6] Incidentally, Frege had two 'logics': a *Satzlogik*, presented axiomatically, in his **BS**, and a *Regellogik*, in **Grundgesetze der Arithmetik 1**, 1893 [**GGA**]. On the latter, see, however, Note 10 below and, possibly, Rezuş 2009. For Lukasiewicz, see now Rezuş 2017.

[7] The special case $n = 1$ (pointing out to so-called 'monolemmatic' arguments) is also attested in extant Stoic texts, although not explicitly so in Chrysippus. In particular, even though our sources are not very clear on this, entailments of the form $A \vdash A$ (expressing the 'law of identity') were, curiously enough, not accounted for as (valid) 'syllogisms'. (Perhaps on the reason they do not necessarily involve logical connectors.) However, this is a mere terminological detail – after all, Aristotle did not call the monadic entailments 'syllogisms' either, yet he recognised valid *immediate inferences* [later terminology] $A \vdash B$, for specific propositional 'types' A, B – since, given the Stoic way of understanding and explaining entailmens (via rejections / refutations / [logical] conflict), an entailment of the form $A \vdash A$ is to be accounted for as being equivalent with the 'law of [non-] contradiction' stated in terms of polar oppositions. And it did not occur to Chrysippus to deny the validity of the latter or to defend a would-be 'paraconsistent' logic, as in the case of some moderns. (See details below.) Similar remarks apply, *mutatis mutandis*, to entailments of the form $A, B \vdash A$, resp. $A, B \vdash B$ or, more generally, $\Gamma \vdash A_i$ (where $0 < i < n+1$, and Γ is as above).

Like for the moderns, the main concern of logic, according to Chrysippus, consists of sorting out arguments (good vs bad): entailments can be valid or invalid, so that good arguments are expressed by valid entailments, bad arguments by invalid entailments.

Valid entailments can be generated from 'axioms'[8], i.e., primitive valid entailments, called *indemonstrables* ([logoi] anapodeiktoi), by rules of inference (validity-preserving transitions). The logical move goes both ways, since by reversing the rules one can, ultimately, 'reduce' any valid entailment to (valid) indemonstrable entailments – in the guise of 'axioms' – in finitely many steps.[9] Implicitly, there is a claim of completeness behind the Stoic technique, because validity (for entailments) can be characterised, alternatively, by truth conditions for the corresponding conditionals, i.e., by a criterion of the form:

the entailment $\Gamma \vdash C$ is valid iff $(\&\Gamma \to C)$ is true,

where $\&\Gamma$ stands for the conjunction of the elements of Γ (taken in some canonical order, $(...(A_1 \wedge A_2) \wedge ... \wedge (A_{n-1} \wedge A_n)$, $n \geq 3$, by associating to the left, say).[10]

Anyway, since one should not expect a very strict conceptual demarcation between (formal) syntax and semantics in the Stoic logical doctrines, the talk about the (would-be) 'completeness of the Stoic system (of logic)' is a bit pointless and rather unhistorical.

Rejections / refutations / (logical) conflict

In particular, a rejection / refutation (*elenchos*) is an entailment whose conclusion is a contradiction. This is rather tricky, because Chrysippean negation is not exactly a 'primitive' idea, as in the moderns (Frege, Russell, Łukasiewicz, etc.)

What is the (logical) form of the conclusion C, in the latter case? In other words, *how would a Stoic logician express a contradiction?*

Atomic propositions and the Chrysippean negation

The atomic propositions (atoms, for short) are, by definition, so to speak, divided into 'polar' pairs. On the other hand, the atoms can be either 'variable(s)' or constants.

---

[8] In the modern sense (i.e., valid entailments taken as primitive). The Stoics used axiōma as a technical term for "proposition".

[9] For technicalities, see, e.g., the PhD Dissertation of Katerina Ierodiakonou [Analysis in Stoic Logic, London 1990], and Susanne Bobzien's monograph on *Stoic syllogistic*, 1996. Cf. also Bobzien 1999, 2019.

[10] In this sense, the Stoic *distinguo* between entailments and conditionals (expressed here by implicative formulas) is close to contemporary (post-Fregean) conceptual standards, and, in a way, superior to Frege, who did not make such a distinction in his **GGA**. Actually, Frege won't have had understood the point behind the 'deduction theorem' (Tarski 1920–1921).

Proper atoms

The 'variable' atoms, are taken as primitives. By way of example, one has polar pairs of the form: 'It is day' vs 'It is night', 'Kallias is walking' vs 'Kallias is sitting / standing'[11], or else, and better, in English 'John is married" vs 'John is a bachelor', or 'n is odd' vs 'n is even', for any particular n > 0, whence no real need for an 'internal' negation, in order to express (proper) atomic polar oppositions. One might, indeed, think that the Stoic 'atomic' negations make up a (semantic) feature of the natural language, they are not *indicators of logical form*.[12]

Atomic constants

In the case of propositional constants, one can pick up two arbitrary propositions, $\top$ (*verum*) resp. $\bot$ (*falsum*) say, whose truth values do not change according to the circumstances. *Examples:* $\top :=$ 'two is less than three' vs $\bot :=$ 'two is more than three'.[13]

With this notation, a Greek rejection / refutation (*elenchos*) should be of the form $\Gamma \vdash \bot$, i.e., a rejection is an entailment where C [its conclusion] is an arbitrary false proposition (set $C := \bot$).

Formally, if A is an atom, let us write opp(A) for its polar opposite, and define the polar opposite of opp(A) as A := opp(opp(A)).[14] The 'law of double negation' is, thus, 'built-in, semantically', at atomic level, so to speak.[15]

Complex propositions

Complex propositions are built up, inductively, from simples (or atoms), by binary links (binary connectives), called 'connectors' (*sundesmoi*). Formally,

---

[11] Assuming a world without gradual transitions, shadows, dawn, twilight zones and so on.

[12] We can normalise this situation – 'syntactically', so to speak – by fiat, taking 'non' as a formal indicator for polar opposition in "variable" atoms, and write, e.g., non(A), for any 'variable' atom A. Finally, the choice is arbitrary, of course, as we end up with (language) meaning postulates of the form A = non(non(A)), for every such an atom A. (Cf. with the use of literals in recent approaches to proof theory.) See, however, the remarks on 'syntax' vs 'semantics' in Stoic logic, following below.

[13] The Ancient's preferred examples would have rather been of the kind 'the part is equal to the whole', for $\bot$, and 'the part is less than the whole', for $\top$.

[14] Put things on a sphere – or on a circle, for that matter – to see the point behind the ad hoc 'polar'-terminology. If I am living in Western Europe, my Canberran friend, Bob, is my 'polar', from my point of view, and conversely, from his, so that each of us is the 'polar of a polar', etc.

[15] This is much similar to the way some recent proof-theorists would present classical logic. Cf., e.g., Pyotr S. Novikov, Kurt Schütte, or, *mutatis mutandis*, Jean-Yves Girard, in his 'linear' logic.

where # is a connector (*sundesmos*), (A # B) is a proposition, if so are its immediate components, A and B.[16]

The 'method' of polar oppositions is used to classify complex propositions, as well. That is, complex propositions fall into pairs (A ⊕ B) vs (A ⊗ B), where ⊕ resp. ⊗ instantiate a specific connector # (see below).

Example. (A ∧ B) vs (A △ B), where ∧ (**and**) is classical conjunction and △ (**nand**) is its polar opposite (incompatibility, a 'Sheffer functor').[17]

Like in atoms, we have (A ∧ B) = opp(A △ B), resp. (A △ B) = opp(A ∧ B) as meaning postulates for 'opp', whence, again, 'double negation': opp(opp(C)) = C, for C := (A ∧ B), resp. C := (A △ B).

The procees is repeated for the remaining (polar) pairs. There are four of the kind left.

Tabulating as appropriate, in modern notation the Stoic connectors would amount to the following ten symbols, grouped in five polar teams (we provide also due colloquial names in the meta-language):

[1] △ (**nand**) vs ∧ (**and**),
[2] → (**if**, material implication; reading approximatively: 'if... then...') vs **more**, ↛ (its polar opposite: māllon... ē..., in Stoic parlance, a kind of 'rather... than...', in English),
[3] ← (**since**, co-implication, the converse of material implication; colloquial, approximate reading: 'since') vs **less**, ↚ (its polar opposite, in Stoic jargon: ētton... ē..., a kind of 'rather not... than...'),
[4] ∨ (inclusive **or**) vs ▽ (its polar opposite, i.e. the analogous 'Peirce-Sheffer functor' **nor**), and,
[5] ↔ (**iff**, material equivalence) vs ↮ (**xor**, i.e., Boole's exclusive **or**),

and similarly for the corresponding complex propositions.[18]

---

[16] 'Multiary' links, as in the case of ∧ (conjunction) and ∨ (inclusive disjunction) are to be taken, again, as a feature of the natural language and are 'resolved' / analysed into binary links / connectors, in the obvious way.

[17] Incompatibility is exemplified in the third Stoic indemonstrable T3. Incidentally, Charles Sanders Peirce (1839–1914) re-discovered the Stoic **nand** before Henry Maurice Sheffer (1882–1964), but he did not manage to publish his finding.

[18] Given the method of construction (by 'polars'), it is enough to attest a single member of each polar pair in our texts. Only case [4] does not occur explicitly in Chrysippus (it appears in later Stoic textbooks, however). The above correspond to the modern truth-functional (or Boolean) binary connectives (see below). In particular, in case [2] and [3], one must have, semantically, 'conjunctions' (A ↛ B) = (A ∧ opp(B)), and (A ↚ B) = (opp(A) ∧ B), resp., so that the Stoic intended readings māllon A ē B, and ētton A ē B, resp. would have been quite intuitive, in the end, granted the fact that the official Stoic negation could have been defined in terms of polar oppositions, as not(A) := opp(A), for any (complex) A. This terminology has nothing to do with would-be 'comparative' (non-truth-functional) propositions, and the like, as some recent readers of Chrysippus used to speculate. See also below. — Apparently, however, Chrysippus and his followers thought of

Let us call the first four [1]–[4] polar pairs (of connectors, resp. complex propositions) *proper* (polar pairs) and the latter [5] *sub-polar*, or *improper* (polar pairs).

One can easily see that the proper polars are well-behaved semantically: one has (classical) 'disjunctions' on the left (LHS, in the above) and (classical) 'conjunctions' on the right (RHS, above). The corresponding *duals*, in modern terminology, appear, in each case, in alternate pairs, while the sub-polars are *self-dual*.[19] In other words, the proper polars can be analysed back / decomposed into components, on a uniform pattern.

In particular, from this point of view, the first three Stoic indemonstrables [T1–T3] are different in character from the latter two [T4–T5].

Otherwise, the latter two can be eliminated definitionally: explicit definitions of **xor** – and thus **iff** – in terms of proper polars [and opp] can be found in the late Stoic textbook lore. Cf., e.g., Galen's Inst. log., IV.3 – and the comments of John Sprangler Kieffer ad loc. – for the definitional expansion of the Stoic exclusive disjunction [**xor**] in terms of [inclusive] **or** and **nand**, viz. (A ↔ B) =_df ((A ∨ B) ∧ (A △ B)) or, colloquially: '(A or B), but not (both A and B)'.

On this subject, see also Bobzien 1999, p. 111, who reads correctly (i.e., truth-functionally) 'A māllon B' and 'A ētton B', resp. as 'both (either A or B) and A' and 'both (either A or B) and B', resp.[20], but, curiously, omits noticing explicitly the ('material') equivalences:

((A △ B) ∧ A) ↔ (A ∧ opp(B)) ↔ (A ↛ B) ↔ opp(A → B)
[= A māllon ē B, *scilicet*], resp.
((A △ B) ∧ B) ↔ (opp(A) ∧ B) ↔ (A ↚ B) ↔ opp(A ← B)
[= A ētton ē B],

and, thereby, the polarity principle behind the Chrysippean ('semantic') construction.

On the other hand, the fact that the proper polars **if** and **nand** are interdefinable – if negation (resp. opp) is present – is to be handled separately. (See below.)

As to the proper polars, let us call, for convenience, the 'disjunction'-like connectors (△, →, ←, ∨), appearing on the LHS, *additive*, and the 'conjunction'-like connectors (appearing on the RHS), *multiplicative*, and similarly for the corresponding complex propositions. Generic notation: (A ⊕ B) resp. (A ⊗ B).

We have, again, semantically, for each pair (⊕,⊗), meaning postulates of the form:

'intensional' (non-truth-functional) connectors, as well. Our sources are not very illuminating on this subject, though.

[19] The latter can be viewed as 'disjunctions' as well as 'conjunctions', and can be analysed into / reduced to / defined in terms of proper connectors. Otherwise, our texts confirm the corresponding equivalences.

[20] By pondering upon Apollonius Dyscolus Conj. [Schneider GG II.i] 222.25–26. Here, round parantheses are mine – where 'either' might have been left out, by modern standards – i.e., one must have ((A △ B) ∧ A) resp. ((A △ B) ∧ B).

$\mathrm{opp}(A \oplus B) = A \otimes B$, and $\mathrm{opp}(A \otimes B) = A \oplus B$,

that is, in particular, $(A \triangle B) = \mathrm{opp}(A \wedge B)$ resp. $(A \wedge B) = \mathrm{opp}(A \triangle B)$, and so on, whence, in general,

$C = \mathrm{opp}(\mathrm{opp}(C))$, for each $C := (A \oplus B)$, resp. $C := (A \otimes B)$.

In other words, double negation is, again, 'built-in', by construction.

One can understand the equivalences above as meaning postulates for 'opp', whereby (classical) negation can be viewed as a defined notion, by setting, finally, $\mathrm{non}(A) := \mathrm{opp}(A)$.

We can think of the above as a piece of semantics, which we can even formalise as appropriate. As already noted before, this does not mean that the Stoics would have cared to distinguish, conceptually, between bare syntax (as recent formalists would have it) and semantics (or model theory).[21]

Note also that the explanations above do not make any explicit appeal to a truth-value account of the Stoic connectors and of the concept of polar opposition. Whether this was actually the case in Chrysippus and his followers, we cannot tell with ultimate certainty, given the poor state of our sources. One can say, however, for sure, that the Stoics were well aware of the fact that most of the connectors they used to theorise explicitly upon can be characterised by something similar to our truth-tables (as in Peirce, Frege, Russell, Post, and, later on, in Wittgenstein [Tractatus]).

Once more, there is a close parallel to all this in contemporary proof-theoretic work concerning classical logic, mainly. Typically, in order to simplify the syntax of a Gentzen L-system, for instance, and to save repetitions, a proof-theorist would take the atoms to be as above, with only (classical) **and** and (inclusive) **or** as proposition-forming primitives, defining next (classical) negation by the usual Ockham / De Morgan transformations. An equivalent technique consists of manipulating 'signed' formulas, instead.[22] The least thing to say, here, is that the modern / contemporary techniques have been vastly anticipated by Chyrisppus and his followers.[23]

---

[21] A better name for the polar construction – implicit in the Stoic way of understanding logic – would be, perhaps, 'proof-theoretic semantics' (as in recent work of Dag Prawitz and some of his followers), and we can take the method as making up *the right way of justifying classical logic, conceptually*.

[22] Cf., e.g., Schütte *et al.*

[23] A terminological aside: J.-Y. Girard used the terms 'additive', 'multiplicative' and 'polarity' in a different sense.

§ 2 Chrysippean proof semantics

Analysis and polar projections

For 'analytical' purposes, so to speak, let us define, next, *polar projections* left(C), right(C), for each complex proper polar proposition C, separately (leaving the sub-polars aside, for a while), as follows.[24]

For additive C := A ⊕ B:

If C = (A △ B), then left(C) = opp(A), right(C) = opp(B).
If C = (A → B), then left(C) = opp(A), right(C) = B.
If C = (A ← B), then left(C) = A, right(C) = opp(B).
If C = (A ∨ B), then left(C) = A, right(C) = B.

For multiplicative C := A ⊗ B:

If C = (A ∧ B), then left(C) = A, right(C) = B.
If C = (A ↛ B), then left(C) = A, right(C) = opp(B).
If C = (A ↚ B), then left(C) = opp(A), right(C) = B.
If C = (A ▽ B), then left(C) = opp(A), right(C) = opp(B).

With this *schematic notation*, one has, as meaning postulates (semantically thus), equivalences of the form:

A ⊕ B = left(A ⊕ B) ∨ right(A ⊕ B),
A ⊗ B = left(A ⊗ B) ∧ right(A ⊗ B),

i.e., the additives are (classical) disjunctions, while the multiplicatives are (classical) conjunctions, as announced already in the above.

Note that the usual (Boolean) duals are exactly those pairs that agree on polar projections. This is not the case for the sub-polar pair (**iff** vs **xor**), where each complex propositional form can be viewed either as an additive (classical disjunction) or as a multiplicative (classical conjunction). In other words, (A ↔ B) and (A ↮ B) are self-duals.[25]

The Chrysippean logic as a rejection / refutation system

The fact that the overall construction described here corresponds actually to (what we call) classical logic, indeed, is obvious from the Stoic way of defining (valid) entailments (here: Stoic [valid] arguments),

---

[24] This is just a convenient shorthand, to save repetitions.
[25] This does not make, as yet, the underlying logic 'classical', as one might be tempted to think at a first look. Indeed, the construction 'by (classical) polarities', sketched in the above, would also apply to a couple of so-called 'substructural' logics, as well (like the Anderson-Belnap relevance logic $\mathcal{R}$, the non-distributive $\mathcal{R}$, or 'Lattice $\mathcal{R}$' [$\mathcal{LR}$], and, even, Girard's '(classical) linear' logic $\mathcal{LL}$, for instance [by the Anderson / Belnap standards, $\mathcal{LL}$ is *non-distributive*, by the way]).

First, we must remember the fact that entailments of the form $\Gamma, A \vdash C$ are to be analysed (actually: defined) in terms of rejections / refutations / (logical) conflict, by:

($\vdash$) $\Gamma, A \vdash C \Leftrightarrow \Gamma, A, \text{opp}(C) \vdash \bot$,

where Γ is as above, and $\Leftrightarrow$ stands for equivalence (in the meta-language).

Here we may think of $\Gamma \vdash \bot$ as being a primitive monadic (single-place) predicate on sequences Γ. Ad hoc alternative notation: $\Gamma \models$, with intended reading, in modern terms: 'Γ is inconsistent'. So the above equivalence ($\vdash$) can be wiewed as a definition of $\vdash$ in terms of $\models$, viz.

(df $\vdash$) $(\Gamma, A \vdash C) \Leftrightarrow_{df} (\Gamma, A, \text{opp}(C) \vdash \bot)$ [$\Leftrightarrow (\Gamma, A, \text{opp}(C) \models)$],

where Γ, taken as a parameter, might be empty, as a limit case.

In other words, the definition of a (valid) entailment should involve the (genuinely classical) *reductio ad absurdum*, as well.[26]

Explicitly, the basic tenet is that Chrysippean logic is constructed in terms of rejection / refutation (expressing [logical] conflict), taken as a primitive notion, to be further characterized 'axiomatically' so to speak.

In modern terms, this covers an obvious induction, where defined is the (primitive recursive) monadic predicate $\models$.

With $\Rightarrow$ standing for the meta-conditional, and & for conjunction in the meta-language, one has two 'axioms':

($\bot \Vdash$) $\bot \models$,
(cut $\Vdash$) $A, \text{opp}(A) \models$,

or even, more generally,

(cut $\Vdash \Gamma$) $\Gamma, A, \text{opp}(A) \models$,

and transitions (in Frege's terms [**GGA**]) or 'structural' rules of inference (à la Gentzen 1934–1935) of the form:

(dil $\Vdash$) $(\Gamma \models) \Rightarrow (\Gamma, C \models)$,
(prm $\Vdash$) $(\Gamma, A, B \models) \Rightarrow (\Gamma, B, A \models)$,
(CUT $\Vdash$) $(\Gamma, A \models)$ & $(\Gamma, \text{opp}(A) \models) \Rightarrow (\Gamma \models)$,

written in *parametric* form (i.e., keeping Γ as a parameter on both sides of $\Rightarrow$, where appropriate).

Note that a special case of the 'inner' (cut $\Vdash$) is:

(cut $\Vdash \top$) $\top, \bot \models$.

---

[26] With this notation, $\Gamma \Vdash$ means actually $\Gamma \vdash \bot$.

Here, the 'axioms' make up the basis of the induction and the transitions the inductive step.[27]

The 'inner'-cut axiom (cut ⊪) is just a way of expressing the 'law of (non-)contradiction'.

Given the parametric spelling of the (primitive) rules above, from this, one has also, as a derived rule:

(ctc ⊪) $(\Gamma, A, A \models) \Rightarrow (\Gamma, A \models)$,

i.e., the so-called 'contraction' rule (Frege's *Verschmelzung*, in **GGA**).

Here, (dil ⊪), (prm ⊪), (ctc ⊪), and the global (CUT ⊪)-rule stand for the usual 'dilution', 'permutation', 'contraction' and the 'syllogism' rules, resp., in Frege's **GGA**, as well as in Gentzen's *Inauguraldissertation 1934–1935*.[28]

Dilution and the fallacies of relevance

If we have at hand formal means in order to express the fact that a proof of C does not depend on Γ (ad hoc notation: C|Γ), then (dil ⊪) can be reversed, i.e., we have, also:

(dil ⊪ $\Gamma \Leftrightarrow$) $(\Gamma \models) \Leftrightarrow (\Gamma, A \models)$, provided $\perp | A$,

i.e., provided that the proof of the contradiction reached in the refutation ($\perp$, say) does not depend on A.[29]

Note that the reading of dilution (dil ⊪) is quite natural, in this context, and does not involve would-be *fallacies of relevance*, as recent relevance logic defendors might imply, viz.:

> if (the sequence) Γ is inconsistent then, *a fortiori*, so is any proper extension of it[30].

---

[27] The moderns would likely want to have a limit case, i.e., an additional 'axiom' concerning the empty sequence (nil). On technical reasons, it is appropriate, indeed, to identify the empty sequence (nil) with the constant atom ⊤ [sic], having thus ⊤ ⊢ A ⇔ ⊢ A ('theoremhood' for A), as usual in our textbooks.

[28] Gentzen had *Schnitt*, 'cut', instead of Frege's *(Ketten)schlusss*, whence also our current way of speaking in proof theory. Apparently, Gentzen borrowed the idea from Paul Hertz (1881–1940), a former physics student of David Hilbert, not from Frege's **GGA** (I am trusting Paul Bernays on this, who actually supervised Gentzen's Göttingen Dissertation). As regards terminology, in English, dilution (dil) is oft referred to as 'weakening', an inadvertent translation from German, where one has *Verdünnung* (like for wine and/or in chemistry).

[29] In modern logic, the restrictive proviso (C|Γ) can be properly formalised in *witness theory*, by using an explicit λ-calculus notation for 'witnesses' (formal proofs), as, e.g., in N. G. Bruijn's AUTOMATH proof-checking systems, and, in general, in logic systems / calculi based on the 'Curry-Howard Correspondence' [or 'Isomorphism'].

[30] Or else, in plain English: *contradictions are infectious*.

The reversal of (dil-Γ): $(\Gamma, A \models) \Leftarrow (\Gamma \models)$, provided $\bot|A$, reads naturally, again:

> if (the sequence) (Γ, A) is inconsistent and A has never been used in establishing this very fact, then A is redundant and one can safely get rid of it, i.e., Γ is inconsistent, as well.

This way of understanding dilution (dil) and 'redundancy' in Stoic logic contexts leaves to think that Chrysippus and his followers payed little or no attention to what the moderns would call *fallacies of relevance*.[31]

In particular, for $A := \top$, one has

(dil ⊩ ⊤) $(\Gamma, \top \models) \Rightarrow (\Gamma \models)$

anyway, in view of the gobal (CUT ⊩)-rule, so that, with this stipulation, the 'inner' (cut)-'axiom' makes redundant the 'axiom':

$(\bot \Vdash) \bot \models$.

Actually, all 'structural' rules (transitions), except the global (CUT ⊩), can be reversed, i.e., we have also:

(prm ⊩⇔) $(\Gamma, A, B \models) \Leftrightarrow (\Gamma, B, A \models)$,
(ctc ⊩⇔) $(\Gamma, A, A \models) \Leftrightarrow (\Gamma, A \models)$.

The above, taken together with (df ⊢) generate the usual Frege-Gentzen ('structural') rules, as well as the 'law of identity':

(id) $A \vdash A$,
  resp. the 'projection' rule:
(id Γ) $\Gamma, A \vdash A$,
  that can be also written as:
(prj) $\Gamma \vdash A_i$ ($0 < i < n+1$)
  (recall that Γ is a shorthand for $A_1, ..., A_n$).

---

[31] This *contra* Jonathan Barnes [cf. his *Proof destroyed* (1980)], Susanne Bobzien (1996), Marek Nasieniewski (1998) *et al.*, who suggested that Chrysippus' logic should be rather viewed as 'a kind of relevance logic' [sic]. With reference to the discussion of (dil) above, a 'true relevantist' (i.e., a paradigmatic relevance logic defendor) would have likely talked about 'relevantly [in-] consistent' sequences – or sets – of propositions, instead. A correct formal model-theoretical account of the alternative goes, however, far beyond the techniques Chrysippus and his followers would have had at hand. As the alternative reading is quite tempting, I shall examine in detail the Barnes-Bobzien suggestion in a sequel to these notes.

A 'dialectical theorem'

Before leaving the subject, it might be a good idea to pause, once more, on Sextus Empiricus' statement of a so-called *dialectical theorem* [theōrēma dialektikon[32]] of the Stoics (Sextus Adv. Math., VIII, 231, 3–6):

> '[W]hen we know the premisses [lēmmata] Γ which imply a certain conclusion [sumperasma] [C], we know also potentially the conclusion [C] involved in them [Γ] [dunamei kakeĩno en toutois echomen to sumperasma], even though it is not explicitly [kat' ekphoran] stated' (Robert Gregg Bury's English, from the Loeb edition, *ad loc.*).

As a side-remark, the Latin of the industrious Frenchman Gentien Hervet (1499–1584) is more readable: '*Cum habuerimus propositiones ex quibus colligitur aliqua conclusio, vi ac potestate in his habemus illam conclusionem, etiamsi diserte non enuntietur.*' (ed. Johann Albert Fabricius, Leipzig 1718, page 502).

For the Greekless reader, the quote contains a mix-up of Peripatetic and Stoic technical jargon (sumperasma, in place of epiphora, on a par with the lēmmata), so we may also wonder where was actually Sextus copying from.

In the light of the above, the 'dialectical theorem' referred to by Sextus looks, rather, like an inductive stipulation.

Take first Γ modulo arbitrary permutations (i.e., as a multiset; this is implicit in the Chrysippean way of understanding the lēmmata, anyway).

Then define $\Gamma \succ C$ [i.e., C *dunamei en tois* Γ], inductively, by:

[1] C is an element of Γ [basis clause for ($\Gamma \succ C$)],
[2] there is a proposition A[33] such that $\Gamma, A \succ C$, and $\Gamma \succ A$ [inductive step, with same conclusion ($\Gamma \succ C$), via the global (CUT)].

The basis clause (of the induction) [1] is a diluted (id) – i.e. the 'projection axiom' (prj) above – covering (dil), as well, while the inductive step [2] covers the remaining 'structural' rules of inference, viz. (CUT) and (ctc) [as well as (prm), in fact, given the multiset assumption on Γ's].

If the over-argutious Sextus was actually quoting Chrysippus or a genuine Stoic source[34], one can only admire this concise intuitive phrasing of a basical logical idea that took about twenty two centuries to be retrieved.[35]

---

[32] A 'logical principle', as, e.g., Ierodiakonou 1990, II.2.2 had it, while insisting on the absence of the *definite* article (before theōrēma) in the Sextan text (which I took seriously).
[33] Expressed by a 'CUT-formula', in Gentzen's terms.
[34] Quite unlikely, but why not?
[35] By Gerhard Gentzen, via Paul Hertz, namely.

Proper logical rules

What about the proper 'logical' rules (transitions), i.e., those involving connectors?

With the projective shorthand-notation above, one has, for each polar pair $(\oplus, \otimes)$, equivalences of the form:

($\otimes$) multiplicative case, where $C := (A \otimes B)$,
$(\Gamma, \text{left}(C), \text{right}(C) \models) \Leftrightarrow (\Gamma, C \models)$,
($\oplus$) additive case, where $C := (A \oplus B)$,
$(\Gamma, \text{opp}(\text{left}(C)) \models) \,\&\, (\Gamma, \text{opp}(\text{right}(C)) \models) \Leftrightarrow (\Gamma, C \models)$,

where the left-to-right transition is a corresponding Gentzen-rule (an 'introduction' rule), while a right-to-left transition is the associated resolution-rule (as in the Beth-Hintikka-Smullyan tableaux, etc.; an 'elimination' rule, thus).

Example. Case of this, taking $\otimes$ to be $\wedge$ and $\oplus$ to be $\triangle$:

($\wedge$) $(\Gamma, A, B \models) \Leftrightarrow (\Gamma, (A \wedge B) \models)$,
($\triangle$) $(\Gamma, \text{opp}(A) \models) \,\&\, (\Gamma, \text{opp}(B) \models) \Leftrightarrow (\Gamma, (A \triangle B) \models)$,

and, since one has defined $\text{not}(A) := \text{opp}(A)$, in the latter case, one has:

($\triangle$) $(\Gamma, \text{not}(A) \models) \,\&\, (\Gamma, \text{not}(B) \models) \Leftrightarrow (\Gamma, (A \triangle B) \models)$.

Putting (df $\vdash$) at work yields:

($\wedge$) $(\Gamma, A, B \vdash C) \Leftrightarrow (\Gamma, (A \wedge B) \vdash C)$ ['confusion'].
($\triangle$) $(\Gamma, (\text{not}(A) \vdash C) \,\&\, (\Gamma, \text{not}(B) \vdash C) \Leftrightarrow (\Gamma, (A \triangle B) \vdash C)$,

with also, in view of (df $\vdash$), derived rules:

(adj) $(\Gamma \vdash A) \,\&\, (\Gamma \vdash B) \Rightarrow (\Gamma \vdash (A \wedge B))$
[the 'adjunction rule' for $\wedge$],

as well as the expected $\wedge$-'projections', i.e., the 'simplification' or 'elimination' rules for $\wedge$:

(fst $\wedge$) $(\Gamma \vdash (A \wedge B)) \Rightarrow (\Gamma \vdash A)$,
(snd $\wedge$) $(\Gamma \vdash (A \wedge B)) \Rightarrow (\Gamma \vdash B)$.

Note that the ($\triangle, \wedge$)-team above generates also the derived rule ($\triangle$-introduction on the right):

(split) $(\Gamma, A, B \vdash \bot) \Rightarrow (\Gamma \vdash A \triangle B)$.

On the other hand, the meaning postulates for the built-in negation, yield derived equivalences of the kind:

$(\Gamma, A \vdash \bot) \Leftrightarrow (\Gamma \vdash \text{not}(A))$,
$(\Gamma, \text{not}(A) \vdash \bot) \Leftrightarrow (\Gamma \vdash A)$,
$(\Gamma, \text{not}(\text{not}(A)) \vdash C) \Leftrightarrow (\Gamma, A \vdash C) \Leftrightarrow (\Gamma, A \vdash \text{not}(\text{not}(C)))$, etc.

In the end, by the standards above, a Gentzen sequent 'multiple on the right' (i.e., something of the form $\Gamma_1 \vdash \Gamma_2$, where Γ_i (i := 1,2) are finite sequences of propositions) is just a Stoic rejection / refutation (i.e., of the form: $\Gamma_1, \text{opp}(\Gamma_2) \vdash \bot$, where $\text{opp}(\Gamma_2) := \text{opp}(B_1), ..., \text{opp}(B_m)$, for $\Gamma_2 := B_1, ..., B_m$, m > 1).[36]

All this is very redundant, of course. As a matter of fact, a single (proper) polar pair $(\oplus,\otimes)$ is sufficient in order to get full classical (propositional) logic.

Chrysippus' logic is classical logic

Let us call, for further reference, **Ch** ['Ch' for **Chriyppus**] this (global) formulation of classical logic. More precisely, we may refer to it as **Ch**[⊩], in order to stress the fact that the rejection predicate [⊩] has been taken as a (semantical) primitive. The proof that **Ch** is classical (propositional) logic, indeed, is straightforward. [*Hint.* Construct, first, an algebra **Ch-seq**, say, on finite sequences of propositions, satisfying the 'Chrysippean' conditions stated earlier. Show next that **Ch-seq** is a Boolean algebra [BA]. Finally, concoct a Stone-like argument to the effect that every BA can be so represented. Whence 'classical completeness'.[37]]

Modulo provisions already alluded to in the above, regarding the distinction syntax *vs* semantics in Ancient logic, **Ch** might be thought of as being a kind of genuine 'semantical' justification of (what we understand, nowadays, by) classical logic (as opposed to intuitionistic logic, say, based on the BHK [Brouwer-Heyting-Kolmogorov] interpretation of the Brouwer-Heyting logic).[38] Note also that the explanations above did not make appeal explicitly to a truth-functional account of the Chrysippean connectors and of the concept of polar opposition.[39] Incidentally, one can also realise the fact that, once valid entailments are characterised on the proposed pattern, the predicates **True** and **False** – as applied to propositions – can be defined explicitly in terms of

---

[36] The reader has already realised, by now, that the polar statement of the proper logical rules (those involving connectors) above amounts to a *nearly verbatim copy* of the Gentzen **LK**-rules [for classical logic], taken modulo (⊢), and including reversals (i.e., as equivalences, instead of unidirectional meta-conditionals).

[37] Note that we have both tautology- as well as (classical) rule-completeness, thereby.

[38] Even if the polar construction suggested here would have no historical support in the Stoic texts, the *conceptual justification of classical logic* should rest on the very same principle, in the end, a rather simple fact that has been overseen to my knowledge – by, virtually, all defenders of classical logic, so far. As I have gotten the idea by paying attention to Chrysippus and his followers, in the first place, I'd better credit him with the finding: the present remarks make up just a piece of (historical) data-retrieval, indeed. In fact, the entire construction is based on a *very Greek idea* (in the ancient sense), there is nearly nothing to wonder about. Yet, *fallait-il y penser !*

[39] Whether this was actually the case in Chrysippus and his followers, we cannot tell for sure: the supporting positive evidence in the extant (ancient) texts is rather scarse.

monadic entailments, by **True**(A) iff $\top \vdash A$ resp. **False**(A) iff $A \vdash \bot$. (See also the previous remarks on classical completeness.)

Derivable rules

Summing up, given (df $\vdash$), one can easily derive, from **Ch**[⊩], the following 'structural' bi-transitions, in terms of $\vdash$ alone:

($\vdash$ dil ⇔) $\Gamma \vdash C \Leftrightarrow \Gamma, A \vdash C$, if C does not depend on A,
($\vdash$ prm ⇔) $\Gamma, A, B \vdash C \Leftrightarrow \Gamma, B, A \vdash C$,
($\vdash$ ctc ⇔) $\Gamma, A, A \vdash C \Leftrightarrow \Gamma, A \vdash C$,

together with the following forms of global (CUT):

(CUT $\vdash \otimes$) $(\Gamma, A \vdash C) \& (\Gamma \vdash A) \Rightarrow (\Gamma \vdash C)$
   ['parametric'],
(CUT $\vdash \oplus$) $(\Gamma_1, A \vdash C) \& (\Gamma_2 \vdash A) \Rightarrow (\Gamma_1, \Gamma_2 \vdash C)$
   ['cumulative'].

As already mentioned above, the 'inner' (cut)-'axiom' (the 'law of (non-) contradiction') is tantamount 'the law of identity':

(id) $A \vdash A$,

in this setting, whence the 'projection' rule:

(prj) $\Gamma, A \vdash A$

follows by (repeated) dilutions.

On the other hand, in the case of the polar pair of connectors [∧,△], the proper 'logical' rules of inference (those involving connectors), yield:

($\vdash \wedge \Leftrightarrow$) $(\Gamma, A, B \vdash C) \Leftrightarrow (\Gamma, (A \wedge B) \vdash C)$ ["confusion"],
($\vdash \triangle \Leftrightarrow$) $(\Gamma, \text{opp}(A) \vdash C) \& (\Gamma, \text{opp}(B) \vdash C) \Leftrightarrow \Gamma, (A \triangle B) \vdash C$,

whereas, from the latter, we get, as noted before:

($\triangle \vdash \Leftrightarrow$) $(\Gamma, A, B \vdash \bot) \Leftrightarrow (\Gamma \vdash (A \triangle B))$,
($\wedge \vdash \Leftrightarrow$) $(\Gamma \vdash A) \& (\Gamma \vdash B) \Leftrightarrow (\Gamma \vdash (A \wedge B))$.

i.e., the 'split'-rule (△-'introduction' on the right), resp. the usual 'adjunction'- and 'projection'-rules for ∧.

The other three polar pairs of connectors yield the expected bi-transitions (in terms of $\vdash$).

In particular, the [↠,→]-pair gives:

($\vdash \twoheadrightarrow \Leftrightarrow$) $(\Gamma, A, \text{opp}(B) \vdash C) \Leftrightarrow (\Gamma, (A \twoheadrightarrow B) \vdash C)$
   [a case of 'confusion'],
($\vdash \rightarrow \Leftrightarrow$) $(\Gamma, \text{opp}(A) \vdash C) \& (\Gamma, B \vdash C) \Leftrightarrow (\Gamma, (A \rightarrow B) \vdash C)$,

whereas the latter two yield (in this order):

$(\rightarrow \vdash \Leftrightarrow)$ $(\Gamma, A \vdash B) \Leftrightarrow (\Gamma \vdash (A \rightarrow B))$,
$(\nrightarrow \vdash \Leftrightarrow)$ $(\Gamma \vdash A)$ & $(\Gamma, B \vdash \bot) \Leftrightarrow (\Gamma \vdash (A \nrightarrow B))$, resp.
$(\nrightarrow \vdash \Leftrightarrow)$ $(\Gamma \vdash A)$ & $(\Gamma \vdash \text{opp}(B)) \Leftrightarrow (\Gamma \vdash (A \nrightarrow B))$.

An easy excercise shows that the definitional stipulations (equivalences):

$(A \rightarrow B) = (A \vartriangle \text{opp}(B))$,
$(A \nrightarrow B) = (A \wedge \text{opp}(B))$

allow the derivation of the $[\nrightarrow,\rightarrow]$-equivalences from the corresponding $[\wedge,\vartriangle]$-equivalences, listed previously, whereas the (definitional) stipulations:

$(A \vartriangle B) = (A \rightarrow \text{opp}(B))$,
$(A \wedge B) = (A \nrightarrow \text{opp}(B)) = \text{opp}(A \rightarrow \text{opp}(B))$

guarante the corresponding derivations in the opposite direction.

As expected, the remaining proper polar pairs ($[\nleftrightarrow,\leftarrow]$, resp. $[\triangledown,\vee]$) yield analogous equivalences (bi-transitions) in terms of $\vdash$ alone, whereas the appropriate definitional equivalences allow the derivation of the latter teams from either one of those mentioned previously, and, ultimately, from those generated by the $[\wedge,\vartriangle]$-pair alone, for instance.

Let us call the overall [re-] construction of **Ch**[⊩] in terms of $\vdash$, **Ch**[$\vdash$]. It is obvious that the latter formulation (in terms of entailments) is equivalent to the former one (in terms of refutations)[40]. In other words, both 'semantic' formulations – **Ch**[⊩] and **Ch**[$\vdash$] – turn out to be equivalent, in the sense they allow deriving the same set of rules [transitions], modulo (df $\vdash$). Moreover, the discussion above shows that we can eventually *fibrate* **Ch** 'semantically' *along* a single proper polar pair of connectors, i.e., *cut off* would-be *redundancies*, on this pattern, in four distinct ways.[41]

From a *genuinely Chrysippean* point of view, the most interesting and (historically) relevant 'semantic' fibration appears to be the one along the $[\wedge,\vartriangle]$ polar pair of connectors.[42]

A historical aside

One might argue that Chrysippus could have borrowed the so-called 'hypothetical syllogisms' involving conditionals and exclusive disjunctions (as mentioned

---

[40] Just define $\Gamma \Vdash \Leftrightarrow_{df} (\Gamma \vdash \bot)$, in **Ch**[$\vdash$].
[41] There is no need to be very formal on the meaning of 'fibration', here: the term points out to a specific form of analysis (meant to eliminate conceptual redundancies). Technically, given a redundant formulation of a formal system $\mathcal{S}$, fibrating $\mathcal{S}$ along a given choice of primitives amounts to reformulating $\mathcal{S}$ in non-redundant terms, relative to the given choice.
[42] In view of what has been said before, it is a simple exercise to 'reconstruct semantically' the full **Ch** along the $[\wedge,\vartriangle]$-fibration, say.

explicitly in the Stoic indemonstrables T1–T2, and T4–T5, resp.) from late Peripatetic sources (Theophrastus, Eudemus, and 'some other of Aristotle's associates [hetairoi]', as Alexander of Aphrodisias has occasionally implied[43]). Putting aside the – quite obvious – fact that neither the Great Aristotle nor his lesser hetairoi – or later (Peripatetic) followers, for that matter – had the slightest idea about what is a logical connector [proposition-forming binary connective][44], even the astute, bright and partinic (!) *connoisseur* Alexander did not attempt to assign the discovery of the **nand**-connector to any one of his fellow (Peripatetic) predecessors who lived in the shadow of the Great Master.[45]

What follows in the next section concerns 'syntactic' constructions, based on (the intented semantics of) **Ch**, whose association to the actual Stoic ideas are, historically speaking, rather conjectural. They are, however, *in the Chrysippean spirit*, so to speak.

§ 3 Redundancy, fibrations and formal systems

As noted above, in view of the well-known interdefinability of classical connectives, **Ch** is very redundant. Like, *mutatis mutandis*, Gentzen's **LK** – the sequent-version of classical logic, in his 1934–1935 – actually.[46] Except for the fact that, unlike in Gentzen's **LK**, the 'Ancient Logic' **Ch** – as presented here – has an implicit *conceptual justification*, as well as an explicit *criterion of construction* (which, otherwise, Gentzen missed).

It is instructive to examine subsystems or fragments of **Ch**, say, based on functionally complete sets of classical (Boolean) connectives.

---

[43] See, e.g., In Anal. Pr. 389,31–390,19 = Hülser **1137** and **1083**, resp.

[44] By Chrysippean standards, the Peripatetics used to deal only with atoms [atomic propositions], so to speak. — As an aside, in this guise, the generous idea of some of the *recentiores* to correct Łukasiewicz 1957 (who, otherwise, confused *use* and *mention*, in his heroic endeavours of 'reconstructing' Aristotle 'from the standpoint of modern formal logic'), on the basis of a would-be genuine, alternative idea of 'natural deduction' is, at best, a terminological *quiproquo*, since what we usually call *natural deduction* in contemporary logic – the Jaśkowski-Gentzen-Fitch-Prawitz-etc. approach – consists, *prima facie*, of an attempt to characterise, theoretically, the behaviour of the *logical connectives*, things unheard of in Aristotle and badly mishandled in the later Perpipatetic lore, as well as in the learned ruminations of 'historians' of logic à la Carl [von] Prantl, Eduard Zeller and the like.

[45] Cf. the (rather deceiving) comments of Jonathan Barnes on this, in Barnes 1985.

[46] Gentzen's **LK** is based on an ad hoc choice of primitives, reflecting, most likely, his casual interest in the Heyting logic, which, reputedly, is *an empirical construction*. (See the discussion of the first-order Heyting logic, viewed as a proof-theoretic fragment of classical logic – in witness-theoretical terms [$\lambda\gamma$-calculus] – as appearing in Rezuş 1991, 1993.) It never occurred to any intuitionist logician – or mathematician – to *justify* (*conceptually*) the choice of the primitives in the 'standard' propositional intuitionistic signature [$\neg, \to, \wedge, \vee$], or else in the 'reduced' one [$\bot, \to, \wedge, \vee$]. Indeed, why not an intuitionistic primitive **nand** [incompatibility], or an intuitionistic **nor**, for that matter? (The *empirical presuppositions* of Brouwer – and Heyting – in logic are discussed, in proper terms, elsewhere.)

Syntactic fibrations

Leaving the sub-polar pair (**iff**, **xor**) aside, one can *prima facie* fibrate 'semantically' the overall construction [**Ch**], in four distinct ways, as suggested above, by choosing a single specific (proper) polar pair of connectors ($\oplus, \otimes$), as a primitive setting, while still tinkering on (the status of) negation.

Further, one can specialise this choice, by taking only one of the connectors, (classical) negation, and a constant ($\top$ or $\bot$) as primitives. This yields eight possible (syntactic) fibrations of **Ch**, so to speak.

In two cases (those involving **nand** resp. **nor** as primitives), negation is redundant (as a primitive), because we have:

$\text{not}(A) = (A \triangle A) = (A \triangledown A),$

as meaning postulates.

In these cases, redundancy is 'built-in' so to speak, since at least one propositional constant is necessary for functional (Boolean) completenesss, while, granted $\bot$, say, we can define, alternatively, an inferential negation (à la Peirce 1885), by setting:

$\text{not}(A) := (A \to \bot)$

(as material implication is definable in terms of $\triangle$ resp. $\triangledown$ alone, without using propositional constants), etc.

On the other hand, the other two cases (involving $\wedge$ resp. $\vee$ as primitives), one needs a primitive negation, as well, whence again, 'built-in' redundancy, since one also needs at least one constant, as above, in order to guarantee functional completeness.

Now, in the remaining four cases (where either $\to$ or its converse $\leftarrow$ is present) the only non-redundant choices consist of taking:

(1) either [$\bot, \to$] resp. [$\bot, \leftarrow$] as primitive signatures, with negation defined 'inferentially', by:

$\text{not}(A) := (A \to \bot)$, as above, or by:
$\text{not}(A) := (\bot \leftarrow A)$, resp.

(2) or [$\top, \not\to$] resp. [$\top, \not\leftarrow$] as primitive signatures, with negation defined 'co-inferentially', by:

$\text{not}(A) := (\top \not\to A)$, or by:
$\text{not}(A) := (A \not\leftarrow \top),$

and it is relatively easy to see that any other one of the remaining choices has 'built-in' redundancy, too.

Some reflection on this elementary combinatorics shows that the most economic – and conceptually clean – choices of primitives are those involving an additive connector ($\triangle, \to, \leftarrow, \vee$).

On the other hand, it is obvious that there is no (conceptual) profit in using $\leftarrow$ instead of $\to$, resp. $\vee$ instead of $\triangle$, as primitives (the latter behave, proof-theoretically, exactly in the same way, modulo trivial transformations: (A ← B) = (B → A), on the one hand, resp. (A ∨ B) = (not(A) △ not(B)), by Ockham / de Morgan, on the other hand), so that we end up with just two, conceptually distinct, non-redundant strategies, viz. with the primitive signatures [⊥,△] and [⊥,not,→], resp. This yields, essentially, two conceptually distinct ways of formulating (syntactically) the 'Ancient Logic' of Chrysippus.

An Ancient Logic, Formulation 1

(*The* [⊥,△]-*case*) Let us introduce, first, the fragment **Ch**[⊥,△] of **Ch**, based on the primitive (propositional) signature [⊥,△] alone.

Taking △ as a primitive, together with ⊥, allows defining the remaining classical connectives (sic – including thus classical negation, understood as a [modern] connective, this time) on a familiar, well-known pattern, by setting (definitionally), e.g.:

not(A) := (A △ A),
(A ∧ B) := not(A △ B),
(A → B) := (A △ not(B)),
(A ← B) := (not(A) △ B),
(A ∨ B) := (not(A) △ not(B)),

etc., so that, for instance, on the primitive signature above, granted the appropriate team of 'structural' rules for ⊢ (including the global (CUT) for ⊢), the following rules of inference are sufficient for full classical (propositional) logic:

(i-cut) (Γ ⊢ non(A)) & (Γ ⊢ A) ⇒ (Γ ⊢ ⊥) ['the law of (non-) contradiction'],
(red) (Γ, non(A) ⊢ ⊥) ⇒ (Γ ⊢ A) [reductio ad absurdum],
(fst) (Γ ⊢ A ∧ B) ⇒ (Γ ⊢ A),
(snd) (Γ ⊢ A ∧ B) ⇒ (Γ ⊢ B),
(adj) (Γ ⊢ A) & (Γ ⊢ B) ⇒ Γ ⊢ (A ∧ B) ['adjunction'],

Remember that, in this setting, conjunction [∧] is a defined notion, i.e., (A ∧ B) := not(A △ B). Here, we just ignore the (two) remaining ways of defining negation ("inferentially").

An Ancient Logic, Formulation 2

(*The* [⊥,not,△]-*case*) The most preferred modern formulation of classical logic (Frege, Church, Łukasiewicz, Jaśkowski, etc.) relies on a primitive [not,→]-signature. The latter is, actually, a functionally incomplete set, since the propositional constants are missing.[47]

---

[47] Without further additions in the primitive syntax, they can be recovered only by an algebraic trick, as in groups, by proving, e.g., first that 'all zeroes are equal' [unlike in groups, in a Boolean algebra we have two 'zeroes', i.e., ⊤ and ⊥, not a single one], i.e. by something like: (1) define first ⊤[C] := (C → C) and (2) show next that the rules imply: ⊤[A] = ⊤[B], for all A, B, in the sense of material equivalence **iff**, and analogously for ⊥[C].

On the other hand, adding a primitive propositional constant (pick up $\bot$, for instance) yields 'built-in' redundancy as noted before, in view of the inferential definition(s) of negation, à la Peirce. Nevertheless, as in the case of Formulation 1 above, there is no reason to bother about, since one can, simply, ignore the inferential alternative(s).

The [$\bot$,not,$\to$]-fragment of **Ch**, **Ch**[$\bot$,not,$\to$] say, consists of the following rules of inference:

(i-cut) $(\Gamma \vdash \text{not}(A))$ & $(\Gamma \vdash A) \Rightarrow \Gamma \vdash \bot$ ['the law of (non-) contradiction'],
(red) $\Gamma, \text{not}(A) \vdash \bot \Rightarrow \Gamma \vdash A$ [*reductio ad absurdum*],

exactly as for **Ch**[$\bot$,Δ] above, meant to handle the proof-theoretical behaviour of negation (and $\bot$), with, moreover, a 'replication' of the (i-cut)-(red)-team:

($\to$-cut) $(\Gamma \vdash A \to B)$ & $(\Gamma \vdash A) \Rightarrow \Gamma \vdash B$ [*modus ponens*, $\to$-elimination]
(abs) $\Gamma, A \vdash B \Rightarrow \Gamma \vdash A \to B$ ['the deduction theorem', $\to$-introduction],

meant to handle the proof-theoretic behaviour of (minimal, as well as material) implication.

§ 4 Alternative Formulations and witness theory

Some variations and/or improvements on the 'syntactic' fibrations mentioned earlier are still possible.

Inferential Formulations

From Formulation 2, **Ch**[$\bot$,not,$\to$], we can get a non-redundant Formulation 2i, **Ch**[$\bot$,$\to$] – with 'i' short for 'inferential' – by defining not(A) 'inferentially', as above, whereby (i-cut) becomes a special case of ($\to$-cut) = (*modus ponens*).

Double-negation enrichments

On the other hand, from both Formulation 1 and 2, one can obtain slightly redundant Formulations 1[DN] and 2[DN] say, on the same primitive propositional signatures, by adding two double-negation [DN] rules, that are actually redundant in both Formulation 1 and Formulation 2 (resp. 2i),

(∇) $\Gamma \vdash A \Rightarrow \Gamma \vdash \text{not}(\text{not}(A))$ [double-negation introduction],
(Δ) $\Gamma \vdash \text{not}(\text{not}(A)) \Rightarrow \Gamma \vdash A$ [double-negation elimination].

The latter Formulations are significantly more efficient, in practice, as well as much cleaner, conceptually.

As a matter of fact, the conceptual profit of a DN-formulation is visible only if we are interested in a 'witness theoretic' presentation of classical logic, by also 'formalizing the proofs themselves', so to speak. In the DN-cases, the 'witnessed' DN-rules are supposed to obbey obvious inversion-principles, governing witnesses / proofs, of the form:

[∇] $\Gamma \vdash \nabla(a) : \text{not}(\text{not}(A))$, if $\Gamma \vdash a : A$,
[Δ] $\Gamma \vdash \Delta(c) : A$, if $\Gamma \vdash c : \text{not}(\text{not}(A))$,

for all witnesses a : A, resp. c : not(not(A)) relative to Γ, subjected to explicit (witness / proof) isomorphisms making up an inversion:

$(\beta\Delta) \vdash \nabla(\Delta(c)) = c : \text{not}(\text{not}(A))$,
$(\eta\Delta) \vdash \Delta(\nabla(a)) = a : A$,

resp. (where = stands for proof-conversion or proof-isomorphism, this time). It is easy to see that in the DN-less formulations at least one of the $(\beta/\eta\Delta)$-conditions would normally fail (in the appropriate λ-calculus), whence the idea of taking ∇ and Δ as primitive (proof-) operators (resp. rules of inference).

Witness theory

The equational 'witness theories' (i.e., typed λ-calculi) matching the natural deduction formulations mentioned above along a suitably modified *Curry-Howard Correspondence* (Rezuş 1990, 1991, 1993, Rehof & Sørensen 1994, Sørensen & Urzyczyn 2006, etc.), are as follows:

Formulation 1 **Ch**[⊥,Δ] coresponds to the λπ-calculus **λπ** (λ-calculus with 'surjective pairing'), typed as appropriate.
Formulation 2i **Ch**[⊥,→] corresponds to the typed λγ-calculus (Rezuş, cca 1987). Cf., e.g., Rezuş 1990, 1991, 1993, building upon the pioneering work of Dag Prawitz, PhD Dissertation Stockholm 1965. A similar λ-calculus-based construction (λμ-calculus) has been proposed by Michel Parigot, around 1991.
Formulation 2 **Ch**[⊥,not,→] and the 'redundant' DN-formulations are variations on the above. The corresponding (typed) λ-calculi are replications [here, just duplications] of pure λ-calculus and, actually, proper subsystems of λπ-calculus, even at undecorated ['type-free'] level, since, unlike the pure λ-calculus, **λπ** contains infinitely many distinct, non-trivial copies of itself.

There are many more such, but the [sub-] systems listed in the above are very close to the original Stoic system, as well as to the 'Chrysippean' way of justifying classical logic.

§ 5 Bibliographical notes

As regards the relevant textual sources, the edition of Hans van Arnim 1903–1905, 1924 [SVF] is also available in recent reprints. It has been partially translated in several modern languages (see, e.g., Dufour 2004 [French], Baldassari 1984, and Radice 1989 [Italian]). The new, comprehensive collection of Hülser 1987–1988 contains also a German translation (otherwise not always very inspired). Although included in the fragment-collections listed above, other specific sources (Alexander of Aphrodisias, Sextus Empiricus, and Galenus) have been listed separately. *Pace* Charles S. Peirce, the only (more or less professional) mathematician referring explicitly to Chrysippus, I know of, is Girolamo Cardano (1501–1576); cf. Cardano 1570 (reprinted in Cardano 1663), and, possibly, Rezuş 1991, Bellissima & Pagli 1996. For the 'traditional' views on Stoic logic, see Prantl 1855 (vol. 1) and Zeller 1879 (German original), 1892 (English translation). For modern technical discussions and/or 'reconstructions' of Chrysippus' logic, see Łukasiewicz 1934, Mates 1948, 1953, Becker 1957, Kneale & Kneale 1962; Egli 1967, 1978, 1979, 1993, 2000 (mainly, on 'Stoic quantifiers'), Frede 1974, Gould 1974, Corcoran 1974, Mueller 1974, 1978, 1979, Brunschwig 1980, Ierodiakonou 1990, Mignucci 1993, Bobzien 1996, 1999, 2003, 2019, Nasieniewski 1998, O'Toole & Jennings 2004 (this item contains too many errors to be useable; not only tyographical, unfortunately), and, last but not least, the John Locke Oxford Lectures (Summer 2004) of Jonathan Barnes 2007. (See also Barnes 1980, 1985, 1996, 1999.) On Frege's **BS** (1879) *vs* **GGA** (1893) – i.e., his *Regellogik* – see also Rezuş 2016. For Charles S. Peirce, see Peirce 1880, 1902 (as a precursor of Sheffer 1913) and Peirce 1885 (for the definition of 'inferential' negation, etc.). For details on Henry M. Sheffer, see Scanlan 2000, and Urquhart 2012. For the origins of natural deduction and sequent logic, and for recent technicalities on the subject, see Hertz 1922, 1923, 1928, 1929, 1929a, Legris' (2012) introduction to his translation of Hertz 1922, Jaśkowski 1927, 1934 (work of 1926–1927), Gentzen 1932 (on Hertz), and his *Inauguraldissertation* 1934–1935, Fitch 1952, Prawitz 1965, 1971, 1973, 1974, 1977, 1979, 1981, Rezuş 1981, 1990, 1991, 1993, 2015a, Indrzejczak 1998, 2016, Pelletier 1999, 2001, Barendregt & Ghilezan 2000, Hazen & Pelletier 2012, 2014, von Plato 2017, etc. (On the behaviour of the 'Peirce-Sheffer functors' – i.e., **nand** and **nor** – in this context, cf., e.g., Price 1961, von Kutschera 1962, Gagnon 1976, Read 1999, Zach 2015, etc.) For λ-calculus and type-theories based on λ-calculus, see Barendregt 1981, 1984^2, Rezuş 1981, 1986, 1986a, Hindley & Seldin 1986, 2008, Hindley 1997, Barendregt *et al.* 2013, and, possibly, Rezuş 2015 (a review of the latter item). On the Curry-Howard Correspondence ('proposition as types') for classical logic, see Rezuş 1990, 1991, 1993, Rehof & Sørensen 1994, and Sørensen & Urzyczyn 2006 (containing also a brief description of Parigot's $\lambda\mu$-calculus, mentioned earlier). Nicolaas G. de Bruijn's AUTOMATH proof-systems are documented in de Bruijn 1980, Rezuş 1983, and Barendregt & Rezuş 1983. For modern 'polar' proof-theoretical / semantical constructions, see Novikov 1941, Schütte 1977 and the references appearing in Girard's Rome 2004-lectures, issued now also in English, as Girard 2011. Finally, the early history of the tableaux-systems [Beth-Hintikka] can be recovered from the Amsterdam PhD Dissertation of Paul van Ulsen 2000, while the basics on tableaux can be retrieved from the monograph Smullyan 1968 and the survey of D'Agostino 1999. Readers interested in the computer-science counterpart of the same story (resolution) might profit from perusing the booklet of John Alan Robinson (Robinson 1979), the father of resolution.

12 December 2016

Im Buchstabenparadies (2009)

§ 1 The Many Names of *The Truth*

> *non idem est si duo dicunt idem*
> MIHAI EMINESCU[1]

The Originary Symbolic Dichotomy (*Zeichenarten*)

On the first page of his **Begriffsschrift**[2], Gottlob Frege begins the «explanation of designations» (*Erklärung der Bezeichnungen*) appearing in his Concept[ual]

---

[1] This is the way most Romanians – even high school teenagers – would eventually quote the Latin saying, following a poem of their national poet, Mihai Eminescu: «Noi amândoi avem același dascăl, // Școlari suntem aceleiași păreri... // Unitul gând oricine recunoască-l. // Ce știi tu azi, eu am știut de ieri. // De-aceleași lucruri plângem noi și râdem... // *Non idem est si duo dicunt idem.*», etc. [We, both of us, have gotten the same teacher, // Both you and me are pupils of the same opinion hence... // We're one-in-thought, let this be known to everybody ever. // Nevertheless, what you today know was my yester knowledge once. // Together crying, laughing at same things, and both in tandem... // Well, friend of mine: *non idem est si duo dicunt idem.* (Eminescu 1879; my translation, AR)] — Pace the poetical inversion, Eminescu is, actually, misreading Terence, here, together with a later learned – likely Jesuit – paremiological tradition: '*duo cum faciunt idem, non est idem*', '*si duo faciunt idem, non idem est*', etc. (The Jesuits were thereby equipping a former slave with subtle – their own – political views.) Cf. Publius Terentius Afer, **Adelphoi**, V.3, 821–825: MICIO *multa in homine*, DEMEA, *signa insunt, quibus ex coniectura facile fit, duo quom idem faciunt, saepe ut possis dicere 'hoc licet inpune facere huic, illi non licet', non quo dissimilis res sit, sed quo is qui facit.* In more recent times, the Latin saying had '*dicere*' for '*facere*' ('*si duo dicunt idem, non idem est*') and classics scholars would usually render the modified variant by 'If two *languages* say the same thing, it is not the same thing.'. So, unlike in the Jesuit reading, the modified Latin saying was about [two] *languages*, not about *people* (as in Eminescu's poem and as currently understood and/or oft quoted, colloquially, by most Romanians). See, e.g., Jon R. Stone **The Routledge Book of World Proverbs**, Routledge, London 2006, etc.

[2] **Begriffsschrift**, *Eine der arithmetischen nachgebildete Formelsprache des reinen Denkens*, Halle a/S, Verlag Louis Nebert 1879; henceforth referred to as **BS,** followed by paragraph and page number, in this edition. Unless specified explicitly otherwise, I will use my own translations, they are usually free renderings of the German original. As a rule, German terms are mentioned, in parentheses, only the first time an English translation occurs in the text. – There are two English translations of **BS**: **Concept Script**, [translated] by Stefan Bauer-Mengelberg, in: Jean van Heijenoort (ed.), **From Frege to Gödel:** *A Source Book in Mathematical Logic, 1879–193*, Harvard University Press, Harvard MA 1976, and **Conceptual Notation and Related Articles**, with a biography and introduction, [translated] by Terrell Ward Bynum, Clarendon Press, Oxford UK and Oxford University Press, New York NY 1972 [*Oxford Scholarly Classics*].

Script (*Begriffsschrift*)³ with a *distinguo* he estimates to be a «fundamental idea» (*Grundgedanke*) and claims that he intends to make it useful for the Domain – or Realm – of Pure Thinking ([*das*] *Gebiet des reinen Denkens*).

Paying attention to his colloquialisms, as well as to the formal terminology, the basic *distinguo* states a kind of symbolic proto-dichotomy, *Ur-zwiespalt*, at the root of the mathematics⁴, and runs as follows:

There are only two kinds (*Arten*) of symbols (*Zeichen*) in mathematics: the first kind consists of letters (*Buchstaben*) serving to express generality (*Allgemeinheit*), the second one consists of symbols that have a definite sense ([*solche*] *die einen ganz bestimmten Sinn haben*).

From his examples and further explanations, we are first inclined to render the distinction, in modern terms, as one between *variables* and *constants*. This is not exactly the case.

The first kind contains *Buchstaben* only, and covers actually what the moderns would call *variables*, as well as *meta-variables* (perhaps even *meta-meta-variables*).

The second kind covers what we mean nowadays by *constants*, in a large sense, though. Some of Frege's 'designations' would *translate* to our 'constants', indeed (i.e., they would become 'constants' – in our sense – after translation). Yet, we don't have exact equivalents for all of them. In the end, this should not cause serious problems, because we can always invent some notation, in order to provide would-be equivalents.

The real trouble appears while attempting to translate the first kind of symbols (i.e., Frege's 'letters') into modern terms. This does not work smoothly.

Unlike most of our logic fellows – among the *recentiores*, those of the last fifty years, say – Frege has a veritable *Buchstabenparadies*:

- Greek, Latin, and German letters (for short: Grk, Lat, Ger),
- the *case distinction* matters, for him, too: *uppercase*, *lowercase* (UC, LC), and, sometimes, even
- the distinction between *vowels* and *consonants* (V, C) is going to be relevant.

So we have combined syntactic types: Grk-UC-C, Grk-LC-V, Lat-UC, Lat-LC, Ger-LC-V, etc. We might call them *sub-kinds*, for instance. Fortunately, not all combinations⁵ would be actually used. For instance, the V/C distinction won't really matter in Latin type.

There is more, however.

If, as regards the *vowels*, only two sub-kinds are reserved for specific purposes, viz. Grk-LC-V, and Ger-LC-V, some of the *consonants* are reserved for some uses, while some other are meant for different uses.

---

³ As well as in **BS**, the Begriffsschrift (the book) itself.
⁴ *Grössenlehre* in **BS**, but this covers, actually, nearly all of what would have counted as 'mathematics' in his times.
⁵ Here 18 = 12 + 6, because one can distinguish among distinctions, too, so that we can have, in principle, Grk-UC-C, and Grk-UC-L, on a par with Grk-UC, all of them as distinct sub-kinds, after all.

The overall design – of the sub-kinds in *Buchstaben* – is governed by what we might call 'semantic criteria' and/or, even, 'ontological presuppositions'.

In order to understand the full meaning of Frege's Conceptual Script we must revert to a later work, namely to his Grundgesetze der Arithmetik[6]. This work makes use of the mature Fregean distinctions: *Sense vs Denotation* (*Sinn / Bedeutung*), *Object vs Function* (*Gegenstand / Function*), and *Function vs Concept* (*Function / Begriff*)[7].

Ontology first

Frege's ontological equipment is rather minimalistic: there are only two categories of entities in the world ('out there'): Objects and Functions. Ultimately, the Senses (of the words) are 'out there', too, because they seem to be objective. Yet, in as far as the notation itself (i.e., the Conceptual Script) is concerned, only the former two would actually matter.

The Ojects are *saturated* (*gesättigt*) or *complete* – self-sufficient to themselves[8] – while the Functions are not so.

---

[6] Grundgesetze der Arithmetik, *begriffsschriftlich abgeleitet*, I. Band, Jena, Verlag von Hermann Pohle 1893; II. Band, 1903 [reprinted by Olms Verlag, Hildesheim 1962], henceforth **GGA**:1, **GGA**:2. (Added in proof [15 October 2019]. The original edition of Hermann Pohle has been reset in modern notation as: Grundgesetze der Arithmetik, *begriffsschriftlich abgeleitet*, Band I und II, In moderne Formelnotation transkribiert und mit einem ausführlichen Sachregister versehen von Thomas Müller, Bernhard Schröder und Rainer Stuhlmann-Laeiz, MENTIS, Paderborn 2009 [ISBN: 978-3-89785-692-9].) As a matter of fact, only the first volume would matter for present purposes. The previous remarks on the **BS**-translations apply to **GGA**, too. — **GGA** has been partially translated into English in: The Basic Laws of Arithmetic, [translated by] Montgomery Furth, University of California Press, Berkeley and Los Angeles CA 1964. There is also a ('very rough') translation of **GGA**:2 (§§ 53 to the end), by Richard G. Heck [Jr.] and Jason Stanley, 2004 (online at Brown University, Providence RI), as well as a group at *Arché*, St Adrews UK, working on a complete translation, to be likely published by Oxford University Press. (Added in proof [15 March 2016]. The *complete* English translation of **GGA** has appeared in 2013: Basic Laws of Arithmetic. *Derived using concept-script*, Volumes I & II [in one], Translated and edited by Philip A. Ebert & Marcus Rossberg, with Crispin Wright, Oxford University Press, Oxford 2013.)

[7] Frege's *Function* is not a 'function' in our sense – not even in the sense of his mathematical contemporaries – so I might be tempted to leave *Function* untranslated. However, in order to make things look uniform in Middle English, I would always capitalize Frege's Objects, and Concepts, on a par with his *Functionen*, and equip the latter with an English plural, so we can read colloquially, anyway.

[8] The Mediaeval's *Nihil* (✠, say), the Arab's Zero (0), and the Modern's Empty Set (∅), – even the Unicorns (∅, ◯, and ⊙) – are best examples in point. This is, actually, Frege's Paradigm of an (Individual) Object, because Numbers, People, The Earth, The Moon, The Sun, The Stars, and whatever else admitting of a concrete or abstract (mind-) pointing to... (*Meinen*) are democratically equal first-level Citizens

There are, in particular, two very special Objects 'out there', entities Self-sufficient to Themselves as every Object, endowed with a Biblical status, more or less, called '*das Wahre*' and '*das Falsche*', in Frege — *The Truth* and *The Falsehood*, in Middle English, or *verum* and *falsum*, in Mediaeval Latin. We shall reserve special (Proper) Names for the latter: ⊤ and ⊥, for instance[9].

By Originary Ontological Dichotomy, the Functions are seen (and declared) to be *ungesättigt, non-saturated, incomplete*, and *in need of completion* or 'hungry' (*ergänzungsbedürftig*), so to speak: they are always ready to eat something (else). Eventually, it would turn out they are able to eat nearly everything, not only Objects, but also Functions, so that Self-Eating, in the guise of Function-Autophagy, is tolerantly allowed in the Script.

The Concepts are special cases of Functions, they are *pure and simple* or *simply hungry* (*einfach ergänzungsbedürftig* – the due meta-conceptual explanation comes in a moment), so they are non-saturated, and in need of completion, too[10].

Names, Fathers, and Sons

Like in the Bible – the Old Testament – more or less, everything can be *named* in Frege's Script[11]. Beware, though: there are Names and Names in Paradise!

The Names of Objects are called *Proper Names* (*Eigennamen*).

Some of them are granted to us, as, e.g., the Name of Socrates, the Name of 0, the Name of 1 (the latter are *Ziffern*, of course), and, in order to distinguish

---

of his World, and are going to have exactly the same ontological status as 0, ∅, ∅, ○, etc.

[9] (Added in proof [15 October 2019]. Unlike in the rest of this book, where I am using, as a rule, in non-Fregean contexts, **v** and **f**, resp., as abbreviations for the Latin *verum* and *falsum*, resp.) These Names are *not used* in the Script, however. They are just Meta-Names, we use (in order) to speak about those Very Special Objects, while speaking about Frege, and his (Conceptual) Script, for instance.

[10] This being said, one can realize, after a while, that Frege was a Left Wing Logical Radical, in fact, because The Hungry Ones – traditionally, a factor of instability in the World – are likely going to eat nearly everything eventually – well, mainly Objects, but also a bit of themselves – in order to make The Realm of Pure Thinking (*das Gebiet des reinen Denkens*) possible.

[11] Including The Truth Itself (*das Wahre*), and Its Eternal Foes, The False [One] (*das Falsche*), in particular. Except, perhaps, for [The] Everything (Itself). There is no Name for Everything – for The Overall Togetherness, I mean – in Frege. At least, this was the initial intention. And this, probably because, otherwise, The Name of Everything would have been a Name of Itself, thereby including Itself, as well as The Name of Its Name, The Name of All Its Names, and so on. — In fact, as Sir Bertrand (Third Earl) Russell was able to show slightly later, Everything *can* be named in the Script Itself (as well as and in **GGA**), because every Fregean (Proper) Name is a Name for everything, and this is already a Theorem of The Script Itself. Let's not anticipate, however.

between the Name of Rose (*sc.* the sister of Miss White – actually Mrs Black[12]) and Rose herself, or else between Lady Di[13] and Her Name, we must use quotes[14]. Single quotes, to be precise. So, as expected, the Name of Rose should be 'Rose'[15], while the Name of The Cat – my cat, the one sleeping *next to* The Mirror, as well as *in* The Mirror – should be 'Diana'[16], even for Gottlob Frege[17].

All this looks, *prima facie*, childish. Incidentally, however, Frege detects a bad use of the Proper Names in the mathematical writings of his very learned contemporary fellows (confusions between *Ziffer* and *Nummer*, for instance) and takes the opportunity to poke some fun on the subject. Consequently, in order to be sure that my very learned contemporary fellow readers won't confuse those many Bostons, out there (in the World) or in here (on Paper / in the Mirror), the Name of 1 (read *Eins* in German) is '1', not 1. To make a long story short, *the numeral* is the name of *the number*, while the number itself is not – and cannot be – a name, even if we are gödelizing Principia Mathematica[18].

---

[12] See the *Appendix*. In the meantime, the former Miss Rose White married Mr Black, the topologist.

[13] My cat, invented for the purpose. See below.

[14] They are not exactly the usual German *Gänsefüßchen*, in Frege; and they do *not* ressemble the French *guillemets* nor the English double / single quotes, either. The Fregean quotes might be called 'German single quotes' (opening-down and closing-up). – Strangely enough, the traditional German *Gänsefüßchen* have become the *official* double quotes in Romanian, although the modern German typography does not seem to use them anymore. – In what follows, I will tacitly translate the Fregean quotes into (contemporary) British & American English *single quotes* (opening-up and closing-up), as in 'Alice', for instance, the Name of Miss White, the twin-sister of Mrs Black. (Added in proof [15 October 2019]. Unlike in the rest of this book, I shall keep using the English [both British and American] double quotes informally, in colloquial contexts, reserving the corresponding single quotes for technical [Fregean] purposes.)

[15] And, possibly, 'Rose Black', according to her passport.

[16] Diana (The Cat, I mean) has gotten a passport of her own, and that's Her Name, indeed, according to the Appropriate Minister of Her Majesty, The Queen. (Added in proof [15 March 2016]. We have gotten a King now, but I wrote the paper in 2009.)

[17] The modern typographical alternative (TEX, LATEX, $\mathcal{A}\mathcal{M}$STEX, etc.), using *quasi-quotes* – a recent invention of Professor Willard van Orman Quine – was not available to *Herr* Hermann Pohle, Frege's Publisher of Jena. We shall, therefore, refrain, in general, of putting The Cat, Lady Di, or other Distinguished Objects between quasi-quotes, as in ⌜Diana⌝.

[18] Some of the learned characters appearing in Jonathan Swift's Travels (1726) would eventually use Objects in place of Names in order to discuss philosophical questions, but this is a slightly different epistemic scenario: those brave people were, by no means, confused. (Imagine, for a moment, the headache – and all that suffering – it would have taken to do [scientific] semantics or [Tarskian] model theory. Or, even, proof theory: no way to draw those nice [Gentzen] trees, nor any kind of pictures, anymore!) On the other hand, if some of my learned readers would think that such

The remaining Fregean Proper Names can be manufactured easily, by using the definite article[s] (and, perhaps, the demonstratives), as in the case of the Names of the Planet Venus[19], 'The Father of Socrates', 'The Square Root of (*Quadratwurzel aus*) 4', and so on.

This is important in what follows, because 'The' (*der / die / das usw*, and its modern translations, i.e., a Name of The Definite Article, no matter in what modern language, be it German, English, Dutch, Romanian, or Arabic / MSA) is a primitive symbol in **GGA**. Yet, beware! 'The' is *not* a Proper Name, in **GGA**.

In modern terms, the latter kind of Fregean Proper Names – the ones we can make with 'The' – are called '*descriptions*'; to Frege they are Proper Names, as well.[20]

Now, all Fregean Proper Names are Names of Objects, even if obtained by composition, like 'The Father of Socrates', for instance.

The Fregean technical jargon is, in this case, as follows: The Father of Socrates (i.e., Sophroniskos) is the *denotation* (*Bedeutung*) of (The Proper Name) 'The Father of Socrates', and, likewise, Socrates himself is the denotation of 'Socrates' etc.[21].

For Venus and her (its?) philosophically beloved Names we have yet another useful *distinguo*: The Proper Names of Venus have the planet itself (herself?) as a denotation; yet all those Names are different, and they differ by their Senses.

So 'The Morning Star' and 'The Evening Star' *express* different Senses, even if they denote / stand for the same Object (The Planet Venus).

And, *mutatis mutandis*, we get the same story, about '5' and '2+3', for instance.

If we extract, now, The Name of Socrates from 'The Father of Socrates' (a Proper Name), we don't get a Proper Name anymore, but a Name of something else, something in need for completion, viz. the Name of an un-saturated entity, i.e., a Function-Name, in Frege's terminology; this one would be 'the Father of...', a Name that *indicates* (refers to, *deutet an*) a 'hungry' thing (it is a Function, and, in particular, a Concept: see below). In this case, Frege would have said that the Object Socrates *falls under* (*fällt unter*) the Concept referred to by the Function-Name 'the Father of...'.[22]

---

confusions are unlikely in modern / contemporary mathematics, I should strongly dissent. Enough to open a recent book on categories, for instance, in order to get the fun.

[19] 'The Morning Star', 'The Evening Star', etc. Plenty of xx-th century philosophers have fallen in love with Her (Its?) Names!

[20] Added in proof [15 October 2019]. In fact, the *distinguo* proper names / descriptions is more recent and counts as the major contribution of Sir Bertrand Russell to logic.

[21] In other words, The Cat is the denotation of Her Name: easy to remember! Of course, on this plan, Lady Di is also the denotation of The Name of The Cat (mine), but this is only because Alice – Miss White – use to call my cat a Lady.

[22] Unlike for most of the moderns, like Freud, for instance – who was able to draw the otherwise unlikely picture of a Self-Satisfied and Self-Sufficient Father, Master

Whence, 'Falling Under' would be a relation – in our terms, and for Frege too – subsisting between an Object and a Concept. We shall come back to this important Relation later.

What about something like: 'Sophroniskos is the Father of Socrates'?

Obviously, we can play the extraction trick, in this case, too: if we extract The (Proper) Name of the Father, we get something looking, more or less, like what we had before, viz. '...[is] the Father of Socrates' – yet another Function-Name[23].

There is more to do, here: we can also extract The Name of the Son from what remained, getting this time '...[is] the Father of...'. According to the convention on saturation / incompleteness, this should be The Name of a Function, too; namely of one that is *twice hungry* (*zwiefach ergänzungsbedürftig*), so to speak.

Now we can say what kind of Function is a Concept: a Concept is an unsaturated entity that is *hungry simpliciter* (*einfach ergänzungsbedürftig*), *once hungry*: Concepts can eat only once, to saturation.

To simplify this kind of (meta-meta-) talk, Frege introduces the generic means to express *generality* we encountered at the very first page of **BS**: *die Buchstaben*.

Willkommen im Paradies

Required to this purpose is the sub-kind said Grk-LC-C, before. The corresponding letters are used to fill in the holes marked '..' in the above.

Actually, there are only two of them, ξ and ζ: the first one, ξ, is used to fill in the single hole of a Concept-Name, while both ξ and ζ are used to fill in – in this order – the two holes occurring in a Function-Name that is 'twice-hungry'!

In either case, both ξ and ζ are used only as *place-holders*, and they are called 'arguments'. As usual in mathematics, in fact. The corresponding holes are said to be *argument-places* or *argument-positions* (*Argumentsstellen*).

So we have Concepts or One-Argument Functions, and Two-Argument Functions, so far.

The Functions have Names of their own, namely *Function-Names* (*Functionnamen*). Like what they name, the Function-Names are hungry, and in need for

---

of Everything, Eating Himself, etc. – the Fathers are always hungry for Frege. As a rule, they are eating just Sons and Daughters, and – in Truth – only their own. Because, again – more or less as in the Bible, the New Testament, this time – One can possibly be – in Truth – only The Father of His Own Son. There is an obvious progress, however, mainly in the direction of the recent Feminist Movements: in Frege, One can have True Daughters of His Own, as well. In the end, the implied Parental Relation is equally patriarchal (a Falling submissively Under), but one should note the fact that the Fregean Mothers are, actually, in the same position of (Conceptual) supremacy.

[23] This is, for sure, a Concept-Name, too, pointing out to (*deutend an*, or referring to) the Concept of a specific Father, the Father of that Brave Old, Wise Man, vividly pictured in the dialogues of Plato, and Xenophon, mocked in the plays of Aristophanes, and so on.

completion, they are *Incomplete Names*, or Names with Holes, as in: 'the Father of...', or in '...[is] the Father of Socrates' (Names of two different Concepts), and '...the Father of...' (a would-be Name of The Relation Subsisting Between Father and Son), resp.

In both cases, the Holes are mere Places or Locations for Arguments (*Argumentsstellen*) and they should disappear in favor of the corresponding Place-Holders, Greek Letters in sub-kind Grk-LC-C: ξ, in the first case, or ξ and ζ – in this order – in the second case.

Now, as we might want to get rid of all those boring examples (with Fathers and Sons, resp. Sophroniskos and Socrates, in particular), and speak in general – *in the Name of Generality*, so to speak – we can also add yet another sub-kind of *Buchstaben*, with, this time, an additional Object / Function *distinguo*.

The additional distinction O / F generates sub-sub-kinds: Grk-UC-C-O and Grk-UC-C-F, say.

To this purpose, four letters in sub-kind Grk-UC-C – e.g., Γ, Δ, and Φ, Ψ – two parentheses, and an inevitable comma[24], should largely suffice, for a while.

---

[24] To be accurate, the Comma of (to be used in) the (Fregean) Script [**BS**] should be different from the comma used to speak about the Script itself. For obvious reasons, the first one – a simple meta-comma – cannot be a (Classical, Old) Greek comma or a Greek semicolon [**?**], it can be only Latin (or German, at most), while the second one – a meta-meta-comma – should be a honest English comma — unless I'd revert to German for the rest of this paper, like at the very beginning, in the title (this would look, however, strange, and rather impolite to most of my readers, I suspect). Should I use, perhaps, invisible [meta-meta-] quasi-quotes for the last one, and claim it's in (Middle) English? Hard to say... Noting – for the record – the fact that this points out to a serious oversight in the current practice of (both German and American) mathematics and (mathematical) logic – as well as in the design of TeX, and $\mathcal{AMS}$TeX, for that matter – I will let my friends, the True meta-Quineans of Boston – and Cambridge – MA, the Distinguished Officers of the AMS Governance – http://www.ams.org/about-us/governance/governance, etc. – the Council Members, Committees, and Representatives of the ASL – http://www.aslonline.org/info-governance.html, http://www.aslonline.org/info-council.html – Professor Donald E. Knuth – http://www-cs-faculty.stanford.edu/~uno/vita.html – and, possibly, the *Bundestag* – http://www.bundestag.de/ – to decide about, in this intricate affair, promising to introduce all due corrections in the next version of this paper, once they have reached, democratically, *the final solution* (according to the Constitution of the Association for Symbolic Logic [ASL] – http://www.aslonline.org/info-about-constitution.html – the fundamental text Quine 1940 [revised edition 2003], and, possibly, the *Grundgesetz für die Bundesrepublik Deutschland* [sic] – http://www.gesetze-im-internet.de/bundesrecht/gg/gesamt.pdf, etc.). As for the common (round) parentheses, I will pretend they're international (see, e.g., the current usage in the Charter of the United Nations – http://www.un.org/en/documents/charter/ [in English, with links to the MSA, Chinese, French, Russian, and the Spanish versions, on the same page] – if in doubt), so they won't ultimately deserve quoting – nor translation – anyway.

9 Im Buchstabenparadies (2009)

- Grk-UC-C-O for Object meta-variables: Γ, Δ, instead of 'Socrates', 'Sophroniskos', and the like,
- Grk-UC-C-F for Concept meta-avariables: Φ, instead of 'the Father of...', or '... [is] the Father of Socrates',
- Grk-UC-C-F for [Two-Argument] Function meta-avariables: Ψ, instead of '...[is] the Father of...'.

Remember: all of them – *die Buchstaben* thus – are 'means to express generality', as said before.

So we get now things like

- Δ, Γ,
- $\Phi(\ )$, $\Psi(\Delta,\ )$, $\Psi(\ ,\Gamma)$,
- $\Psi(\ ,\ )$,

as well as $\Phi(\Delta)$, $\Psi(\Delta,\Gamma)$ (no holes, as in the case of Δ and Γ).

These things are *not* in the (Holy) Script Itself, they are only used *to speak about* the Script. That's why we call the corresponding letters 'metavariables', nowadays. To Frege they are *Buchstaben*, pure and simple.

In modern parlance, we use to say that

- Δ, Γ *range over* Proper Names,
- Φ *ranges over* Concept-Names, and
- Ψ *ranges over* [Two-Place] Function-Names,

but, although he has something as *the range of a letter* (to us: metavariable) in **BS** and **GGA**, Frege would not allow us to write down things like Φ and Ψ, without Holes – markers for non-saturation or incompleteness – and recommends filling in the Holes of $\Phi(\ )$ and $\Psi(\ ,\ )$ with the Grk-LC-C-letters reserved for Place-Holders, ξ and ζ resp., like in: $\Phi(\xi)$ and in $\Psi(\xi,\zeta)$, resp.

On the other hand, since I introduced already a (meta-meta-) notation (sic) for sub-kinds, there is no need to copy the original Greek (meta-) notation from **BS** and **GGA**, anymore; we can just 'declare metavariables' (and their 'syntactic sub-kinds') of our own, in the usual way.

Moreover, I shall conveniently adopt a so-called 'autonymous usage' for metavariables (confusing thus, deliberately, Boston and 'Boston' – or ⌜Boston⌝, perhaps – at meta-level). This is just to spare on (meta-) quotes, of course.

Metavariable declarations. We use the convenient (meta-meta-meta-) notation '... :: ...' in order to put things on (meta-) paper.

Let x, y :: Grk-LC-C, a, b :: Grk-UC-C-O, and F, G :: Grk UC C F.

We can have F(x) and G(x,y), thus, in place of things with Fathers, Sons, and Holes, and spare on Greek letters, with the same occasion[25].

Samples of autonymous (meta-meta-) usage:

- the letter F stands for an arbitrary Concept (a Function with a single Argument-Place) indicated by the letter x, as in F(x);

---

[25] So that we can use them, for other purposes perhaps, later on.

- the letter G stands for an arbitrary Function with two Argument-Places, indicated by x and y, resp., in this order, as in G(x,y).

With an additional specification on Locatives, says Frege: we are only allowed to write F(x), not a mere F, and similarly, for G's: we are only allowed to write G(x,y), not G or G(x).

Yet, there is no need to (meta-meta-) quote the (meta-) letters F, G, x, and y, this time.

Proto-Substitution. From the previous explanations, we know already that we are allowed to insert a for x, and a, b for x, y resp. in the corresponding Argument-Places / Positions (that is: fill in the holes).

In such cases, we can also record on (meta-) paper what we have done, by using yet another piece of (meta-meta-) notation, as a shorthand for would-be instructions:

- 'put the Object-Letter a in place of the Place-Holder x in a Concept-Name F(x)', and
- 'put the Object-Letters a, b – in this order – in place of the Place-Holders x, y in a Function-Name G(x,y)', resp.,

whereupon we can possibly think of them as instructions to perform operations on Letters (*Buchstaben*, to us: metavariables).

Such (meta-meta-) notations together with the corresponding results can be then written as *inductive definitions*:

- $F(x)[x:=a] =_{def} F(a)$, and
- $G(x,y)[x:=a, y:=b] =_{def} G(a,b)$, resp.,

where the outcomes F(a), G(a,b) can be called 'result of substituting a for x in F(x)', and 'result of substituting a for x and b for y in F(x,y)', resp., taking also ('inductively') into account the fact that we have specified already the corresponding *ranges* of all the Letters therein involved.

'Sophroniskos is the Father of Socrates' should illustrate both cases.

This is a modern, rather recent, habit, however. The corresponding would-be operation is usually called 'substitution', resp. 'simultaneous substitution', and the suffix '...[x:=a]' reads colloquially 'x becomes a in...', etc.

Frege does not bother to invent such (meta-meta-) notational subtleties: too obvious, of course[26]. The basic idea (*Proto-Substitution*, or *Substitution for Object-Arguments*) is *controlled semantically*, so to speak, from his point of view, and follows from the explanations, anyway.

But wait! What kind of thing is the outcome – F(a), and G(a,b) – after all?

Obviously, 'Sophroniskos is the Father of Socrates' is something written down on paper (it is an expression called 'proposition', usually, and a Name, in Frege's terms), because it appears between (Frege) quotes. Yet, according to

---

[26] He would have, certainly, poked a lot of fun on our curious notational [meta-meta-] habits.

the Proto-Wisdom (*Ur-weisheit*) referred to earlier, there are only *two* Kinds of Names on Earth[27]: Names of Objects and Names of Functions[28]. And the Name above does not seem to fit either of those Kinds.

Frege's answer to the question is at least surprising, at a first look: the Name 'Sophroniskos is the Father of Socrates' is a Name for *The Truth* (*das Wahre*).

For the Truth is an Object, and so should be The Falsehood (*das Falsche*). Both $\top$ and $\bot$ are Objects 'out there', for Frege, remember?

Whence a True Proposition is a Name for The Truth, the Object $\top$, and, symmetrically, a False Proposition is a Name for *das Falsche*, the Object $\bot$.

Propositions are Proper Names, thus, like 'Sophroniskos', 'Socrates', 'Abū ᶜAlī al-Ḥussayn ibn ᶜAbd-Allāh ibn Sīnā' ('Avicenna', for short), 'William Shakespeare', 'Gottlob Frege', 'Rose', 'Alice', 'Diana', 'Venus', and, even '40 Eridani A', as well as '0', and '1', actually, except for the fact that Propositions are denoting The Truth ($\top$) or The Falsehood ($\bot$), instead of people, cats, planets, stars, or numbers.

This way of manufacturing Propositions from Proper Names and Function-Names works with 0 and 1, as well, in place of Sophroniskos and Socrates (or Mr White Sr. and Mrs Rose Black – the sister of Miss Alice White – for a change): for instance '$0 = 0$' is going to be a Name of The Truth ($\top$), and so is '$1 = 1$', no need for Fathers, Sons or Daughters[29]. And, on the same line of thought, '$0 = 1$' should be a Name of The Falsehood ($\bot$), of course.

Note the indefinite article ('a') before 'Name', in the above[30]: like my *Blue Rose*[31], in fact, Frege's Truth (*das Wahre*, *The Truth*, or $\top$) has many, many (Proper) Names.

---

[27] *Pace* his frequent references to (The Planet) Venus, Frege does not really care about Naming Conventions on other planets in our solar system, nor about those that might be in force on would-be other planets, located in systems like Canopus, Sirius, and 40 Eridani A, B, C, for that matter. (This is, likely, one of the reasons why Mr Isaac Asimov, Ms Doris Lessing, and Mr Gene Roddenberry never referred to him, alas.)

[28] The Kinds (*Arten*) Themselves need *not* be on Earth. Nor on (Meta-) Paper, for that matter. At least for a while.

[29] Nor for Lady Di.

[30] The name of a Name is a not a mere name, of course. For instance, a (Proper) Name of 'Name' is ' 'Name' ' – possibly ⌜⌜Name⌝⌝ – not 'Name' itself.

[31] Miss Alice White, *scilicet*.

§ 2 The Logic of *Grundgesetze*

> Die Frage nun, warum und mit welchem Rechte
> wir ein logisches Gesetz als wahr anerkennen,
> kann die Logik nur dadurch beantworten,
> dass sie auf andere logische Gesetze zurückführt.
> GOTTLOB FREGE Grundgesezte i, 1893, Vorwort, p. xvii

The New Rules of Logic

Unlike the Script of his booklet[32] of 1879 [**BS**], the Conceptual Script of the **Grundgesetze** [**GGA**] is supposed to record the inferential structure of *arithmetic*.

Although arithmetic *is* logic, for Frege, the primitive setting of **BS** (axiomatics with *modus ponens*, substitution, etc. as rules of inference) is not best suited to this purpose.

Instead of *deriving* the would-be rules of inference required for this purpose from the **BS**-system, Frege starts afresh with a new set of inferential rule-schemes.

On this reason, the **GGA**-*Logik* appears to us as a *logic of rules* (*Regellogik*), not as a mere *proposition*[*al*] *logic* (*Satzlogik*).

If we take the trouble to manufacture the appropriate notation, the Fregean design of 1893 becomes strangely familiar to the modern reader[33].

In order to make things more transparent, we consider first the quantifier-free fragment of the **GGA**-logic[34].

---

[32] Frege's word (*Büchlein*).

[33] The parallel with the so-called 'natural deduction' systems and/or the sequent-presentations of logic (Paul Hertz 1921–1929, Stanisław Jaśkowski 1926–1934, Gerhard Gentzen 1933–1934) has been already noticed by several authors in print (as, e.g., Tichý 1988, von Kutschera 1996, Schroeder-Heister 1997, 1999, etc.). In retrospect, it is amazing that neither Jaśkowski, nor Hertz – not even Gentzen – would refer to **GGA**. Jaśkowski 1934 quoted only **BS**, and obliquely so, just in order to show that two of the axiom-schemes of **BS** – i.e., (K) and (S) below – are equivalent to one of his 'supposition rules' (conditionalisation / the deduction theorem), if *modus ponens* is present. His 'supposition calculus' – a system of 'natural deduction' (Gentzen's term) – is, in fact, different from Frege's **GGA**-logic, in that he makes from the very beginning distinctions Frege won't have recognized as legitimate. On the other hand, although Gentzen 1934–1935 did not mention Frege either, his sequent presentation of logic is based, obviously, on **GGA** (by pondering also on Hertz 1922, 1923, 1928, 1929, 1929a, perhaps).

[34] Technically, both the **BS**- and the **GGA**-logic include the so-called 'quantifiers'. The word – although not the concept – comes from Charles S. Peirce, cca 1880.

To begin with, we fix the terminology and some notation, keeping always an open eye on would-be equivalences subsisting between the current way of speaking in logic and Frege's own terms.

Propositions containing implication (Frege's *Bedingungstrich*) will be printed horizontally (A → B).

A Fregean *conditional* is a proposition (*Satz*) of the form

$$A_1 \to .A_2 \to ... \to .A_n \to C, [n \geq 1],$$

with a *succedent* (called *Oberglied*) C, and one or more *antecedents* (its *Unterglieder*) $A_1, ..., A_n$[35]. Obviously, in the propositional fragment, the *Oberglied* can be only an *atom* (a *propositional variable*), in our terms, or a negation (printed ¬A, here).

We collect conveniently the *Unterglieder* in a sequence $\bar{\Gamma} := (A_1, ..., A_n)$, so that the above becomes

$$\bar{\Gamma} \to C \equiv (A_1, ..., A_n \to C), [n \geq 1]$$

(where ≡ stands for syntactic identity). On practical reasons, we might also need an additional convention to the effect that $\bar{\Gamma} \to C \equiv C$, for n = 0, in the above.

For asserted conditionals, $\vdash \bar{\Gamma} \to C$ is *the same thing as* $\vdash (A_1 \to ... \to .A_n \to C)$, so the conventions above are meant to save printing →'s, and parentheses (and/or separating dots), mainly.

A Fregean *transition* (*Übergang*) is a meta-conditional of the form

$$\vdash \bar{\Gamma}_1 \to C_1, ..., \vdash \bar{\Gamma}_k \to C_k \Rightarrow \vdash \bar{\Gamma} \to C$$

with, usually, k := 1, 2, corresponding to what we call *rules of inference*.

In order to state the Fregean transitions in pure 'syntactic' terms, we shall introduce a few systematic abbreviations.

Notation (*Shuffling*). Where $\bar{\Gamma}$ is a finite sequence (here: a sequence of propositions) and i is a non-negative integer, $\bar{\Gamma} *_i A$ reads 'insert A at place i in $\bar{\Gamma}$'.

This operation can be defined by an obvious induction[36], as follows:

(0,i)   If $\bar{\Gamma}$ is empty (length 0), then $\bar{\Gamma} *_i A \equiv A$, for any i, else,
(n,i)   let $\bar{\Gamma} := (A_1, ..., A_n)$, n ≥ 1 ($\bar{\Gamma}$ has length n); then

(n,0) $\bar{\Gamma} *_0 A \equiv (A, A_1, ..., A_n)$,
(n,i<n), $\bar{\Gamma} *_i A \equiv (A_1, ..., A_i, A, A_{i+1}..., A_n)$, for ≤ 1 ≤ i ≤ n–1, and
(n,i≥n), $\bar{\Gamma} *_i A \equiv (A_1, ..., A_n, A)$, for i ≥ n.

---

[35] In a horizontal arrangement, the corresponding terms would have been, likely, *Vorder-* and *Hinterglied*, resp.

[36] Specifically, by induction on pairs (m,i), for any two non-negative integers m, i.

In other words, location 0 is 'in front of' the elements of $\bar{\Gamma}$, while location $i \geq 1$ is 'after' the i-th element of $\bar{\Gamma}$, up to $i \leq n{-}1$, and 'after' its last element, if $i \geq n$.

One can also extend this notation to pairs of finite sequences $\Gamma_1 *_\sigma \Gamma_2$, for an appropriate (finite) sequence of non-negative integers, $\sigma := (i_1, \ldots i_m)$, where m is the length of Γ_2[37].

We shall apply the notation above to asserted Fregean propositions,

$$\vdash \bar{\Gamma} \to C \ (\equiv A_1 \to \ldots \to .A_n \to C), \ [n \geq 1],$$

where C is the succedent (*Oberglied*), and the A_i's (elements of $\bar{\Gamma}$) are the antecedents (*Unterglieder*) of the proposition $\bar{\Gamma} \to C$, by writing, e.g.,

$$\vdash \bar{\Gamma} *_i A \to C,$$

for the asserted proposition obtained from $\bar{\Gamma} \to C$, by inserting A at 'place' i in $\bar{\Gamma}$:

$$\vdash \bar{\Gamma} *_i A \to C \ (\equiv A_1 \to .A_2 \ldots \to .A \ldots \to .A_n \to C),$$

and analogously for $\vdash \bar{\Gamma}_1 *_\sigma \bar{\Gamma}_2 \to C$, for an appropriate shuffle-subscript σ.

Conventions. If $\bar{\Gamma}$ is as above, then $\bar{\Gamma}_\Pi$ stands for an abitrary permutation of $\bar{\Gamma}$, while $\bar{\Gamma}_W$ stands for the sequence obtained from $\bar{\Gamma}$ by deleting all duplicates of its elements, if any (so that only first occurrences are retained).

These conventions are supposed to apply, *mutatis mutandis*, to asserted Fregean propositions $\vdash \bar{\Gamma} \to C$, as well.

Notation. Finally, in order to save some (meta-) talk, while re-stating the Fregean transitions called *Wendung(en)*, – our usual *contraposition* rules – we set $A^\perp := \neg C$, if $A \equiv C$, and $A^\perp := C$, if $A \equiv \neg C$, where $\neg C$ stands for (the primitive Fregean) negation[38].

We can now turn to the proper logical ingredients of **GGA**[39].

There is, first, an axiom (scheme), and a neutral group of transitions (ruleschemes in fact, analogues of what most people would call, *mutatis mutandis*, 'structural rules', nowadays):

(id) $\vdash \bar{\Gamma} *_i A \to A,$

---

[37] Intuition: although not recommended in the game of poker and the like, cardshuffling can be performed sequentially, by inserting one card at a time, *at an arbitrary place*, in a half-deck of cards. In particular, suppressing the i- and σ-subscripts on the shuffle-operation * amounts, ultimately, to the fact that sequences so described are to be taken modulo arbitrary permutations (i.e., if 'context-free', they would become so-called *multisets*).

[38] This is mainly because the Fregean *Wendung* is stated as a single rule (-scheme) in terms of truth-values in **GGA**:1, § 48, p. 61. The notational convention above yields the closest 'syntactic' schematic analogue of Frege's (meta-) statement *ad locum*: a single (meta-) statement in place of four; see below.

[39] On obvious reasons, the order of exposition following below is not that of **GGA**.

(prm) $\vdash \bar{\Gamma} *_i A \to C \Rightarrow \vdash \bar{\Gamma} *_j A \to C,$
(ctc) $\vdash \bar{\Gamma} *_i A *_j A \to C \Rightarrow \vdash \bar{\Gamma} *_k A \to C,$

for any sequence $\bar{\Gamma} := (A_1, ..., A_n)$, $n \geq 0$, and all i, j, k $\geq$ 0.

As regards the terminology, the rule-scheme (prm) – said, more or less, 'permutation' or 'exchange' in modern terms – is called *Vertauschbarkeit der Unterglieder* in Frege[40], while the (ctc)-scheme – our 'contraction', more or less – is called 'fusion' (*Verschmelzung gleicher Unterglieder*)[41].

In matters of propositional axioms, Frege states actually[42] only two instances of (id), viz. those obtained by instantiating (1) $\bar{\Gamma} := (A, B)$, with i = 0, and (2) $\bar{\Gamma} := (A)$, with i = 0, resp., i.e.,

(K) $\vdash A \to (B \to A)$, and
(I) $\vdash A \to A,$

as would-be laws (*Grundgesetze*), in contrast with the mere *Regeln*, stated and discussed at length separately, but we can easily obtain the rest, including the more general *dilution* principle (called also 'weakening', or 'thinning', or even 'irrelevance', by recent writers on allied topics),

(dil) $\vdash \bar{\Gamma} \to C \Rightarrow \vdash \bar{\Gamma} *_i A \to C,$

for any sequence $\bar{\Gamma} := (A_1, ..., A_n)$, $n \geq 0$, and all i ≥ 0[43].

From the above, one can easily obtain, with our global (meta-) notation for permutations and fusions,

(prm*) $\vdash \bar{\Gamma} \to C \Rightarrow \vdash \bar{\Gamma}_\Pi \to C,$
(ctc*) $\vdash \bar{\Gamma} \to C \Rightarrow \vdash \bar{\Gamma}_W \to C,$

for any sequence $\bar{\Gamma} := (A_1, ..., A_n)$, $n \geq 2$.

In general, Frege's (propositional) *Gesetze*, as well as his transitions are to be taken modulo (prm*) [exchanges/permutations] and (ctc*) [fusions/contractions].

Worth noting separately are the following instances of (prm*) and (ctc*), 'limit cases', so to speak:

[C] $\vdash A \to .B \to C \Rightarrow \vdash B \to .A \to C,$
[W] $\vdash A \to .A \to C \Rightarrow \vdash A \to C,$

where [C] is the one-premiss rule-form of the (otherwise redundant) axiom-scheme (C) of **BS**[44], and [W] the one-premiss rule-form of (W), also obtained, at least implicitly, in **BS**[45]:

---

[40] **GGA**:1, § 14, 26, and § 48, p. 61, Rule (2).
[41] **GGA**:1, § 15, p. 29, and § 48, p. 61, Rule (4).
[42] **GGA**:1, § 18, p. 34, § 47, p. 61.
[43] Obtain first the single-premiss rule-form [K], of (K), with *modus ponens* – an instance of (cut) below – substitute, and apply then (prm) in order to get (dil). The fact that Frege meant something like (id), in general, instead of (K), is obvious from **GGA**:1, § 18, p. 34, for instance, where (I) is said to be *ein besonderes Fall* of (K).
[44] **BS**, § 16, proposition (8).
[45] In **BS**, § 16, proposition (11). Actually, (11) is $\vdash (A \to .A' \to C) \to (A' \to C)$.

(C) ⊢ (A → .B → C) → (B → .A → C),
(W) ⊢ (A → .A → C) → (A → C).
Next, there are two more transitions called 'inferences' (*Schlüsse, Schluss-weisen*):
(cut) ⊢ $\bar{\Gamma} *_i$ A → C, ⊢ $\bar{\Gamma}'$ → A ⇒ ⊢ $\bar{\Gamma} *_\sigma \bar{\Gamma}'$ → C,
(syl) ⊢ $\bar{\Gamma} *_i$ B → C, ⊢ $\bar{\Gamma}' *_j$ A → B ⇒ ⊢ $\bar{\Gamma} *_\sigma \bar{\Gamma}' *_k$ A → C,
for any two sequences $\bar{\Gamma} := (A_1, ..., A_n)$, $\bar{\Gamma}' := (B_1, ..., B_m)$, with m, n ≥ 0, i, j ≥ 0, and an arbitrary shuffle-index σ.

The first transition[46] corresponds, more or less, to our 'cut' (*Schnitt*, following Gentzen). The special case of (cut) with m = n = 0 is just *modus ponens*, of course. Repeated applications of (cut) yield an obvious 'multicut', a rule-scheme with k+1 premises, for k ≥ 2.

The second transition (syl) corresponds to the traditional syllogism, and is stated explicitly modulo (ctc*) in **GGA**[47]. One can generalise (syl) to a would-be 'polysyllogism', corresponding to the traditional *Kettenschluss*, a rule-scheme with k+1 premises, for k ≥ 2, as for (cut), but Frege does not seem to appreciate such complex transitions: his rules of inference have at most two premises[48].

All this is redundant, of course, since one can easily have (syl) from the general form of (cut) above. However, Frege seems to prefer a primitive (cut)-scheme with empty $\bar{\Gamma}'$, so that (syl) must be stated separately.

If we intend a more economical primitive setting, one can choose only (id), (dil) and a special case of (cut), with $\bar{\Gamma} \equiv \bar{\Gamma}'$, incorporating already the effect of (ctc):
(cut$_W$) ⊢ $\bar{\Gamma} *_i$ A → C, ⊢ $\bar{\Gamma}$ → A ⇒ ⊢ $\bar{\Gamma}$ → C,
wherefrom one can obtain easily (ctc), as well as (syl)[49]. But economy is not a concern in **GGA**, anyway.

---

[46] Cf. *6. Schliessen (a)*, in **GGA**:1, § 48, p. 62, etc.

[47] Cf. *7. Schliessen (b)*, in **GGA**:1, § 48, p. 62.

[48] The general 'syllogism' appears first in Hertz 1922–1923 – where it is wieved as a generalisation of *modus* BARBARA – together with (dil) and (id). Incidentally, Hertz never refers to Frege's logic and to Frege in general (except once, to the Grundlagen der Arithmetik, but this *not* in a logic paper, and the reference has no bearing to logic, anyway).

[49] Gentzen 1933 noticed that Hertz's general form of (syl) can be obtained from (id), (dil), and (cut$_W$). His main concern was different from that of Hertz, however, so he preferred, to have, *mutatis mutandis*, a primitive setting with (I), and the 'structural rules' [Gentzen's term] (dil), (prm), (ctc), and (cut). Like Jaśkowski, nearly a decade before (most of Jaśkowski 1934 is based on results obtained in 1926), Gentzen 1934–1935 then distinguished between deductions or inferences and conditional propositions, written as implications (so that he could also state conditionalisation / the *deduction theorem*), supplied the [now] missing 'logical rules' and then went to show that (cut) is redundant, etc. This would have likely seemed strange to Frege.

In view of would-be further simplifications, one might note the following special instances of (syl):

[C∘BB] $\vdash A \to .B \to C, \vdash A' \to B \Rightarrow \vdash A \to .A' \to C,$
[S] $\vdash A \to .B \to C, \vdash A \to B \Rightarrow \vdash A \to C,$
[B] $\vdash B \to C, \vdash A \to B \Rightarrow \vdash A \to C,$
[CB] $\vdash A \to B, \vdash B \to C \Rightarrow \vdash A \to C,$

and the fact that [S] follows from [C∘BB] and [W] above.

Notation (*Witness grammar*). Instead of the global Fregean separators (*Abzeichen*) marking applications of (cut) and (syl), we shall use next a more specific witness notation, described as follows:

(0) if X is a witness for $(A \to C)$, and Y is a witness for A, we write $X \triangleright Y$, for the result of applying *modus ponens* to X and Y, i.e., for C;

(1) if X is a witness for $(B \to C)$, and Y is a witness for $(A \to B)$, we write $X \circ Y$, for the result of applying rule [B] to X and Y, i.e., for $(A \to C)$,

and, analogously,

(2) if X is a witness for $(A \to .B \to C)$, and Y is a witness for $(A \to B)$, we write $X \square Y$, for the result of applying rule [S] to X and Y, i.e., for $(A \to C)$.

(We also assume association to the left for the defined binary operations.) This notation is to be taken taken modulo uniform substitutions[50].

The 'witness operations' corresponding to *modus ponens*, [B], and [S] can be thus viewed as binary operations on appropriate witness labels X, Y.

In particular, where (S) is a witness for proposition (2) of **BS**, and $\triangleright$ is as above,

(S) $\vdash (A \to .B \to C) \to (A \to B \to .A \to C),$
[▷] $\vdash A \to C, \vdash A \Rightarrow \vdash C,$

the two-premiss rule [S] amounts to $[S](X,Y) := X \square Y = (S) \triangleright X \triangleright Y$, of course.

Likewise, writing [K] – resp. [K](X) – for the single-premiss rule obtained with *modus ponens* from (K),

[K] $\vdash A \Rightarrow \vdash B \to A,$

i.e., $[K](X) := (K) \triangleright X$, one has *by definition*, for appropriate X and Y, $[B](X,Y) := X \circ Y = [K](X) \square Y$, as well as $[CB](X,Y) := Y \circ X$, and $[B](X,Y) := [CB](Y,X)$, modulo (prm), etc.

Examples.

(1) Given the rule [S] and the axiom-scheme (K) we can also have (I), that is $\vdash A \to A$, above, by applying [S] to appropriate substitution instances in (K), i.e., $(I) = (K) \square (K)$.

---

[50] The corresponding 'operations' on witnesses represent thus rules of inference (modulo substitution), and are order-sensitive, of course. The above amounts to an applied 'combinatory logic' notation for proofs / derivations. We don't need the 'characteristic equations' of the witnesses (I), (K), (S), (B), (CB), however; we have just a 'witness grammar', a convenient notational tool, à la Carew A. Meredith, say. Cf., e.g., Rezuş 1982.

(2) Analogously, sparing on square brackets and parentheses (by associating to the left, say) – with K(X) for [K](X) = (K) ▷ X, for instance, as above – the rather lengthy Łukasiewicz 1934 derivation of the **BS** axiom (C) [= **BS**, proposition (8)] from (K) and (S), with *modus ponens* and substitution, simplifies to (C) := K(CBK) □ (S), where (CBK) := K(S) □ K □ K(K), condensing thus half a page (in print) to a single line[51].

So, in general, deductions become *explicit definitions*.

Remark (*Conditionalisation*). The careful reader has already noticed, by now, the fact that, with this minimal equipment – i.e., granted (K) and rule [S] – one can already obtain the usual *conditionalisation* rule, or the *deduction theorem*[52]. On doctrinal reasons, this rule cannot be stated in the Script, however. For Frege, a statement ⊢ (A → B) is, in fact, *the same thing* as *the deduction of* B *from the assumption* A, and a *law* (*Gesetz*) *of logic is just a codification of a rule of inference* (or else a codification of a package of such rules).

Finally, there are two other transitions, serving to manipulate negation, the first one, called *Wendung*[53], corresponds to *contraposition*, and the last one to a form of classical *dilemmatic reasoning*[54]:

(ctp) $\vdash \bar{\Gamma} *_i A \to C \Rightarrow \vdash \bar{\Gamma} *_j C^\perp \to A^\perp$,
(abs) $\vdash \bar{\Gamma} *_i \neg A \to C, \vdash \bar{\Gamma} *_j A \to C \Rightarrow \vdash \bar{\Gamma} \to C$.

Of course, (ctp) amounts to four distinct *contrapositions*, in our terms[55]:

(ctp B) $\vdash \bar{\Gamma} *_i A \to C \Rightarrow \vdash \bar{\Gamma} *_j \neg C \to \neg A$,
(ctp B̂) $\vdash \bar{\Gamma} *_i \neg A \to \neg C \Rightarrow \vdash \bar{\Gamma} *_j C \to A$,
(ctp C)) $\vdash \bar{\Gamma} *_i A \to \neg C \Rightarrow \vdash \bar{\Gamma} *_j C \to \neg A$,
(ctp Ĉ)) $\vdash \bar{\Gamma} *_i \neg A \to C \Rightarrow \vdash \bar{\Gamma} *_j \neg C \to A$.

Our (meta-) notation with ...$^\perp$'s yields a 'syntactic' transcription of Frege's semantic formulation of the rule-scheme called *Wendung*, in terms of truth-values (*Wahrheitswerthe*).

The rule-scheme (abs) is stated, again, modulo (ctc), in **GGA**.

---

[51] The original derivation of Łukasiewicz 1934 – a typical piece of (proto-) 'Polish logic' – amounts, actually, to (C) := (S) ▷ K(S) ▷ (K) ▷ (CBK) ▷ (S), i.e., it is not '(S)-(K)-normal', so to speak. As a matter of historical detail, Carew A. Meredith, the main promotor of the ('condensed') proof-style exemplified here, has learnt this kind of 'witnessing' proofs / derivations by *modus ponens* (and substitution) from his mentor, Jan Łukasiewicz, while attending the Pole's lectures in Dublin, some time after the Second World War, around the early nineteen fifties. Cf., e.g., Meredith 1977, for bio-bibliographical details on C. A. Meredith, and Kalman 1974, 1983, Rezuş 1982, Hindley & Meredith 1990, Hindley 1997, etc., for further information on the so-called 'condensed detachment' operator of C. A. Meredith.
[52] In combinatory logic, the λ-abstractor can be defined in terms of (K) and (S). In fact, one needs only (I) and two operations: a first one corresponding to rule [K], the one-premiss rule from (K), and a second one corresponding to [S].
[53] Cf. *3. Wendung*, in **GGA**:1, § 48, p. 61, cf. also § 14, p. 27, etc.
[54] Cf. *8. Schliessen (c)*, in **GGA**:1, § 48, p. 62, etc.
[55] Cf. **GGA**:1, § 14, p. 27, etc.

It yields the cognate rule *consequentia mirabilis* of Girolamo Cardano (1570), also known as *the Rule of Clavius*[56]:

(clv) $\vdash \bar{\varGamma} *_j \neg A \rightarrow A, \Rightarrow \vdash \bar{\varGamma} \rightarrow A$.

In fact, given (id), on the one hand, and (ctp Ĉ), (syl), on the other, (abs) and (clv) turn out to be equivalent.

Whence, ultimately, the following list contains a very compact, non-redundant, and complete set of rules for classical (propositional) logic:

- a single axiom-scheme: (K),
- two (cut) rules: *modus ponens*, and
- rule [S] (alternatively, [S] can be replaced by [C∘BB] and [W]),
- two contrapositions: (ctp C), and (ctp Ĉ),
- rule (abs) (or, alternatively (clv), as noted above),

and it is easy to see that one cannot make any further reductions, i.e., one has already a non-redundant rule-system, indeed.

Otherwise, there are many possible variations on this theme. Yet, as already mentioned above, the economy of means is not an issue in **GGA**.

Consistency has been already established in **GGA** (where the rules are first justifed 'semantically', so to speak).

For completeness it is enough to derive the postulates (axioms and rules) of a system for which we do already have the result (!). In order to obtain the axiom system of **BS**, for instance, one should remember the Remark on conditionalisation, above, and notice the fact that the double negation rules:

---

[56] Apparently, Girolamo Cardano (1501–1576) got it on his own, and he was rather proud of this. (See, e.g., his treatise **De proportionibus**, Basle 1570 [reprinted in Cardano 1663, 4], where this proof-pattern is qualified as being *res admirabilior quae inventa sit ab urbe condito. . . , longe majus Chrysippaeo Syllogismo* [*Et est res admirabilior quae inventa sit ab urbe condito, scilicet ostendere aliquid ex suo opposito, demonstratione non ducente ad impossibile* [*et*] *ita, ut non possit demonstrari ea demonstratione nisi per illud suppositum quod est contrarium conclusioni, velut si quis demonstraret quòd Socrates est albus quia est niger,* [*et*] *non posset demonstrare aliter,* [*et*] *ideo est longè maius Chrysippeo Syllogismo.*], Cardano 1570, p. 231 = Cardano 1663, 4, p. 580.) The Jesuit Kristoph Klau – Christophorus Clavius SJ (1537–1612) or *Pater Clavius*, in Latin – noticed, however, a prior usage of the *mirabilis argumentandi modus* in Euclid's **Elementa** [*etiam usus est Euclides*, a comment *ad* Eucl., IX.12], and – by confusing use and mention (!), as well as some other things – dismissed the priority claim of the Pavian (cf. Clavius **Opera mathematica I**, Mayence 1611, *ad loc.*, in 1.2, page 11; and, possibly, his comments *ad* Theodosius, in *ibid.*, I.12). Subsequent learned Jesuit propaganda induced scholars to credit Clavius with the corresponding 'Law', later on. As a matter of fact, *Pater Clavius* missed other antecedents in this respect, as well, e.g., Aristotle, in his **Protrepticus**, now lost, and the Stoics, as reported by Sextus Empiricus **Adv. math.**, VIII, 281–2, 466 *et sq.* For further details on *consequentia mirabilis* and the like, see Kneale 1957, Kneale & Kneale 1971, Miralbell 1987, Nuchelmans 1991, 1992, Bellissima & Pagli 1996, etc.

$[\Delta^\perp]$ ⊢ ¬¬A ⇒ ⊢ A,
$[\nabla^\perp]$ ⊢ A ⇒ ⊢ ¬¬A,

can be obtained from (I) and (ctp C), (ctp Ĉ), whence also (ctp B), and (ctp B̂). Finally, closing under conditionalisation yields:

(K) = **BS** (1)   ⊢ A → .B → A,
(S) = **BS** (2)   ⊢ (A → .B → C) → (A → B → .A → C),
($B^\perp$) = **BS** (28)   ⊢ A → C → .¬C → ¬A.
($\Delta^\perp$) = **BS** (31)   ⊢ ¬¬A → A,
($\nabla^\perp$) = **BS** (41)   ⊢ A → ¬¬A,
(*modus ponens*) ⊢ A → C, ⊢ A ⇒ ⊢ C

i.e., the full propositional logic of **BS**. As completeness (for **BS**) has been already established by Łukasiewicz 1931, we have completeness for the **GGA** *Regellogik*, as well[57].

---

[57] Here, as observed by Jan Łukasiewicz 1934, (C) [= **BS**, proposition (8)] is redundant. A somewhat simpler derivation has been already mentioned above, as an example of 'witnessing'. Moreover, the resulting axiom system – without (C) – is independent, as shown, for instance, by Christian Thiel, in his Erlangen Dissertation (1965). Cf. the translation Thiel 1968, p. 21.

§ 3 Appendix. *The Name of a Rose*

> *Damit ein solches Unternehmen Erfolg haben könne, müssen natürlich die Begriffe, deren man bedarf, scharf gefasst werden.*
> GOTTLOB FREGE Grundgesezte i, 1893, *Einleitung*, p. 1

...*a rose by any other name would smell as sweet...*

So Juliet, «Daughter to» Capulet, and William Shakespeare: a Rose is a rose – *as sweet as any other Rose* – even if (she is) called by any other name.

And what if she is a young lady called Rose? 'The name of a Rose, if called by her own name, is Rose!' – my logic textbook says.

What if she's called by any other name? 'The name of Rose, if called by any other name than her own (name), is still Rose!' – insists the book.

What about *Roses Called by Something Else Than Names*? What about a *Blue Rose*, for instance? My neighbour, if she's *Dressed in Blue*, a Rose by her own (name). Or else *Blue Rose*, for instance, a fancy white rose to me, she is (a) White, indeed, although no Rose, in fact, as she's called so *by her blue eyes* alone. So she is a *White Rose*, too, this even by her (own) name, the fancy rose, *Blue Eyes* and so.

*

In front of The Mirror, near Diana – that's Lady Di, The Cat, my cat – there is a bunch of sweet RED Roses.

Just got them fresh, this morning, from Alice Blue Eyes, the sister of my neighbour, Rose.

*

Miss White and Mrs Black are twins. They like colors and roses most of all (the father is a painter, the mother a fashion-designer — this is not a reason, though).

They used to be both White, exactly like their parents – like any (other pair of) twins 'round the world, in fact – but Rose married Dr. Black, the mathematician, not too long ago.

Alice got blue eyes, Rose's are black. (So, while in school, and later, among friends and so, they were able to fool people only from far.)

Since they were kids, I used to call them by names of roses, and we played the Game of Names. As Alice was and still is no rose, but only a name, I called her White Rose or Blue Rose. Her sister was, sure, a White Rose by her own name, but most of the time, Black Rose, in view of her eyes, mainly. Anyway, Rose was Rose White, too, in all kind of scripts, including her passport (the girls were not inscribed in their parents' passports, just in case), but this was not too entertaining, for us, because she could not read by then.

After a while, the twins became experts in the Game of Names, and even tried to fool me by asking tricky questions, using also the talents of their mother to reinforce the effect, by dressing themselves colourfully for example, in blue or white: "What is my name, if I am dressed in blue?", asked Alice. Or else, the other one, while dressed in white, likewise. "White Rose in Blue", "Black Rose in White", I answered undisturbed.

To make things a bit more interesting (this was before the girls were able to write), I invented the Game of the Name in the Mirror. For example: Alice in the Mirror (look at her, in there!) is the name of the name of White Rose. Question: What is the name of Alice in the Mirror? Alice in the Mirror or Blue Rose?

This was not so simple, nor very easy anymore, mainly because each girl had a Mirror of her own, and they were able to have a look at their own names in the name of the other etc.

On obvious reasons, Mrs White, the fashion-stylist, was very angry with me at first, during a week or so – the girls were making a lot of noise around their new Mirror-Names – but, soon after, the daughters caught her in the Mirror, too – in the new Game, I mean – and she ended by indulging herself in endless Mirror-debates, together with the two young ladies, aged – each – six and a half. So that, after another while, her husband – my friend, the painter – became suspicious, and inquired, cautiously nonetheless, about 'that new mirror story you invented for the benefit of my family'.

Once the girls have learnt to write (and count), there was no need for Mirrors anymore: they could use quotes! This was not easy at first, but they got the point eventually. The Game of Names reduced itself to a matter of counting quotes.

*

"*The Name of a Rose* – Alice expertly comments on Names, this morning – is *not* the name of a Rose, my friend[58]. It's the name of a book – added White Rose, Blue Eyes, in Black, for my logical comfort – by somebody else than my Rose, sweet Rose — she's *Dressed in Blue*, my Rose, sweet-rose, indeed. And *Red Rose* is *not* the name of any Rose! It is an Amsterdam café, remember? The one where we used to drink, once in a while, a coffee, and *frühstück*, last year. But don't call me, please, *White Rose*. It is *not* my name. It's even *not* the name of Rose. She's a Black now, even if a Rose, and *Dressed in White*. And I am still a White, even if no Rose, and *Dressed in Black*, this morning. *Blue Eyes*, you said? Who cares? I won't, anymore! So, don't call me *Blue Eyes, in Black*, either. It is *not* my name. It's not even a Name! Nor is *Black Eyes, in Blue* a name of Rose, if *Dressed in Blue*, even if she has gotten black eyes.

---

[58] Denying thereby, to my surprise – as I was unable to *hear* first her (quasi-) quotes – a Law called «A is A» or even «⌜Boston⌝ = ⌜Boston⌝» in my logic books (mainly in those printed in Boston MA).

Alice-in-Black, would do, perhaps, once in a while, for me. This morning, for instance. Like *Rose-in-Blue*, for Rose, today[59]."

*

What's the Name of (a) Rose? The Name of the One White Rose called Alice in my story, for example[60].

And, if it is *true* that the Name of a Rose is not the name of a rose – as Alice claimed – then what should we Name this Truth?

And, before anything else: *What is a Name of Truth?* Is *The Truth* a Name of (the) Truth? Is it (the) One? – *The* Truth, perhaps? – Is it the only One?[61]

13 June 2009 – 15 October 2019

---

[59] I wonder, how should I have printed *Red Rose*, in $\mathcal{A}_{\mathcal{M}}\mathcal{S}T_{\rm E}$X, here? With Boston quasi-quotes, perhaps? I'm not sure of this. Miss White was right! *Red Rose* is *not* a name at all, but a café.

[60] The reader should perhaps notice the fact that I did not ask "*Where is the Name of Rose?*" (or Alice)! As a matter of fact, the latter kind of questions is uninteresting 'from a logical point of view', so to speak. — On the other hand, the idea that the Name of Mrs Rose Black (or of Miss Alice White, for that matter) is, ultimately, *on paper* (possibly after printing a copy of this paper, say) is a very wrong one – a mere way of speaking, at best – because even the Name of my Cat – Diana or Lady Di – is, after all, an Abstract Entity, and all what we can actually put 'on paper', while printing (or writing things 'down'), are molecules of ink (or else graphite / 'lead', allotropes of carbon). Same thing, *mutatis mutandis*, about blackboards, chalk and so (including the more recent case of so-called *e-paper* – and *e-ink* –; although technologically more sophisticated, the latter are stubbornly remaining 'out there', in the so-called 'physical realm'). In short, like Ideas, Concepts, and Thoughts, the Names cannot be found 'out there' either. Among logicians, the Stoics and the so-called 'formalists' – even some early otherwise famous Polish pioneers (like Jan Łukasiewicz, Stanisław Leśniewski, Alfred Tarski), as well as Professor Willard van Orman Quine, in more recent times – were, certainly, a bit in a hurry while locating the 'syntax' next to the real Cats (and Dogs), or else at the same level as would-be traces of ink on paper, chalk on blackboards and so on...

[61] *Endnote* [13 June 2009] The rest of this section has been lost, mainly because of (the former) Miss Alice White. My friend is not to be blamed, nevertheless. The fact is that Miss White married Mr Green, an ET with green eyes, two years ago and, eventually, lost interest in Terran Logic, and in my lectures on Frege, (Blue) Roses and Co. She moved recently (together with Mr Green, of course) to Blue Vegas – a nice little ecological village located on a small theoretical planet in Alpha Centauri – and took her shorthand notes (*Gabelsberger Kurzschrift*) with her. As communication with Blue Vegas is rather difficult nowadays, I was unable to ask her about the whereabouts of my Fregean ruminations, so far.

10

'Begriffsschrift' as *Beweisschrift* (2019a)

The notes following below concern the conceptual *heritage* of Gottlob Frege. My comments are focused on the least exciting part of Frege's published writings, namely on his *logic proper*, a term by which I understand *proof theory*. This is a venerable, bi-millennial *theoretical* endeavour meant to elucidate the main concept of the *science of logic*, viz., the concept of (a) *proof*.

Introduction

First born some twenty-two centuries ago in Ancient Greece – due, mainly, to the endeavours of Chrysippus of Sol[o]i, a Greek allochtone native of *Cilicia Campestris*, currently in contemporary Turkey – classical logic is re-born afresh, from the head of Gottlob Frege, about 140 years ago – in 1879, to be precise – as allegedly, *olim*, Minerva from the head of Jupiter.

> *Pace* the rather intricated details concerning the gestation of the Ancient Goddess, we are, even nowadays, in the dark as regards the actual birth of 'Frege's logic'. The fact is that Frege did not use to acknowledge sources. On this matter, we can make educated guesses, at most. Like Leibniz, he had, apparently, pasigraphic *penchants*. Notably, his informed, recent biographers did not fail to notice the relation to the Grassmanns and, even, a *grammatical* connection in the family. The father, Karl Alexander (1809–1866), was a school-teacher and issued, by the time the son was born, a grammar-book for German-speaking children aged 'from 9 to 13' [Hülfsbuch zum Unterrichte in der deutschen Sprache für Kinder von 9 bis 13 Jahren], with a second edition, 'durchaus umgearbeitet', in 1850, and a third one [Druck und Verlag der Hinstorff'schen Hofbuchhandlung, Wismar & Ludwigsluft], in 1862.[1] See,

---

[1] As to the intended, *German*-speaking audience of his father, worth noting is the fact that Frege's birthplace, Wismar, was under Swedish law during 1648–1871. It became part of Germany in 1871, though Sweden officially renounced its claims to the city only in 1903.

e.g., Kreiser 1995, 2001, and Grattan-Guinness 2000, **4.5**, p. 199. This does not explain, however, why did Frege take, as a starting point in *classical proof theory*, (1) two applied *combinators* – properly invented, about forty years later (by Moses Schönfinkel, in 1920, in print 1924) – (2) other two in order to account for the Stoic double-negation isomorphism, and, finally, (3) equivalents of the first two Chrysippean *anapodeiktoi* [indemonstrables] – the Stoic *modi*, *ponendo ponens* and *tollendo tollens* – in his axiomatics of 1879.

Like many other inventions of the Western European XIX-th century, Frege's new-born – full record in his **Begriffsschrift** [**BS**] (1897), a booklet (*Büchlein*, in his terms) with nice figures and weird symbols of about 100 pp. in print (actually, x + 88) – was unconventional in appearance – to say the least – and not entirely *sine naevo*. *Habent sua fata libelli*: with a few exceptions, the immediate, contemporary readearship of the booklet did yield a poor and rather irrelevant feedback on matters logical. Nonetheless, putting aside some minor – and negligible – practical inconveniences[2], the **BS**-booklet delivered the birth certificate of what we are, nowadays, calling '(first-order) classical logic'.

A few *preliminary remarks* are in order. Frege's **BS**-logic was a *Satzogik* (Rezuş 2017, **§1**, footnote 2): it focused exclusively on formalising propositions resp. propositional schemes – as expressed by formulas – and ignored

(1) the so-called rules of inference and, more significantly,
(2) the proofs themselves, i.e., the main objects of concern in logic.

[1] As regards (1), Frege's **BS**-logic was essentially *formularian*, the main concept to be taken into account in **BS** being the concept of *provability*, a property of formulas (or else, of *propositions* resp. *propositional schemes*, by semantic delegation).[3] In fact, the term *Formelsprache* ('formula[rian] language') was programmatically present in the (sub-) title of **BS**. Otherwise, Frege mentioned explicitly, in subsequent *pro domo* comments to **BS**, the fact that his aim was

---

[2] As perspicuoulsy illustrated by the new German edition (Paderborn 2009) of the **GGA**, the 'Japanese' features of Frege's original *Begriffsschrift* – i.e., the vertical diplay of formulas that has upset some of his learned contemporary readers – are of no consequence – except perhaps for typographers and publishers – since its 'modernisation' (i.e., the transcription in 'linear', Peanian **Formulaire**-like equivalents, say) is, more or less, a matter of mechanical transformation. Incidentally, this shows – even at formularian level – that *conceptual* matters are, ultimately, independent of their (visual) *representation*. Unfortunately, the original Fregean 'concept-script' reflects just *formula-structures*, not *proof-structures*.

[3] In witness theory – a philosophically neutral mathematical discipline – we are not – and should not be – concerned with the metaphysical status of the entities called *propositions* (Frege's *Gedanken*, corresponding, more or less, to the Stoic *lekta*, to Bolzano's *Sätze an sich*, etc.). Whatever its would-be epistemic relevance, the subject – although discussed *ad nauseam* by subsequent authors (mainly by Bertrand Russell, cf., e.g., Church 1984, Urquhart & Pelham 1994, Pehlam 1996, Landini 1998, etc.), causing a lot of headache to analytic philosophers and to philosophers of logic – is, simply, not a logic subject.

that of formulating, in the footsteps of Leibniz, a *lingua characteristica* – i.e., a conceptual scheme[4] – meant to formalise in detail the *propositions* of classical mathematics (classical analysis, more or less).

The stock of primitive rules of inference in Frege's **BS** was rather limited. Worth noting, however, is the fact that Frege came back, a few decades later, in 1893 [**GGA** 1], with a full-fledged *Regellogik*, meant to be used in everyday mathematics.[5]

Enough to say that, in both 'logics' (**BS**, as well as **GGA**), the Fregean concept of (a) *rule of inference* is rather parochial from a modern point of view.

Specifically, Frege considered only 'flat' rules of inference, of the form:

($\flat$) $\vdash A_1, ..., \vdash A_n \Rightarrow \vdash C$ (n a natural number),

where the turnstile $\vdash$ would read (in our terms): '... is provable', and A_i, with $0 < i < n+1$, C are formulas, ignoring the general case, where the premisses of ($\flat$) might have been hypothetical, i.e., of the form $B_1, ..., B_{m(i)} \vdash A_i$ ($m(i) > 0$).

Roughly speaking, in our terms, Frege identified *asserted conditionals* (i.e., formulas) of the form:

($\rightarrow$) $\vdash A_1 \rightarrow (A_2 \rightarrow ... \rightarrow (A_n \rightarrow C)...)$, (n a natural number),

with *valid 'entailments'* or *'sequents'*, i.e., finite sequences of formulas, of the form:

($\vdash$) $A_1, ..., A_n \vdash C$ (n a natural number),

were exactly one (here: C) is tagged (here: with $\vdash$) as a conclusion.

> Actually, by witness-theoretical standards, the concept of (an) *entailment* is a *pseudo-concept*, a mere notational expedient, but this becomes visible only after cleaning up the traditional conceptual mess (**§ 1.12**). — In formularian terms, Chrysippus and his followers were more *advanced* – in a sense – than Frege, since they distinguished carefully among (1) true conditionals, (2) valid entailments – or 'arguments' (*logoi*) – and (3) transformation recipes (*schemata*) used to obtain valid entailments from valid 'primitive' or 'indemonstrable' entailments (*anapodeiktikoi logoi*). Cf. Rezuş 2016. (For an alternative 'modern' reading of Chrysippus' conceptual scheme, see Bobzien 2019.)

Technically speaking, in modern terms, Frege's – likely deliberate – confusion amounts to the fact that one has to assume a kind of meta-rule that allows passing from ($\flat$) to ($\rightarrow$), so that, in the presence of a specific 'flat' rule – viz. [*modus (ponendo) ponens*] [PP] for $\rightarrow$, below – ($\flat$) and ($\rightarrow$) are deductively equivalent, a thing hardly worth mentioning, in the end.

In general, in parametric variant, an inference rule is of the form:

---

[4] Cf. van Heijenoort 1967.
[5] As I have already discussed the basics of the **GGA**-logic in some detail (Rezuş 2009), there is no point in insisting, once more, on the subject. In what follows, I shall only casually refer to **GGA**.

$(\bar{\Gamma}\flat)$ $\bar{\Gamma} \vdash A_1$; ...; $\bar{\Gamma} \vdash A_n \Rightarrow \bar{\Gamma} \vdash C$ (n a natural number),

where $\Rightarrow$ stands for the meta-conditional and the parameter, i.e., the sequence $\bar{\Gamma} := [A_1, ..., A_m]$ is to be viewed as an (elliptic) proof-context, containing (elliptic) assumptions A_i ($0 \leq i \leq m$). In particular, $\bar{\Gamma}$ might be empty: the nil-sequence (the Fregean case).

So, among other things, Frege would not have seen the point behind the *Deduction Theorem* (Tarski 1921), i.e., the so-called [→-introduction] rule:

(λ) $\bar{\Gamma}, A \vdash B \Rightarrow \bar{\Gamma} \vdash A \to B$,

where $\bar{\Gamma} := [A_1, ..., A_m]$ is as above (or empty).[6]

[2] As regards (2), the oversight concerning *the proofs themselves* is not characteristic for Frege nor for the other pioneers (Peirce, Peano, Russell, and, somewhat later, all those bright *Göttinger* around Hilbert, and the – Lvov-Warsaw – Poles around Leśniewski and Łukasiewicz, or else the casual Romanians, and/or the *Wiener*-Austrians, incidentally, active in the area since the late nineteen twenties and the early thirties, etc.).

Until very recently – about 50 years ago – the so-called 'mathematical' or 'symbolic' logic inherited of this kind of *professional blindness*. In fact, the mainstream logic research is, even nowadays, *essentially formularian*.[7]

And Frege was only in part responsible for this kind of *conceptual blunder*.

§ 1 Frege's 'Begriffsschrift'-logic

Frege's *Satzlogik* of the 'Begriffsschrift' [**BS**] (1879) is a piece of axiomatics for classical (first-order) logic.

Using the modern familiar notation, Frege's **BS**-logic relies on the following abstract grammar:

individual variables :: u, v, w, ...
individual terms :: r, s, t, ... ::= u | ...
atomic formulas :: P[**t**]
formulas :: A, B, C, ... ::= P[**t**] | A → B | ¬A | ∀u.A,

where P ranges over predicate letters, **t** stands for a finite sequence of individual terms, while → stands for the material implication, ¬ for the (classical) negation, and ∀ for the (first-order) universal quantifier.[8]

We ignore quantifiers for a while (see, however, § **6**), and use propositional variables p, q, r, ... in the spelling of quantifier-free formulas, with round parentheses for parsing. Formally, we rely, first, on the reduced grammar:

---

[6] But see Resnik 1980, p. 174, and Tichý 1988, Chapter XIII, for instance.
[7] Being a formularian, qua pioneer, in 1879 or around 1893 – and even half a century later, during the nineteen thirties – has a certain *historical justification*, no doubt. There is no *reasonable* excuse for this a hundred years afterwards, however.
[8] The inner structure of individual terms may be ignored in 'pure' logic.

atomic formulas = propositional variables :: p, q, r, ...
formulas :: A, B, C, ... ::= p | A → B | ¬A.

Further, we have six axiom(scheme)s:

(K)   ⊢ K[A,B] : A → (B → A)
     [Simp = Simplification, Irrelevance],
(S)   ⊢ S[A,B,C] : (A → (B → C)) → ((A → B) → (A → C))
     [Frege = Selfdistribution on the major],
(C)   ⊢ C[A,B,C] : (A → (B → C)) → (B → (A → C))
     [Commutation],
(CB$^\rho$)   ⊢ CB$^\rho$[A,B] : (A → B) → (¬B → ¬A)
     [modus (tollendo) tollens, qua tautology] [TT],
(∇)   ⊢ ∇[A] : A → ¬¬A
     [Double Negation introduction] [DN-i],
(Δ)   ⊢ Δ[A] : ¬¬A → A
     [Double Negation elimination] [DN-e],

and a single inference rule:

(▷)   :: ⊢ A → B, ⊢ A ⇒ ⊢ B
     [modus (ponendo) ponens = detachment = →-elimination] [PP].

1.1 We used formula-parameters on primitive witnesses in the axiom-schemes:

(LABEL)   ⊢ LABEL[formula-parameters] : [AXIOM],

writing decorations (formula-parameters) on labels, in place of:

(LABEL)   ⊢ [AXIOM].

> This is mainly to simplify typography and to avoid sub- and/or superscripts. Normally, one would want to write the list of [formula-parameters] as a (long) sub-script on a witness-LABEL, as, e.g., $S_{A,B,C}$, in place of S[A,B,C], etc.

§ 1.2 Frege (1879) had also a primitive *uniform substitution rule*, assumed tacitly, more or less, and corresponding axioms, as, e.g., in:

(K)   ⊢ K[p,q] : p → (q → p),
(S)   ⊢ S[p,q,r] : (p → (q → r)) → ((p → q) → (p → r)),

etc. In view of a side-remark of John von Neumann, we can have axiom-schemes instead and ignore the substitution rule (for propositional variables).

§ 1.3 Incidentally, the [Commutation]-scheme is redundant (Łukasiewicz 1929, 1934) and one can also derive easily:

(I)   ⊢ I[A] : A → A [Identity],

from (K), (S), using (▷) alone (cf. **§ 1.9** and **§ 1.11**, below).

§ 1.4 In view of a casual remark of Tarski (1921), the axiom-schemes (K) [Simp] and (S) [Frege] can be replaced by the *Deduction Theorem* (λ), also known as [→-introduction]-rule in 'natural deduction' formulations of classical logic:

(λ) :: $A \vdash B \Rightarrow \vdash A \to B$ [Deduction Theorem = $\to$-introduction] [DT].

In fact, the pair (K), (S) together with the rule ($\triangleright$) [*modus (ponendo) ponens*] is tantamount the pair of rules (λ), ($\triangleright$), as regards provability resp. derivability. (Cf. **§ 1.8**.)

Given this, we can also replace, equivalently, the tautologies (CB^ρ), (∇), (Δ) by the corresponding (two-, resp. single-premiss) rules:

($\bar{\triangleright}$) :: $\vdash A \to B, \vdash \neg B \Rightarrow \vdash \neg A$ [*modus (tollendo) tollens*] [TT],
(∇) :: $\vdash A \Rightarrow \vdash \neg\neg A$ [Double Negation introduction] [DN-i],
(Δ) :: $\vdash \neg\neg A \Rightarrow \vdash A$ [Double Negation elimination] [DN-e].

More definitely, using parametric proof-contexts $\bar{\Gamma} := [A_1, ..., A_n]$ (n a natural number), we shall write:

($\bar{\triangleright}$) :: $\bar{\Gamma} \vdash A \to B, \bar{\Gamma} \vdash \neg B \Rightarrow \bar{\Gamma} \vdash \neg A$,
(∇) :: $\bar{\Gamma} \vdash A \Rightarrow \bar{\Gamma} \vdash \neg\neg A$,
(Δ) :: $\bar{\Gamma} \vdash \neg\neg A \Rightarrow \bar{\Gamma} \vdash A$,

as well as:

(λ) :: $\bar{\Gamma}, A \vdash B \Rightarrow \bar{\Gamma} \vdash A \to B$,

where, in the limit case, the sequence $\bar{\Gamma}$ can be empty.

§ 1.5 Everywhere here, the mnemonic labels on rules and/or on specific tautologies are, more or less, those used in traditional formularian logic textbooks. The Latin inferential *modi* ($\triangleright$) and ($\bar{\triangleright}$) were so identified in a late peripatetic tradition conflating Aristotelian and Stoic terminology.

> Note that the usual colloquial talk about 'introduction / elimination' (int-elim) rules of inference – derived from Gentzen 1934–1935 – is also based of formularian thinking. After disambiguation, every logic inference rule is, in fact, supposed to 'introduce' the provability behaviour of a specific proof operator, while 'elimination' may refer, in general, only to the elimination of a specific proof-détour, in Gentzen's terms, resp. to the elimination of a 'redex', in λ-calculus jargon (**§ 6**). Incidentally, 'Double Negation introduction / elimination' would actually correspond to détour-eliminations only in the equational proof DN-normalisation approach discussed in **§ 4**.

§ 1.5.1 *Contrapositions and rule cognates.* In particular, (CR^ρ) stands for a case of *Contraposition*.

In classical logic, we have four of the kind, viz.:

(CB^ρ) $\vdash \mathsf{CB}^\rho[A,B] : (A \to B) \to (\neg B \to \neg A)$,
(C^ρ) $\vdash \mathsf{C}^\rho[A,B] : (A \to \neg B) \to (B \to \neg A)$,
(CB^∂) $\vdash \mathsf{CB}^\partial[A,B] : (\neg A \to B) \to (\neg B \to A)$,
(C^∂) $\vdash \mathsf{C}^\partial[A,B] : (\neg A \to \neg B) \to (B \to A)$.

Incidentally, the first two are valid in the *Minimalkalkül*, while the other two are genuinely classical.

The corresponding single-premiss rules are (cf. Frege's transitions in **GGA 1, § 14**, etc., and Rezuş 2009):

(CB^p) :: $\bar{\Gamma} \vdash A \to B \Rightarrow \bar{\Gamma} \vdash \neg B \to \neg A,$
(C^p) :: $\bar{\Gamma} \vdash A \to \neg B \Rightarrow \bar{\Gamma} \vdash B \to \neg A,$
(CB^∂) :: $\bar{\Gamma} \vdash \neg A \to B \Rightarrow \bar{\Gamma} \vdash \neg B \to A,$
(C^∂) :: $\bar{\Gamma} \vdash \neg A \to \neg B \Rightarrow \bar{\Gamma} \vdash B \to A,$

whence also the two-premiss rule (▷) [TT] and its cognates:

(▷) :: $\bar{\Gamma} \vdash A \to B, \bar{\Gamma} \vdash \neg B \Rightarrow \bar{\Gamma} \vdash \neg A,$
(▷̈) :: $\bar{\Gamma} \vdash A \to \neg B, \bar{\Gamma} \vdash B \Rightarrow \bar{\Gamma} \vdash \neg A,$
(▷$^\partial$) :: $\bar{\Gamma} \vdash \neg A \to B, \bar{\Gamma} \vdash \neg B \Rightarrow \bar{\Gamma} \vdash A,$
(▷̈$^\partial$) :: $\bar{\Gamma} \vdash \neg A \to \neg B, \bar{\Gamma} \vdash B \Rightarrow \bar{\Gamma} \vdash A.$

Here (CB^p) yields (▷), while the other three two-premiss rules – as well as the single-premiss Contrapositions mentioned above – can be obtained from (▷), using (∇), (Δ) and (▷) [*modus* (*ponendo*) *ponens*] [PP] alone.

§ 1.6 '*Proof-contexts*'. This notation is, in fact, elliptical. In what follows, we use an explicit *witness-theoretical notation* for rules, viewed as (proof-) operators, more or less as in abstract algebra. In this guise, a so-called 'proof-context' is to be understood as a finite sequence of witness-assumptions, i.e., as a list of decorated ('typed'), pairwise distinct, free proof-variables:

(Γ) $\Gamma := [x_1:A_1, ..., x_n:A_n]$ (n a natural number),

instead of a bare list of formulas $\bar{\Gamma} := [A_1, ... A_n]$, where, in $\Gamma \vdash c : C$, the (decorated, 'typed') variables $x_1:A_1, ..., x_n:A_n$ may occur (as free variables) in the witness c of C. We show this explicitly by writing $c \equiv c[x_1, ..., x_n]$. So, a proof-context is nothing but a *finite sequence of pairwise distinct* (decorated, 'typed') *proof-variables*.

Where $\Gamma := [x_1:A_1, ..., x_n:A_n]$, the extension $\Gamma [x:A]$ means that $[x:A]$ does not occur in the sequence Γ.

For convenience, in order to simplify the witness-theoretical specification of the proof operators, we allow *void occurrences of free variables* in proof-terms (so, where $c \equiv c[x_1, ..., x_n]$, some of the x_i's, $0 < i < n+1$, may 'occur voidly' in c, so to speak).

By way of example, single-premiss rules are to be written as (prefix) singulary ('unary') operators, like in:

(∇) :: $\Gamma \vdash a : A \Rightarrow \Gamma \vdash \nabla[A](a) : \neg\neg A,$
(Δ) :: $\Gamma \vdash c : \neg\neg A \Rightarrow \Gamma \vdash \Delta[A](c) : A,$

two-premiss rules are going to be written as (infix) binary operators, like in:

(▷) :: $\Gamma \vdash f : A \to B, \Gamma \vdash a : A \Rightarrow \Gamma \vdash f \triangleright a : B,$

($\tilde{\triangleright}$) :: $\Gamma \vdash f : A \to B,\ \Gamma \vdash c : \neg B \Rightarrow \Gamma \vdash f\tilde{\triangleright}c : \neg A$,

while (λ) [DT] will be written as a decorated ('typed') abstraction-operator, with term-form $\lambda x{:}A.b[x]$ [: $A \to B$], like in:

(λ) :: $\Gamma\ [x : A] \vdash b[x] : B \Rightarrow \Gamma \vdash \lambda x{:}A.b[x] : A \to B$.

We shall encounter more abstraction operators as we go, later on.

§ 1.6.1 The notation $\Gamma \vdash c : C$ is, in fact, redundant, and the 'concept' of a *proof-context* (denoted by Γ, here) is a *pseudo-concept*, since Γ can be obtained from the structure of the witness c, while the intended decoration ('type') C of c can be inferred from the rules of inference, provided the proof-term [witness] c is correctly contructed. Whence $\Gamma \vdash c : C$ says, simply, that the witness c of C is *correctly constructed*, i.e., it is enough to write $\vdash c$, instead. Roughly speaking, the triple consisting of a proof-context Γ for a witness-term a, taken together with the very witness-term a and the formula A witnessed thereby, is just an *annotated* (correctly constructed) *proof-term* for A. The (meta-theoretical) annotation amounts to a *local explicitation* of the 'official' proof-/ witness-notation, and may be ultimately absent.

> Essentially, proving a formula A, in logic, amounts to the explicit construction of a proof-term (witness) a for A. In the colon-notation (for 'typing'), 'a : A' is to be understood as a single name 'a^A' (cf., for instance, with 'Socrates, the Greek'), where 'a' is the label for a witness-term, with super-scripted 'A' as its (unique) decoration (or 'type'). With minor exceptions, the lazy, 'type-free' notation should make sense, however – at least equationally – more or less as in category theory, where the equations concern morphisms and can be oft written down without drawing 'commuting' diagrams, the latter being only meant – at least initially – to help intuition.

§ 1.7 *Derivable rules.* In this setting, a *derivable* (or *derived*) rule of inference would correspond, notationally, to at least one (*explictly*) definable operator, called *proof* or *witness operator*, here.

Notation. We spare on round parentheses by associating to the left, omitting oft the outermost pair. Put also, for convenience,

($\triangleright_1$) $\Gamma \vdash F(X) := F \triangleright X : B$,
  if $\Gamma \vdash F : A \to B$, and $\Gamma \vdash X : A$,

and, in general,

($\triangleright_n$) $\Gamma \vdash F(X_1) \ldots (X_n) := F \triangleright X_1 \ldots \triangleright X_n : B$,
  if $\Gamma \vdash F : A_1 \to (A_2 \ldots \to (A_n \to B)\ldots),\ \Gamma \vdash X_i : A_i$
  ($0 < i < n+1,\ n \geq 0$).

If F and the X_i's are (witness-) variables, and there is no danger of confusion, we shall even omit round parentheses and write $xy_1\ldots y_n$ for $x(y_1)\ldots(y_n)$, $n > 0$.
  Set also

(□) $\Gamma \vdash F \square G := \mathsf{S}[A,B,C](F)(G) : A \to C$,
    if $\Gamma \vdash F : A \to (B \to C)$, and $\Gamma \vdash G : A \to B$.

We have then *derived rules* (viz., *explicitly definable proof operators*):

(k) :: $\Gamma \vdash b : B \Rightarrow \Gamma \vdash \mathsf{k}_A(b) := \mathsf{K}[B,A](b) : A \to B$,
(□) :: $\Gamma \vdash f : A \to (B \to C), \Gamma \vdash g : A \to B \Rightarrow \Gamma \vdash f \square g : A \to C$,
    where $f \square g := \mathsf{S}[A,B,C](f)(g): A \to C$,

and, analogously,

(▷) :: $\Gamma \vdash f : A \to B, \Gamma \vdash c : \neg B \Rightarrow \Gamma \vdash f \triangleright c := \mathsf{CB}^\rho[A,B](f)(c) : \neg A$.

etc. This kind of shorthand can be also combined, as, e.g., in:

(∘) $\Gamma \vdash F \circ G := \mathsf{B}[A,B,C](F)(G) : C \to B$,
    if $\Gamma \vdash F : A \to B$, and $\Gamma \vdash G : C \to A$, where
(B) $\Gamma \vdash \mathsf{B}[A,B,C](F)(G) := \mathsf{k}_B(F) \square G : A \to C$, with F, G as above,

and so on.

> Once we shall state the equational behaviour of the corresponding proof operators, it should be obvious that this notation makes sense, in general, without decorations (i.e., 'type-free', so to speak).

§ 1.8 *The Deduction Theorem* [DT]. The proviso on void occurrences of free variables is an artifice meant to unify the notation for λ-abstractions (while witnessing the Deduction Theorem) and abstraction operators in general.

Indeed, the derivation of (λ) [DT] amounts to an *inductive argument*, with:

(λI) :: $\Gamma [x : A] \vdash b[x] : B \Rightarrow \Gamma \vdash \lambda x{:}A.x := \mathsf{I}[A] : A \to A$,
    if $b[x] \equiv x$ (and thus $A \equiv B$),
(λK) :: $\Gamma [x : A] \vdash b : B \Rightarrow \Gamma \vdash \lambda x{:}A.b := \mathsf{K}[B,A](b) : A \to B$,
    if x is not free in b,

as a basis of the recursion, and, as inductive steps:

(λ_ext) :: $\Gamma [x : A] \vdash b[x] := f \triangleright x : B \Rightarrow \Gamma \vdash \lambda x{:}A.b[x] := f : A \to B$,
    if x is not free in f,
(λS) :: $\Gamma [x : A] \vdash b[x] := f \triangleright c : B \Rightarrow \Gamma \vdash \lambda x{:}A.b[x] := \hat{f} \square \hat{g} : A \to B$,
    otherwise,

where $\Gamma [x : A] \vdash b[x] := f \triangleright c, \Gamma [x : A] \vdash f : C \to B, \Gamma [x : A] \vdash c : C$, with also $\Gamma \vdash \hat{f} := \lambda x{:}A.f : A \to (C \to B)$, resp. $\Gamma \vdash \hat{g} := \lambda x{:}A.c : A \to C$, and $\hat{f} \square \hat{g} := \mathsf{S}[A,C,B](\hat{f})(\hat{g})$.

§ 1.9 Once we have gotten (λ) [DT], the primitive witnesses (K), (S) can be written down explicitly with [DT] and [*modus (ponendo) ponens*] [PP] alone:

(K) $\vdash \mathsf{K}[A,B] := \lambda x{:}A.\lambda y{:}B.x : A \to (B \to A)$,

(S) ⊢ **S**[A,B,C] := λx:A→(B→C).λy:A→B.λz:A.((x▷z)▷(y▷z))
     : (A→(B→C)) → ((A→B) → (A→C)).

Incidentally, one can also avoid a direct derivation of the otherwise redundant [Commutation] (C) – à la Łukasiewicz 1929, 1934 (cf. **§ 1.11**) – in the original **BS**-system, noting that:

(1) [Commutation] has not been used in the derivation of (λ) above, and that:
(2) one can obtain it by using only (λ) [DT] and (▷) [modus (ponendo) ponens] [PP], as, e.g., in:

(C) ⊢ **C**[A,B,C] := λx:A→(B→C).λy:B.λz:A.((x▷z)▷y) : (A→(B→C))→((B→(A→C)).

§ 1.10 As mentioned already, in both **BS** (1879) and in **GGA** (1:1893, 2:1903), Frege allowed only 'flat' rules of inference, of the form:

(♭) :: ⊢ A_1, ..., ⊢ A_n ⇒ ⊢ C

corresponding to n-ary 'algebraic' operators:

(♭) :: Γ ⊢ a_1 : A_1, ..., Γ ⊢ a_n : A_n ⇒ ⊢ c ≡ ♭(a_1,...,a_n) : C,

confusing thus asserted conditionals, of the form:

(→) ⊢ A_1 → (A_2 → ... → (A_n → C)...)

and 'entailments', of the form:

(⊢) A_1, ..., A_n ⊢ C.

The latter are to be disambiguated, witness-theoretically, as in:

(Γ⊢) Γ ≡ [x_1 : A_1, ..., x_n : A_n] ⊢ c ≡ c[x_1,..., x_n] : C.

So, as observed earlier, Frege would not have seen the point behind (λ) [DT], whence, in particular, he would not have been able to invent a compact proof-notation of the kind suggested here.

Technically, this yields a notation equivalent to the term-syntax of the pure 'typed' λ-calculus **λ**[→].

The appropriate equational conditions governing the behaviour of the primitives (λ) and (▷) will be discussed later.

§ 1.10.1 Actually, Frege's systematic 'confusion' between (⊢) and (⊢) – in **GGA**, say – can be easily accommodated in the present notational setting, by writing, for instance,

⊢ $\bar{\Gamma}$ → C := A_1 → (A_2 → ... → (A_n → C)...)

everywhere, in place of our $\bar{\Gamma} \vdash C$, where $\bar{\Gamma} := [A_1, ..., A_n]$ – cf., e.g., the ad hoc 'shuffling operations' described formally in Rezuş 2009 – although, in this case, the corresponding witness-theoretical disambiguation would require some additional formal provisions, as regards the 'shuffling' of λ-prefixes in λ-abstracts, i.e., in terms of the form $\lambda x_1. ... .\lambda x_n.c$, resp. the 'shuffling' of the 'long cut-forms' $f(a_1)...(a_n)$, involved in witnessing his 'transitions' [*Übergänge*] corresponding to the 'structural' rules of Gentzen. (Cf. **1.13**.)

§ 1.11 *Consensed detachment.* Basically, for Frege, a proof is a *finite sequence of formulas*, where, inductively, each element of the sequence is either (1) an axiom or else (2) it follows from preceding elements of the sequence by applying the rules of inference. So, in fact, there was hardly anything to invent, since the witness-theoretical notation exemplified here amounts, roughly, to a bare formalisation of the comments in a proof, taken in Frege's sense.

By way of example, the derivation (proof) of (I), from (K) and (S) with (▷) [*modus (ponendo) ponens*] is the sequence:

1 ⊢ S[A,B,A] : (A → (B → A)) → ((A → B) → (A → A))
   :: a substitution instance of (S),
2 ⊢ K[A,B] : A → (B → A)
   :: (K),
3 ⊢ S[A,B,A] ▷ K[A,B] : (A → B) → (A → A)
   :: apply (▷) to 1 and 2 [in this order],
4 ⊢ S[A,B→A,A] ▷ K[A,B→A] : (A → (B → A)) → (A → A)
   :: a substitution instance of 3,
5 ⊢ (S[A,B→A,A] ▷ K[A,B→A]) ▷ K[A,B] : A → A
   :: apply (▷) to 4 and 2 [in this order],

whereby we have, verbosely:

(I) ⊢ I[A] := (S[A,B→A,A] ▷ K[A,B→A]) ▷ K[A,B] : A → A,

and one can even ignore the formula-decorations, by writing, elliptically:

(I) ⊢ I := S(K)(K) : A → A,

instead (where $F(X) := F \triangleright X$, as above), since – as long as we intend to obtain 'most general' results – the latter can be restored uniquely up to a uniform relettering of the propositional variables.

> This follows from a well-known theorem of John A. Robinson on *unifiers* (Robison 1965; cf. the 'Unification Theorem', in Robinson 1979, Chapter 11, p. 192), while performing uniform substitutions on pure implicational formulas.

Without abstractors – here: (λ), for the moment being – this amounts to a notational variant of the so-called 'condensed detachment' [CD] operator of Carew A. Meredith.[9]

---
[9] Cf. Prior 1962² [*Appendix 2*], Kalman 1974, 1983, Rezuş 1982, 2010, Hindley & Meredith 1990, McCune & Wos 1992, Ulrich 2001, etc.

With a slightly more involved example – and ignoring, again, formula-decorations – the Łukasiewicz 1929, 1934 derivation of (C) from (K) and (S) amounts to:

(C) ⊢ C := (B □ K(K)) ∘ S : (A → (B → C)) → (B → (A → C)),

where we have set, for convenience, B := K(S) □ K, and F ∘ G := K(F) □ G, with F □ G := S(F)(G), as before.

> Frege obtained (I) from (K) and (C), with (▷) alone, as I := C(K)(K), in **BS**, **§ 16** (27). As a notable idiosyncrasy, in general, Frege did not use derived rules, in **BS II**, in order to shorten proofs / derivations. This would have brought him to invent some more – picturesque – bi-dimensional 'shorthand'.

The accommodation of assumptions – resp. of decorated ['typed'] λ-abstractions – in this setting yields a notational variant for the 'natural deduction' [ND] formulation of the minimal logic, as in Jaśkowski 1927, 1934; the principle is exactly the same.[10]

Of course, if we had additional primitive rules, we should have modified the witness notation as appropriate.

§ 1.12 In view of the above, if we have at hand the appropriate proof notation, the 'concept' of (an) *entailment* [finite sequence of formulas where exactly one is tagged as a conclusion, the remaining ones counting as assumptions or as 'hypotheses'] is a *pseudo-concept*, too: in the disambiguated version of (⊢) above, the conditional witness $c \equiv c[x_1, ..., x_n]$ of C contains already the assumptions $x_1 : A_1, ..., x_n : A_n$. So, in fact, we need such expedients only if we ignore the proof-information, as, e.g., in a colloquial, elliptic spelling of the rules of inference.

§ 1.13 *'Structural rules'*. The elliptic spelling of the rules is inducing *auxiliary stipulations* (called 'structural rules' in Gentzen's Dissertation) meant for the formal manipulation of the (elliptic) proof-contexts. The latter correspond, more or less, to some of Frege's 'transitions' [*Übergänge*] in **GGA**.

Elliptically – were $\bar{\Gamma}$ stands for a sequence of formulas – Frege's transitions would read, in our (i.e., Gentzen's) terms:

(dil) :: $\bar{\Gamma}$ ⊢ C ⇒ $\bar{\Gamma}$, A ⊢ C
  [*Verdünnung*, Dilution],
(prm) :: $\bar{\Gamma}$, A, B ⊢ C ⇒ $\bar{\Gamma}$, B, A ⊢ C
  [*Vertauschbarkeit*, Interchange or Permutation],
(cto) :: $\bar{\Gamma}$, A, A ⊢ C ⇒ $\bar{\Gamma}$, A ⊢ C
  [*Verschmelzung*, Fusion, in Frege; Contraction, in Gentzen],

except for the fact Frege would have had ⊢ $\bar{\Gamma}$ → C where we have $\bar{\Gamma}$ ⊢ C (cf. **§ 1.10.1**).

---

[10] See Rezuş 2017 for a detailed discussion of Jaśkowski's original approach.

The Fregean [Dilution] is also known as 'Weakening', in popular Middle English (the American version) logic books, a mis-translation due, apparently, to Stephen C. Kleene. As noted elsewhere, the German 'verdünnen' refers rather to bare dilutions, as, e.g., in (inorganic) chemistry, or else in improper oenological manipulations (diluting a noble wine with water, and the like). The 'weakening' of the resulting solution is a side-effect. This is a bad simile, in the end, because, actually, in logic, 'dilutions' can be reversed (**§ 1.14**).

Analogously, in the limit case, one has:

(id) :: $A \vdash A$ [The 'Law' of Identity]

or, more generally, modulo (dil), with $\bar{\Gamma} := [A_1, ..., A_n]$, for $0 < i < n+1$,

(prj) :: $\bar{\Gamma} \vdash A_i$ [Projection].

For Gentzen (prj) – resp. the special case (id) – would not have counted as a 'structural' rule, but rather as an axiom (scheme).

Besides, in **GGA**, Frege had also, *mutatis mutandis*, a special transition called *Schluss* [Syllogism], known to the moderns as [CUT] [*Schnitt*], from Gentzen's Göttingen *Inauguraldissertation*:

(CUT) :: $\bar{\Gamma}, A \vdash C; \bar{\Gamma} \vdash A \Rightarrow \bar{\Gamma} \vdash C$ [*Schluss*, Syllogism; *Schnitt*, CUT].

If properly disambiguated, with $\Gamma := [x_1:A_1, ..., x_n:A_n]$ this time, one has

(prj) :: $\Gamma \vdash x_i : A_i$, with $0 < i < n+1$, resp. (for $n = 1$),
(id) :: $x : A \vdash x : A$,

where $n > 0$, and, for $\bar{\Gamma}$ arbitrary (including $\bar{\Gamma} \equiv$ nil):

(dil) :: $\Gamma \vdash c : C \Rightarrow \Gamma [x : A] \vdash c : C$, if x is not free in c,
(prm) :: $\Gamma [x : A] [y : B] \vdash c[x,y] : C \Rightarrow \Gamma [y : B] [x : A] \vdash c[x,y] : C$,
(ctc) :: $\Gamma [x : A] [y : A] \vdash c[x,y] : C \Rightarrow \Gamma [z : A] \vdash c[z,z] : C$,
where c[z,z] stands for the simultaneous substitution of x and y for z in c[x,y], and, finally,
(CUT) :: $\Gamma [x : A] \vdash c[x] : C; \Gamma \vdash a : A \Rightarrow \Gamma \vdash c[x:=a] : C$,
where c[x:=a] (reading: 'x becomes a in c') stands for the substitution of a for x in c[x].

The concept of *substitution* – for proof- / witness-tems – can be defined in the obvious way (by induction 'on the structure of the proof-term c[x]').

It is easy to see that:

(1) [Contraction] (ctc) is a *special case* of [CUT] (as applying to 'atomic' substitutions) and that
(2) except for the [*Schnitt*] [CUT], the 'structural rules' (dil), (prm), and (ctc) can be *reversed* (see also Rezuş 2016).

§ 1.13.1 Given (λ) and ($\triangleright$), the corresponding Fregean **GGA**-transitions are, obvioulsy, tantamount Gentzen's 'structural' rules mentioned above (cf. Rezuş 2009).

> Gentzen had, in fact, a different idea of a 'proof context', as his 'sequents' (*Sequenzen*) are finite sequences of formulas of the form $\bar{\Gamma}_1 \vdash \bar{\Gamma}_2$, with $\vdash$ as an ad hoc separator (a punctuation mark), where, the *length* of $\bar{\Gamma}_i$ (i := 1, 2) is $\mathsf{lh}(\bar{\Gamma}_i) \geq 0$ [sic]. In detail, where $\bot$ stands for an arbitrary *false* proposition (*falsum*), the case $\mathsf{lh}(\bar{\Gamma}_2) = 0$ amounts to $\bar{\Gamma}_1 \vdash \bot$, the case $\mathsf{lh}(\bar{\Gamma}_2) = 1$ amounts to the usual 'entailments' $\bar{\Gamma}_1 \vdash C$, and applies only to intuitionistic logic (resp. to *Minimalkalkül*), while the case $\mathsf{lh}(\bar{\Gamma}_2) > 1$, applying to classical logic, consists of a kind of hypocritical way of hiding a classical *surface* negation (i.e., the outermost $\neg$), and amounts to $\bar{\Gamma}_1, \neg(\bar{\Gamma}_2) \vdash \bot$, where $\neg(\bar{\Gamma}) := [\neg A_1, ..., \neg A_n]$, for $\bar{\Gamma} \equiv [A_1, ..., A_n]$. Among other things, the net effect of this piece of notational, formularian schizophrenia – there is no 'left' and 'right' in logic – amounts to the duplication of the number of the 'structural' rules (in Gentzen's Dissertation, we have also analogous primitive 'structural' rules that are 'multiple on the right', so to speak: the case $\mathsf{lh}(\bar{\Gamma}_2) > 1$, above). While reasoning in formularian terms, this does not pose special problems, as the proof-information is absent, anyway (we may think of the a Gentzen 'sequent' $\bar{\Gamma}_1 \vdash \bar{\Gamma}_2$ as a shorthand for $\bar{\Gamma}_1, \neg(\bar{\Gamma}_2) \vdash \bot$, for instance), but, upon due witness-theoretical disambiguation, this formularian notational artifice is not so innocent, as the Gentzen 'structural' rules counting as 'multiple on the right' are, in fact, instances of *logical* rules, governing special uses of *genuinely classical proof operators*. For instance, (ctc) [Contraction] 'on the right' of the turnstile would amount to a case of *consequentia mirabilis*, etc. Needless to say that, for Frege, the 'sequents' of Gentzen – even those 'singular on the right' – would have been sheer nonsense, or else ad hoc notational conventions resting on bare *conceptual confusions*.

§ 1.13.2 Gentzen's 'structural rules' are, actually, *pseudo-rules of inference*: as presented above, they are just *notational expedients* meant to accommodate a *conceptually defective* approach to proof theory.

> Whence, in particular, the endemic formularian references to 'CUT-admissibility' and the recent popular talk about so-called 'substructural' logics are conceptually inadvertent – to say the least – since the concept of *admissibility* – à la Paul Lorenzen and Haskell B. Curry – applies to proper ('logical') rules of inference, while Gentzen's (CUT)-rule refers to *global substitutions* in witness terms, whereas the alleged 'substructurality' of a logic – the absence of (dil) and/or (ctc), for instance – refers to restrictions on witness term construction, and, specifically, on inference rules represented by abstraction operators, as, e.g., *strictness*, as in relevance logics, or *linearity*, as in the case of the BCK- resp. BCI-logics. (Actually, both 'strictness' and 'linearity' applied originally to λ-calculus since around 1978; cf. Rezuş 1981, 1982, Hindley 1989, 1997.)

§ 1.13.3 Whether Frege's **GGA** presentation of classical logic – using 'flat' rules (exclusively) – could be accounted for as being *conceptually superior* to Gentzen's ad hoc manipulation of (pseudo-) rules (dubbed 'structural', for the occasion) is a matter of otiose speculation, in the end, since both approaches are *underspecified*, proof-theoretically.

> Putting aside his roundabout way of coping with the [DT], in some cases, Frege was, incidentally, mixing-up model- and proof-theoretic considerations while stating *some* logical rules of inference – cf., e.g, his way of introducing the Contraposition-rules in **GGA** – and – in as far as we can tell – he was unaware of the distinction between rule-derivability and rule-admissibility, a matter on which – although not brought explicitly into discussion – Gentzen was quite accurate, at least in as far as his specific views on the matter are concerned, on what is to be accounted for as a rule of inference. (The *Hauptsatz* of Gentzen 1934–1935 says that the CUT – i.e., substitution in witness-terms – is admissible for classical – resp. intuitionistic – logic. Witness-theoretically, this information can be obtained by *replacing systematically* the *global substitution operator*, i.e., Gentzen's CUT [an inductive definition, in fact], by *local substitutions*, associated to each primitive proof operator, appearing in the system.)

> As far as we know, Gentzen did *not* read **GGA**. On the authority of his actual PhD supervisor, Paul Bernays, Gentzen borrowed his 'structural' rules (mainly the CUT), from Paul Hertz (1881–1940) – a former student and collaborator of Hilbert, on theoretical physics, as well as an admirer and a friend of Einstein – and elaborated, independently, upon.[11].

> In retrospect, it might have been interesting to know Gentzen's would-be assessment of Frege's *Regellogik* of **GGA**. (The quantifier-free segment of the **GGA**-logic is equivalent, qua provability, to the quantifier-free logic of **BS**, and thus consistent.)

---

[11] See, for instance, Gentzen 1933, and Bernays 1965, pp. 3–5. Paul Hertz (cf. Bernays 1969, www.linkfang.de/wiki/Paul_Hertz_(Physiker), etc.) and Paul Bernays were colleagues at *Augusta*, in Göttingen, and close to each other (the latter has also reviewed several publications of Hertz, including his Springer book, Hertz 1923a), mainly because Bernays was also a neo-Kantian philosopher, and an active member of a *Neu-Fries'sche Schule*, while Hertz was – among other things – keenly interested in epistemological problems and in the philosophy of science, otherwise matters on which he lectured for a while in Göttingen. Although a physicist by training, philosophy (and logic) was more than a *violon d'Ingres* for Hertz: he had strong connections with the *Wiener Kreis* and the Berlin neopositivists (cf. Haller 1993), he also edited, together with Moritz Schlick, the epistemological papers of Hermann von Helmholtz (Julius Springer, Berlin 1921), translated Russell's **Problems of Philosophy** in German (Weltkreis-Verlag, Erlangen 1926), etc. Except for casual post-cards, the Bernays *Nachlass* – now at ETH Zürich – has not preserved any Bernays-Hertz correspondence [I am indebted to the late Ernst Paul Specker (ETH Zürich), for this information], and, apparently, this is also the case for the Hertz *Nachlass*. To my knowledge, a significant part of the *Nachlass* of Paul Hertz – including some unpublished MS-material – has been donated by one of his sons – Mr. Rudolf Hertz, a former banker of New York City – to the Archives of Scientific Philosophy of the Pittsburgh University, Pittsburgh PA. My earlier plans (Rezuş 1988) to edit the relevant *logico-epistemological papers* of Paul Hertz did not substantiate, due to lack of interest among publishers, both Western European and American. See, however, Abrusci 1982, Schroeder-Heister 2002, and Legris 2012.

§ 1.14 (1) After disambiguation, the reversal of (dil) [Dilution] reads:

(rev-dil) :: Γ [x : A] ⊢ c : C ⇒ Γ ⊢ c : C, if x is not free in c [Reversed Dilution],

and amounts, intuitively, to the trivial observation that, in actual [witness] derivations, we can always get rid of the 'assumptions that have never been used', so to speak (Rezuş 2016). Of course, without a proper proof-notation, we cannot state (rev-dil) formally.

(2) Here, the stipulations corresponding to Fregean transitions have been written with parametric proof-contexts. In current notation, we can – and we shall actually – leave out the Γ-parameters.

(3) The two-premiss transition [*Schluss, Schnitt*] has an equivalent, *cumulative* version:

(CUT⊕) :: Γ_1 [x : A] ⊢ c[x] : C; Γ_2 ⊢ a : A ⇒ Γ ⊢ c[x:=a] : C [cumulative CUT],

where the sequences Γ_1 and Γ_2 do not share (decorated) proof-variables and Γ := Γ_1, Γ_2 (sequence concatenation).

(4) Analogously, any derived n-premiss (logical) rule (n > 1) has an equivalent cumulative version as well.

§ 1.15 Summing up – and ignoring the 'structural rules' – we have shown that Frege's original axiom system of **BS** is equivalent, qua provability – resp. rule-derivability – to the following (primitive) logical inference rules:

(λ) = [→-introduction, the Deduction Theorem] [DT],
(▷) = [→-elimination, *modus (ponendo) ponens*] [PP],
(▷̄) = [*modus (tollendo) tollens*] [TT],
(∇) = [Double Negation introduction] [DN-i],
(Δ) = [Double Negation elimination] [DN-e].

> We can also show *non-redundancy* – i.e., the fact that the five inference rules above are independent as regards derivability – but the argument would likely distract the reader from the main line of thought, so we'd better leave it as an exercise.

§ 1.15.1 Here, only (Δ) stands for a genuinely classical rule of inference; the remaining ones are already minimal, i.e., derivable in the [implication-negation]-fragment of Johansson's *Minimalkalkül* Johansson 1937, and so, a fortiori, intuitionistically valid. (This shows already independence for (Δ), by the way.)

§ 1.16 So far, we have established a witness-theoretical notation for (the quantifier-free segment of) Frege's **BS**-logic, an alternative *Begriffsschrift*, so to speak. Specifically, in contrast with the original Fregean formula(rian) *script*, we have a *proof formalism* – a *proof-script* [*Beweisschrift*] – instead, where the main objects of concern are not the formulas (expressing propositional schemes) anymore, but rather their *proofs* (technically: *witnessess*).

In the initial, axiomatic variant of **BS**, the tautologies taken as axioms (here: as axiom-schemes) are witnessed by *primitive* proof-objects (witnesses)

(K), (S), (CB$^\rho$), (∇), (Δ) – decorated with appropriate formula-parameters, as, e.g., in K[A,B], S[A,B,C], etc. – while the proofs of the tautologies derived from axiom(scheme)s by [*modus (ponendo) ponens*] [PP] are represented by complex expressions constructed from primitive witnesses and a binary operation ($\triangleright$) – a *proof operator on witnesses* – meant to represent the primitive inference rule [PP], so that we have an abstract proof-grammar:

>primitive witness-terms :: c := K, S, CB$^\rho$, ∇, Δ
>witness-terms :: a, b, c, d, e, f ::= c | c$\triangleright$a,

where the intended decorations – also called 'types' sometimes – are given by (K), (S), (CB$^\rho$), (∇), (Δ), and ($\triangleright$) above.

> Here, the witness for Frege's redundant (C)-axiom can be taken to be a shorthand, as in **§ 1.11**, matching, e.g., the derivations of Łukasiewicz 1929, 1934.

In the revised formulation of **BS** – as summarised in **§ 1.15** – we have been left with:

(1) a stock of decorated ('typed') *assumption-variables* (witness-variables) x:A, y:B, z:C, ..., etc., and
(2) *primitive inference rules* [DT], [PP], [TT], [DN-i], [DN-e][12], represented by appropriate witness operators (λ), ($\triangleright$), ($\bar{\triangleright}$), (∇), (Δ), resp.,

whereby the actual proofs are represented by *witness-terms* built up, inductively, from *witness variables* and *witness operators*, according to the abstract proof-grammar:

>witness-variables :: x, y, z, ...
>witness-terms :: a, b, c, ... ::= x | λx.b | c$\triangleright$a | c$\bar{\triangleright}$a | ∇(a) | Δ(c)

where decorations ('types') are to be supplied, inductively, again, by (1), as a basis of the recursion, and by (2) – namely (λ), ($\triangleright$), ($\bar{\triangleright}$), (∇), and (Δ), resp. – as inductive steps.

This is a minor achievement, because, at this point, we have *only* an explicit proof-*notation* – a *proof-grammar*, at most.[13] – as the corresponding witness operators are *not yet defined* (in the algebraic sense). This must be done separately, by subjecting them to *explicit equational conditions*.[14]

---

[12] So, we have, in fact, an equivalent 'natural deduction' formulation of **BS**, à la Jaśkowski 1927, 1934, and Gentzen 1934-1935, as regards provability, resp. rule-derivability.

[13] In other words, the 'derivation' rules of the formularian logician are – proof-theoretically speaking – just 'term-formation' or 'well-formedness' rules.

[14] *Mutatis mutandis*, this kind of requirement is the everyday, bread and butter tool of the ('universal') algebraist, and goes back to Leopold Kronecker, at least. (Remember, for instance, his insistence on 'defining' everything that was 'non-arithmetical' to his mind – irrationals and complex numbers included – by explicit equational conditions.)

The reader has likely realised, so far, the fact that the idea of *defining a proof operator*, in witness theory, is *bi-dimensional* so to speak, *mutatis mutandis* as in category theory. Namely, in any such a definition, one has (1) a *provability* component, given by explicit decorations ('typing': the object-part, in categories) and (2) an *equational* component, given by explicit 'characteristic' equational conditions (the morphism-part, taken care of by 'commuting' diagrams, in categories). So, in principle, one can proceed 'elliptically' in two distinct ways: [1] by ignoring the decorations / 'typing' and by doing 'proof theory without formulas' (!) – here: just 'type-free' λ-calculus – or else [2] by ignoring the explicit proof-notation and by doing 'proof theory without (mentioning) the proofs themselves', as in axiomatics, 'natural deduction', Gentzen sequent systems and the like, in the old-fashioned formularian tradition. The first kind of elliptic thinking is more informative, in a way, but lacks *intuitive support*, so that the would-be expert in 'type-free' λ-calculus behaves, more or less, like *Monsieur* – and, possibly, *Madame* – *Jourdain*, who was unaware of the fact (s)he was speaking *in prose*. On the other hand, the traditional formularian thinker relies on *rather vague* and *badly tutored intuitions* about what is to be meant by *proving in logic / mathematics* and is bound to handle the proofs themselves by proxy, i.e., by ad hoc (figurative or visual, even 'geometrical') expendients – sequences, multisets or sets of formulas, trees of such sequences or multisets, etc., block structures and the like – (pseudo- and para-) mathematical ingredients that give her / him the illusion (s)he knows what (s)he is talking about. *Quod non*, the formularian expert is *radically confused* on rather *elementary conceptual issues* (in logic, at least): think of the vast literature on 'Gentzenisations' – the industry of so-called CUT-*elimination* – or else of the comedy of errors populating our logic text-books with ad hoc 'natural deduction'-systems and so on (cf., e.g., Pelletier 1999, 2001, Hazen & Pelletier 2014).

One can do this in several distinct ways. On Frege's primitive (formularian) signature [→,¬], and with a choice of primitive rules of inference (resp. proof operators) as above, the corresponding equational stipulations are slightly involved, however, and non-transparent, intuitively.

Mutatis mutandis, I have been confronted with an analogous situation while analysing, in witness-theoretical terms, the Łukasiewicz-Jaśkowski axiomatics – resp. 'natural deduction' – for classical propositional logic (Łukasiewicz 1929, Jaśkowski 1927, 1934), based on [→,¬], in Rezuş 2017.

We prefer to do this (i) by extending, first, Frege's primitive signature, and (ii) by analysing, next, the rather complex primitive rule [*modus (tollendo) ponens*] [TT] – resp. the corresponding proof operator (▷) – into simpler components.

§ 2 Restoring the logical constants: *verum* and *falsum*

At this point, one might want to note another Fregean oversight. The primitive signature [→,¬] is (*expressively*) *incomplete* (in algebraic jargon, the corresponding Boolean set of connectives or logical constants is said to be 'functionally incomplete'). In other words, we cannot define explicitly the propositional constants $\top$ (*verum*) and $\bot$ (*falsum*) from the shown primitives, so, given a primitive negation $\neg$, we must have at least one of them, either $\top$ or $\bot$, as a primitive.

> Intuitively, $\top$ (resp. $\bot$) is supposed to stand for an *arbitrary* true (resp. false) proposition. – One might speculate about the *pure logical character* of the 'logical' constants. For Frege, they won't have posed special problems. (Later, he even allowed the corresponding 'objects' [*Gegenstände*] in his semantics: *das Wahre* resp. *das Falsche* as denotations [*Bedeutungen*] for true resp. false propositions, not to be imported into the conceptual script itself, though.) – In fact, as the **BS**-system was supposed to contain at least Classical Arithmetic (in our terms: *non-logical* symbols, although for Frege, arithmetic was *pure logic*), as in Peano Arithmetic [**PA**], and (a first-order) equality $\doteq$, one could also define the *falsum*-constant by $\bot := (0 \doteq 1)$. In the presence of the (first-order) equality and the universal quantifier, there is even no need for arithmetic constants, since one can set $\top := \forall u.(u \doteq u)$, etc. Of course, one could also define both $\bot$ and $\top$ by using *propositional quantifiers*, but, unlike Peirce 1885 and Russell 1906, Frege did not mention explicitly such things in **BS**. Otherwise, the familiar 'algebraic' procedure (like, *mutatis mutandis*, in groups, say) – which consists, roughly, of introducing, first, a relative *verum* $\top_p$ (resp. *falsum* $\bot_p$), by using an arbitrary tautology containing a single propositional variable, as, e.g., $(p \to p)$, with $\top_p := (p \to p)$, etc. and by showing, next, $\top_p \leftrightarrow \top_q$, for any two p, q, so that one could, ultimately, set $\top := \top_p \leftrightarrow \top_q$, modulo material equivalence $\leftrightarrow$ – involves a *hidden use* of propositional quantifiers. — In the **GGA**-system the problem does not occur, as the formalism contains plenty of additional (non-logical) constants.

We choose to add $\bot$ (*falsum*) as a new primitive to the pure logical signature $[\to, \neg]$, and define:

($\top$) $\top := \neg \bot$ [*verum*],

setting:

($\sim$) $\sim A := A \to \bot$ [inferential negation; Peirce 1885][15].

§ 2.0.1 Here, one might note the fact that this extension works for the intuitionistic logic and the *Minimalkalkül*, as well. Moreover, the choice of the primitive propositional constant is arbitrary, so that one has both $\top = \neg \bot$, and $\bot = \neg \top$, even minimally so, whence also (propositional) equivalences $\top = \neg \neg \top$, $\bot = \neg \neg \bot$ (with = for definitional equivalence). In particular, the tautologies $\nabla[A]$, $\Delta[A]$ are both minimally (and so intuitionistically) valid for $A := \top, \bot$.

> Any classical quantifier-free tautology containing only $\top$ and $\bot$ is *minimally valid*, and, by this very token, the propositional quantifers $\forall^p$ and $\exists^p$ are legitimate in a (purely formal) intuitionistic setting, since one can define them

---

[15] Note that $\top$ and $\dagger := \sim \bot \equiv \bot \to \bot$ are *distinct* propositions.

explicitly by using the minimal (resp. intuitionistic) **and** resp. **or** and substitution: $\forall^P p.A[p] := (A[p:=\top]$ **and** $A[p:=\bot])$, resp. $\exists^P p.A[p] := (A[p:=\top]$ **or** $A[p:=\bot])$. In *Minimalkalkül* one must be careful, however, because the usual quantifier rules for $\forall^P$ would automatically yield validity for the *intuitionistically characteristic* rule *ex falso*, for $\bot := \forall^P p.p$. In classical logic, one could call them 'truth-value quantifiers', perhaps (Nuel D. Belnap Jr.). So, in fact, propositionally quantified classical logic – also called 'extended propositional logic / calculus' sometimes, a subject derived from Peirce (1885), the young Russell (1906), Leśniewski, Tarski, and Łukasiewicz – is *not more* than the bare classical 'propositional' logic. Cf. Rezuş 2017, **§ 3**. Alternative definitions of *falsum* – and thus classical negation – in propositionally quantified classical logic have been considered by Russell 1906 (without reference to Peirce 1885, though; see also Anellis 1995). Otherwise, on the addition of the propositional constants (and inferential negation), propositionally quantified (classical) logic – à la Peirce and/or Russell – is, at least in as far provability is concerned, the same thing as 'propositional' classical logic, à la Frege. If, however, one uses a primitive negation instead, as in Frege's **BS**, the outcome is witness-theoretically different from what can be obtained with inferential negation alone.

§ 2.0.2 In classical logic, the presence of the propositional constants in the primitive setting amounts to the fact that the *intended*, truth-value semantics of the logic is implicit in the bare syntax, i.e., we have the intended truth-value behaviour already available at a pure proof-theoretical level, modulo material equivalence ($\leftrightarrow$).

Consider $\vdash \bot \to \bot \leftrightarrow \top, \vdash \bot \to \top \leftrightarrow \top, \vdash \top \to \bot \leftrightarrow \bot, \vdash \top \to \top \leftrightarrow \top$, etc., where $\leftrightarrow$ is to be defined in the expected way.

This makes, ultimately, (formularian) consistency – and completeness – matters (the usual Lindenbaum-Tarski industry of the early thirties) rather trivial (Cf. Pogorzelski & Wojtylak 2008, etc.).

§ 2.1 On the enriched signature, $[\bot, \to, \neg]$, Frege's original axiom system (**BS**) is incomplete, of course: we cannot even prove the famous (minimal) tautology:

$(\Omega) \vdash \Omega : \top$,

let alone more interesting trivia, like the transforms:

$(\bar{\tau}) \vdash \bar{\tau}[A] : {\sim}A \to \neg A$

and its converse:

$(\tau) \vdash \tau[A] : \neg A \to {\sim}A$ [the Law of Non-Contradiction, qua tautology],

or else the intuitionistically characteristic 'Law':

$(\dot{\varpi}) \vdash \dot{\varpi}[A] : \bot \to A$ [i.e., the Mediaeval *ex falso quodlibet (sequitur)*],

as well as the otherwise very basic (and also minimal, in the sense above) rule of inference:

($\cdot$) :: $\vdash \neg A, \vdash A \Rightarrow \vdash \bot$ [the Law of Non-Contradiction, as a rule],

and so on.

§ 2.1.1 Note that the genuinely classical tautology:

(E$^\gamma$) $\vdash$ E$^\gamma$[A] : ($\sim$A $\rightarrow$ A) $\rightarrow$ A
    [the Law of Clavius = inferential *consequentia mirabilis*, qua tautology]

is a proper substitution instance in the genuinely classical tautology:

(P) $\vdash$ P[A,B] : ((A $\rightarrow$ B) $\rightarrow$ A) $\rightarrow$ A
    [Peirce's Law; cf. Peirce 1885, and Russell 1906]

(with $\bot$ for B), but, as the $\bot$-constant is not primitively available in **BS**, we cannot prove the inferential [Law of Clavius] either. Actually, in Frege's original script of **BS** one can only prove the non-inferential version of [Clavius], viz.:

(E) $\vdash$ E[A] : ($\neg$A $\rightarrow$ A) $\rightarrow$ A
    [the non-inferential Law of Clavius, *consequentia mirabilis*].

> The cognate rule $\neg A \vdash A \Rightarrow \vdash A$ – corresponding to the proof operator (ϵ), mentioned below – was first spotted by Girolamo Cardano (1501–1576), and appears explicitly in his famous *opus* De proportionibus (Sebastian Henricpetri [Petri], Basle 1570). The traditional reference to Clavius [viz., to Kristoph Klau SJ (1538–1612), 'the Euclides of the XVI-th century'], in this context, is due to the fact that the Jesuit noticed (Clavius 1611) the use of this reasoning pattern in ancient Greek mathematics texts (Euclides and Theodosius). In fact, the same reasoning pattern could have been found in Aristotle's (now lost) early treatise Protrepticus. One should not confuse – like Father Clavius – *use* and *mention*, however: unlike his venerable predecessors, the Pavian was also *aware* of this specific proof-pattern. Subsequently, another famous Jesuit mathematician, Giovanni Girolamo Saccheri (1667–1733), used systematically the 'Law of Clavius' in his ruminations on Aristotle's syllogistic (Logica demonstrativa, 1697), as well as in his 'defense' of the Euclidean geometry (Euclides ab omni naevo vindicatus, 1733), a vindication whereby he discovered a couple of *non-Euclidean geometries* somewhat *avant la lettre*. The modern mathematicians would also use endemically Cardano's *mirabilis consequentia*, as, e.g., while showing, at undergraduate level, that a group has a single identity element (assume that are two of them, distinct, $\neg$A thus, and conclude – using the group-axioms – that they are identical: A thus; whence A). Notably, the corresponding single-premiss, 'flat' rule, $\vdash \neg A \rightarrow A \Rightarrow \vdash A$, appears in Frege's **GGA 1**, as well. For details, see Kneale 1957, Miralbell 1987, Rezuş 1990, 1991, 1993, 2009, 2017, 2017a, 2017b, Nuchelmans 1991, 1992, Bellissima & Pagli 1996, and the references given there.[16]

---

[16] As an aside, worth mentioning perhaps is also the fact that Brouwer owned a copy of Clavius 1611 (5 large-sized in-folio's, a quite expensive piece of of *rara mathematica*) in his personal library, now the 'L. E. J. Brouwer Collection', in the Provincial Library of Friesland, Leeuwarden, the Netherlands (the collection – Brouwer's testamentary donation – contains many other rather rare books on – or relevant to – the history of mathematics).

Even superficial a reflection on the matter shows that we need at least the axiom (Ω) and the (single-premiss) rule-variants of the transforms $(\bar{\tau})$ and (τ) in order to pass from the primitive signature $[\to, \neg]$ to the enriched signature $[\bot, \to, \neg]$ and conversely.

§ 2.2 *The $\lambda\gamma$-calculi I.* Before anything else, a side-remark on the *economy of means* should be appropriate. On the non-redundant (and expressively complete) signature $[\bot, \to]$, with inferential negation defined, à la Peirce (1885), by $(\sim)$, as above, we need (K), (S) and $(\triangleright)$ – or else (λ) and $(\triangleright)$ – and the inferential version of (Δ), viz.:

$(\Delta^\gamma) \vdash \Delta^\gamma[A] : {\sim}{\sim}A \to A$,

resp. its single-premiss rule-variant:

(Δ^γ) :: $\Gamma \vdash c : {\sim}{\sim}A \Rightarrow \Gamma \vdash \Delta^\gamma[A](c) : A$,

or else a new (monadic) abstraction-operator γ (term-form: $\gamma z{:}{\sim}A.e[z]$ [: A]), with:

(γ) :: $\Gamma\,[z : {\sim}A] \vdash e[z] : \bot \Rightarrow \Gamma \vdash \gamma z{:}{\sim}A.e[z] : A$
[the inferential *reductio ad absurdum*],

in order to obtain the full (quantifier-free) classical logic.

In this case, (Δ^γ) – resp. the rule (Δ^γ) – and the (γ)-rule are inter-derivable (resp. inter-definable, qua proof operators) by:

$(\Delta^\gamma_=)\ \Gamma \vdash \Delta^\gamma[A] := \lambda z{:}{\sim}{\sim}A.\gamma x{:}{\sim}A.(z{\triangleright}x) : {\sim}{\sim}A \to A$, resp.
$(\gamma_=)\ \Gamma \vdash \gamma x{:}{\sim}A.e[x] := \Delta^\gamma[A](\lambda x{:}{\sim}A.e[x]) : A$, for $\Gamma\,[x{:}A]\,|\text{-}\,e[x] : \bot$.

This economical strategy is unobjectionable, and, if we impose appropriate equational conditions on the three primitives – (λ), $(\triangleright)$, and (γ) – we get a proper equational witness theory for the (quantifier-free) classical logic, viz. a 'typed' $\lambda\gamma$-calculus (Rezuş 1990, 1991, 1993, 2017c).

§ 2.2.1 There are several distinct ways of playing the economical game, even in a 'type-free' (undecorated) setting.

The simplest variant does not go beyond the *pure* λ-calculus – based on (λ) and $(\triangleright)$ – with, as 'characteristic' equational conditions, the familiar stipulations:

$(\beta\lambda) \vdash (\lambda x.b[x]) \triangleright a = b[x{:=}a]$, and
$(\eta\lambda) \vdash \lambda x.(f{\triangleright}x) = f$, if x is not free in f,

including the expected monotony conditions for both (λ) and $(\triangleright)$.

In 'type-free' terms, set $<a> := \lambda x.(x{\triangleright}a)$, where x is not free in a, and define:

$(\gamma)\ \gamma x.e[x] := \lambda x.e[x{:=}<x>]$

(this would be 'the Glivenko (γ)'; cf. Rezuş 2017a, 2017b, 2017c). We have then, in the pure λ-calculus,

(Cl$_=$) ⊢ <a> ▷ c = c ▷ a, whence also:
($\hat{\beta}\gamma$) ⊢ γx.(x▷(γy.e[x,y]) = γz.e[x,y:=z] [surface diagonalisation],
($\eta\gamma$) ⊢ γx.(x▷a) = a, x not free in a [extensionality for inferential *reductio*],

as well as:

($\hat{\xi}\gamma$) ⊢ γx.c = γx.d ⇒ ⊢ γx.(f▷c) = γx.(f▷d)

(Rezuş 1990, 1991, 1993, 2017a, 2017c), and we can decorate ('type') both primitives, λ and ▷, as intended – by (λ) and (▷) – with, in particular, for [the inferential *reductio*]:

(γ) Γ ⊢ γx:∼A.e : A, if Γ [x : ∼A] ⊢ e : ⊥,

taken as a (proof-) primitive.

The resulting equational theory, **λγ**$_G$[⊥,→] (with 'G' for Glivenko), based on (λ), (γ), and (▷) alone, supposed to satisfy the conditions (βλ), (ηλ), ($\hat{\beta}\gamma$), ($\eta\gamma$), and ($\hat{\xi}\gamma$), is (Post-) consistent (since its 'type-free' cognate, **λγ**$_G$, is already contained in the pure λ-calculus **λ**), so the decorated ('typed'), now primitive, proof operators (λ), (γ) and (▷) are defined equationally: we don't have a bare proof-notation, as in 'natural deduction' and the like. Here, ($\hat{\beta}\gamma$) ['surface diagonalisation'], and ($\hat{\xi}\gamma$) can be replaced equivalently (Rezuş 1990, 1991, 1993) by:

($\bar{\beta}\gamma$) ⊢ γx.f[x]▷(x▷(γy.e[x,y]) = γz.f[x:=z]▷e[x,y:=z] [full diagonalisation].

§ 2.2.2 The full ('type-free') extensional **λγ**-calculus, **λγ**, is more involved and the corresponding consistency proof goes beyond pure **λ**. We shall go into it later (**§ 5.7**).

> Full extensional **λγ** is, actually, a proper subsystem of the **λπ**-calculus; cf. Rezuş 1990, 1991, 1993, 2017a, 2017b. Incidentally, one can also simulate *inside* the pure **λ** an 'intensional', β-only **λγ**-calculus, the Kolmogorov **λγ**$_K$, discussed briefly in Rezuş 2017c. The latter lacks (ηλ), but has an 'inferential' β-analogue of the (ζ)-condition (ζγ); in fact, one has **λγ** = **λγ**$_K$ + (ηλ). Cf. **§§ 5.6–5.7**. All **λγ**-calculi can be 'typed' on the signature [⊥,→] with negation (∼) defined inferentially, and make up proof-systems for the (quantifier-free) classical logic.

Notably, the decorated ('typed') version **λγ**$_G$[⊥,→] admits of a *definitional extension* that consists of:

(1) restricting the equational (γ)-conditions to 'atomic' γ's, i.e., to witness-terms of the form γx:∼A.e, where A is atomic (here, a propositional variable), and
(2) defining, γx:∼A.e inductively, for 'complex' γ's (by induction 'on the structure of A'):

as in:

($\gamma_\perp$) $\Gamma \vdash \gamma z{:}\dot{\top}.e[z] := e[x{:=}\lambda x.x] : \perp$,
    if Γ [x:$\dot{\top}$] $\vdash$ e[z] : $\perp$, with $\dot{\top} := {\sim}\perp$,
($\gamma_\rightarrow$) $\Gamma \vdash \gamma z{:}{\sim}(A{\rightarrow}B).e[z] := \lambda x{:}A.\gamma y{:}{\sim}B.e[z{:=}\mu(x,y)] : A \rightarrow B$,
    if Γ [z:(A→B)] $\vdash$ e[z] : $\perp$, Γ [x:A] [y:$\sim$B] $\vdash \mu(x,y) := \lambda z{:}A{\rightarrow}B.y\triangleright(z\triangleright x)$.

> See also Prawitz 1965, Rezuş 1990, 1991, 1993, and the (ζ)-conditions discussed in § **5.7**. Prawitz 1965 had no (η)-conditions, only an analogue of ($\bar\beta\gamma$) ['full diagonalisation'] – i.e., a single reduction (détour elimination) 'rule' for the 'inferential' *reductio ad absurdum*, here (γ) – and *no formal proof-syntax* (he used 'natural deduction', instead, and colloquial explanations in order to cope with abstraction operators). Moreover, Prawitz would not have allowed witness-terms of the form $\gamma z{:}\dot{\top}.e[z]$, and $\gamma z{:}{\sim}{\sim}A.e[z]$ (the latter were represented by λ-abstracts, i.e., by witness-terms of the form $\lambda z{:}{\sim}A.e[z]$), he had also – redundantly – a primitive classical ∧ (**and**), etc. — The connection with Glivenko 1928, 1929 should be obvious from Rezuş 1990.

§ **2.2.3** Although mathematically correct, there is a conceptual inconvenience in all this. Namely, if we think, more or less traditionally, in terms of rules of inference, we can realise that the inferential (Peirce) definition of negation ($\sim$) amounts to the fact that the Rule of Non-Contradiction [LNC] is simulated, in this setting, by a proper substitution instance of ($\triangleright$) [*modus* (*ponendo*) *ponens*] [PP], as a kind of special case of it. This kind of 'specialisation' is *counter-intuitive*, to say the least.[17] Technically, there is no real inconvenience, however, because the mathematics behind the game 'works'. It does not work *as intuitively intended*, however.

> *Mutatis mutandis*, with a famous simile, the well-known Kuratowski definition of an ordered pair in **ZF** – as a mini-set of the form {x, {x,y}} – 'works' for all practical purposes, although the definitional identification is rather ad hoc and the very basic idea of 'ordering' – otherwise essential in everyday mathematics – is thereby lost. Think of grounding our usual set-theoretical intuitions of *ordinals* on this kind of, apparently innocent, conceptual reduction.

§ **2.3** Obviously, if we are going to add a primitive negation ¬ to the signature [$\perp,\rightarrow$], we are also supposed to subject the would-be additional proof operators to appropriate, 'characteristic equations'.

> Algebraically, we must specify an algebraic *(sub-) variety*, à la Birkhoff-Mal'cev (i.e., as in 'universal algebra'), by stating *explicit equational conditions*.

§ **2.3.1** Since the required substitutivity conditions (Leibnizian replacement of equals by equals, *salva veritate*) are already ensured[18], this concerns, essentially, the $(\bar\tau,\tau)$-pair of rules corresponding to – and, in view of ($\triangleright$), derivable from – the pair $(\bar\tau,\tau)$ of combinators mentioned previously (§ **2.1**):

---

[17] Putting aside French – and Dutch – linguistic idiosyncrasies, potatoes are hardly special cases of apples!

[18] All proof operators are suposed to be 'algebraic', satisfying the expected monotony conditions.

$(\bar{\tau})\ \vdash \bar{\tau}[A] : \sim A \to \neg A,$
$(\tau)\ \vdash \tau[A] : \neg A \to \sim A$
  [the Law of Non-Contradiction, qua tautology],

namely,

$(\bar{\tau}) :: \Gamma \vdash f : \sim A \Rightarrow \Gamma \vdash \bar{\tau}[A](f) : \neg A,$
$(\tau) :: \Gamma \vdash c : \neg A \Rightarrow \Gamma \vdash \tau[A](c) \sim A$
  [the Law of Non-Contradiction, qua rule].

The latter are minimally valid, i.e., derivable in Johansson's *Minimalkalkül*, on the signature $[\bot,\to,\neg]$, and are supposed to act, as regards provability / rule-derivability, more or less, like a kind of 'casting' operators in programming.[19]

The *natural* equational conditions on the the $(\bar{\tau},\tau)$-pair would consist of making them one-one, i.e., by assuming they make up an *inversion*. Ignoring decorations ('types'), this amounts to:

$(\beta\bar{\tau})\ \vdash \tau(\bar{\tau}(f)) = f,$
$(\eta\bar{\tau})\ \vdash \bar{\tau}(\tau(c)) = c.$

As ever, $(\bar{\tau})$ and (τ) are supposed to be 'algebraic' (one assumes they 'respect' equality, i.e., that the due monotony conditions hold).

Set, in the appropriate syntax, $\boldsymbol{\lambda\bar{\tau}}[\bot,\to,\neg] := \boldsymbol{\lambda}[\bot,\to,\neg] + (\beta\bar{\tau}) + (\eta\bar{\tau})$.

§ 2.3.2 Alternatively, we may attempt to disambiguate the 'special' cases of (λ) [DT] and ($\triangleright$) [*modus (ponendo) ponens*] [PP], by adding primitive proof operators on the enriched signature $[\bot,\to,\neg]$, as, e.g., in:

$(\rho) :: \Gamma\ [x : A] \vdash e[x] : \bot \Rightarrow \Gamma \vdash \rho x{:}A.b[x] : \neg A\ [\neg\text{-introduction}],$
$(\cdot) :: \Gamma \vdash c : \neg A;\ \Gamma \vdash a : A \Rightarrow \Gamma \vdash c{\cdot}a : \bot\ [\text{LNC}]\ [\neg\text{-elimination}],$

(both are minimally valid, i.e., already derivable in Johansson's *Minimalkalkül*, on $[\bot,\to,\neg]$, as well), thereby regimenting the rule / proof operator $(\cdot)$ [LNC] in its proper ('intuitive') category of things.

On this plan, the new proof operators (ρ) and $(\cdot)$ are supposed to satisfy equational conditions analogous to $(\beta\lambda)$ and $(\eta\lambda)$, viz.,

$(\beta\rho)\ \vdash (\rho x.e[x]) \cdot a = e[x{:=}a],$ and
$(\eta\rho)\ \vdash \rho x.(f{\cdot}x) = f,$ if x is not free in f,

including, as expected, monotony for both (ρ) [$\neg$-introduction] and $(\cdot)$ [LNC].

Set, in the appropriate syntax, $\boldsymbol{\lambda\rho}[\bot,\to,\neg] := \boldsymbol{\lambda}[\bot,\to,\neg] + (\beta\rho) + (\eta\rho)$.

§ 2.3.3 One can establish equivalence $\boldsymbol{\lambda\bar{\tau}}[\bot,\to\neg] \simeq \boldsymbol{\lambda\rho}[\bot,\to\neg]$, as regards provability / rule-derivability as well as equationally.

Reciprocal derivability is immediate, since we have, as minimally valid principles:

---

[19] Professional 'working' mathematicians – as well as other logic outsiders – would usually take the 'casting'-trick as granted and trivial, but logicians should be more cautions.

(ρ) $\Gamma \vdash \rho$x:A.e := $\bar{\tau}$[A](λx:A.e[x]) : ¬A, if Γ [x:A] $\vdash$ e[x] : $\bot$.
($\cdot$) $\Gamma \vdash$ c·a := τ[A](c) $\triangleright$ a, if $\Gamma \vdash$ c : ¬A and $\Gamma \vdash$ a : A,

in $\boldsymbol{\lambda\bar{\tau}}[\bot,\to,\neg]$, as well as,

($\bar{\tau}$) $\Gamma \vdash \bar{\tau}$[A](f) := ρx:A.(f$\triangleright$x) : ¬A, if $\Gamma \vdash$ c : ∼A,
(τ) $\Gamma \vdash \tau$[A](c) := λx:A.(c·x) : ∼A, if $\Gamma \vdash$ c : ¬A,

in $\boldsymbol{\lambda\rho}[\bot,\to,\neg]$.

For the equational part, one can even proceed without decorations ('type-free'). Indeed, let, in the appropriate ('type-free') syntax, $\boldsymbol{\lambda\bar{\tau}} := \boldsymbol{\lambda} + (\beta\bar{\tau}) + (\eta\bar{\tau})$, and $\boldsymbol{\lambda\rho} := \boldsymbol{\lambda} + (\beta\rho) + (\eta\rho)$. Define, in $\boldsymbol{\lambda\bar{\tau}}$, ρx.e[x] := $\bar{\tau}$(λx.e[x]) and c·a := τ(c)(a). This yields ($\beta\rho$) and ($\eta\rho$). Conversely, define, in $\boldsymbol{\lambda\rho}$, $\bar{\tau}$(f) := ρx.f(x) and τ(c) := λx.c·x. This yields ($\beta\bar{\tau}$) and ($\eta\bar{\tau}$).

§ 2.3.4 (1) Obviously, Frege's (primitive) Contraposition-combinator (**CB**$^\rho$), and the corresponding two-premiss rule:

($\triangleright$) :: $\vdash$ A $\to$ B, $\vdash$ ¬B $\Rightarrow$ $\vdash$ ¬A [*modus (tollendo) tollens*] [TT]

are derivable in $\boldsymbol{\lambda\rho}[\bot,\to,\neg]$ resp. in $\boldsymbol{\lambda\bar{\tau}}[\bot,\to,\neg]$, by using both ($\rho$) and ($\cdot$), say, as in:

($\tilde{\triangleright}_=$) $\Gamma \vdash$ f $\tilde{\triangleright}$ c := ρz:¬A.c·(f$\triangleright$z) : ¬A, if $\Gamma \vdash$ f : A $\to$ B, and $\Gamma \vdash$ c : ¬B.

(2) Moreover, $\boldsymbol{\lambda\rho}[\bot,\to,\neg] \simeq \boldsymbol{\lambda\bar{\tau}}[\bot,\to,\neg]$ yields also the ($\boldsymbol{\Omega}$)-combinator mentioned earlier (**§ 2.1**), as $\vdash \boldsymbol{\Omega} := \rho$x:$\bot$.z : $\top$, and a deductively equivalent version of Frege's ($\boldsymbol{\nabla}$), as well as the corresponding operator / rule (∇), which we write next, for convenience, with distinguished typography, as:

($\dot{\boldsymbol{\nabla}}$) $\vdash \dot{\boldsymbol{\nabla}} := \lambda$x:A.$\rho$z:¬A.z·x : A $\to$ ¬¬A,
($\dot{\nabla}$) $\Gamma \vdash \dot{\nabla}$(a) := ρz:¬A.z·a : ¬¬A, if $\Gamma \vdash$ a : A.[20]

(3) Note, however, that, on the signature $[\bot,\to,\neg]$, the combinators ($\boldsymbol{\Omega}$), and ($\boldsymbol{\nabla}$) – as well as the operator / rule (∇) – are not available from Frege's Contraposition-combinator (**CB**$^\rho$) – resp. the operator / rule ($\triangleright$) [TT] – with (λ) and ($\triangleright$) alone. In fact, granted (λ), ($\triangleright$), and ($\tilde{\triangleright}$), one needs a primitive ($\boldsymbol{\Omega}$) or something similar in order to obtain ($\bar{\tau}$) – and thus (ρ) – and (at least a special instance of) the ($\boldsymbol{\Delta}$)-combinator, in order to obtain ($\cdot$), and thus (τ). Cf. **§ 3.1**.

§ 2.3.5 From the above we have a proof-system for the Johansson minimal logic, on the signature $[\bot,\to,\neg]$. Although this is not immediately visible, the uncomfortable point in the extended game – the one with both $\bot$ and ¬ as primitives – consists of the fact that we get a rather big bag of *infinitely many, provably non-equivalent proof operators* corresponding to [LNC] thereby: formally pairwise distinct ($\cdot_n$)'s (n a natural number).

This remark – on which I am going to ponder briefly next – applies to Frege's **BS**-logic, as well.

---

[20] The dotless symbol is reserved for a *primitive* notion.

§ 3 Frege and the Law of Non-Contradiction

We may wonder: *How many Laws of Non-Contradiction are, out there?* Historically speaking, the so-called Law of Non-Contradiction [LNC] has been first spotted about twenty-three centuries ago by Aristotle – Metaph., Γ – in the shadow of Parmenides (and Heraclitus). In this section, we shall end up with *infinitely many* [LNC]'s.

> As everything is 'constructive' in witness theory – Rezuş 2017a, 2017b – we can have only *countably many* such things around, though.

The main thing worth exploiting is ($\bar{\triangleright}$) [*modus (tollendo) tollens*] [TT].
Let us forget about the minimal $\lambda\rho[\bot,\to,\neg] \simeq \lambda\bar{\tau}[\bot,\to,\neg]$, for a moment, and asume that, on the extended primitive signature $[\bot,\to,\neg]$,

(1) we have (λ), ($\triangleright$) – i.e., the primitives of pure λ – (Ω), ($\bar{\triangleright}$), and (Δ), and that
(2) we have already established their equational behaviour.

> This amounts to the fact that we are already supposed to deal with a complete set of *proof operators* for classical logic based on $[\bot,\to,\neg]$, although not necessarily on the minimal route mentioned under **§ 2.3**. See, specifically, **§ 2.3.4** (3).

On this basis, we shall proceed by a chain of (explicit) definitions of proof operators in terms of (λ), ($\triangleright$), (Ω), ($\bar{\triangleright}$), and a *minimally* valid instance of (Δ):

§ 3.1 *Definition.*

($\bar{\tau}$) $\Gamma \vdash \bar{\tau}[A](f) := f \bar{\triangleright} \Omega : \neg A$, if $\Gamma \vdash f : \sim A$,
(ρ) $\Gamma \vdash \rho x{:}A.e[x] := \bar{\tau}[A](\lambda x{:}A.e[x]) : \neg A$, if $\Gamma\,[x:A] \vdash e : \bot$,
($\cdot$) $\Gamma \vdash c \cdot a := \Delta[\bot]((\lambda x{:}\top.a) \bar{\triangleright} c) : \bot$, if $\Gamma \vdash c : \neg A$, $\Gamma \vdash a : A$,
whence also:
(τ) $\Gamma \vdash \tau[A](c) = \lambda x{:}A.(c \cdot x) : \sim A$, if $\Gamma \vdash c : \neg A$, and
($\dot{\nabla}$) $\Gamma \vdash \dot{\nabla}(a) := \rho x{:}\neg A.(x \cdot a) : \neg\neg A$, if $\Gamma \vdash c : A$.

Here, we have used, for convenience, the same notation as in **§ 2.3**. Note, however, that, even though they have the intended, minimal *provability* behaviour, the operators *defined* in **§ 3.1** might not be the same as those already mentioned in **§ 2.3**, qua *equational* behaviour. Indeed, in **§ 2.3**, the pairs (ρ,·), resp. ($\bar{\tau}$,τ) were taken as *primitives*, with *postulated* equational behaviour, whereas, in **§ 3.1**, the latter three operators have been defined explicitly in terms of a *genuinely classical* operator (Δ.)

In particular, on the basis of the explicit definitions in **§ 3.1**, the (ρ,·)-pair may not satisfy the (postulated) equational conditions of $\lambda\rho$, so that both (·) and ($\dot{\nabla}$), say, defined as in **§ 3.1**, may have an erratic equational behaviour as regards the *minimality criteria* of $\lambda\rho$, etc.

Let us pause, parenthetically, on analogous situations going *beyond minimality*.

§ 3.1.1 Incidentally, the transform ($\bar{\tau}$) can be viewed as a special case of ($\bar{\triangleright}$) [modus (tollendo) tollens] [TT], modulo (Ω).

On a similar line of thought, its converse (τ) yields a special case of the characteristic intuitionistic rule:

(CO) :: $\Gamma \vdash c : \neg A,\ \Gamma \vdash a : A \Rightarrow \Gamma \vdash$ CO[A,B](c)(a) : B
[ex contradictione quodlibet].

Indeed, granted the intuitionistically characteristic:

($\dot{\varpi}$) $\vdash \dot{\varpi}$[A] : $\bot \to$ A [ex falso quodlibet, qua tautology]

or else the corresponding single-premiss rule:

(ϖ) :: $\Gamma \vdash e : \bot \Rightarrow \Gamma \vdash \varpi$[A](e) : A [ex falso quodlibet, qua rule of inference]

and ($\cdot$), i.e., the minimal [Law of Non-Contradiction] [LNC], we have the (genuinely) intuitionistic tautology (CO), derived as in:

(CO) :: $\vdash$ CO[A,B] := λz:$\neg$A.λx:A.ϖ[A](z·x) : $\neg A \to (A \to B)$,

resp. the (genuinely) intuitionistic rule:

(CO$_{A,B}$) :: $\Gamma \vdash c : \neg A,\ \Gamma \vdash a : A \Rightarrow \Gamma \vdash$ CO[A,B](c)(a) = ϖ[B](c·a) : B,

and, in particular, its substitution instance,

(CO$_{A,\bot}$) :: $\Gamma \vdash c : \neg A,\ \Gamma \vdash a : A \Rightarrow \Gamma \vdash$ CO[A,$\bot$](c)(a) = ϖ[$\bot$](c·a) : $\bot$,

so that one could have, alternatively:

(τ) :: $\Gamma \vdash c : \neg A,\ \Gamma \vdash a : A \Rightarrow \Gamma \vdash \tau$[A](c)(a) = ϖ[$\bot$](c·a) : $\bot$.

§ 3.1.2 Intuitionistic insiders would likely claim that they have a (genuinely intuitionistic) proof-isomorphism:

($\varpi_\bot$) $\Gamma \vdash \varpi$[$\bot$](e) = e : $\bot$, if $\Gamma \vdash e : \bot$,

(see **§ 4.2.1**), so that one could have also obtained [LNC], as regards provability / derivability, by:

($\cdot$) $\Gamma \vdash$ c·a := CO[A,$\bot$](c)(a) [$\equiv \tau$[A](c)(a) $\equiv \varpi$[$\bot$](c·a)] : $\bot$,
if $\Gamma \vdash c : \neg A,\ \Gamma \vdash a : A$,

directly from (CO).

The fact is that the intuitionistically characteristic [ex falso]-rule (beyond Minimalkalkül) is nothing but a *degenerated case* of a genuinely classical pattern of reasoning (sic) – as shown below – so I'd rather prefer to have both ($\bar{\tau}$) and (τ), as well as (ρ) and ($\cdot$) introduced 'minimally'.

A side-remark (for experts in *Minimalkalkül* and intuitionism). Recall that $\Delta[\bot]$ – a proper substitution instance of the genuinely classical tautology $\Delta[A]$ – is also a minimally – and thus intuitionistically – valid tautology (**§ 2.0.1**). Whence so is the rule ($\cdot$) [the Law of Non-Contradiction] [LNC]. Anyway, a minimalist intuitionist – Kolmogorov or Johansson, for instance – would not have denied the *minimal validity* of this *specific instance* of that famous Law of Logic.

A historical remark. My side-remarks on *Minimalkalkül* and (logical) intuitionism might be useful, here, mainly in order to fix the ideas, in retrospect. – The fact that Frege did not bother to read the Dutch *proefschrift* of the young Brouwer (1907) or his 1908a is immaterial, of course. – As to minute historical details, Kolmogorov 1925 was actually pondering mainly on separation matters appearing in Hilbert's axiomatics, during the early twenties – including the so-called 'positive' logic – not on Frege's **BS**. So, in the end, my insistence on separation matters [*Minimalkalkül versus* intuitionism *versus* classical (Fregean) logic] is, rather, unhistorical, because Frege did not – and could not – make such distinctions. Incidentally, he also ignored Peirce 1885 in **GGA**. (Russell's 1906 – the immediate precursor of Principia Mathematica on formularian technicalities – appeared too late in the picture, anyway.)

§ 3.2 Next, we shall use the full *provability* strength of the genuinely classical (Δ)-rule – leaving its equational behaviour unspecified, for a while – in order to simulate a *genuinely intuitionistic* proof-principle (beyond the *Minimalkalkül*).

§ 3.2.1 *Definition.* Set, in $\lambda\rho[\bot,\to,\neg] + (\Delta)$:

(∂) $\Gamma \vdash \partial x{:}\neg A.e[x] := \Delta[A](\rho x{:}\neg A.e[x]) : A$, if $\Gamma\ [x : \neg A] \vdash e[x] : \bot$, so ($\partial$) stands for the genuinely classical [*reductio ad absurdum*], and

($\star$) $\Gamma \vdash c \star a := \dot{\nabla}[A](a) \cdot c : \bot$, if $\Gamma \vdash c : \neg A$ and $\Gamma \vdash a : A$ (yet another [LNC]).

§ 3.3 We can now analyse (∂) – the genuinely classical [*reductio ad absurdum*] – into components, by defining (deriving) the specific instances:

(ϖ) $\Gamma \vdash \varpi[A](e) := \partial x{:}\neg A.e : A$, if $\Gamma\ [x : \neg A] \vdash e : \bot$, x not free in e [ex falso qodlibet][21], and

(ϵ) $\Gamma \vdash \epsilon z{:}\neg A.a[x] := \partial x{:}\neg A.(x\star a[x]) : A$, if $\Gamma\ [x : \neg A] \vdash a : A$ [Cardano's *consequentia mirabilis*].

Recall that, as regards provability – resp. rule-derivability – the [*ex falso*]-rule (ϖ) is *genuinely intuitionistic* (over and beyond the *Minimalkalkül*), while the Cardano rule (ϵ) is *genuinely classical* (over and beyond the Glivenko-Heyting

---

[21] In view of the previous remark **§ 1.14** (1) on (rev-dil) [Reversed Dilution], the proviso on (ϖ) reads 'if $\Gamma \vdash e : \bot$'; so, anyway, x is fresh for e.

intuitionistic logic).[22] Conversely, given appropriate equational conditions on (ϖ) and (ϵ), the abstraction operator (∂) – standing for the genuinely classical [reductio ad absurdum] – can be recovered from (ϖ) [ex falso] and (ϵ) [consequentia mirabilis], by setting:

($\partial_=$) $\quad \Gamma \vdash \partial$x:¬A.e[x] := ϵx:¬A.ϖ[A](e[x]) : A, if Γ [x : ¬A] $\vdash$ e[x] : $\bot$.

§ 3.3.1 The genuinely intuitionistic proof operator (ϖ), standing for the [ex falso]-rule, is thus, in this context, a specific instance – a *degenerated case* – of the genuinely classical (∂) [*reductio ad absurdum*].

> As an aside, formularian – Kolmogorov-Glivenko-Heyting style – intuitionism is absent-mindedly *inadvertent* – to say the least – since, for Brouwer, the genuinely classical tautology (Δ[A]) was supposed to be intuitionistically valid for any atomic, *decidable* A, i.e., not only for A := $\bot$, $\top$, resp. for A := ¬B, with B atomic, etc. (as, e.g., in Glivenko 1928, 1929, and Kolmogorov 1925). In other words, formularian (logical) intuitionism is *intuitionistically defective* [sic] unless one is able to formalise (propositional) *decidability* in the *very same* formal setting. In retrospect, the early Kolmogorov-Glivenko-Heyting (formularian) ruminations on 'Brouwer's logic' turned out to be epistemically useful only in as far our understanding of *classical* logic is concerned. See also van Atten 2017.

§ 3.4 The promised *infinitistic* ingredient comes now.

> The long preliminary discussion above is motivated by the fact I intended to insist on minimality, à la Kolmogorov 1925 and Johansson 1937.

Assume that we have already established the equational behaviour of ($\cdot$) [LNC][23] and (∇).
Define next, inductively:

§ 3.4.1 *Definition*.

($\cdot_0$) $\quad \Gamma \vdash$ c $\cdot_0$ a := c . a : $\bot$, if $\Gamma \vdash$ c : ¬A, and $\Gamma \vdash$ a : A, and
($\cdot_{n+1}$) $\quad \Gamma \vdash$ c $\cdot_{n+1}$ a := ∇[A](a) $\cdot_n$ c : $\bot$, if $\Gamma \vdash$ c : ¬A, and $\Gamma \vdash$ a : A
(n a natural number).

We have thus plenty of *equationally distinct proof operators* corresponding to the Law of Non-Contradiction [LNC].

---

[22] Cf. Kezuş 1990, 1991, 1993, 2017, 2017a, 2017b, 2017c and the references given there. Around the early fifties, H. B. Curry isolated (Curry 1950, 1952, 1963) a proper, non-minimal fragment of classical logic – he used to call it 'logic of strict negation' – with (λ), ($\triangleright$), (ϵ), as primitives, but without (ϖ) [ex falso]. Notably, in presence of the intended rules for the minimal **or**, Curry's logic contains Brouwer's foe, [*tertium non datur*], although the genuinely classical [*reductio ad absurdum*] (∂) and the (Δ)-rule are not derivable in it.
[23] Or else ($\star$), for that matter (cf. **§ 3.2.2**).

Note that this is, possibly, the case *in a minimal setting*, although, in principle, this *might* happen only if we are not careful while manipulating minimal proof-isomorphisms, i.e., if minimality is not correctly defined equationally. See also **§ 4.1**.

As an aside, such – apparently hair-splitting, legitimate though – details remain, usually, un-noticed in standard proof-theoretical treatements of logical intuitionism – cf., e.g., Troelstra 1973, van Dalen 1980, 2013[5] or Schwichtenberg & Troelstra 1996, 2000[2] – mainly because (minimal resp. intuitionistic) negation is taken to be a defined notion there, i.e., one relies on *inferential* negation ($\sim$) instead. Needless to say – while hiding such difficulties *by omission* – the corresponding 'theoretical' attempts to provide neat *proof semantics*, for minimal / intuitionistic logic, in terms of 'constructions' etc. – based on the familiar Brouwer-Heyting-Kolmogorov [BHK] 'explanations' of the minimal / intuitionistic $\to$ and $\bot$ (minimal / intuitionistic 'absurdity') – go into unnecessary metaphysical complications involving a lot more hairs to split (as, e.g., rather weird speculations on 'constructions that do – and can – not exist', presupposing a would-be additional, underlying *modal logic of constructions*, and so on).

Here, the elementary observation is that, in (classical) logic, even a very basic fact like the Law on Non-Contradiction [LNC] is, actually, *infinitistic* in nature. See also **§ 4.2**.

§ 4 The equational proof-normalisation of Double Negation

We have left the equational behaviour of the Double Negation pair (Δ,∇) aside, so far.

§ 4.1 Note that, with the ($\cdot$)- and (∇)-operators introduced as above – in terms of a primitive ($\bar{\triangleright}$) [TT] – there is, a priori, no 'natural' way of collapsing the hierarchy of the [LNC]-operators ($\cdot_n$) (n a natural number), by imposing appropriate equational conditions on the (initial) proof operators ($\cdot$) and (∇) – i.e., ultimately, on ($\bar{\triangleright}$) [TT] – unless we make the pair ($\rho,\cdot$) satisfy ($\beta\rho$), for instance, and obtain (∇) from ($\rho,\cdot$), qua ($\dot\nabla$) as in:

($\dot\nabla$) $\Gamma \vdash \dot\nabla(a) := \rho x{:}\neg A.(x\cdot a) : \neg\neg A$, if $\Gamma \vdash c : A$.

So, in a minimal setting, we can get a *definable* ($\dot\nabla$)-operator – resp. a *derivable* inference rule ($\dot\nabla$) – by using the minimal (ρ)-rule [$\neg$-introduction], say, as, e.g., in ($\dot\nabla$) above, whereupon, the minimal equational conditions:

($\beta\rho$) $\vdash (\rho x.e[x]) \cdot a = b[x{:=}a]$, and
($\eta\rho$) $\vdash \rho x.(f{\cdot}x) = f$, if x is not free in f,

on a primitive pair ($\rho,\cdot$), would eventually identify equationally the [LNC]-operators ($\cdot_n$) (n a natural number).

However, with ($\cdot$) defined from ($\bar{\triangleright}$) [*modus (tollendo) tollens*] [TT], by:

(·) $\Gamma \vdash $ c·a := $\Delta[\bot]((\lambda$x:⊤.a$) \mathbin{\bar{\rhd}} $ c$) : \bot$, if $\Gamma \vdash $ c $: \neg$A and $\Gamma \vdash $ a $:$ A [LNC],

as earlier, the equational conditions $(\beta\rho)$ and $(\eta\rho)$ – understood as 'natural' conditions on a would-be primitive (ρ)-operator – would look rather un-natural, to say the least.

> Here, such conditions are to be ultimately understood as *characteristic conditions* on the $(\bar{\rhd})$-operator, i.e. on [*modus (tollendo) tollens*] [TT], which, as mentioned before, stands for a *minimally valid* rule of inference.

Let alone the fact that one could have provided, alternatively, a (definable) $(\bar{\nabla})$-operator by something like:

$(\bar{\nabla})$ $\Gamma \vdash \bar{\nabla}[A](a) := \rho$x$:\neg$A.(x$\star$a) $: \neg\neg$A, if x is not free in a, and $\Gamma \vdash$ a $:$ A,

where

(ρ) $\Gamma \vdash \rho$x:A.e[x] := ∂x:$\neg\neg$A.e[x:=Δ[A](x)] $: \neg$A, if Γ [x:A] $\vdash$ e $: \bot$, and
(Δ) $\Gamma \vdash \Delta[A](c) := \partial$x:$\neg$A.(c$\star$x) $:$ A, if $\Gamma \vdash$ c $: \neg\neg$A

are supposed to be defined in terms of a primitive pair $(\partial, \star)$ – to be specified equationally as appropriate – and generate, next, an analogous hierarchy of [LNC]-operators from this kind of $(\bar{\nabla})$, and an initial (primitive) [LNC], with c·a taken to mean c$\star$a, this time, etc.

> Putting things colloquially – and anticipating slightly – we are going to be *in trouble* – al least by common, formularian standards – if we are damaging, some way or another, the *equational behaviour* of the minimal logic segment based on $[\bot, \to, \neg]$. On the other hand, we can easily recover, in a classical setting, the *intended* equational proof-behaviour of minimal implication (and minimal negation), by assuming a *primitive* pair (∇, Δ) of one-one proof operators, behaving *as intended classically* (more or less as in Chrysippus, the actual founder of [classical] logic).[24]

§ 4.2 A reasonable *way out* from the rather intricated formalistic *quid pro quo*'s above would consist in ignoring the various $(\bar{\nabla})$'s already available definitionally – while still respecting provability, resp. rule-derivability, one way or another – and impose equational conditions on a *primitive* [sic] pair (∇, Δ), where the corresponding operators are supposed to make up an *inversion*, i.e., to satisfy:

$(\beta\Delta)$ $\Gamma \vdash \nabla[A](\Delta[A](c)) = $ c $: \neg\neg$A, if $\Gamma \vdash$ c $: \neg\neg$A, and
$(\eta\Delta)$ $\Gamma \vdash \Delta[A](\nabla[A](a)) = $ a $:$ A, if $\Gamma \vdash$ a $:$ A,

If joined to a primitive minimal $(\rho, \cdot)$-pair supposed to satisfy $(\beta\rho)$ and $(\eta\rho)$ – i.e., to the minimal $\lambda\rho$-calculus, called $\boldsymbol{\lambda\rho}$ here – this yields the appropriate $(\beta\partial)$ and $(\eta\partial)$ conditions for the $(\partial, \star)$-pair, as in:

$(\beta\partial)$ $\vdash$ c $\star$ $(\partial$x.e[x]$)$ = e[x:=c],

---

[24] On might have here room for a *historical speculation*: Why did Frege choose – in the spirit of Chrysippus – both (∇) and (Δ) as *primitives* in **BS**?

$(\eta \partial) \vdash \partial x.(x \star a) = a$, if x is not free in a,

and conversely: if joined to a $(\partial,\star)$-pair satisfing $(\beta\partial)$ and $(\eta\partial)$, the $(\beta$-$\eta\Delta)$-conditions would yield the convenient minimal $(\beta\rho)$ and $(\eta\rho)$-conditions of $\boldsymbol{\lambda\rho}$. In other words, we are able to recover (equationally) the minimal $\boldsymbol{\lambda\rho}[\bot,\to,\neg]$, by using pure $\boldsymbol{\lambda}$ and *genuinely classical* postulates (on *reductio ad absurdum* and the [LNC].

As a bonus, this works in a decoration-free ('type-free') setting. Besides, (Post-) consistency for $\boldsymbol{\lambda\rho}$ is straightforward on this plan, as the resulting 'type-free' systems are, in fact, proper subsystems of the $\lambda\pi$-calculus (Rezuş 2017b).

The *equational proof DN-normalisation* makes official the fact that the inference rule corresponding to the Law of Non-Contradiction [LNC] is *systematically ambiguous* and that it remains so, even if we attempt to make explicit its equational behaviour qua proof operator. Moreover, this *basic ambiguity* is *all-pervasive*, in a way, since any witness (here: 'typed' combinator) of a (classical, resp. intuitionistic or minimal) tautology A using explicitly [LNC] in its derivation won't actually correspond to a single proof-object (witness), but rather to an *infinite set of pairwise non-isomorphic* witnesses of A. The same situation obtains for the witness operators defined explicitly in terms of the '[LNC]-hierarchy' (and, possibly, other operators).

§ 4.2.0 *Examples.* With (ϖ) taken to be a primitive – or else to be defined 'classically' by $\Gamma \vdash \varpi[A](e) := \partial z{:}\neg A.e$ (z not free in e), for $\Gamma \vdash e : \bot$, as above – we have distinct (proof-theoretically non-equivalent) witnesses for the typical intuitionistic tautology $\neg A \to (A \to B)$:

(CO.) $\vdash$ CO[·][A,B] $:= \lambda z{:}\neg A.\lambda x{:}A.\varpi[A](z{\cdot}x) : \neg A \to (A \to B)$, and
(CO$\star$) $\vdash$ CO[$\star$][A,B] $:= \lambda z{:}\neg A.\lambda x{:}A.\varpi[A](z{\star}x) : \neg A \to (A \to B)$,

already in a pure intuitionistic proof-setting, and, in fact, with the $(\cdot_n)$'s defined minimally, as above, infinitely many such pairwise distinct [proof-theoretically non-equivalent], intuitionistically valid proof-objects (witnesses):

(CO·$_n$) $\vdash$ CO[·$_n$][A,B] $:= \lambda z{:}\neg A.\lambda x{:}A.\varpi[A](z \cdot_n x) : \neg A \to (A \to B)$
(n a natural number).

The use of the genuinely intuitionistic (ϖ) [ex falso] is *not essential* here, since one could have had, instead, a package of apparent substitution instances of (CO[·$_n$][A,B]) of the form:

$\vdash$ CO[·$_n$][A,$\neg$B] $:= \lambda z{:}\neg A.\lambda x{:}A.\rho y{:}B.(z \cdot_n x) : \neg A \to (A \to \neg B)$
(n a natural number).

§ 4.2.1 As an aside, in order to make things well-behaved equationally, the accommodation of substitution instances in witness-theoretical (ϖ)-manipulations (in intuitionism) would incidentally require additional – rather ad hoc – conversion-conditions of the form:

($\varpi_\bot$) $\Gamma \vdash \varpi[\bot](e) = e : \bot$, if $\Gamma \vdash e : \bot$,
($\varpi_\neg$) $\Gamma \vdash \varpi[\neg A](e) = \rho y{:}A.e : \neg A$, if $\Gamma \vdash e : \bot$, y not free in e,
($\varpi_\rightarrow$) $\Gamma \vdash \varpi[A{\rightarrow}B](e) = \lambda x{:}A.\varpi[B](e) : A{\rightarrow}B$, if $\Gamma \vdash e : \bot$, x not free in e,

etc., supposed to be intuitionistically valid.

> As they pay attention to specific information about the formula-structure built-in (within) the proof operator (here ϖ) itself, the latter would hardly count as *proper proof-détours*, by the standards of Gentzen 1934–1935 (no 'justifying' int-elim talk therein involved). With a would-be primitive minimal **or** – resp. a minimal *first-order existential quantifier* – in the picture, the required, additional equational patching – in a proof-system for the Glivenko-Heyting-style intuitionism – is even more intricated. See, e.g., Rezuş 1991 for a rather annoying inventory of such accidents. The least thing to say, here, is that the Glivenko-Heyting logic can be hardly characterised – proof-theoretically – by *proper* proof-détours alone, so that, in particular, Gentzen's plan – reductive, in essence – would eventually fail, even on the *Minimalkalkül*.

§ 4.3 *Retrieving minimality.* We can already *revise* our list of required proof operators on the extended primitive signature $[\bot, \rightarrow, \neg]$, i.e. ($\lambda$), ($\triangleright$), ($\bar{\triangleright}$), ($\nabla$), and ($\Delta$), resp., by replacing the rather complex – though minimal – ($\bar{\triangleright}$)-operator (in traditional terms, a case of Contraposition), *either*

> $\boldsymbol{\lambda\rho}[\bot, \rightarrow, \neg]$: with the pair ($\rho, \cdot$) – corresponding to the pair of minimal rules: [$\neg$-introduction] and [LNC] – subjecting it to the expected (β-$\eta\rho$)-conditions of $\boldsymbol{\lambda\rho}[\bot, \rightarrow, \neg]$,

or else, equivalently (**§ 2.3.3**),

> $\boldsymbol{\lambda\bar{\tau}}[\bot, \rightarrow, \neg]$: with the pair of minimally valid transforms ($\bar{\tau}, \tau$) – where (τ) stands for a minimal [LNC], once more – subjecting it to the analogous (β-$\eta\bar{\tau}$)-conditions of $\boldsymbol{\lambda\bar{\tau}}[\bot, \rightarrow, \neg]$.

This kind of revision – essentially minimal, qua provability, resp. rule-derivability – yields a definable operator ($\dot{\nabla}$), sharing the provability behaviour of the generic (∇), but leaves the genuinely classical (Δ)-operator to be further specified equationally.

§ 4.4 *Equationally proof DN-normalised $\lambda\partial$-calculi.* As suggested above, we can, however, retrieve minimality (equationally) on a genuinely classical route, on the top of pure λ, in which case the genuinely classical alternative based on equational proof DN-normalisation takes automatically care of both a primitive (Δ) and a would-be new primitive (∇)-operator (sic), whereby we can do some combinatorial postulate-chopping.

Let us introduce some convenient *shorthand* first. This is based on our previous notational practice. Consider the following ($\boldsymbol{\beta, \eta}$)-pairs:

$$\boldsymbol{\lambda} := (\beta\lambda) + (\eta\lambda), \text{ resp.}$$
$$\boldsymbol{\rho} := (\beta\rho) + (\eta\rho), \ \boldsymbol{\bar{\tau}} := (\beta\bar{\tau}) + (\eta\bar{\tau}),$$
$$\boldsymbol{\partial} := (\beta\partial) + (\eta\partial), \ \boldsymbol{\Delta} := (\beta\Delta) + (\eta\Delta)$$

(expected monotony conditions included)[25], and define combinations: $\boldsymbol{\lambda\rho}$, $\boldsymbol{\lambda\bar{\tau}}$, $\boldsymbol{\lambda\rho\partial}$, $\boldsymbol{\lambda\bar{\tau}\partial}$, $\boldsymbol{\lambda^{\Delta}\rho}$, $\boldsymbol{\lambda^{\Delta}\bar{\tau}}$, resp., as undecorated ('type-free') equational theories (here: λ-calculi) extending the pure λ-calculus $\boldsymbol{\lambda}$.

With $T_1 \prec T_2$ standing for 'T_1 is a (proper) equational sub-system of T_2', and $T_1 \simeq T_2$ standing for 'T_1 and T_2 are equationally equivalent', we have, from § 2.3.3 (resp. § 4.3) and § 4.2:

$$\boldsymbol{\lambda} \prec \boldsymbol{\lambda\rho} \simeq \boldsymbol{\lambda\bar{\tau}} \prec \boldsymbol{\lambda\rho\partial} \simeq \boldsymbol{\lambda\bar{\tau}\partial} \simeq \boldsymbol{\lambda^{\Delta}\rho} \simeq \boldsymbol{\lambda^{\Delta}\bar{\tau}}.$$

As all systems are interpretable in – resp. are fragments of – $\boldsymbol{\lambda\pi}$, (Post-) consistency is straightforward.

> The corresponding *proper* extension(s) of $\boldsymbol{\lambda\rho}[\bot,\to,\neg] \simeq \boldsymbol{\lambda\bar{\tau}}[\bot,\to,\neg]$ is / are (Post-) consistent, because the latter 'typed' λ-calculi have the same 'type-free' equational cognate that is a fragment (an equational sub-system) of the $\lambda\pi$-calculus ('type-free' $\boldsymbol{\lambda\pi}$ contains *infinitely many equationally distinct inversions*, cf. Rezuş 2017b).

While decorating ('typing') accordingly, on the signature $[\bot,\to,\neg]$, we obtain the analogous proof-systems:

$$\boldsymbol{\lambda\rho}[\bot,\to,\neg], \ \boldsymbol{\lambda\bar{\tau}}[\bot,\to,\neg],$$

for minimal logic, resp.

$$\boldsymbol{\lambda\rho\partial}[\bot,\to,\neg], \ \boldsymbol{\lambda\bar{\tau}\partial}[\bot,\to,\neg], \ \boldsymbol{\lambda^{\Delta}\rho}[\bot,\to,\neg], \ \boldsymbol{\lambda^{\Delta}\bar{\tau}}[\bot,\to,\neg],$$

for classical logic.

In the latter case, the stratification (decoration / 'typing') preserves equational (strict) inclusions resp. equational equivalences.[26] The latter four (pairwise equivalent) systems are based on equational proof DN-normalisation and contain a kind of *built-in redundancy* as regards provability, although not equationally.

---

[25] In the terminology of Rezuş 2017b, $\boldsymbol{\lambda}$, $\boldsymbol{\rho}$ and $\boldsymbol{\partial}$ are mono-folds, while $\boldsymbol{\bar{\tau}}$ and $\boldsymbol{\Delta}$ are inversions. For the sake of conceptual uniformity, one may also think an inversion is a kind of *degenerated* (mono-) fold. On the other hand, if we allow infinitary syntax, a (mono-) fold could be accounted for as an appropriately indexed infinite family of inversions. One can probably make the general setting look more *algebraic*, but this goes beyond our present purposes. Incidentally, I discussed the matter publicly in some detail a few years ago (2015), for a TCS-audience, under the heading 'Is the lambda-calculus algebraic?'. The basic reference is the survey Selinger 2002, but see also Lambek & Scott 1986. Algebraically, one may want to ensure the fact that we are concerned with so-called *algebraic classes* (also known as *algebraic varieties*, à la Birkhoff and Mal'cev). In particular, in logic (proof theory) we are dealing with rather *specific monoids*.

[26] Note that, in general, 'typing' need not preserve equational equivalences.

§ 4.5 *The $\lambda\partial$-calculus* $\lambda\partial[\bot,\to,\neg]$. It is easy to see that we have also at hand a weaker way of postulate-chopping, genuinely classical as well, by using the (primitive) pairs $\boldsymbol{\lambda} \equiv (\lambda,\triangleright)$ and $\boldsymbol{\partial} \equiv (\partial,\star)$, subjected to equational conditions $(\beta\lambda)$, $(\eta\lambda)$ and $(\beta\partial)$, $(\eta\partial)$, resp.

The latter solution takes care of both (∇) and (Δ) only in as far provability (resp. rule-derivability) is concerned, *without* equational proof DN-normalisation. However, in this case, the minimal $\rho \equiv (\rho,\cdot)$-team is badly mishandled, equationally speaking, since we loose the conditions $(\beta\rho)$, $(\eta\rho)$, i.e., the minimal sub-system $\lambda\rho$.

Technically, this yields the poor man's *Curry-Howard Correspondence* [CHC] for (quantifier-free) classical logic on the primitive signature $[\bot,\to,\neg]$, i.e., a witness theory $\lambda\partial[\bot,\to,\neg]$, with 'type-free' cognate $\lambda\partial$ (cf. Rezuş 2016, 2017, 2017a, 2017b and **§ 6**, below).

In other words, $\lambda\partial[\bot,\to,\neg]$ does *not* contain (*equationally*) the minimal proof-calculus $\lambda\rho[\bot,\to,\neg] \simeq \lambda\bar{\tau}[\bot,\to,\neg]$, and analogousy for 'type-free' cognates, although the decorated ('typed') system contains a *definable* (∇)-operator corresponding to the minimally valid (∇)-rule.

By the same token, the latter alternative hides the fact that we have, actually – in classical logic – as regards provability / derivability, infinitely many equationally non-equivalent variants of the Law of Non-Contradiction [LNC], thought of as a rule of inference. Theoretically, if viewed as a rule, [LNC] stands, rather, for a rather large package of pairwise non-equivalent proof operators. The system(s) based on equational DN-normalisation yield(s) a proper equational extension of $\lambda\partial[\bot,\to,\neg]$, and make(s) the said [LNC]-ambiguity explicit, from the very beginning.

> In a way, the λ-calculi based on equational proof DN-normalisation are more efficient in practice (if we are going to implement in sofware the corresponding proof-systems, for instance).
>
> Besides, the equational proof DN-normalisation policy yields a natural way of coping with alternative primitive signatures for the (quantifier-free) classical logic, as, e.g., $[\bot,\vee,\neg]$, where $\vee$ stands for the classical inclusive **or** etc. (Cf., e.g., Rezuş 2017b.)
>
> [Homework] At this point (anticipating slightly the next **§**), the industrious readear may find useful taking a pause, by attempting to formulate the quantifier-free segment of Principia Mathematica – based on the primitive (classical) signature $[\bot,\vee,\neg]$ (sic) – in *witness-theoretical terms*, with equational proof DN-normalisation postulates, such as to be able to retrieve the intended classical (proof-) equivalences.

§ 5 Extending the witness-theoretical picture

Consider the extended λ-calculus $\lambda\partial[\bot,\to,\neg]$, of **§ 4.5**. Without equational proof DN-normalisation we have gotten a witness theory (proof-system) for full classical propositional logic, but we have lost the (derivable) equational properties of the minimal pair $\boldsymbol{\rho} \equiv (\rho,\cdot)$, and, in particular – in as far provability is concerned, since the right kind of (ρ) [¬-introduction, in formularian terms] is absent – a natural way of coping with the [¬-introduction]; one has at hand, thereby, only a *roundabout* way of 'introducing' a primitve negation ¬.

We can, however, decide to enlarge the picture and to exploit the fact that classical conjunction (∧) is already available definitionally, on the primitive signature $[(\bot), \to, \neg]$, by setting A ∧ B := ¬(A→¬B).

> So, classically, we have, as expected, ¬(A→B) = (A ∧ ¬B), modulo Double Negation, whence, in formularian terms, [∧-introduction] is a special case of [¬-introduction], in a way.

§ 5.1 The $\lambda\pi$-calculi.
In order to cope properly with the ∧-connective in a ('typed') λ-calculus setting, we shall use an extension of the pure λ-calculus with primitive pairs.

Let us consider first the 'type-free' $\lambda\pi$-calculus, $\boldsymbol{\lambda\pi}$, i.e., the *λ-calculus with 'surjective' pairing*.

In 'type-free' terms, $\boldsymbol{\lambda\pi}$ is a proper extension of the pure λ-calculus $\boldsymbol{\lambda} \equiv \boldsymbol{\lambda}_{\beta\eta}$ – where the $(\lambda,\triangleright)$-pair satisfies the $(\beta\lambda)$ and $(\eta\lambda)$-conditions – which is also equipped with an extensional ('surjective') pairing $\boldsymbol{\pi} := (\pi, \pi_1, \pi_2)$, i.e., with primitive pairs $\prec a, b \succ := \pi(a,b)$ and projections $\mathbf{j}(c) := \pi_j(c)$, j := 1, 2, satisfying the conditions:

$(\beta_j \pi) \vdash \mathbf{1}(\prec a, c \succ) = a, \vdash \mathbf{2}(\prec a, c \succ) = c$ (j := 1,2), and
$(\eta \pi) \vdash \prec \mathbf{1}(f), \mathbf{2}(f) \succ = f$,

the expected monotony conditions for the primitives included.

The extensional $\lambda\pi$-calculus $\boldsymbol{\lambda\pi}$ is known to be (Post-) consistent (Dana S. Scott 1969 [by models], Christian Støvring 2005 ['constructively', by conservativity]) and equationally equivalent to a *C-monoid*, in the sense of Joachim Lambek (Lambek & Scott 1986).

> Essentially, a C-monoid is what we obtain from a cartesian closed category [CCC] by forgetting the (categorical) object-part (the terminal object included). In other words, a C-monoid is just the 'type-free' counterpart of a CCC, i.e., a monoid equipped with a (lifted) extensional pairing and an inversion.

As such, $\boldsymbol{\lambda\pi}$ contains at least one *inversion*, i.e., generically, a pair $\updownarrow := (\uparrow, \downarrow)$ with:

$(\beta \uparrow) \vdash \downarrow(\uparrow(a)) = a,$
$(\eta \uparrow) \vdash \uparrow(\downarrow(c)) = c,$

expected monotony conditions for $\uparrow$ and $\downarrow$ included.

> To see this, pick up, e.g., the usual (curry-ing, un-curry-ing)-pair of $\lambda\pi$-combinators, and take the corresponding λ-lifted terms in the guise of a representative for a paradigmatic inversion.

Of course, the 'type-free' $\lambda\pi^\uparrow := \lambda\pi + (\beta\uparrow) + (\eta\uparrow)$ — consisting of the primitive pairs $\lambda \equiv (\lambda,\triangleright)$, $\uparrow \equiv (\uparrow,\downarrow)$, and of the triple $\pi \equiv (\pi,\pi_1,\pi_2)$ supposed to satisfy the conditions $(\beta\lambda)$, $(\eta\lambda)$, and $(\beta\uparrow)$, $(\eta\uparrow)$, as well as $(\beta_j\pi)$, $(\eta\pi)$, j :=1, 2, resp. (all monotony conditions included) — is, equationally, tantamount $\lambda\pi$.

At 'type-free' level, the relevant facts about $\lambda\pi$ amount to the observation that (Rezuş 2017b):

(1) $\lambda\pi$ contains infinitely many — pairwise non-equivalent — copies of the pure (extensional) λ-calculus λ (and so infinitely many — pairwise non-equivalent — copies of itself),
(2) the extended λ-calculus obtained by joining together two distinct copies of the pure (extensional) λ is consistent — because it is a (proper) subsystem of $\lambda\pi$ — and contains infinitely many — pairwise non-equivalent — inversions.[27]

§ 5.1.1 Incidentally, this yields (Post-) consistency for all calculi ('formal systems') of concern here.

§ 5.2 *The $\partial\pi$-calculi.* We use, next, ad hoc *typographical variants* of $\lambda\pi^\uparrow$ and $\lambda\pi$, called $\partial^\Delta\pi$ and $\partial\pi$, resp., where we impose specific decorations ('typing') on the primitives. The typographical variance consists of the fact we identify the $\lambda \equiv (\lambda,\triangleright)$-pair with the $\partial \equiv (\partial,\star)$-pair (reversing the order, for convenience, while writing down the corresponding 'cuts'; here: $\triangleright$ vs $\star$) and the $\uparrow \equiv (\uparrow,\downarrow)$-pair with the $\Delta \equiv (\Delta,\nabla)$-pair resp., leaving the triple $\pi \equiv (\pi,\pi_1,\pi_2)$ unchanged. (Cf. Rezuş 2017b, for details.) Under the corresponding *art déco* ('typing'), we loose equational equivalence: the decorated ('typed') $\partial^\Delta\pi$ becomes a *proper extension* of the analogously decorated ('typed') $\partial\pi$. In general, however, the initial decoration ('typing') is inherited in subsystems of $\partial^{(\Delta)}\pi$.

In $\partial^{(\Delta)}\pi[\bot,\to,\neg]$, the decorated ('typed') version of the underlying ('surjective') pairing is given by:

$(\pi) \vdash \prec a,c \succ\ :\ \neg(A \to B)$, if $\vdash a : A$ and $\vdash c : \neg B$,
$(\pi_1) \vdash \mathbf{1}(f) : A$, if $\vdash f : \neg(A \to B)$,
$(\pi_2) \vdash \mathbf{2}(f) : \neg B$, if $\vdash f : \neg(A \to B)$,

and is supposed to satisfy the analogous equational conditions $(\beta_j\pi)$, (j := 1, 2), and $(\eta\pi)$:

$(\beta_j\pi) \vdash \mathbf{1}(\prec a,c \succ) = a, \vdash \mathbf{2}(\prec a,c \succ) = c : D$, if $\vdash a : A, \vdash c : \neg B$ (j := 1, 2),
$(\eta\pi) \vdash \prec \mathbf{1}(f), \mathbf{2}(f) \succ\ = f$, if $\vdash f : \neg(A \to B)$,

---

[27] Here, we mean *equational* (non-)equivalence, of course. In the terminology of Rezuş 2017b, (1) a cartesian fold contains an n-fold, for each n > 1, and (2) a 2-fold (di-fold) contains infinitely many distinct inversions (and so infinitely many distinct cyclic mono-folds).

where we have added, for convenience, the intended decorations ('types').

§ 5.2.1 Granted both $\boldsymbol{\lambda} \equiv (\lambda, \triangleright)$ and $\boldsymbol{\partial} \equiv (\partial, \star)$, in $\boldsymbol{\lambda\partial}[\bot, \to, \neg]$, we can even ignore the primitive projections $\mathbf{j}(f)$ ($j := 1, 2$), for a while, and subject directly the pair-construct to the condition:

($\beta\mathbf{p}$) $\vdash \prec a,c \succ \star f = c \star (f \triangleright a) : \bot$,
  where $\vdash a : A$, $\vdash c : \neg B$, $\vdash f : A \to B$, so $\vdash \prec a,c \succ\ : \neg(A \to B)$,

already derivable in $\boldsymbol{\partial\pi}[\bot, \to, \neg]$.

§ 5.2.2 Of course, in $\boldsymbol{\partial\pi}[\bot, \to, \neg]$, one has also:

($\eta\mathbf{p}$) $\vdash g \star f = \mathbf{2}(g) \star (f \triangleright \mathbf{1}(g)) : \bot$, where $\vdash f : A \to B$, and $\vdash g : \neg(A \to B)$.

In view of ($\eta\pi$), the conditions ($\beta\mathbf{p}$) and ($\eta\mathbf{p}$) are equationally equivalent in $\boldsymbol{\partial\pi}$.

§ 5.2.3 Ignoring decorations ('types'), the resulting ('type-free') $\lambda\partial$-calculus, $\boldsymbol{\lambda\partial p} := \boldsymbol{\lambda\partial} + (\beta\mathbf{p})$, is (Post-) consistent since it is a subsystem of $\boldsymbol{\partial\pi}$. From ($\beta\mathbf{p}$), we get:

($\triangleright$) $\vdash f \triangleright a := \partial z{:}\neg B.(\prec a,z \succ \star f) : B$,
  if $\vdash f : A \to B$, $\vdash a : A$, where z [$: \neg B$] is not free in a and f.

§ 5.3 Define now a ('typed') dyadic abstractor $\int[x{:}A,y{:}\neg B].e[x,y]$ [$: A \to B$], where x, y are distinct ('typed') variables – a *split-operator*, so to speak – by:

($\int$) $\vdash \int[x{:}A,y{:}\neg B].e[x,y] := \lambda x{:}A.\partial y{:}\neg B.e[x,y] : A \to B$,
  if $[x{:}A, y{:}\neg B] \vdash e[x,y] : \bot$.

§ 5.3.1 This yields, for ($\int$):

(λ) $\vdash \lambda x{:}A.b[x] = \lambda x{:}A.\partial y{:}\neg B.(y \star b[x]) = \int[x{:}A,y{:}\neg B].(y \star b[x]) : A \to B$,
  if $[x{:}A] \vdash b[x] : B$ (y not free in $b[x]$),

and characteristic equational conditions:

($\beta\int$) $\vdash \prec a,c \succ \star (\int[x{:}A,y{:}\neg B].e[x,y]) = e[x{:}a,y{:=}c] : \bot$,
  if $[x{:}A,y{:}\neg B] \vdash e[x,y] : \bot$, $\vdash a : A$, $\vdash c : \neg B$,
($\eta\int$) $\vdash \int[x{:}A,y{:}\neg B].(\prec x,y \succ \star f) = f$ (with x, y not free in f),
  if $\vdash f : A \to B$,

so, among other things, both (λ) and ($\triangleright$) are available in $\boldsymbol{\partial\pi}$, i.e., $\boldsymbol{\partial\pi}$ contains the usual, pure extensional λ-calculus $\boldsymbol{\lambda}$, while its 'typed' version, $\boldsymbol{\lambda}[\to]$, inherits of the intended decorations from the original 'typing' of $\boldsymbol{\partial}^{(\Delta)}\boldsymbol{\pi}[\bot, \to, \neg]$.

§ 5.3.2 The resulting ('typed') λ-calculus, $\boldsymbol{\lambda\partial p}[\bot, \to, \neg]$, is (Post-) consistent (as so is its 'type-free' version $\boldsymbol{\lambda\partial p}$), and a witness theory for classical propositional logic (based on $[\bot, \to, \neg]$), as well (Rezuş 2017b).

§ 5.4 *The $\partial\!\int$-calculus.* The 'type-free' version of the latter λ-calculus admits of an alternative formulation, without primitive (λ) and ($\triangleright$), since we obtained the pair $\boldsymbol{\lambda} \equiv (\lambda,\triangleright)$, with ($\beta\lambda$) and ($\eta\lambda$), from its subsystem based on $\boldsymbol{\partial} \equiv (\partial,\star)$, and $\boldsymbol{\int} \equiv (\int,\pi)$ as primitives, subjected to the corresponding equational conditions ($\beta\partial$), ($\eta\partial$), ($\beta\!\!\int$), ($\eta\!\!\int$) and ($\beta\mathbf{p}$) (expected monotony conditions included).

Let $\boldsymbol{\partial\!\int}$ be the alternative formulation. Formally, the 'type-free' $\partial\!\int$-calculus is based on the abstract grammar:

witness variables :: x, y, z, ...
witness terms :: a, b, c, d, e, f ::= x | ∂z.e | c$\star$a | $\int$[x.y].e | $\prec$a,c$\succ$

and, *mutatis mutandis*, the equational conditions ($\beta\partial$), ($\eta\partial$), ($\beta\!\!\int$), ($\eta\!\!\int$) above (monotony for the primitives included).

Defining λ, and $\triangleright$ resp. in $\boldsymbol{\partial\!\int}$, by ($\lambda$) and ($\triangleright$) resp. – as in **§ 5.3.1** and **§ 5.2.3**, resp. – yields ($\beta\lambda$) and ($\eta\lambda$), as well as ($\beta\mathbf{p}$), so $\boldsymbol{\lambda\partial p}$ and $\boldsymbol{\partial\!\int}$ are equationally equivalent.

§ 5.4.1 (Post-) consistency for $\boldsymbol{\partial\!\int}$ is straightforward, since it can be interpreted in $\boldsymbol{\partial\pi}$, upon defining $\int$, using projections, in $\boldsymbol{\partial\pi}$, by:

($\int$) ⊢ $\int$[x,y].e[x,y] := ∂z.e[x:=**1**(z),y:=**2**(z)].

§ 5.4.2 Now, $\boldsymbol{\partial\!\int}$ can be decorated ('typed') alternatively, on the primitive classical propositional signature $[\bot,\triangle]$, where $\triangle$ stands for the Peirce-Sheffer **nand**-connective (the negation of conjunction), and one defines, as usual, negation by:

$\sim_\triangle A := A \triangle A$.

Putting (in this subsection only) for convenience, $\neg A \equiv \sim_\triangle A$, one has:

$A \wedge B := \neg(A \triangle B)$, $A \to B := A \triangle \neg B$, $A \leftarrow B := \neg A \triangle B$,
$A \vee B := \neg A \triangle \neg B$, $A \leftrightarrow B := (A \to B) \wedge (A \leftarrow B)$, etc.

The resulting 'typed' calculus $\boldsymbol{\partial\!\int}[\bot,\triangle]$ is a witness theory for classical propositional logic, too, and so is, *mutatis mutandis*, its proper extension $\boldsymbol{\partial\pi}[\bot,\triangle]$ – where we had additional primitive projections – a typographical variant of the $\lambda\pi$-calculus, appropriately decorated ('typed').

§ 5.4.3 Notably, in $\boldsymbol{\partial\!\int}[\bot,\triangle]$, we can simulate projections $\mathbf{j}_\star(c)$, satisfying, *mutatis mutandis*, conditions ($\beta_{\mathrm{j}\star}\pi$) analogous to ($\beta_\mathrm{j}\pi$) above, j := 1,2 (cf. Bonus 2017b). Of course, ($\eta\pi$) is not available for the latter kind of pairing.

§ 5.4.4 In particular, in this setting, the Double Negation reads (definitionally) $\neg\neg A := (A \wedge A)$, whereas the Double Negation tautologies amount to ⊢ $A \to (A \wedge A)$ [DN-introduction] and ⊢ $(A \wedge A) \to A$ [DN-elimination], resp., i.e. to $\wedge$-idempotence, so that the Double Negation rules (∇) and (Δ) are ultimately taken care of by the primitive pair-construct and the definable projections $\mathbf{j}_\star$ (j := 1,2).

As everything is *genuinely classical* from the very beginning, in $\partial\smallint[\bot,\triangle]$, we don't have to worry about minimality and/or intuitionism in as far material implication ($\rightarrow$), say, is concerned, although both (λ) [DT] and ($\triangleright$) [*modus (ponendo) ponens*] [PP] are available definitionally.

§ 5.4.5 Similarly, here, (ϖ) – the genuinely intuitionistic [*ex falso*]-rule – is (as mentioned in **§ 3.3.1**) just a specific instance of the primitive (∂)-operator, whereas the proof-properties of the classical (inclusive) **or** [here: A ∨ B := ¬A △ ¬B] are taken care of by the primitive split-construct ($\smallint$), and it is easy to see that one can simulate trivially, in such terms, the minimal / intuitionistic *injections* and the *proof-by-cases*-construct, as well as the associated proof-properties of the minimal / intuitionistic **or**. Incidentally, with such simulations, extensionality fails for the proof operators associated to the intuitionistic **and** and **or**, in $\partial\smallint[\bot,\triangle]$, and – for the intuitionistic **or** – even in $\partial\pi[\bot,\triangle]$. In fact, the usual presentations of the intuitionistic (resp. minimal) proof theory along the Curry-Howard Correspondence [CHC] ignore such subtleties, and handle only the 'intensional' (β)-cases.

> In proper proof-theoretic terms, extensionality for the proof-operations associated to the intuitionistic – resp. minimal – **or** corresponds to the so-called 'disjunction property'. Otherwise, this is *not characteristic* for logical intuitionism, as the said property holds for many other formal systems located in the *intermediate* regnum, between (formularian) intuitionism and (formularian) classical logic, for instance.

§ 5.5 *The $\partial^\triangle\smallint$- and the $\partial^\triangle\pi$-calculi.* In view of **§ 5.2**, we have a 'type-free' (Post-) consistent extension $\partial^\triangle\smallint$ of $\partial\smallint$, with additional primitives ($\triangle$), (∇) supposed to satisfy an inversion, i.e., a subsystem of $\partial^\triangle\pi$, with $\partial^\triangle\smallint := \partial\smallint + (\beta\triangle) + (\eta\triangle)$.

There are many ways of decorating $\partial^\triangle\smallint$ 'classically'. In fact, on the primitive signature $[\bot,\triangle]$ we have several ways of defining a (classical) negation; e.g., the classical **not**(A) can be viewed either (i) as A △ A or (ii) as A △ ⊤ or (iii) as ⊤ △ A. We may also ignore the alternative definitions and assume – *mutatis mutandis*, as in the case of the redundant $[\bot,\rightarrow,\neg]$-signature – that we have also a *primitive negation* ¬, and decorate $\partial^\triangle\smallint$ on the (extended, redundant) signature $[\bot,\triangle,\neg]$.

The preferred variant of $\partial^\triangle\smallint[\bot,\triangle,\neg]$ has three *parasitary*, definable negations:

$$\sim_\triangle A := A \triangle A, \quad \sim_\rightarrow A := A \triangle \top, \quad \sim_\leftarrow A := \top \triangle A,$$

that might *never* be needed – unless we intend to be very careful as regards the would-be (proof-theoretical) extensions of the game – and a *primitive negation* ¬, used to define the remaining (classical) connectives, *mutatis mutandis*, as usual, by:

$$A \wedge B := \neg(A \triangle B), \quad A \rightarrow B := A \triangle \neg B, \quad A \leftarrow B := \neg A \triangle B,$$
$$A \vee B := \neg A \triangle \neg B, \quad A \leftrightarrow B := (A \rightarrow B) \wedge (A \leftarrow B), \text{ etc.}$$

The fact that the primitive negation $\neg$ – as well as the remaining classical connectives defined in terms of $\neg$ and a primitive **nand**-connective $\triangle$ – are *as intended* (classically, as regards provability, resp. rule derivablility) is ensured by subjecting the proof-theoretical primitives:

$$(\partial), (\star), (\textstyle\int), (\pi), (\triangle), (\nabla)$$

to equational conditions:

$$(\beta\partial), (\eta\partial), (\beta\textstyle\int), (\eta\textstyle\int), (\beta\triangle), (\eta\triangle)$$

resp. (expected monotony conditions included), assuming thus equational proof DN-normalisation.

> Recall that the negation used in decorating ('typing') the proof-theoretical primitives – including $(\triangle)$, (∇), resp. $(\beta\triangle)$, $(\eta\triangle)$ – is the *primitive* negation $\neg$.

Obviously, the only real novelty over $\partial\!\!\int [\bot,\triangle]$ concerns, here, the equational normalisation of the $(\triangle,\nabla)$-pair, advantages and disadvantages included.

In fact – at 'type-free' level – $\partial^{\triangle}\!\!\int$ is a proper subsystem of $\partial^{\triangle}\pi$ (whence, once more, consistency), so that – as regards decorated ('typed') systems – the proper comparison consists of contrasting $\partial\!\!\int[\bot,\triangle,\neg]$ with $\partial^{\triangle}\pi[\bot,\triangle,\neg]$, i.e., with a *proper* extension of it.

> Adding projections π_j to $\partial\!\!\int[\bot,\triangle,\neg]$, with the 'surjective' pairing conditions $(\beta_j\pi)$, $(\eta\pi)$, $j := 1,2$, yields, in fact, an equationally redundant version of $\partial^{\triangle}\pi[\bot,\triangle,\neg]$.

One of the significant technical details to be noticed here consists of the fact that, unlike $\partial\pi$ and its (pseudo-) extension $\partial^{\triangle}\pi$, the $\partial\!\int$-calculi $\partial(^{\triangle})\!\!\int$ do not have a definable 'surjective' pairing.[28]

§ 5.6 ζ-extensions. Note that, in ('type-free') $\partial\pi$, one can derive the ζ-condition (Rezuş 2017b):

$$(\zeta\partial_\rightarrow) \vdash \partial z.e[z] = \lambda x.\partial y.e[z:=\prec x,y\succ] \equiv \textstyle\int[x,y].e[z:=\prec x,y\succ],$$

in view of $(\eta\pi)$, while, in its ('type-free') pseudo-extension $\partial^{\triangle}\pi$, one has also:

$$(\zeta\partial_\neg) \vdash \partial z.e[z] = \rho z.e[z:=\nabla(z)],$$

from $(\eta\triangle)$, by definition, since $\rho z.e[z] := \partial z.e[z:=\triangle(z)]$.

So, on the extended signature $[\bot,\rightarrow,\neg]$, one can decorate, in $\lambda^{\triangle}\partial p[\bot,\rightarrow,\neg] := \lambda\partial p[\bot,\rightarrow,\neg] + (\beta\triangle) + (\eta\triangle)$:

$$(\zeta\partial_\neg)\ \Gamma \vdash \partial z{:}\neg\neg A.e[z] = \rho z{:}\neg[A].e[z:=\nabla[A](z)] : \neg A,$$
for $\Gamma\,[z{:}\neg\neg A] \vdash e\,[x] : \bot$, and

---

[28] In the terminology of Rezuş 2017b, all interesting subsystems of $\partial\pi$ are *subcartesian*, i.e., one can only define a pairing that does not satisfy the $(\eta\pi)$-condition ('surjectivity'). The remark applies to decorated ('typed') systems, as well.

$(\zeta\partial_\rightarrow)$ $\Gamma \vdash \partial z{:}\neg(A{\rightarrow}B).e[z] = \lambda x{:}A.\partial y{:}\neg B.e[z{:}{=}\prec x,y\succ] : A \rightarrow B$,
for $\Gamma [z{:}\neg(A{\rightarrow}B] \vdash e[x] : \bot$,

while, *mutatis mutandis*, on the extended signature $[\bot,\triangle,\neg]$, one can decorate, analogously, in $\partial^\triangle \smallint[\bot,\triangle,\neg]$:

$(\zeta\partial_\neg)$ $\Gamma \vdash \partial z{:}\neg\neg A.e[z] = \rho z{:}\neg[A].e[z{:}{=}\nabla[A](z)] : \neg A$,
for $\Gamma [z{:}\neg\neg A] \vdash e[x] : \bot$, as above, and

$(\zeta\partial_\rightarrow)$ $\Gamma \vdash \partial z{:}(A{\wedge}B).e[z] = \smallint[x{:}A,y{:}B].e[z{:}{=}\prec x,y\succ] : A \triangle B$,
for $\Gamma [z : (A{\wedge}B) \vdash e[z] : \bot$.

§ 5.6.1 Now, 'type-free' equational λ-theories are (consistently) *extendible*, and, in particular, so is $\lambda\pi$ (and its proper subsystems), whereas this is even more frequently the case for their decorated ('typed') cognates.

There is a huge λ-calculus lore on this, in print, inspired essentially by the the previous work of Dana S. Scott (1969 and later) on λ-models, partly documented now in Barendregt *et al.* 2013. For relevant proof-theoretical issues, see the review Rezuş 2015a.

Let us focus, for a moment, on $\partial^\triangle \smallint[\bot,\triangle,\neg]$ (where $\top \equiv \neg\bot$) versus $\partial\smallint[\bot,\triangle]$ (where we had $\top_\triangle := \sim_\triangle \bot \equiv \bot \triangle \bot \not\equiv \top$).

The tautology $\top_\triangle$ has many equationally non-equivalent witnesses in $\partial\smallint[\bot,\triangle]$; at least two, as, e.g., those inherited from its proper extension $\partial\pi[\bot,\triangle]$:

$(\boldsymbol{\Omega}_j)$ $\vdash \boldsymbol{\Omega}_j := \smallint[x_1{:}\bot,x_2{:}\bot].x_j : \top_\triangle$, $(j := 1, 2)$.

On the other hand, under equational DN-normalisation – as, for instance, in $\partial^\triangle \smallint[\bot,\triangle,\neg]$, as well as in $\partial^\triangle \pi[\bot,\triangle,\neg]$ – we have also (ρ), and thus $\boldsymbol{\Omega} := \rho x{:}\bot.x$ – a paradigmatic (minimal) witness for $\top$ – and, as both $\bot$ and $\neg$ count as primitives, in this setting, nothing prevents us to make $\boldsymbol{\Omega}$ unique, in $\partial^\triangle \smallint[\bot,\triangle,\neg]$, by stipulating – beyond 'type-free' λ-calculi – that:

$(\top)$ $\vdash a = \boldsymbol{\Omega}$, if $\vdash a : \top^{29}$,

adding, moreover, something like:

$(\zeta\partial_\bot)$ $\Gamma \vdash \partial z{:}\top.e[z] = e[z{:}{=}\boldsymbol{\Omega}] : \bot$, if $\Gamma [z : \top] \vdash e[z] : \bot$.

Altogether, we can add consistently – to $\partial^\triangle \smallint[\bot,\triangle,\neg]$, for instance – $(\zeta\partial_\bot)$, $(\zeta\partial_\neg)$, $(\zeta\partial_\rightarrow)$, so that we can also get, in particular, their degenerated – intuitionistically valid – counterparts $(\varpi_\bot)$, $(\varpi_\neg)$, $(\varpi_\rightarrow)$, resp., mentioned in **§ 4.2.1**.

---

[29] This is vaguely reminiscent of a definitional stipulation on terminal objects in CCC's.

Alternatively, one can restrict the primitive (∂)-decoration ('typing') to atoms, and take ($\zeta\partial_\bot$), ($\zeta\partial_\neg$), ($\zeta\partial_\to$) as *definitional* stipulations. One can also consider similar ζ-extensions of 'typed' $\boldsymbol{\lambda\partial}\mathbf{p}[\bot,\triangle,\neg]$, of course. See also Rezuş 2017a, 2017b, 2017e. For an analogous strategy, as applied to ('inferential') $\lambda\gamma$-calculi, based on Prawitz 1965, and meant to recover the 'inferential' intuitionistic ϖ–conditions, see Rezuş 1990, 1991.

§ 5.6.2 (1) One can generalise ($\zeta\partial_\to$) in the obvious way. Set first $\mathbf{A} \to \mathbf{B} := A_1 \to (A_2 \ldots (A_n \to B)\ldots)$, for $n > 0$ fixed. Iterate, next, the (primitive) pair-construct to the right:

$\Gamma \vdash \prec x_1, y \succ\, := \,\prec x_1, y \succ\, : \neg(A \to B)$,
for $\Gamma \vdash x_1 : A_1$, $\Gamma \vdash y : \neg B$,
$\vdash \prec x_1, x_2, \ldots, x_n, y \succ\, := \,\prec x_1, \prec x_2, \ldots, x_n, y \succ \succ\, : \neg(\mathbf{A}\to\mathbf{B})$,
for $\Gamma \vdash x_i : A_i$, $\Gamma \vdash y : \neg B$, $0 < i < n+1$,

and let, for $n > 0$ fixed,

$\Gamma \vdash \prec \mathbf{x}, y \succ\, := \,\prec x_1, x_2, \ldots, x_n, y \succ\, : \neg(\mathbf{A} \to \mathbf{B})$,
for $\Gamma \vdash x_i : A_i$, $\Gamma \vdash y : \neg B$, $0 < i < n+1$, and
$\Gamma \vdash \lambda\mathbf{x}{:}\mathbf{A}.b[\mathbf{x}] := \lambda x_1{:}A_1. \ldots, \lambda x_n{:}A_n.b[x_1, \ldots, x_n] : \mathbf{A} \to \mathbf{B}$,
for $\Gamma\, [x_1{:}A_1, \ldots, x_n{:}A_n] \vdash b[x_1, \ldots, x_n] : B$.

Then

($\zeta\partial_\to^n$) $\Gamma \vdash \partial z{:}\neg(\mathbf{A}\to\mathbf{B}).e[z] = \lambda\mathbf{x}{:}\mathbf{A}.\partial y{:}\neg B.e[z{:=}\prec\mathbf{x},y\succ] : \mathbf{A} \to \mathbf{B}$,
 for $\Gamma\, [z{:}\neg(\mathbf{A}\to\mathbf{B})] \vdash e[x] : \bot$,

is derivable in $\boldsymbol{\lambda\partial}\mathbf{p}[\bot,\to,\neg]$.
(2) *Mutatis mutandis*, the 'type-free' variant:

($\zeta\partial^n$) $\vdash \partial z.e[z] = \lambda\mathbf{x}.\partial y.e[z{:=}\prec\mathbf{x},y\succ]$

is derivable in $\boldsymbol{\lambda\partial}\mathbf{p}$, resp. in $\boldsymbol{\partial\pi}$.
(3) Set, in $\boldsymbol{\lambda\partial}\mathbf{p}$, $\epsilon z.a[z] := \partial z.(z\star a[z])$, resp. $\varpi(e) := \partial z.e$, if z is not free in e, and, analogously, for decorated ('typed') witness-terms, in $\boldsymbol{\lambda\partial}\mathbf{p}[\bot,\to,\neg]$:

$\Gamma \vdash \epsilon z{:}\neg A.a[z] := \partial z{:}\neg A.(z\star a[z]) : A$,
for $\Gamma\, [x{:}\neg A] \vdash a[z] : A$, resp.
$\Gamma \vdash \varpi[A](e) := \partial z{:}\neg A.e : A$,
for $\Gamma\, [z : \neg A] \vdash e : \bot$, if z is not free in e.

We have, for $n > 0$:

($\zeta\epsilon^n$) $\vdash \epsilon z.a[z] = \lambda\mathbf{x}.\epsilon y.a[z{:=}\prec\mathbf{x},y\succ](x_1)\ldots(x_n) \equiv$
  $\lambda x_1. \ldots .\lambda x_n.\epsilon y.a[z{:=}\prec\mathbf{x},y\succ](x_1)\ldots(x_n)$, ($\mathbf{x}$ fresh for $a[z]$), resp.
($\zeta\varpi^n$) $\vdash \varpi(e) = \lambda\mathbf{x}.\varpi(e) \equiv \lambda x_1. \ldots .\lambda x_n.\varpi(e)$, ($\mathbf{x}$ fresh for e),

in $\boldsymbol{\lambda\partial}\mathbf{p}$, and

($\zeta\epsilon_\rightarrow^n$) $\Gamma \vdash \epsilon z{:}\neg(\mathbf{A}{\rightarrow}\mathbf{B}).a[z] = \lambda\mathbf{x}.\epsilon y{:}\neg\mathbf{B}.a[z{:}{=}\prec\mathbf{x},y\succ](x_1)...(x_n) \equiv$
$\lambda x_1{:}A_1.\ ...\ .\lambda x_n{:}A_n.\epsilon y{:}\neg\mathbf{B}.a[z{:}{=}\prec\mathbf{x},y\succ](x_1)...(x_n)$
: $\mathbf{A}{\rightarrow}\mathbf{B}$,
for Γ [z${:}\neg(\mathbf{A}{\rightarrow}\mathbf{B})] \vdash a[z] : \mathbf{A}{\rightarrow}\mathbf{B}$, ($\mathbf{x}$ fresh for a[z]), resp.

($\zeta\varpi_\rightarrow^n$) $\Gamma \vdash \varpi[\mathbf{A}{\rightarrow}\mathbf{B}](e) = \lambda\mathbf{x}.\varpi[\mathbf{B}].(e) \equiv \lambda x_1{:}A_1.\ ...\ .\lambda x_n{:}A_n.\varpi[\mathbf{B}].(e) : \mathbf{A}{\rightarrow}\mathbf{B}$,
for Γ [z${:}\neg(\mathbf{A}{\rightarrow}\mathbf{B})] \vdash e : \bot$, ($\mathbf{x}$, z fresh for e),

in $\boldsymbol{\lambda\partial p}[\bot,{\rightarrow},\neg]$.

As noted elsewehere (Rezuş 2017a, 2017b), one can recover the (∂)-operator of $\boldsymbol{\lambda\partial p}$, resp. $\boldsymbol{\lambda\partial p}[\bot,{\rightarrow},\neg]$ – i.e., [*reductio ad absurdum*] – from (ϵ) and (ϖ).

§ 5.7 The $\boldsymbol{\lambda\gamma}$-calculi II: the full calculus $\boldsymbol{\lambda\gamma}$ (Rezuş 1990).

Let, in $\boldsymbol{\partial\pi}$, $\dot{\tau}(c) := \lambda x.(c{\star}x)$ and $\gamma z.e[z] := \partial z.e[z{:}{=}\dot{\tau}(z)]$, resp. $\epsilon z.a[z] := \gamma z.(z{\triangleright}a[z])$. We have thus $\varpi(e) := \partial z.e \equiv \gamma z.e$, if z is not free in e.

It is an easy exercise to obtain the characteristic equational conditions of the (full 'type-free') $\lambda\gamma$-calculus $\boldsymbol{\lambda\gamma}$ (Rezuş 1990, 1991, 1993), viz., *mutatis mutandis*:

($\hat{\beta}\gamma$) $\vdash \gamma x.(x{\triangleright}(\gamma y.e[x,y])) = \gamma z.e[x,y{:}{=}z]$ [surface diagonalisation],
($\eta\gamma$) $\vdash \gamma x.(x{\triangleright}a) = a$, x not free in a [extensionality for inferential *reductio*],
($\hat{\xi}\gamma$) $\vdash \gamma x.c = \gamma x.d \Rightarrow\ \vdash \gamma x.(f{\triangleright}c) = \gamma x.(f{\triangleright}d)$

of **§ 2.2** above, as well as:

($\zeta\gamma$) $\vdash \gamma z.e[z] = \lambda x.\gamma y.e[z{:}{=}\mu(x,y)]$, where $\mu(x,y) := \lambda z.y{\triangleright}(z{\triangleright}x)$,
(x, y fresh for e[z]), and, in general, for n > 0,
($\zeta\gamma^n$) $\vdash \gamma z.e[z] = \lambda\mathbf{x}.\gamma y.e[z{:}{=}\mu(\mathbf{x},y)]$, where $\mu(\mathbf{x},y) := \lambda z.y{\triangleright}(z{\triangleright}x_1...{\triangleright}x_n)$,
($\mathbf{x}$, y fresh for a[z]).

The intended decoration ('typing') of $\boldsymbol{\partial\pi}[\bot,{\rightarrow},\neg]$, yields

(γ) $\Gamma \vdash \gamma x{:}{\sim}A.e : A$, if Γ [x : ${\sim}A$] $\vdash e : \bot$,

for the γ-abstractor, and:

(ε) $\Gamma \vdash \varepsilon x{:}{\sim}A.a : A$, if Γ [x : ${\sim}A$] $\vdash a :A$,
(ϖ) $\Gamma \vdash \varpi[A](e) : A$, if Γ [x : ${\sim}A$] $\vdash e : \bot$, x not free in e,

for the other two operators.

Note that (ε) differs from (ϵ) in the same way (γ) differs from (∂), while (ϖ) is the degenerated variant of both.

In the resulting 'typed' calculus $\boldsymbol{\lambda\gamma}[\bot,{\rightarrow}]$, conditions analogous to ($\zeta\epsilon_\rightarrow$), resp. ($\zeta\epsilon_\rightarrow^n$) obtain, *mutatis mutandis*, for (ε), n > 0.

As in the case of (∂) in $\boldsymbol{\lambda\partial p}$ – resp. in $\boldsymbol{\lambda\partial p}[\bot,{\rightarrow},\neg]$ – one can recover (γ) in $\boldsymbol{\lambda\gamma}$ – resp. in $\boldsymbol{\lambda\gamma}[\bot,{\rightarrow}]$ – from (ε) and (ϖ).

See, in detail, Rezuş 2017a, for the characteristic equational conditions of $\boldsymbol{\lambda\varepsilon\varpi} \simeq \boldsymbol{\lambda\gamma}$ versus $\boldsymbol{\lambda\epsilon\varpi} \simeq \boldsymbol{\lambda\partial}$.

§ 5.8 Although this goes beyond our use of ('type-free') $\partial\pi$ (i.e., ultimately, $\lambda\pi$), one can likely extend equationally the 'typed' calculus $\lambda\partial p[\bot,\to,\neg]$, with something like:

($\zeta\partial_\bot$) $\Gamma \vdash \partial z{:}\top.e[z] = e[z{:}{=}\Omega]$, if $\Gamma\ [z{:}\top] \vdash e[z] : \bot$, where $\vdash \Omega : \top$,

as in § 5.6.1.

The analogous condition for $\lambda\gamma[\bot,\to]$ could have been (cf. Rezuș 1990, 1991):

($\zeta\gamma_\bot$) $\Gamma \vdash \gamma z{:}{\sim}\bot.e[z] = e[z{:}{=}\lambda x{:}\bot.x]$, if $\Gamma\ [z : {\sim}\bot] \vdash e[z] : \bot$.

The latter is, however, *not* a special case of the former, as we don't have, without further provisions,

$$\vdash \dot{\tau}(\Omega) \equiv \lambda x{:}\bot.(\Omega{\star}x) = \lambda x{:}\bot.x,$$

in $\lambda\partial p[\bot,\to,\neg]$.

For the degenerated case – which amounts to the apparently harmless, intuitionistically specific (equational) postulate:

($\varpi_\bot$) $\Gamma \vdash \varpi[\bot](e) = e$, for $\Gamma \vdash e : \bot$

– this is immaterial, as $\Gamma \vdash \varpi[\bot](e) \equiv \partial z{:}\top.(e) \equiv \gamma z{:}\top.(e) : \bot$, for x not free in e, and $\Gamma \vdash e : \bot$.

§ 6 Frege's quantifiers and Hilbert's *epsilons*

Except for the fact that he did not insist with too many formal details on the status of the first-order terms and on substitution of first-order individual variables in first-order terms resp. formulas, Frege's first-order quantification theory of **BS** is pretty much similar to our way of understanding the subject, today: there is, practically, nothing to change.

> Actually, as Gödel incidentally noticed a while ago (1944), Frege's quantification theory is superior to the Russell-Whitehead way of handling the matter in Principia Mathematica, a few decades later.

With syntax specified as under § 1, the atomic formulas are of the form P[**t**] – with **t** := $t_1, ..., t_n$ a finite sequence of individual terms – arising from 'matrices' of the form P[**u**], where P is an n-ary predicate letter and **u** := $u_1, ..., u_n$ is a finite sequence of pairwise distinct individual variables (incidentally, Frege would take only n := 1 or n := 2, in examples), distinguishing notationally between free and bound individual variables in that the former are ranged over by Latin lowercase letters while the latter are represented by German lowercase letters. Metavariables were distinguished accordingly, using uppercase Greek and Latin types. In **GGA**, one encounters more distinctions, a veritable *Buchstabenparadies* (cf. Rezuș 2009), because (i) the underlying 'language' was a higher-order language and, mainly, because, in the meantime, (ii) Frege managed to elaborate heavily on semantics.

Hereafter, we denote *mutatis mutandis*, substitutions for individual variables by a[u:=t] (in witness-terms) and A[u:=t] (in formulas) resp.

§ 6.1 Technically, in his **BS**-logic, Frege had, in our terms, (1) a *Generalisation* rule [∀-i] [∀-introduction], stated informally, and (2) an axiom scheme equivalent – granted the [PP] primitive rule – to the *Instantiation* rule [∀-e] [∀-elimination]).

> In **BS**, the formal status of [∀-i] is much similar to the situation of [DT] [the Deduction Theorem].

After disambiguation, we are left with two distinct proof operators: a monadic abstraction operator (Λ) and a binary 'cut' operator ($\blacktriangleright$), resp. – with term-forms Λu.a[u] and f$\blacktriangleright$t, resp. – matching formally, *mutatis mutandis*, the $\lambda \equiv (\lambda,\triangleright)$-pair.

Let, for convenience, $\boldsymbol{\Lambda} := (\Lambda,\blacktriangleright)$. In 'type-free' terms, the $\boldsymbol{\Lambda}$-pair is supposed to be a replication of the $\boldsymbol{\lambda}$-pair, i.e., to be subjected to equational conditions:

($\beta\Lambda$) ⊢ (Λu.a[u]) $\blacktriangleright$ t = a[u:=t], and
($\eta\Lambda$) ⊢ Λu.(f$\blacktriangleright$u) = f, if u is not free in f,

including the expected monotony conditions for both (Λ) and ($\blacktriangleright$).

As regards consistency-matters, where T[$\boldsymbol{\lambda}$] is an equational extension of $\boldsymbol{\lambda}$, if T[$\boldsymbol{\lambda}$] is (Post-) consistent, so is T[$\boldsymbol{\Lambda}$] := T[$\boldsymbol{\lambda}$] + $\boldsymbol{\Lambda}$ (by an obvious translation argument).

Now, the decorated ('typed') version of T[$\boldsymbol{\Lambda}$] on a primitive signature including [(⊥),→,∀] is obtained from the correspondingly 'typed' T[$\boldsymbol{\lambda}$] by adding the decoration ('typing'):

(Λ) Γ [u] ⊢ a[u] : A[u] ⇒ Γ ⊢ Λu.a[u] : ∀u.A[u],
  provided u is fresh for Γ, [Generalisation] [∀-introduction]
($\blacktriangleright$) Γ ⊢ f : ∀u.A[u] ⇒ Γ ⊢ f $\blacktriangleright$ t : A[u:=t]
  [Instantiation] [∀-elimination].

Let, for convenience, T[$\boldsymbol{\lambda\Lambda}$][⊥,→,¬,∀] be the 'typed' cognate of T[$\boldsymbol{\lambda\Lambda}$]. The 'typed' calculus is (Post-) consistent, since so is it 'type-free' cognate.

As both (Λ) and ($\blacktriangleright$) are minimally valid rules of inference, this takes care of (Post-) consistency for the first-order extensions of all witness theories discussed thus far, including the corresponding fragments of the proof-calculi for the Johansson *Minimalkalkül* and the Glivenko-Heyting intuitionistic logic.

§ 6.1.0 As mentioned earlier, Frege had an axiom scheme in **BS**, viz. something like:

(Cl$_v^{\forall u}$) Γ [v] ⊢ Cl$_v^{\forall u}$[A[u]] : ∀u.A[u] → A[v], v fresh for Γ and A[u],

in our terms, in place of [Instantiation] ($\blacktriangleright$), whence also a 'first-order combinator':

$(Cl^\forall)\ \vdash Cl^\forall[A] : \Lambda v.(\lambda f{:}\forall u.A[u].(f \blacktriangleright v)) : \forall v.\forall u.A[u] \to A[v]$.

§ 6.2 *Witnessing by examples and counter-examples: minimal / intuitionistic pseudo-existence.* In **BS**, Frege defines the (classical) existential quantifier ∃ – with formula-form $\exists u.A[u]$ – classically, by 'duality', i.e., as $\exists u.A[u] := \neg\forall u.\neg A[u]$.

This kind of 'existence' is, of course, available in (first-order) *Minimalkalkül*, as well as in (formularian) Glivenko-Heyting intuitionistic logic.

Proof-theoretically, the minimal / intuitionistic version is weird even by classical standards, a fact noticed – at least implicitly – by Kolmogorov 1925 (he used to speak about *pseudo-concepts* in this context).

David Hilbert addressed the problem, indirectly and somewhat *avant la lettre* – around 1905 or so – as he used to think about (first- resp. higher-order) quantifiers as being *infinitistic* ingredients in logic, while Kolmogorov payed attention to what Hilbert had to say on the matter.

In order to have at hand a clear conceptual distiguo between the minimal / intuitionistic concept and its classical versions, we shall write $\dot\exists$ resp. $\dot\exists u.A[u]$ for the $\mathcal{M}$ resp. the $\mathcal{H}$-variant, keeping in mind that, in both $\mathcal{M}$ resp. $\mathcal{H}$, we have a *primitive* dotless-version, as well.

The fact is that we can witness minimally – resp. intuitionistically – $\dot\exists$-statements is immediate, even at plain provability level.

And we can also do this by genuinely classical means, too.

In such cases, the required witness-theoretical construct is a *mixed* pair-operator, with term-form [t,a], where a is a witness for A[t], for an individual term t, specified in advance, the 'witness-example' in point.

> Classically, one would rather have 'counter-examples', as well as – more significantly – 'no-counter-examples' in point.

Here, we have set $A[t] \equiv A[u{:=}t]$, as ever.

We shall write, next, $\downarrow_t(a)$ ($\equiv [t,a]$) for this kind of pairs, distinguishing among minimal / intuitionistic and classical concepts.

One has, first, a *minimally definable* notion:

$(\downarrow^{\dot\exists})\ \Gamma \vdash \downarrow_t^{\dot\exists}(a) := \rho z{:}\forall u.\neg A[u].((z \blacktriangleright t) \cdot a) : \dot\exists u.A[u] \equiv \neg\forall u.\neg A[u]$,
   if $\Gamma \vdash a : A[t]$, where z is not free in a and t,

in $\lambda\rho\Lambda$, with both (ρ) and ($\cdot$), as well as ($\blacktriangleright$), as primitives. From this, we get, by ($\beta\rho$):

$(\beta\downarrow^{\dot\exists})\ \Gamma \vdash \downarrow_t^{\dot\exists}(a) \cdot f = (f \blacktriangleright t) \cdot a : \bot$,
   for $\Gamma \vdash f : \forall u.\neg A[u]$ and $\Gamma \vdash a : A[t]$.

Next, we have, analogously, *classically definable* notions:

$(\downarrow_n^{\dot\exists})\ \Gamma \vdash \downarrow_t^{\dot\exists}(a) := \rho z{:}\forall u.\neg A[u].((z \blacktriangleright t) \cdot_n a) : \exists u.A[u] \equiv \neg\forall u.\neg A[u]$,
   if $\Gamma \vdash a : A[t]$, where z is not free in a and t,

in $\boldsymbol{\lambda\partial\Lambda}$, with ($\rho$) and ($\cdot_n$), n $\geq$ 0, definable from the classical primitives (∂) and ($\star$), by:

$\Gamma \vdash \rho$x:A.e[x] := ∂x:¬¬:A.e[x:=Δ(x)] : ¬A, for Γ [x:A] $\vdash$ e[x] : $\bot$,
as in **§ 4.1**,
$\Gamma \vdash \nabla$(a) := ρx:¬A.x$\star$a (x not free in a), and
$\Gamma \vdash$ c $\cdot_0$ a := c $\cdot$ a := ∇(a) $\star$ c : $\bot$,
$\Gamma \vdash$ c $\cdot_{n+1}$ a := ∇(a) $\cdot_n$ c : $\bot$, for $\Gamma \vdash$ c : ¬A, $\Gamma \vdash$ a : A,
on the pattern of **§ 3.4.1**.

So, classically, we have infinitely many distinct ways of witnessing 'pseudo-existence', and it is obvious that, without additional provisions in $\boldsymbol{\lambda\partial\Lambda}$, the equational behaviour of the defined witness operators ($\downarrow_n^\exists$), n $\geq$ 0, is rather erratic. (Otherwise, many other variations on this are possible.)

Of course, minimally (and intuitionistically), in $\mathcal{M}$ (resp. in $\mathcal{H}$), we have also a *primitive* notion, corresponding to the minimal / intuitionistic *witness-examples*: we prove $\exists$u.A[u] by displaying an example t such that A[t] holds (i.e., formally, such that one has a witness for A[t]), viz.:

($\downarrow^\exists$) $\Gamma \vdash \downarrow_t^\exists$(a) : $\exists$u.A[u], if $\Gamma \vdash$ a : A[t],

with minimal / intuitionistic $\exists$ as a primitive. This is, however, of limited interest for the moment being: we shall come back to it later.

On the other hand, in a genuinely classical setting, one may also have a *primitive* 'mixed'-pair construct, namely:

($\downarrow$) $\Gamma \vdash \downarrow_t$(a) : ¬$\forall$u.A[u], if $\Gamma \vdash$ a : ¬A[t],

corresponding, intuitively, to the classical reasoning by *counter-examples*: we *disprove* $\forall$u.A[u] by supplying a counter-example to it, namely an individual term t such that ¬A[t] holds (i.e., formally, such that one has a witness for ¬A[t]): old wise Aristotle was already familiar with this proof-pattern.

Note that the special case of the minimal / intuitionistic ($\downarrow^\exists$) – obtained by substituting ¬A for A – yields only a counterexample for $\forall$u.¬¬A[u]. This does not pose special problems to the minimalist, as the said special case yields someting like:

($\downarrow^{\bar\forall}$) $\Gamma \vdash \downarrow_t^{\bar\forall}$(c) := ρz:$\forall$u.A[u].(c$\cdot$(z$\blacktriangleright$t)) : ¬$\forall$u.A[u],
    if $\Gamma \vdash$ c : ¬A[t] (z not free in c and t),

already in $\boldsymbol{\lambda\rho\Lambda}[\bot,\rightarrow,\neg,\forall]$ – with primitive (ρ), ($\cdot$), and ($\blacktriangleright$) – a perfectly unobjectionable minimal / intuitionistic derived notion, whence, by ($\beta\rho$) and ($\beta\Lambda$):

($\beta\downarrow^{\bar\forall}$) $\Gamma \vdash \downarrow_t^{\bar\forall}$(c) $\cdot$ f = c $\cdot$ (f $\blacktriangleright$ t),
    for $\Gamma \vdash$ f : $\forall$u.A[u], and $\Gamma \vdash$ c : ¬A[t], and also
($\beta\downarrow\Lambda$) $\Gamma \vdash \downarrow_t^{\bar\forall}$(c) $\cdot$ Λu.a[u] = c $\cdot$ a[t],
    for Γ [u] $\vdash$ a[u] : A[u], and $\Gamma \vdash$ c : ¬A[t].

So, ultimately, 'reasoning by counter-examples' should be minimally – and thus intuitionistically – 'reliable' (although we might thereby go into – unsuspected – *infinitistic* complications, in the background).

§ 6.3 *Minimal and intuitionistic existence.* Putting things in a (historically) counterfactual prospect, if the minimalist (resp. the intuitionist) could afford herself to use the classical pseudo-concept of existence (here: $\dot\exists$), she won't need going too far beyond $\boldsymbol{\lambda\rho\Lambda}[\bot,\to,\neg,\forall]$. From a genuinely Brouwerian standpoint, there is, however, a subtle point here, namely: rejecting –'constructively', i.e., by explicit witnessing – the evidence for $\neg\neg\forall u.\neg A[u] \equiv \neg\dot\exists u.A[u]$ is *not* the same thing as providing explicit evidence for $\exists u.A[u]$, understood as a genuinely minimal / intuitionistic proposition.

This makes things slightly involved: the minimalist (resp. the intuitionist) would eventually need:

(1) a *specific primitive* proof operator $\downarrow^\exists$, resp. a pair-construct, $\downarrow^\exists_t(a)$, with $\Gamma \vdash \downarrow^\exists_t(a) : \exists u.A[u]$, for $\Gamma \vdash a : A[t]$, in the guise of an 'introduction' rule for the genuinely minimal (intuitionistic) $\exists$, as well as
(2) a *specific* way of 'eliminating' the genuinely minimal (intuitionistic) $\exists$-witnesses from proofs.

Minimal (intuitionistic) '$\exists$-elimination' is done (under CHC; see also Rezuş 2019b) by a genuine witness operator – a complex abstraction operator – $\mathcal{E}$, with term-form $\mathcal{E}[u,x:A].(f,c[u,x])$ [u, x not free in f] such that:

($\mathcal{E}$) $\Gamma \vdash \mathcal{E}[u,x:A].(f,c[u,x]) : C$,
  if Γ [u] [x: A[u]] $\vdash$ c[u,x] : C, and $\Gamma \vdash f : \exists u.A[u]$,

subjected to the obvious restrictions on bound and free variables, and 'characteristc' equational behaviour given by:

($\beta\downarrow^\exists$) $\Gamma \vdash \mathcal{E}[u,x:A].(\downarrow^\exists_t(a),c[u,x]) = c[u:=t,x:=a] : C$,
  if $\Gamma \vdash a : A[t]$, and Γ [u] [x: A[u]] $\vdash$ c[u,x] : C.

> There is also an (η)-condition, ($\eta\downarrow^\exists$), which will be ignored here, corresponding to the so-called 'existence property', in formularian terms. (Otherwise, the extensionality fails in the classical simulation discussed below, a situation similar to the failure of extensionality for the simulated proof-primitives associated to the minimal / intuitionistic **or**-primitive – corresponding to the so-called 'disjunction property', mentioned in passing above.[30])

The latter two minimal / intuitionistic witness operators are explicitly definable, classically, in $\boldsymbol{\lambda^\Delta\partial\Lambda}[\bot,\to,\neg,\forall]$, say, by:

---

[30] Incidentally, this is true about the completely analogous $\lambda\gamma$-simulations of Rezuş 1991, pointing out to an oversight, *ad loc*. One can eventually recover classically such 'properties' only in a *very extensional* version of the $\lambda\pi$-calculus, by additional equational postulates (cf., e.g., the scalar Π-extensions of Rezuş 2017b and **§ 6.4** below).

($\downarrow^\exists$) $\Gamma \vdash \downarrow_t^\exists(a) := \rho z{:}\forall u.\neg A[u].((z\blacktriangleright t)\cdot a) : \exists u.A[u] \equiv \neg\forall u.\neg A[u]$,
if $\Gamma \vdash a : A[t]$, provided z is not free in a,
i.e., *mutatis mutandis*, as above, and

($\mathcal{E}$) $\Gamma \vdash \mathcal{E}[u,x{:}A].(f,c[u,x]) := \partial z{:}\neg C.f \cdot (\Lambda u.\rho x{:}A[u](z\cdot c[u,x]) : C$,
if $\Gamma [u]\ [x{:}A[u]] \vdash c[u,x] : C$, and $\Gamma \vdash f : \exists u.A[u] \equiv \neg\forall u.\neg A[u]$,
provided u is not free in f and z is not free in f and c[u,x].

> Note that one must be able to use the minimal ($\beta\rho$) and ($\beta\Lambda$)-conditions, as well as the genuinely classical ($\eta\partial$) in order to derive ($\beta \downarrow^\exists$), whence the need for the ($^\Delta$)-component of the classical $\lambda\partial\Lambda$-calculus. Alternatively, one can do the same thing in the equivalent $\lambda\rho\partial\Lambda$-calculus.[31]

§ 6.4 Hilbert's epsilons.
Recall that a *scalar pairing* (Rezuș 2017b, § 1.4 and § 2.2) is, essentially, what is usually meant by *Hilbert's epsilon symbol* (Hilbert 1922, 1923, 1928).

> Roughly speaking, the construction following below is the 'type-free' counterpart of the so-called *no-counterexample interpretation* of the universal quantifier, popular in the Hilbert school during the nineteen twenties and later, a subject recovered by Georg Kreisel in 1949–1950.[32]

§ 6.4.1 Scalar theories.
For equational theories based on scalar syntax (Rezuș 2017b), a *scalar* (or a *mixed*) *pairing* is a triple $[\Pi, \Pi_1, \Pi_2]$ – where Π is a binary operation and Π_1, Π_2 are singulary [single-place] operations, with term-forms [w-terms] $\downarrow_t(c) := \Pi(t,c)$, $\theta(c) := \Pi_2(c)$, and [scalars] $\mathsf{\i}(c) := \Pi_1(c)$ – satisfying the conditions:

($\beta\Pi_1$) $\vdash \mathsf{\i}(\downarrow_t(c)) \doteq t$ (where $\doteq$ stands for scalar equality), and
($\beta\Pi_2$) $\vdash \theta(\downarrow_t(c)) = c$,
($\eta\Pi$) $\vdash \downarrow_t(\theta(c)) = c$, if $\mathsf{\i}(c) \doteq t$,

together with the expected monotony conditions for the primitive operations. An equational theory is *scalar* if it is equipped with a scalar pairing.

§ 6.4.2 Scalar folds.
Let $\boldsymbol{\lambda\partial\Pi}$ be a scalar (extensional) [di-] fold based on $[\lambda, \triangleright]$ and $[\partial, \star]$, with abstract grammar:

scalars :: s, t := u | $\mathsf{\i}$(a) | ...
w-terms:: a, b, c, d, e, f := x | λz.e | c $\triangleright$ a | ∂z.e | c $\star$ a | $\downarrow_t$(c) | θ(c)

satisfying, thus, the conditions:

($\beta\lambda$) $\vdash (\lambda z.b[x]) \triangleright a = b[x{:=}a]$,
($\eta\lambda$) $\vdash \lambda x.(f\triangleright x) = f$ (x not free in f),

---

[31] Analogous (β)-simulations are available in the appropriate $\lambda\gamma(\Lambda)$-calculus (cf. Rezuș 1990, 1991).
[32] Cf. Ackermann 1940, Kreisel 1951, 1952. See also Kreisel 1958 and Feferman 1996.

($\beta\partial$) $\vdash$ c $\star$ (∂z.e[z]) = e[z:=c],
($\eta\partial$) $\vdash$ ∂z.(z$\star$a) = a (z not free in a),

($\beta\Pi_1$) $\vdash$ $\backslash$($\downarrow_t$(c)) $\doteq$ t,
($\beta\Pi_2$) $\vdash$ θ($\downarrow_t$(c)) = c,
($\eta\Pi$) $\vdash$ $\downarrow_t$(θ(c)) = c, if $\backslash$(c) $\doteq$ t,

and the expected monotony conditions for the primitives. Consistency for $\lambda\partial\Pi$ can be established by a simple translation argument (to $\lambda\partial\pi$ [sic], for instance). Define now, in $\lambda\partial\Pi$,

(df Σ) Σ(u,x).e[u,x] := ∂z.e[u:=$\backslash$(z),x:=θ(z)], z fresh for a[u,x],
(df Λ) Λu.a[u] := Σ(u,z).(z$\star$a[u]), z fresh for a[u],
(df ▶) c ▶ t := ∂z.($\downarrow_t$(z)$\star$c), z not free in c.

As ever, this yields:

($\beta\Lambda$) $\vdash$ (Λu.a[u]) ▶ t = a[u:=t],
($\eta\Lambda$) $\vdash$ Λu.(c▶u) = c, (u not free in c),

($\Sigma\Lambda\partial$) $\vdash$ Σ(u,x).e[z] = Λu.∂x.e[u,x],

($\beta\Lambda^*$) $\vdash$ $\downarrow_t$(c) $\star$ (Λu.a[u]) = c$\star$(a[u:=t]),
($\eta\Lambda^*$) $\vdash$ Λu.∂z.($\downarrow_u$(z) $\star$ c) = c (u, z not free in c),
($\zeta\Lambda^*$) $\vdash$ ∂z.e[z] = Λu.∂x.e[z:=$\downarrow_u$(x)] (u, x fresh for e[z]),

($\beta\Sigma$) $\vdash$ $\downarrow_t$(a) $\star$ (Σ(u,x).c[u,x]) = c[u:=t,x:=a]),
($\eta\Sigma$) $\vdash$ Σ(u,x).($\downarrow_u$(x) $\star$ c) = c (u, x not free in c),
($\zeta\Sigma$) $\vdash$ ∂z.e[z] = Σ(u,x).e[z:=$\downarrow_u$(x)] (u, x fresh for e[z]),

and monotony conditions for the defined operations.[33]

§ 6.4.3 *Basic rules of inference.* On the primitive signature [⊥,¬,→,∀], the *basic witness operators* – i.e., the *primitive rules of inference* – of the corresponding witness theory $\lambda\partial\Pi$[⊥,¬,→,∀], the decorated ('typed') cognate of $\lambda\partial\Pi$, are:

(λ) $\Gamma \vdash \lambda$x:A.b[x] : A→B, if Γ[x:A] $\vdash$ b[x] : B
[the 'deduction theorem'],
(▷) $\Gamma \vdash$ (c ▷ a) : B, if $\Gamma \vdash$ c : A→B, and $\Gamma \vdash$ a : A
[modus ponens],
(∂) $\Gamma \vdash \partial$x:¬A.e[x] : A, if Γ[x:¬A] $\vdash$ e[x] : **f**
[reductio ad absurdum],
($\star$) $\Gamma \vdash$ c $\star$ a : ⊥, if $\Gamma \vdash$ c : ¬A, and $\Gamma \vdash$ a : A
[the 'law of non-contradiction'],
(Π) $\Gamma \vdash \downarrow_t$(a) : ∃u.¬A[u] [≡ ¬∀u.A[u]], if $\Gamma \vdash$ a : ¬A[u:=t]
[counterexample],
(Π_1) $\Gamma \vdash \backslash$(c) ≡ t, if $\Gamma \vdash$ c : ∃u.¬A[u]
[counterexample-scalar: 'some u such that ¬A[u]' (Hilbert's epsilon)],

---

[33] We don't actually need the mono-fold λ (based on [λ,▷]), for this.

(Π_2) $\Gamma[v] \vdash \theta(c) : \neg A[v]$, if $\Gamma \vdash c : \exists u. \neg A[u]$ (v fresh for c and Γ)
[counterexample-witness: a witness for $\neg A[v]$, for some $v = \imath(c)$].

Formally, one should have indexed the decorated pairing on formulas, writing, e.g., $\imath_A(c)$ and $\theta_A(c)$, for the appropriate scalar 'projections'. In particular, 'Hilbert's epsilon' – here: $\imath_A(c)$ – corresponds to the *indefinite article* (in German: *ein v so daß c ist ein Beweis für* $\neg A[v]$). Hilbert 1922, 1923, 1929 had no explicit θ-operator because he had no proof / witness-notation at hand, but the idea is implicit in his informal explanations.

In the end, following Hilbert – and, possibly, Chrysippus (cf., e.g., the references to Urs Egli, in Rezuş 2012) – the proof-behaviour of the (*classical*) first-order quantifiers can be derived from the *semantics of the indefinite article* in natural languages.

§ 6.4.4 Derivable rules of inference.
The witness operators (Σ), (Λ), and $(\blacktriangleright)$ defined in terms of the above correspond to *derivable rules of inference* and are *inheriting* of decorations from the basic ones. The reader can easily check the following (as derived from the explicit definitions given above):

(Σ) $\Gamma \vdash \Sigma(u,x{:}\neg A[u]).e[u,x] : \forall u.A[u]$, if $\Gamma[u][x{:}\neg A \vdash e[u,x] : \mathbf{f}$
[indirect generalisation from no counterexamples],
(Λ) $\Gamma \vdash \Lambda u.a[a] : \forall u.A[u]$, if $\Gamma[u] : a[u] : A[u]$
[direct generalisation],
($\blacktriangleright$) $\Gamma \vdash c \blacktriangleright t : A[u{:=}t]$, if $\Gamma \vdash c : \forall u.A[u]$
[direct instantiation].

§ 6.4.5 'Type-free' scalar extensions of $\lambda \partial$.
Let $T := \lambda \partial$, and define, for the appropriate syntax, scalar extensions:

$T\Lambda := T + (\beta\Lambda) + (\eta\Lambda)$
$T\Lambda^* := T + (\beta\Lambda^*) + (\eta\Lambda^*)$
$T\Lambda^*_\zeta := T + (\beta\Lambda^*) + (\eta\Lambda^*) + (\zeta\Lambda^*)$
$T\Sigma := T + (\beta\Sigma) + (\eta\Sigma)$
$T\Sigma_\zeta := T + (\beta\Sigma) + (\eta\Sigma) + (\zeta\Sigma)$,

including the due monotony conditions.

With, as earlier, $T_1 \prec T_2$ standing for 'T_1 is a (proper) equational subsystem of T_2', and $T_1 \simeq T_2$ standing for 'T_1 and T_2 are equationally equivalent', we have, *mutatis mutandis*, as in Rezuş 2017b,

$$T \preceq T\Lambda \preceq T\Lambda^* \simeq T\Sigma \preceq T\Lambda^*_\zeta \simeq T\Sigma_\zeta \preceq \lambda\partial\Pi.$$

Note that, in the above, the inclusions are *strict*.

There are many possible variations on this. One can consider, for instance, systems containing an inversion $[\Delta, \nabla]$, taken as a primitive.

In fact, we can add (conservatively) a scalar pairing $[\Pi, \Pi_1, \Pi_2]$ to any one of the subsystems contained in $\lambda\pi$ already considered in the above (or else in Rezuş 2017a, 2017b).

§ 6.4.6 *Witness theories for first-order classical logic.* We leave as an exercise to the industrious reader the task of defining the *first-order witness theories for classical logic* corresponding to the ('type-free') equational theories discussed here (by interpreting – in the shadow of David Hilbert – scalars as first-order notions).

Whether the resulting proof-formalisms might shed some light on Hilbert's original – and longstanding – concerns with *finitism* [*die finite Einstellung*, resp. *das finites Standpunkt*] is matter for a different – rather philosophical – debate.[34]

At any rate, as shown in **§ 3** above, if considered witness-theoretically, classical logic include *infinitistic features* already at a quantifier-free level.

16 November 2018 – 26 October 2019

---

[34] See the literature mentioned in the references under David Hilbert, Paul Bernays, Georg Kreisel, Wilfried Sieg, and William Tait.

11
Łukasiewicz, Jaśkowski and natural deduction (2017)

In these notes[0] we are tracing back the origins of the so-called 'natural deduction' to the Łukasiewicz Warsaw logic seminars of 1926, reconstructing Jaśkowski's proof system for classical logic [1926–1927, in print: 1934] in *witness-theoretical* terms.

§ 1 Introduction

Sometime during 1926, while still an undergraduate mathematics student at the Warsaw University, Stanisław Jaśkowski (1906–1965) presented, in the local (i.e., Warsaw) Logic Seminar of his teacher, Jan Łukasiewicz (1878–1956), a 'natural deduction' formulation of classical (two-valued) propositional (including propositional quantifiers), first and second-order logic.

Let us pause, first, on *historico-bibliographical* details. Apparently, the work was done at the instigation of Jan Łukasiewicz[1]. As to terminology, the phrase 'natural deduction' (German: 'natürliche Schliessen'), still in common use in logic today, appeared in print first in Gentzen 1934–1935, although the idea was already clear in the motivation of Jaśkowski's research: he meant, on the authority of his teacher, to design a logic of rules (a 'Regellogik', so to speak), close to the 'natural' mathematical reasoning, as opposed to the 'Satzlogik' of Łukasiewicz himself, i.e., an axiomatic presentation – as in the lectures of Łukasiewicz 1929 – in the footsteps of Frege1879, Peirce 1885, Russell 1906, and Whitehead & Russell's *Principia Mathematica* (1910–1913).[2]

---

[0] The paper has first appeared in print as Łukasiewicz, Jaśkowski and Natural Deduction (*Curry-Howard for Classical Logic*), in: Massoud Pourmahdian, and Ali Sadegh Daghighi (eds.), Logic Around the World, AFJ – Andisheh & Farhang-e Javidan Publishing, Tehran 2017 [ISBN: 978-600-6386-99-7], pp. 89–138. The published version differs typographically from the present reprint.

[1] Stanisław Jaśkowski graduated from high-school at eighteen, in 1924, so he was about twenty, by then.

[2] The contrast 'Satz-' vs 'Regellogik' – roughly: 'sentence / proposition logic' [sic] vs 'rule logic' – current in the German logico-philosophical literature, mainly after

The phrase 'deduction theory' [Polish: 'teorja dedukcji'] was, initially, Łukasiewicz's own term for 'propositional logic', including, possibly, propositional and/or first- and second-order quantifiers). Cf. the introductory lines and other occasional side-remarks appearing in Jaśkowski 1934.³

As to publication matters, Jaśkowski's results were promptly announced, as Jaśkowski 1927, at the First Congress of the Polish Mathematicians, held in

---

Gentzen, goes back to Frege (1893) and is meant to stress a difference of approach: pace Frege, the pioneers were mainly concerned with the formal study of *propositions* and /or *propositional schemes*, as expressed by *formulas*, and the properties thereof (like, e.g., *provability* in a given 'logistic' system ['Satzsystem'], etc.), while Frege and, subsequently, Gentzen payed also attention to the *rules of inference* and to their properties (like, e.g., *derivability* and/or *admissibility* in a given system [of rules]). With a suggestive term, we may refer to the former approach – and to its defenders / practitioners – as *Formularian* (with implicit allusion to Peano's various editions of his 'Formulaires', mere collections of [formalised] formulas). Roughly speaking, for a Formularian, a logic is a *set of provable formulas*, and a provable formula ['thesis' or, even, 'theorem'] is, at best, the codification of a [bunch of] rule[s] of inference. The Formularian approach has been effective in the early development of 'algebraic logic' and, later, in model theory, but is, conceptually speaking, rather inadvertent, since two distinct 'logics' may share exactly the same set of 'tautologies' [provable formulas], while still differing as to the corresponding derivable rules. The alternative *rule-oriented approach*, suggested by Łukasiewicz in his Warsaw Seminar (1926), was motivated in terms of 'naturality', by reference to the actual mathematical reasoning and this was, apparently, also the case for Gentzen, somewhat later. Technically speaking, the distinction between *rule-admissibility* [closure of a set of propositions / formulas under a given rule of inference] and *explicit [rule-] derivability*, already implicit in Gentzen (1934–1935), comes rather late to the attention of the logical theorists; to my knowledge, it is due to Paul Lorenzen (1915–1994) – cf. Lorenzen 1955 – and to Haskell B. Curry (1900–1982), slightly later. The first *explicit* counterexample to the (Formularian) tenet that *a logic = a set of provable formulas*, is due to Henry Hiż (*circa* 1957–1958), who described a [three-valued] 'logic', $\mathcal{H}_3$ say, containing all classical, two-valued tautologies as provable formulas, where not all classically valid rules of inference are derivable. See Hiż 1957, 1958, 1959 and, possibly, Nowak 1992, for a model-theoretical account of $\mathcal{H}_3$.

³ For bio-bibliographical and historical details on Jaśkowski and his system(s) based on 'supposition rules' [Polish: 'oparta na dyrektywach założeniowych'], i.e., proper / general rules of inference, using 'assumptions' in their premises, viz., entailments, see, e.g., Dubikajtis 1967, 1975 – a reliable informant on the logical whereabouts of Stanisław Jaśkowski, Lech Dubikajtis (1927–2014) was a former PhD student and a research assistant of Jaśkowski at the Warsaw National Institute for Mathematical Sciences (currently, the Institute of Mathematics of the Polish Academy of Sciences) – as well as Kotas & Pieczkowski 1967, Orłowska 1975, Indrzejczak 1998, 2016, Piętka 2008, and, possibly, the textbook Borkowski & Słupecki 1963. On the subsequent history of 'natural deduction' – a comedy of conceptual errors, indeed – see, e.g., Pelletier 1999, 2000, and Hazen & Pelletier 2012, 2014. General information on the 'Lvov-Warsaw school' can be found in Woleński 1985, 2015, Jadacki 2006, Wybraniec-Skardowska 2009, Garrido & Wybraniec-Skardowska 2018, etc.

Lvov, 7–10 September 1927. See, e.g., the Congress [PPZM] Proceedings (1929) and, specifically, the references of Lindenbaum 1927. Due to circumstances unknown to me (as well as to other, better informed people, apparently), the final paper appeared actually in print only eight years later (in a projected logic collection edited by Jan Łukasiewicz himself, later to become an international logic journal), as Jaśkowski 1934, more or less simultaneously with Gerhard Gentzen's Göttingen *Inauguraldissertation*, Gentzen (1934–1935).

Before going into the proper details of the subject announced in the title, a few more technical and historical remarks on the material available in print – or otherwise – to Jaśkowski, around 1926, are in order.

Modern logic – also called 'mathematical' or 'symbolic' – was (re-) born by the end of the XIX-th century (around 1879, in print, with a sequel, in 1893), in two instalments, authored by Gottlob Frege, viz. Frege 1879 [**BS**] and Frege 1893 [**GGA**:1]. There was an intermediate episode, due to Charles S. Peirce, Peirce 1885 – that Frege ignored – equally worth noting, which, although sketchy, was, in some respects, conceptually superior to Frege's **BS**-account. Both Frege and Peirce had a venerable predecessor, more than twenty one centuries before they were born, in the work of Chrysippus of Sol[o]i (this was an obscure place in *Cylicia Campestris*, nowadays in modern Turkey), a Phoenician emigrant to Athens, the father of Stoic logic and the grandfather of [classical] logic *tout court*. The latter (historical) fact was first noted by Jan Łukasiewicz, sometime during the early 1920.[4]

A less known (historical) fact is that Frege came out with two 'logics', not with a single one: a *Satzlogik* (1879), and a *Regellogik* (1893). Both were 'axiomatic', by modern standards. The latter one was meant to be closer to actual 'mathematical thinking' (a kind of formal counterpart of 'natural' deduction, as occurring in mathematical texts), and was vastly anticipating, among other things, Gentzen 1934–1935, for instance.

The other relevant (historical) fact is that neither Gentzen, nor any other *Göttinger* – David Hilbert (1862–1943) or Paul Hertz (1880–1940), for that matter[5] – had ever read Frege's **GGA**:1, (1893) [sic].

In this matter, I cannot, however, speak for the Lvov-Warsaw Poles – those active in logic before cca 1935 – because we lack the right kind of (historical) documentation. Both Łukasiewicz 1929 and Jaśkowski 1934 mention only Frege's axiomatic system of **BS**, while, curiously enough, the young Alfred Tajtelbaum [aka Tarski] (1901–1983) did not refer to Frege's 'logics' at all.[6]

On the other hand, the main trouble with Frege 1879, 1893 was in the fact he did not recognise the *general* concept of a *rule of inference*. Specifically, with [material] implication, [classical] negation and the [classical] universal quantifier

---

[4] Circa 1923. See the final outcome in Łukasiewicz 1934 and, possibly, Rezuş 2016, for the main claim.

[5] On the authority of Paul Bernays (1888–1977), Gentzen borrowed his 'structural' rules from Hertz, a former physics student of Hilbert in Göttingen.

[6] See, e.g, the *Bibliography* and *Index of Names and Persons* of the collection Tarski 1956.

as primitives, he only acknowledged 'flat' rules (more or less like the algebraic operations), of the form:

(♭) $\vdash \alpha_1, ..., \vdash \alpha_n \Rightarrow \vdash \beta$,

rejecting, implicitly, the (old-fashioned) idea of *entailment* (= finite sequence of propositions, with exactly one being tagged *qua* 'conclusion'):

(⊢) $\alpha_1, ..., \alpha_n \vdash \beta$

as a legitimate – and otherwise essential – logic concept.[7] The young Bertrand Russell (1872–1970) – the only (more or less competent) person who did actually read Frege 1893 in the epoch[8] – was less interested in such absconse distinctions[9], so he missed the point, as well, and sticked to axiomatics, in the shadow of Frege 1879 and Peirce 1885[10].

As another aside, I was, so far, unable to date exactly the event as such, in the moderns, viz. the identification of the concept of a *general rule of inference*.[11] The earliest date I am able to quote is 1921, when Alfred Tarski (Leśniewski's only PhD student) noticed the 'Deduction Theorem' (DT)[12], i.e., the implication-introduction rule of Gentzen (1934–1935), an obvious case of 'non-flat' inferential rule, with an entailement (including 'assumptions') as a premiss:

(DT) $\alpha \vdash \beta \Rightarrow \vdash C\alpha\beta$,

Certainly, Łukasiewicz was aware of such details, if not in 1921, at least sometime before 1926, when he assigned his [very] young student (Jaśkowski) the homework leading to Jaśkowski 1927, 1934. Anyway, the implication int-elim rules (the Deduction Theorem and the famous *modus ponens* / detachment rule) appear explicitly in Jaśkowski's early home-work, and so, *a fortiori*, in Łukasiewicz's Warsaw Seminar, sometime during 1926.[13]

---

[7] In this respect, Frege was behind Chrysippus (and, even, Aristotle, in a way). On this, see, e.g., Rezuş 2009, 2016.

[8] As I could gather from the newest Russell expertise, this happened sometime around 1902–1903.

[9] As noticed, in passing, by Kurt Gödel in his 1944, Russell was even later quite confused about the general concept of a *logical rule of inference*.

[10] Although he did not acknowledge the latter source. Cf. Russell 1906, and Anellis 1995.

[11] One should perhaps read, once more, carefully, the rather vast output of Stanisław Leśniewski, on this. Cf. the collections Leśniewski 1992, 2015, and Srzednicki et al. 1984, Srzednicki & Stachniak 1998.

[12] See Axiom 8* in Tarski 1930, and the footnote on p. 32, in the collection Tarski 1956, for references. Some authors used to credit Herbrand with the discovery. However bright, Jacques Herbrand (1908–1931) was a teenager, just 13 years old, in 1921, so it is unlikely he spotted errors in Frege's [German] texts nobody used to read by then, even in Germany. See also Porte 1982, Pogorzelski 1968, 1994, etc.

[13] At a quantifier-free level, Jaśkowski had a third rule – of the same kind as (DT), actually – yet a less inspired choice I will go into later on.

Whence a question: 'Why has not Jan Łukasiewicz solved the problem [the one assigned to young Jaśkowski] himself – sometime before 1926 – and presented the outcome is his famous lectures Łukasiewicz 1929?' Because he had at hand the (conceptual and technical) means to do it, anyway. Which is *what I mean to show next*. In order to do this, in proper terms, I need a *conceptual revision* of the received views on proving (in logic) and some appropriate notation and terminology.

§ 2 Rules of inference as witness operators

A typical case of the 'flat' rule ($\flat$) above is the *modus ponens* or the *detachment* rule, in logics with implication [here, C], either primitive or defined:

($\triangleright$) $\vdash C\alpha\beta, \vdash \alpha \Rightarrow \vdash \beta$,

Now, in axiomatic presentations of a given logic (classical, two-valued logic, for instance), the premises α_i ($0 < i < n + 1$) of a 'flat' rule of the form ($\flat$) are taken to hold 'unconditionally', without further assumptions, they are provable formulas (expressing propositions / propositional schemes), 'theorems' or 'theses' (in the jargon of the early Polish 'school'); alternatively, they are, semantically, *true* (or else two-valued '*tautologies*', in the classical case).

In fact, any particular axiomatics amounts to an *inductive definition* of the predicate 'provable' (expressed notationally by $\vdash$) applying to formulas (expressing propositions or propositional schemes): the axioms are paradigmatically provable (the basis of the induction), while any (primitive) 'flat' rule of inference carries this property – *provability* – from premises to conclusion (inductive step).

As long as we have only primitive rules of the form ($\flat$) around, 'proving axiomatically' amounts to a piece of *algebraic notation*: a 'flat' *primitive rule of inference* with n premises ($n > 0$) looks like a usual algebraic n-ary operation[14], whereupon a *derivable rule of inference* is just an *explicit definition* of an operation in terms of 'primitive' operations (here, axioms and primitive rules of inference).

'Operations on what?', one might wonder. A first – approximate – answer could be: 'On formulas'.[15] A slightly better one would amount to an additional piece of formalism, to be justified, intuitively, as follows:

---

[14] In the limit case ($n = 0$), the axioms may be thought of as null-ary operations, if we want full generality.

[15] This was actually the case, historically speaking: the idea came first – exactly in these terms – to a later (Irish) student of Łukasiewicz, in the '(logical) Polish quarter' of Dublin, during the early fifties. [After the WWII, 'being unwilling to return to [...] Poland [...], Łukasiewicz looked for a post elsewhere. In February 1946 he received an offer to go to Ireland. On 4 March 1946 the Łukasiewiczes arrived in Dublin, where they were received by the Foreign Secretary and the Taoiseach Eamon de Valera. In autumn 1946 Łukasiewicz was appointed Professor of Mathematical Logic at the Royal Irish Academy (RIA), where he gave lectures

Proving something – a proposition expressed by formula α, say – amounts to *providing a reason* – or 'grounds' – for α, or else, as in court, to displaying a *witness* a for α. Formal notation: $\vdash a : \alpha$.

With this minimal formal equipment, in axiomatic presentations, the axioms are to be witnessed by *primitive constants* (possibly parametric, in the case of axioms schemes), whence 'witnessing' a 'flat' rule of the form ($\flat$) would amount to providing an operation (operator) $\flat$ and a piece of explicit formal notation $\flat(a_1, ..., a_n)$, such that

$$[\,\flat\,] \vdash a_1 : \alpha_1, ..., \vdash a_n : \alpha_n \Rightarrow \vdash \flat(a_1, ..., a_n) : \beta;$$

so, in particular, 'witnessing' a 'flat' rule like *modus ponens*, for instance, would consists of using a binary operation $\triangleright$, say, to the effect that

$$[\,\triangleright\,] \vdash f : C\alpha\beta, \vdash a : \alpha \Rightarrow \vdash (f \triangleright a) : \beta.$$

Summing up, a 'flat' rule of inference is just an *algebraic operation*, in this view. Note, however, that, as long as we do not *define explicitly* the 'operations' $\flat$, we have only a *witness notation*, at most. In other words, in order to have a *witness theory* – as a formal counterpart of (axiomatic) proving – we must be able to *characterise the witness operations* first. Usually, we can do this, as in algebra, by *equational conditions*, expressing *witness-* or *proof-isomorphisms*.

The general case is obtained from the 'flat' case by 'parametrisation' so to speak, where the parameters are finite (possibly empty) sequences of formulas (expressing propositions, resp. propositional schemes) $\bar{\Gamma}_i := [\beta_{i,1}, ..., \beta_{i,m_i}]$ ($0 < i < n+1$, $m_i \geq 0$), resp. $\bar{\Gamma} := [\beta_1, ..., \beta_m]$ ($m \geq 0$), called 'assumption contexts' (alternatively: *witness-contexts* or *proof-contexts*). Every premiss of a general rule is thus an entailment of the form $\bar{\Gamma}_i \vdash \alpha_i$ ($0 < i < n+1$), while the rule has a conclusion of the form $\bar{\Gamma} \vdash \beta$, i.e., one has

$$(\flat\vdash)\ \bar{\Gamma}_1 \vdash \alpha_1, ..., \bar{\Gamma}_n \vdash \alpha_n \Rightarrow \bar{\Gamma} \vdash \beta,$$

with 'witnessed' counterpart of the form

$$[\,\flat\vdash\,]\ \hat{\Gamma}_1 \vdash a_1 : \alpha_1, ..., \hat{\Gamma}_n \vdash a_n : \alpha_n \Rightarrow \hat{\Gamma} \vdash b : \beta,$$

where $\hat{\Gamma}_i := [x_{i,1} : \beta_{i,1}, ..., x_{i,m} : \beta_{i,m_i}]$, the 'witnessed' counterpart of $\bar{\Gamma}_i$ (and analogously for $\hat{\Gamma}$ and $\bar{\Gamma}$) contains 'decorated – or typed – witness-variables', allowing us to manipulate the witness-contexts.[16]

In particular, in the limit case, a null-premiss rule of inference is just an entailment (considered valid). Examples in point:

---

at first once and then twice a week.' (Simons 2014)] See the references to the 𝔇-operator, the 'condensed detachment' operator, of Carew A. Meredith, below, and, possibly the notes Meredith 1977 – by David Meredith, the American cousin of Carew, also a logician – for further historical details.

[16] Besides, one must have additional rules, called 'structural rules', in the current proof-theoretic terminology borrowed from Gentzen 1934–1935, that are rather trivial, and remain un-expressed, formally, in usual presentations of 'natural deduction'.

[id] $\alpha \vdash \alpha$,

or more generally,

[prj] $\bar{\Gamma} \vdash \alpha_i$, for $\bar{\Gamma} \equiv [\alpha_1, ..., \alpha_n]$, $(0 < i < n+1)$,
[▷] $C\alpha\beta, \alpha \vdash \beta$ (*modus ponens*, viewed as a valid entailment),

etc., and analogously for the witnessed variants:

[id⊢] $x : \alpha \vdash x : \alpha$,

resp.

[prj⊢] $\hat{\Gamma} \vdash x_i : \alpha$, for $\hat{\Gamma} \equiv [x_1 : \alpha_1, ..., x_n : \alpha_n]$, $(0 < i < n+1)$,
[▷⊢] $z : C\alpha\beta,\ x{:}\ \alpha \vdash (z \triangleright x) : \beta$.

Here, in [▷⊢], the witness b, appearing in the conclusion, must be of the form $\flat(\flat_1(a_1),...,\flat_n(a_n))$, where the prefixes $\flat_i$ $(0 < i < n+1)$ are either empty (nil) or specific variable-binding operations, called 'abstraction operators', acting on finite sequences $\mathbf{x}_i \equiv [x_{i,1}, ..., x_{i,m_i}]$, $(m_i > 0)$, of *pairwise distinct witness-variables*) and the associated 'body' a_i. In each case, a witness-variable is decorated (or 'typed') by an associated formula thereby witnessed 'hypothetically'.

Of course, if every $\flat$-prefix is empty, we have a 'flat' rule, the 'degenerated' case. E.g., in particular, the most general forms of *modus ponens*, viewed as a rule of inference, should be

[▷⊢⊗] $\hat{\Gamma} \vdash f : C\alpha\beta,\ \hat{\Gamma} \vdash a : \alpha \Rightarrow \hat{\Gamma} \vdash (f \triangleright a) : \beta$ ['parametric'] or
[▷⊢⊕] $\hat{\Gamma}_1 \vdash f : C\alpha\beta,\ \hat{\Gamma}_2 \vdash a : \alpha \Rightarrow \hat{\Gamma}_1, \hat{\Gamma}_2 \vdash (f \triangleright a) : \beta$ ['cumulative'],

where $\hat{\Gamma}_1, \hat{\Gamma}_2$ stands for sequence concatenation.

In general, however, a rule of inference can be arbitrarily complex, so that the identification

(general) rule of inference ≃ witness operator

goes beyond the conventional views on 'algebraic operators'.[17] In order to accommodate, formally, the terminology – and the notation – one can use the

---

[17] The so-called 'abstraction operators' – and, in general, the variable-binding mechanisms – are not welcome in (abstract) algebra, indeed. This on historical reasons, likely. Nicolaas G. de Bruijn observed once, in conversation, that abstraction operators do not occur in pre-xix-century mathematics. This explains, in a way, the initial lack of interest in such phenomena among algebraists. In recent times, when confronted incidentally with such cases – first-order quantifiers, for instance – they made appeal to elaborated *local* solutions in order to cope with the problem. Paul Halmos and Alfred Tarski invented specific constructions – *polyadic algebras*, resp. *cylindric algebras* – in order to algebraise first-order logic with quantifiers, resp. quantifiers and equality. In more general situations, modelling abstraction operations in mathematical terms requires specific category theoretic methods and constructs that go far beyond the traditional algebraic way of thinking about 'opera-

idea of a *generic arity* (gen-arity, for short), viewed as a finite sequence of *non-negative integers*, to be associated to an arbirary operator, taken in the new sense. In this setting, the algebraic [null-ary] constants would get gen-arity nil [= the empty sequence], the usual n-ary algebraic operations would get gen-arity [0, ..., 0] (n times 0, n > 0), the n-adic abstractors would get gen-arity [n] (n > 0), so that the monadic λ-abstractor, as well as the usual quantifiers, for that matter, must have gen-arity [1], the dyadic abstractor *split* [∫], mentioned incidentally below, has gen-arity [2], and so on. In particular, the 'mixed' operators ∫(gen-arity [k_1, ..., k_n], k_i > 0) can be handled as 'flat' (algebraic) n-ary operations acting on k_i-adic abstractors (0 < i < n + 1).

In practice, however, we rarely, if ever, encounter complex rules corresponding to 'mixed' operators; we are normally confronted with 'flat' (ordinary algebraic) operators or with n-adic abstractors with n := 1,2 (operators of gen-arity [1] or [2]), at most, so that, in the end, the talk about gen-arities amounts to a piece of empty generality.[18]

§ 3 The Łukasiewicz Warsaw Lectures

Essentially, the Łukasiewicz Warsaw Lectures of 1928–1929, Łukasiewicz 1929, contain a very detailed axiomatic presentation of

(1) [classical] propositional logic, based on the signature [N,C] (*classical negation* and *material implication*, in Łukasiewicz notation)[19], and

(2) a mild – yet very clean – version of the 'extended propositional [classical] logic', i.e., the [classical] propositional logic with propositional quantifiers, à la Peirce 1885, Russell 1906 and Tarski (doctoral dissertation, Warsaw 1923, under Leśniewski), or else Leśniewski's *protothetic*, for that matter.[20]

---

tions' and 'operators'. Moreover, there are genuine phenomena, occurring frequently in computer science – like, e.g. the *non-local control* (typically, jumps), the *side effects*, or the so-called *continuations* – that can be easily described in terms of abstraction operations, but whose behaviour resist algebraisation, as understood in traditional terms. The (general) logical rules of inference fall within the same category.

[18] Like in the usual algebraic case, in fact, as we do not encounter 13-ary or 17-ary operations in current mathematical practice, either. — In logic, an exception can be encountered in the usual presentation of intuitionistic propositional logic, where the so-called *or*-elimination rule (case-analysis) is a witness operator of gen-arity [0,1,1], as well as in the case of Jaśkowski's rule χ (a witness operator of gen-arity [1,1]) to be discussed below.

[19] Completeness is shown in Łukasiewicz 1929, Chapter III, **§ 22**. Cf. also Łukasiewicz 1931a.

[20] Specifically, Chapter IV of Łukasiewicz 1929 is based 'in great part' [see the Preface of the first edition] on Tarski's previous work. Cf. also Łukasiewicz & Tarski 1930, **§ 5**. For Leśniewski and protothetic, see Sobociński 1939, 1949, 1967, and, more recently, Srzednicki et al. 1984, Srzednicki & Stachniak 1998, Leśniewski 1992, 2015, Jadacki 2016, etc.

Now, except for a minor detail, the latter one is not more than the former, because we can define explicitly the [classical] propositional quantifiers in terms of [classical] connectives K [*and*], A [*or*] (in Łukasiewicz notation), and propositional constants **v** [*verum*] and **f** [*falsum*], anyway (just 'truth value quantifiers', as Nuel Benap Jr. would have had them[21]).

The 'minor detail' refers, here, to the fact that the primitive [N,C]-signature is not *functionally complete*: we cannot obtain the propositional constants **v** and **f** from [N,C] alone. This does not affect our discussion of (2) below, as the signature [N,C,Π], with Π for the *universal propositional quantifier*, is functionally complete and even redundant, since one can define **f** and N à la Peirce 1885 by **f** := Πp.p and Nα := Cα**f**, resp.[22] In an axiomatic quantifier-free setting – as in Łukasiewicz 1929, Chapter II – the absence of the propositional constants might, however, affect the translation of the axioms in terms of rules of inference (and conversely). The point is that we need a primitive **f** [*falsum*] in order to express something as simple as the 'law of (non-) contradiction', for instance, in inferential (entailment-like) terms.[23] By adjoining a *falsum*-constant **f** to [N,C], the Łukasiewicz original axiom system is, however, *incomplete* as it stands. We cannot even prove, from the Łukasiewicz axioms, the 'thesis' ⊢ **v** [≡ N**f**], for instance.[24]

Recall that the Łukasiewicz 1929 quantifier-free axioms, with *modus ponens* and substitution as only 'rules of inference', are (in Łukasiewicz – 'Polish' – spelling):

(CB) ⊢ CB[p,q,r] := 1 : CCpqCCqrCpr
[transitivity of implication: 'suffixing'] - axiom in Peirce 1885

(E) ⊢ E[p] := 2 : CCNppp
[the *consequentia mirabilis* of Girolamo Cardano (1570) or the *Law of Clavius*, viewed as a 'thesis']

(O) ⊢ O[p,q] := 3 : CpCNpq
[*ex contradictione quodlibet*, 'explosion'].

To this team, we add, for reasons discussed above:

(Ω) ⊢ Ω : **v**.

---

[21] Cf., e.g., *mutatis mutandis*, Anderson & Belnap 1992, 2, § 33.4.
[22] As actually done in Łukasiewicz 1929, Chapter IV, § 24.
[23] The 'algebraic' alternative – which consists of [1] defining first a 'relative' *falsum* by **f**[α] := NC$\alpha\alpha$, say, and [2] proving next E**f**[α_1]**f**[α_2], for any two formulas α_1, α_2, etc. – induces unnecessary formal complications. A similar remark applies to Jaśkowski's 'natural deduction' quantifier-free system, based originally on [N,C] alone.
[24] It turns out that all we need is just a single new axiom ⊢ **v**, for this purpose. For completeness, see, for instance, Wajsberg 1937, I, § 5, resp. 1939, II, § 2, and the remark (below) that *ex falso quodlibet* can be obtained from the Łukasiewicz axiom ⊢ O[p,q] : CpCNpq and a paradigmatic proof of **v**, like ⊢ Ω : **v**.

§ 3.1 Proof-combinators

Taking CB, E and O – with the appropriate propositional parameters (here: [p,q,r,] [p], and [p,q], resp.) – as well as Ω, in the guise of primitive 'witnesses' for the corresponding axioms, detachment / *modus ponens* can be viewed as a binary (algebraic) operation $\mathfrak{D}$, from 'detachment' (to be defined properly – i.e., equationally – later on) acting on witnesses, to the effect that:

($\triangleright$ = *modus ponens*) if f is a witness for C$\alpha\beta$ and a is a witness for α, then $\mathfrak{D}$fa is a witness for β.

We write, for convenience, f(a) = (f $\triangleright$ a) := $\mathfrak{D}$fa. This is to be understood modulo arbitrary uniform substitutions, with the proviso that one must take most general substitutions into account (here, substitutions are endomorphisms of the corresponding [free] algebra).[25]

For the record, the formal grammar (for formulas, resp. *witness terms* [*proof-terms* or *w-terms*, for short]) is:

propositional variables :: p, q, r, ...
formulas :: α, β := p | Nα | C$\alpha\beta$
w-variables :: x, y, z, ...
w-terms :: a,b,c,d,e,f := x | Ω | CB | E | O | c$\triangleright$a.

Where α is a formula and a is a w-term, we write, as ever, $\vdash$ a : α, for the fact that a is a witness (actually, a w-term) for α.

So, we have, in particular, *derived* rules (here, *definable witness operators*):

$\vdash$ g $\circ$ f := CB[p,q,r](f)(g) : Cpr, if $\vdash$ f : Cpq and $\vdash$ g : Cqr,
$\vdash$ E[p](f) : p, if $\vdash$ f : CNpp,
$\vdash$ O[p](a)(c) : q, if $\vdash$ a : p and $\vdash$ c : Np.

Examples. Ignoring propositional parameters on witnesses, as well as explicit substitutions[26]:

$\vdash$ 4 := CB $\triangleright$ CB : CCCCqrCprsCCpqs
$\vdash$ 5 := 4 $\triangleright$ 4 = (CB $\triangleright$ CB) $\triangleright$ (CB $\triangleright$ CB) : CCpCqrCCsqCpCsr
$\vdash$ 6 := 4 $\triangleright$ 1 = (CB $\triangleright$ CB) $\triangleright$ CB : [exercise]
$\vdash$ 7 := 5 $\triangleright$ 6 : [exercise]
$\vdash$ 8 := 7 $\triangleright$ 1 : [exercise]
$\vdash$ 9 := 1 $\triangleright$ 3 = CB $\triangleright$ O : [exercise]

...

$\vdash$ I := 16 = 9 $\triangleright$ 2 = (CB $\triangleright$ O) $\triangleright$ C . Cpp - axiom in Peirce 1885

---

[25] Technically speaking, the $\mathfrak{D}$-operator is the 'condensed detachment' operator of Carew A. Meredith (1904–1976). Notably, the Irishman attended Łukasiewicz's lectures in Dublin, during the early 1950. See, e.g., David Meredith's bio-bibliographical note, Meredith 1977, and, possibly, Rezuş 1982, 2010, Kalman 1983, and Hindley & Meredith 1990, for details.

[26] The latter can be uniquely restored (modulo alphabetic variants) by the Robinson unification algorithm. Cf. Robinson 1965, 1979, and Rezuş 1982.

and so on.

Further, Łukasiewicz meticulously obtained

⊢ K := 18 : CqCqp ['the law of simplification'] –
axiom in Frege **BS**
⊢ CI := 20 : CpCCpqq
['assertion' or internalised *modus ponens*]
⊢ C := 21 : CCpCqrCCqCpr
['the law of commutation'] – axiom in Frege **BS**, as well as in Peirce 1885 [*Note* by Łukasiewicz (cca 1925): superfluous in Frege **BS**, it can be already obtained from K and S.]
⊢ B := 22 = C ▷ CB : CCqrCCpqCpr
[transitivity of implication: 'prefixing']
⊢ P := 24 : CCCpqpp
['the Law of Peirce'] - axiom in Peirce 1885
⊢ W := 30 : CCpCpqCpq
['Hilbert' or 'contraction']
⊢ S := 35 : CCpCqrCCpqCpr
['Frege' or 'selfdistribution on the major'] - axiom in Frege **BS**
⊢ Δ := 39 : CNNpp
['law of double negation' (elim)] - axiom in Frege **BS**
⊢ ∇ := 40 : CpNNp
['law of double negation' (intro)] - axiom in Frege **BS**
...
[46–49: 'the laws of transposition' (or 'contraposition')]
⊢ 46 : CCpqCNqNp - axiom in Frege **BS**
⊢ 47 : CCpNqCqCNp
⊢ 48 : CCNpqCNqp
⊢ 49 : CCNpNqCqp

etc.

This amounts to a 'typed' (stratified, decorated) *combinatory logic notation*, where one manipulates formulas in the guise of *principal type schemes*.[27]

Now, as Tarski should have known (in 1921), in presence of *modus ponens*, the Deduction Theorem (DT) – or implication-introduction – can be obtained from K and S alone. This yields the λ-calculus counterpart of the same story.

---

[27] Cf. Hindley 1969, 1997, Hindley & Seldin 1986, and Hindley & Meredith 1990. As a matter of fact, here, one has a 'rigid' typing, à la Church and de Bruijn, instead. For the difference, see Hindley 1997, Barendregt *et al.* 2013 and the review Rezuş 2015. We could have had a (typed) *combinator theory* – a 'combinatory logic' – as well, but, since the equational constraints on the primitive combinators are rather non-transparent, I prefer to skip the details. Otherwise, they can be recovered from remarks following below.

§ 3.2 (DT) and the λ-abstraction-algorithm

I have argued at length in Rezuş 2010 that Tarski must have been, likely, familiar with *some form* of (typed) λ-calculus – or (typed) combinatory logic or both – during the early 1920, knowledge that enabled him to prove some tricky axiomatisability results around 1925.[28]

Indeed, there is, essentially, a *single way* of proving (DT): the proof amounts to a simple *inductive argument*.

The reasoning can be repeated in any (propositional) logic – with substitution and *modus ponens*, as only primitive rules of inference – that contains the (witnessed) 'theses':

(K) ⊢ K[p.q] : CpCqp, and
(S) ⊢ S[p,q,r] : CCpCqrCCpqCpr.

Note first that, in such cases, one has, as derived rules:

[K] $\hat{\Gamma}$ ⊢ f : p ⇒ $\hat{\Gamma}$ ⊢ K[p,q](f) : Cqp, and
[S] $\hat{\Gamma}$ ⊢ f : CpCqr, $\hat{\Gamma}$ ⊢ g : Cpq ⇒ $\hat{\Gamma}$ ⊢ f □ g : Cpr,

where f □ g := S[p,q,r](f)(g), as well as the (witnessed) 'thesis':

(I) ⊢ I[p] : Cpp

[ignoring propositional parameters, the latter is available as S(K)(K)].

Suppose that we have obtained a proof b[x] of β from the assumption that we have a proof x of α (so that b[x] depends possibly on [x:α]). Then (DT) states that we must have a proof λx:α.b[x] := λ([x:α](b[x])) of Cαβ, that does not depend on the proof [x:α], *ceteris paribus*.[29] That is to say, formally,

(λ) $\hat{\Gamma}$ ⊢ λx:α.b[x] : Cαβ, if $\hat{\Gamma}$, [x:α] ⊢ b[x] : β,

for an appropriate assumption-context $\hat{\Gamma}$, as a parameter in the argument.

The induction pays attention to the form ('structure') of b[x]. To save repetitions, set e ≡ λx:α.b[x]. There are only three cases to examine:

(1) b[x] ≡ [x : α]; so α ≡ β; set e := I[α] : Cαα;
(2) b[x] : β does not actually depend on [x : α]; set e := K[β,α](b) : Cαβ;
(3) b[x] ≡ (f ▷ a) : β; where f : Cα'β and a : α'; then the (IH) guarantees $\hat{f}$:= (λx:α.f) : CαCα'β and $\hat{a}$:= (λx:α.a) : Cαα'; set e := $\hat{f}$ □ $\hat{a}$: Cαβ.[30]

---

[28] See, e.g., Rezuş 1982 and the discussion appearing by the end of Rezuş 2010.
[29] The *ceteris paribus* clause refers to the fact that the argument can be taken relative to a parameter $\hat{\Gamma}$ ≡ [x₁:α₁ ... xₙ:αₙ] (n > 0).
[30] This is the so-called 'bracket abstraction algorithm' obtained first in terms of combinators – and, rather late, in this form – by Haskell B. Curry (in 1948–1949) and, independently, by Paul C. Rosenbloom 1950, 2005ᴿ, that is about thirty years after Tarski. See also Rosser 1942, 1953 and Curry & Feys 1958, **6**S.1, etc. One can improve on the last clause (3), by processing first the subcase a ≡ x : α ≡ α', while setting e := f : Cα'β ≡ Cαβ.

§ 3.3 A natural deduction system for classical logic

As in (decorated / 'typed') λ-calculus, we can thus write (ignoring everywhere proof-context parametrisations):

(λ) $\vdash \lambda$x:p.b[x] : Cpq, if [x:p] $\vdash$ b[x] : q,

on a par with the usual 'cut'-condition [*modus ponens*]:

($\triangleright$) $\vdash$ f $\triangleright$ a : q, if $\vdash$ f : Cpq and a : p.

As one might already guess, this makes up the first step in a would-be attempt meant to replace the Łukasiewicz axioms – i.e., the primitive combinator team {CB, E, O} – together with Ω, on the signature [**f**, N,C], by appropriate witness operators (representing rules of inference).

Set now

(ϵ) $\vdash \epsilon$x:Np.a[x] := E(λx:Np.a[x]) : p, if [x:Np] $\vdash$ a[x] : p.

The latter derived rule (definable witness operator) is the *consequentia mirabilis* of Girolamo Cardano (1501–1576) or the *Rule of Clavius*, viewed as a single-premiss rule of inference[31].

As announced before, we adjoin the propositional constant **f** (*falsum*), with **v** := N**f** (*verum*), and a single additional (witness) axiom:

(Ω) $\vdash \Omega$: **v**

and set[32]

(ϖ) $\vdash \varpi$[p](e) := O[**f**,p](e)(Ω) : p, if $\vdash$ e : **f**,

with, finally

($\star$) $\vdash$ c $\star$ a := O[p,**f**](a)(c) : **f**, if $\vdash$ a : p, and $\vdash$ c : Np
    ['inner cut' or the 'rule / law of (non-) contradiction'].

Conversely, E and O can be obtained as

(E) $\vdash$ E[p] := λf:CNpp.ϵx:Np.(f$\star$x) : CCNppp,
(O) $\vdash$ O[p,q] := λx:p.λy:Np.ϖ[q](y$\star$x) : CCpCNpq.

---

[31] Cf. Cardano 1570, Lib. V, Prop. 201, p. 231, resp. Cardano 1663, 4, p. 580. For *pater* Clavius [Christophorus Clavius, aka Christoph Klau, SJ (1537–1612)], cf. Clavius 1611, 1.1, pp. 364–365 [comments *ad* Euclid **Elementa** IX.12], as well as 1.2, p. 11 [comments *ad* Theodosius **Sphaerica** I.12]. See, also Rezuş 1991 pp. 4, 23, 46, and Bellissima & Pagli 1996, *passim*, for details. Notably, Łukasiewicz was familiar with the references above, as well as with the medieval anticipations of his O-axiom (the 'Law of Duns Scotus'). Cf., e.g., Łukasiewicz 1929, 1930, 1934, 1957.

[32] This is the only place where we actually need Ω in derivations.

As is well-known, the rules (λ), ($\triangleright$), (ϵ), (ϖ), and ($\star$), with the additional axiom (Ω), suffice to yield full classical [propositional] logic.[33]

On the other hand, if ($\star$) is present, the rules (ϵ) and (ϖ) of the [**f**,N,C]-signature, taken together, are equivalent, in this context, to *reductio ad absurdum*, (∂), viewed as a single-premiss rule

($\partial \vdash \partial$x:Np.e[x] : p, if [x:Np] $\vdash$ e[x] : **f**.

Indeed, one has

(∂) $\vdash \partial$x:Np.e[x] := ϵx:Np.ϖ[p](e[x]) : p, if [x:Np] $\vdash$ e[x] : **f**,

and, conversely,

(ϵ) $\vdash \epsilon$x:Np.a[x] := ∂x:Np.(x$\star$a[x]) : p, if [x:Np] $\vdash$ a[x] : p, and
(ϖ) $\vdash \varpi$[p](e) := ∂x:Np.e, if [x:Np] $\vdash$ e : **f** (x not free in e)[34],

so that, finally, classical [propositional] logic can be based on

(1) the axiom (Ω), and the four rules:
(2) (λ) [the 'Deduction Theorem', implication-introduction],
(3) ($\triangleright$) [*modus ponens*, implication-elimination],
(4) (∂) [*reductio ad absurdum*], and
(5) ($\star$) ['the law of (non-) contradiction'].

The axiom is, in fact, redundant, since, in this case, one can define explicitly:

(df Ω) $\vdash \Omega := \partial$x:N**v**.$\partial$y:**v**.(x$\star$y) : **v**.

We shall keep, however, Ω around for a while, mainly for the sake of comparison with the Jaśkowski 1934 version of 'natural deduction'.

The 'natural deduction' system above is easily seen to be equivalent to the axiomatics of Łukasiewicz 1929, modified as above such as to fit the primitive [**f**,N,C]-signature. As the rules have been already seen to be derivable from the axioms, this amounts to writing down the explicit definitions of the witnesses (here, *combinators*) CB, E, and O in terms of the proof operators contained in the 'basis' $\{\lambda, \triangleright, \partial, \star\}$.

The corresponding (extended) λ-calculus is discussed next. It turns out that – if we forget about the constant Ω – one can even formulate it in a decoration-free ('type-free') setting. This allows us establishing its (Post-) *consistency* in a straightforward way, using only some basic λ-calculus facts.

---

[33] If the basis consists only of rules, as here, the axiom (Ω) is redundant. See below.
[34] No need for Ω, here. Cf., e.g., Rezuş 1990, 1991.

§ 4 $\lambda\partial$ – an extended λ-calculus

On the primitive [**f**,N,C]-signature, the minimal setting above – consisting of (Ω) [otherwise redundant], (λ) [implication-introduction], ($\triangleright$) [implication-elimination], (∂) [*reductio ad absurdum*], and ($\star$) ['the law of (non-) contradiction'] – can be viewed as an extension of the basic ('simple') 'typed' λ-calculus λ[C], obtained by replicating its pure (λ)-($\triangleright$)-part.

Formally, the decoration-free ('type-free') syntax of the resulting $\lambda\partial$-calculus – $\lambda\partial(\Omega)$, say – is given by:

witness-variables :: x, y, z, ...
witness terms :: a,b,c,d,e,f := x | λx.b | f$\triangleright$a | ∂x.e | c$\star$a.

In the resulting equational system, one has the usual $\beta\eta$-conditions for (λ) and ($\triangleright$) [decoration-free spelling]:

($\beta\lambda$) $\vdash$ (λx.b[x]) $\triangleright$ a = b[x:=a],
($\eta\lambda$) $\vdash$ λx.(c$\triangleright$x) = c (x not free in c),

as well as the analogous $\beta\eta$-conditions for (∂) and ($\star$):

($\beta\partial$) $\vdash$ c $\star$ (∂z.e[z]) = e[z:=c],
($\eta\partial$) $\vdash$ ∂z.(z$\star$a) = a (z not free in a),

together with the expected rules of monotony (compatibility of equality – here, *conversion* – with the operations).

This extension of pure λ can be easily seen to be consistent by interpreting it in the ['type-free'] $\lambda\pi$-calculus, $\lambda\pi$, for instance.[35] Alternatively, one can choose to equip the resulting calculus with an appropriate *notion of reduction* and establish *confluence* [via a Church-Rosser theorem] first.

The *intended* decoration (typing) is given by the conditions (λ), ($\triangleright$), (∂) and ($\star$). In view of the above, if considered as a (decorated / 'typed') λ-theory, the outcome – the $\lambda\partial$-calculus $\lambda\partial$[**f**,N,C] – is a *witness theory for classical logic*.

This yields the simplest *Curry-Howard Correspondence for (propositional) classical logic* I know of.[36] (Cf. **§ 7.4**, below.)

§ 5 The Jaśkowski $\lambda\chi\Omega$-calculus

In his 1927, 1934, Jaśkowski chose to hide the applications of the 'inner cut' ($\star$) – which, as noted above, would have required the additional propositional atom **f** (*falsum*) – and expressed *reductio ad absurdum* in the form of a more complex rule, viz. by the Medieval *ex contradictione quodlibet* ['explosion'] principle, viewed as a rule of inference:

---

[35] This is possible since, unlike the pure λ-calculus λ, the extensional $\lambda\pi$-calculus $\lambda\pi$ contains infinitely many nontrivial copies of itself. [Added in proof. Cf. now Rezuş 2017b.]

[36] See also Rezuş 1990, 1991, and Sørensen & Urzycyn 2006.

(χ) [z:Np] ⊢ c[z] : Nq, [z:Np] ⊢ a[z] : q ⇒ ⊢ χz:Np.(c,a) : p.

Upon adjoining the atom **f** and the 'hidden' rule ($\star$), the complex Jaśkowski rule (χ) can be obtained as:

(df χ) ⊢ χz:Np.(c,a) := ∂z:Np.(c$\star$a) : p, if [z:Np] ⊢ c : Nq, and [z:Np] ⊢ a : q,

while, conversely, one can have:

(∂) ⊢ ∂z:Np.e[z] := χz.($\mathit{\Omega}$, e[z]) : p, if [z:Np] ⊢ e[z] : **f**,

in terms of (χ) and ($\mathit{\Omega}$).

The 'hidden' rule ($\star$) – i.e., the 'law' of non-contradiction – however, can be obtained explicitly from Jaśkowski's (χ) only by an ad hoc contextual artifice, setting, e.g.,

($\star$) ⊢ c $\star$ a := χz:**v**.(c,a) : **f**, if ⊢ c : Np, and ⊢ a : p
[z fresh for c, a].[37]

As an aside, on ultimate formal grounds, I should have rather written down the Jaśkowski rule (χ), as:

[x:Np] ⊢ c[x] : Nq; [y:Np] ⊢ a[y] : q ⇒ ⊢ χ(x,y:Np).(c[x],a[y]) : p,

---

[37] In retrospect, it is difficult to say why Jaśkowski did prefer the complex (χ)-rule (a kind of 'mixed' abstractor, in witness-theoretic terms, like the rather complex *case*-construct [*or*-elimination] in intuitionism), as a primitive rule of inference, in place of the 'elementary' *reductio ad absurdum* (∂) [here, a monadic abstractor, like (λ)] and the 'hidden' rule / operator expressing the 'law of [non-] contradiction' ($\star$). *Prima facie*, I would suspect the choice was a matter of economy. Although there was an even more drastic economy in sight, that both Łukasiewicz and Jaśkowski were, apparently, well aware of, viz. by adopting the 'inferential' definition of negation, à la Peirce 1885, ∼p := Cpf, in which case the primitive rule (χ) could have been replaced by an 'inferential' variant of *reductio ad absurdum* (γ, a monadic abstractor, with ⊢ γz:∼p.e[z] : p, for [z:∼p] ⊢ e[z] : **f**, in decorated / 'typed' version, etc.). Cf. Rezuş 1990, 1991. As a matter of fact, in the latter case, the witness-theoretic properties (as regards proof-conversion resp. proof-reduction [= détour elimination]) of the γ-operator are more involved that those presupposed by the 'natural' [(∂),($\star$)]-pair, but Łukasiewicz and Jaśkowski did not think in such terms, anyway. Even Gentzen (1934–1935) was slightly confused as to the would-be proof-détours that could – and should – be associated to a genuine classical negation. It took us some thirty years, at least, until we were able to reach a clean conceptual insight on the matter. See, e.g., Prawitz 1965 for a solution, applying to the 'inferential' case and the combinator resp. λ-calculus variants, described in Rezuş 1990, 1991 [$\lambda\gamma$-*calculi*]. Besides, it took us about other twenty years, in order to get something as simple as the $\lambda\partial$-calculus sketched under **§ 4** above (Rezuş, cca 1987), corresponding to what the pioneers – Frege, Peirce, Russell, Łukasiewicz, Leśniewski, Tarski etc. – might actually have had in mind.

but, as the two premisses of (χ) are independent, I am probably allowed to use (a subtle form of meta-) α-conversion in this context.[38]

From this, the reader can easily reconstruct by herself the witness theory corresponding to Jaśkowski's natural deduction system for classical logic, i.e., a would-be *Jaśkowski* $\lambda\chi\Omega$-*calculus* – $\lambda\chi\Omega[\mathbf{f},N,C]$, say – (equationally) equivalent to $\lambda\partial\Omega[\mathbf{f},N,C]$ above.

One might also note the fact that the original system of Jaśkowski (1927, 1934) – without $\mathbf{f}$ and (Ω), $\lambda\chi[N,C]$, say – was just a *notational device* (no proof-conversion, resp. proof-reduction rules). Moreover, it was constructed on a *functionally incomplete* propositional signature (as noted before, we cannot retrieve the constants $\mathbf{f}$, $\mathbf{v}$, definitionally, from N and C alone), whence the attempt to associate appropriate conversion-conditions to the Jaśkowski χ-primitive could only yield a *proper subsystem* of $\lambda\partial(\Omega)[\mathbf{f},N,C]$.

§ 5.1 Jaśkowki graphical representation: block structures

Worth mentioning is also the fact that Jaśkowki proposed a perspicuous *graphical representation* of his proof-primitives in the original paper of 1927 – a kind of block-structure, meant to isolate intuitively *sub-proofs* of a given proof (actually, sub-terms in the corresponding λ-calculus description) – that was perfected by Frederic Brenton Fitch (1908–1987) *et alii*, later on.[39]

Otherwise, the tedious and rather non-transparent formal description of the 'supposition rules' in Jaśkowski (1934) can be easily re-shaped, equivalently, in terms of assumption contexts and (witnessed) entailments as already suggested in the above. It is relatively easy to see that the usual 'structural' rules of Gentzen are implicit in Jaśkowski's description. Actually, Gentzen's L-system for classical logic is just a disguised form – namely, *a special case* – of natural deduction.[40]

---

[38] Viewed abstractly, the witnessed entailments are, actually, a kind of *meta-combinators*, or *closed meta-terms*, in the end. — Incidentally, with the terminology mentioned earlier, the Jaśkowski witness operator χ should have gen-arity [1,1], not gen-arity [2] (sic), whence the alternative spelling above.

[39] Cf. Fitch 1952 and Anderson & Belnap 1975, 1992, for applications to intensional logics. Notably, a similar representation was invented and used, later – independently – by Hans Freudenthal (1905–1990), in didactic presentations of classical logic, as well as by Nicolaas G. de Bruijn (1918–2012), in his work on AUTOMATH [automated mathematics] and on the so-called 'Mathematical Vernacular' [WOT = Wiskundige Omgangstaal, in Dutch]. On this, see, mainly, de Bruijn's lectures on Taal en structuur van de wiskunde [The language and structure of mathematics], given at the Eindhoven Institute of Technology, Department of Mathematics and Computing Science, during the Spring Semester 1978, and summarised subsequently [in Dutch], in Euclides 44 (1979–1980), as well as Rezuş 1983, 1990, 1991, for further references.

[40] A Gentzen L-sequent 'multiple on the right', $\alpha_1, ..., \alpha_m \vdash \beta_1, ..., \beta_n$ (m, n $\geq$ 0), is a specific entailment of the form $\alpha_1, ..., \alpha_m, \bar{\beta}_1, ..., \bar{\beta}_n \vdash \mathbf{f}$ – where $\bar{\beta}_i$ is a kind of 'surface negation' of β_i (for $0 < i < n+1$), a rather confusing idea based on an ad hoc piece of ideography – also known as *rejection* or *refutation* (elenchos, in the

§ 5.2 (DN) extensions

Equally worth recording here is the (redundant) extension on the same primitive propositional signature [**f**,N,C], mentioned by the end of Rezuş 2016, which consists of adding the double-negation (DN) rules:

(Δ) $\vdash \Delta[p](c) : p$, if $\vdash c : NNp$ [double-negation! elimination].
(∇) $\vdash \nabla[p](a) : NNp$, if $\vdash a : p$ [double-negation! introduction],

In the latter case, the (DN) witness operators [rules of inference] (∇) and (Δ) are supposed to obey inversion principles of the form

($\beta\Delta$) $\vdash \nabla(\Delta(c)) = c : NNp$,
($\eta\Delta$) $\vdash \Delta(\nabla(a)) = a : p$.

As earlier, the resulting extension (decorated / 'typed' λ-calculus, $\lambda\partial^\Delta$, say) can be shown to be consistent by interpreting its 'type-free' variant in the (undecorated) $\lambda\pi$-calculus.[41]

It is easy to see that, in the formulation without primitive (DN)-rules, at least one of the ($\beta/\eta\Delta$)-conditions would normally fail, whence the idea of taking ∇ and Δ as primitive proof operators (rules of inference).

§ 6 Propositional and first-order quantifiers

The extensions to quantifiers (either propositional or first- resp. second order) are straightforward.[42]

Illustrated next is the extension to *propositional quantifiers* on the (otherwise redundant) signature [**f**,N,C,Π], with Π standing for the universal quantifier, as in Łukasiewicz 1929 and Jaśkowski 1934. As above, α, β, ..., possibly with sub- and/or superscripts are used as metavariables ranging over formulas. If the propositional variable p occurs free (even fictitiously so) in a formula α, we write α[p] in order to make this visible. Substitutions are mentioned accordingly: a[p:=α], and β[p:=α] resp. (read 'p becomes α in a, resp. in β').

For the extended witness-syntax there are required two more proof operators (rules of inference), corresponding to *Generalisation* (Λ) and *Instantiation* ($\blacktriangleright$) resp. The new pair [(Λ),($\blacktriangleright$)] is analogous to the [(λ),($\triangleright$)]-pair above.

Greek of Aristotle and Chrysippus), about two millennia before both Jaśkowski and Gentzen were born. As a matter of fact, *mutatis mutandis*, Chrysippus' conceptual setting was cleaner. Cf Rezuş 2010 for details.

[11] So, once more, consistency can be established already at undecorated / 'type-free' level. In fact, $\lambda\partial^\Delta$ is redundant: if the [(Δ),(∇)]-pair is present, we can leave out either the [(λ),($\triangleright$)]-pair or the [(∂),($\star$)]-pair, viz. $\lambda\partial^\Delta$ is, ultimately, equivalent [in a 'type-free' setting] with each of its 'halves', λ^Δ, resp. ∂^Δ. See, *mutatis mutandis*, § 7.5.

[42] See, e.g., Rezuş 1990, 1991 for first-order quantifiers, and possibly, Rezuş 1986 for the 'extended propositional calculus' (i.e., classical logic with propositional quantifiers), as well as for the second-order case.

We present here a version close to Jaśkowski 1934, leaving to the reader the task of showing equivalence with the corresponding formulation of Łukasiewicz 1929. As above, the construction admits of a decoration-free description.

The decoration-free ['type-free'] syntax of the resulting system ($\lambda\partial\Lambda$) is:

witness-variables :: x, y, z, ...
witness terms :: a,b,c,d,e,f := x | λx.b | f$\triangleright$a | ∂x.e | c$\star$a | Λp.a | f$\blacktriangleright\alpha$.

The additional conversion rules are (in 'type-free' spelling):

($\beta\Lambda$) ⊢ (Λp.a[p]) $\blacktriangleright$ α = a[p:=α],
($\eta\Lambda$) ⊢ Λp.(c $\blacktriangleright$ p) = c, if p is not free in c[43],

together with the corresponding monotony conditions for (Λ) and ($\blacktriangleright$), meant to make equality (conversion) compatible with the operations.[44]

The decoration ['typing'] is, as expected, relative to an arbitrary assumption context (i.e., a finite list Γ of decorated witness variables, omitted below). We have (λ), ($\triangleright$), (∂), and ($\star$), like before, as well as the new rules (for α, β arbitrary formulas):

(Λ) ⊢ (Λp.a[p]) : Πp.α[p], if [p] ⊢ a[p] : α[p],
($\blacktriangleright$) ⊢ (f $\blacktriangleright$ α) : β[p:=α], if ⊢ f : Πp.β[p].

The first- (resp. second-) order case is completely analogous. In each case, the corresponding λ-calculi can be shown to be consistent by simple translation arguments.

§ 7 'Chysippean' re-formulations of the $\lambda\partial\Lambda$-calculus

The careful reader might have noticed a *general principle of construction* behind the *witness theory* (proof-system) $\lambda\partial\Lambda$ above, viz. the fact that the primitive witness / proof operators come in pairs [(abs),(cut)], where (abs) is a (monadic) *abstraction operator* and (cut) is a *'cut'-operator*, i.e., an operation meant to 'eliminate' its associated abstractor (abs). Moreover, each such a pair is supposed to *characterise* the associated rules of inference as *operators*, by *equational stipulations* (here, $\beta\eta$-conditions), i.e., more or less, algebraically, by stipulating their 'characteristic behaviour'. One could thus notice a *uniform introduction-elimination pattern* (of construction), provided *one thinks in terms of witnesses* (here: proofs), *not in terms of* bare *formulas* (expressing propositions / propositional schemes).

---

[43] Exactly as in Girard's 'System F' (PhD Dissertation, Paris 7, 1972). Of course, the latter λ-calculus is the [(∂),($\star$)]-free fragment of $\lambda\partial\Lambda$, i.e., $\lambda\Lambda$, by present notational standards. Cf. Rezuş 1986 for details on the Girard-Reynolds λ-calculus.

[44] Since I have omitted, everywhere in the above, any reference to proof-contexts, the usual provisoes on p-variables are also tacitly assumed.

Technically, it is also possible to describe a *proper extension* of the *witness theory* (proof-system) $\boldsymbol{\lambda\partial\Lambda}$, based on an idea that goes back to the founder of classical logic, the Stoic philosopher Chrysippus of Sol[o]i, twenty-two centuries before. The extension is a λ-theory, i.e., a consistent λ-calculus, as well (both 'type-free', and decorated / 'typed' as above). Of course, I will not credit the famous Phoenician with the details, but the reader should be certainly able to recognise *the Chrysippean spirit* behind the construction.[45]

Writing down things in 'Polish' – i.e., in Łukasiewicz notation, as everywhere here – I will use the same propositional signature as before, viz. $[\mathbf{f}, N, C, \Pi]^{46}$, but choose a slightly different team of primitive witness operators, while leaving ∂, $\star$ and λ unchanged, add two kinds of 'pairs', namely $\pi(...,...)$ and $\downarrow(...,...)$, as well as a (mixed) dyadic abstraction-operator Σ, [writing, conveniently, $\prec a,f \succ \equiv \pi(a,f)$ and $\downarrow_\alpha(a) \equiv \downarrow(\alpha,a)$], and $\Sigma(p,x).c[p,x]$, resp., for proof- / witness-terms a, c[p,x], f and formulas α.

Whence the expected formal grammar [at a decoration-free / 'type-free' level], with p, q, r, ..., as (meta-variables for) propositional variables, as ever:

formulas :: $\alpha, \beta := p \mid \mathbf{f} \mid N\alpha \mid C\alpha\beta \mid \Pi p.\alpha$
w-variables :: x, y, z, ...
w-terms :: $a,b,c,d,e,f := x \mid \partial x.e \mid c{\star}a \mid \lambda x.b \mid \prec a,f\succ \mid \Sigma(p,x).a \mid \downarrow_\alpha(a)$.

The (decoration-free / 'type-free') equational theory – called $\boldsymbol{\partial\lambda^*\Sigma}$, for convenience – consists of

$(\beta\partial) \vdash c \star (\partial z.e[z]) = e[z{:=}c]$,
$(\eta\partial) \vdash \partial z.(z{\star}a) = a$ (z not free in a),

as before, in $\boldsymbol{\lambda\partial(\Lambda)}$, and the following 'polar' conditions:

$(\beta\lambda^*) \vdash \prec a,f\succ \star (\lambda x.b[x]) = f{\star}(b[x{:=}a])$,
$(\eta\lambda^*) \vdash \lambda x.\partial y.(\prec x,y\succ \star c) = c$ (x, y not free in c), as well as
$(\beta\Sigma) \vdash \downarrow_\alpha(a) \star (\Sigma(p,x).c[p,x]) = c[p{:=}\alpha, x{:=}a]$,
$(\eta\Sigma) \vdash \Sigma(p,x).(\downarrow_p(x) \star c) = c$ (p, x not free in c),

together with the expected monotony constraints on the primitive witness operators.

The *intended* decoration ('typing') is given by

---

[45] Cf. Rezuş 2016, for technical – and historical – evidence supporting the claim. The extension works for the system with (DN)-primitives, too.

[46] We may want to abbreviate, for convenience, $\bar{C} := NC$ and $\bar{\Pi} := N\Pi$ (so that $\bar{C}$ is marked as the 'polar [opposite]' of C, and $\bar{\Pi}$ as the 'polar [opposite]' of Π, resp.), but the Łukasiewicz notation makes this superfluous. One might also note the fact that, by the standards of Rezuş 2016, $\bar{C}$ and $\bar{\Pi}$ would have counted as *Chrysippean connectives*. Specifically, $\bar{C}$ corresponds to the Chrysippean connector (binary connective) *more*, i.e., māllon... ē..., a kind of *rather... than...*, in English, while the quantifier-free part of the extended calculus [with (DN)-primitives] – to be described next – corresponds exactly to the semantic $(C-\bar{C})$-fibration of 'Chrysippean logic' **Ch**.

(∂) ⊢ ∂x:Nα.e[x] : α, if [x:Nα] ⊢ e[x] : **f**,
($\star$) ⊢ c$\star$a : **f**, if ⊢ c : Nα and ⊢ a : α,
(λ) ⊢ λx:α.b[x] : C$\alpha\beta$, if [x:α] ⊢ b[x] : β,

as in the case of **$\lambda\partial(\Lambda)$**, with moreover,

(π) ⊢ ≺a,f≻ : NC$\alpha\beta$, if ⊢ a : α, and ⊢ f : Nβ, and
(Σ) ⊢ Σ(p,x:Nα).c[p,x] : Πp.α, if [p] [x:Nα] ⊢ c[p,x] : **f**,
($\downarrow$) ⊢ $\downarrow_\alpha$(c) : NΠp.α, if ⊢ c : Nα

– with the expected restriction on (Σ) – so that the classical 'polarities' become obvious.[47] Note that there is no primitive *modus ponens* ($\triangleright$), in **$\partial\lambda^*\Sigma$**. (See, however, **§ 7.2**.)

A few more (technical) comments are in order.

§ 7.1 An equationally equivalent system

Setting, in **$\partial\lambda^*\Sigma$**,

(df Λ) Λp.a[p] := Σ(p,z).(z$\star$a[p]), z not free in a[p],

we get conditions analogous to ($\beta\lambda^*$) and ($\eta\lambda^*$), viz.

($\beta\Lambda^*$) ⊢ $\downarrow_\alpha$(f) $\star$ (Λp.a[p]) = f$\star$(a[p:=α]),
($\eta\Lambda^*$) ⊢ Λp.∂z.($\downarrow_p$(z) $\star$ c) = c (p, z not free in c),

as well as monotony for the defined operator.

Note that the defined operator *inherits* (of) the *intended* decoration ('typing') from the primitive decoration of the *definientia*.

Conversely, let us replace the primitive dyadic abstractor Σ, of **$\partial\lambda^*\Sigma$**, with a primitive monadic abstraction operator Λ, subjected to the conditions ($\beta\Lambda^*$) and ($\eta\Lambda^*$) above, including monotony for Λ, and call the resulting system **$\partial\lambda^*\Lambda^*$**. Defining, in the latter system,

(df Σ) Σ(p,z).e[p,z] := Λp.∂z.e[u,z],

one has, by easy calculations, ($\beta\Sigma$) and ($\eta\Sigma$), so that, ultimately, **$\partial\lambda^*\Sigma$** and **$\partial\lambda^*\Lambda^*$** turn out to be equationally equivalent.

---

[47] Formally, ∂ looks, in the end, like a kind of degenerated Σ (sic). The informed reader has already realised the fact that the (Σ-$\downarrow$)-rules are just (undecorated / 'type-free') analogues of the usual *intuitionistic* ∃-*rules*. Cf. Rezuş 1986, 1991, etc. — On the historical side, if I am not very mistaken, I remember having encountered something similar to the 'polar' pair [(λ),(π)] in work of Dag Prawitz, going back to the late nineteen sixties and the early seventies (although with no reference to the Stoic lore and/or to would-be [classical] proof-isomorphisms, i.e., to proper proof-conversion rules). Whence, ultimately, the basic idea behind the construction of **$\partial\lambda^*(\Sigma)$** should be, very likely, accounted for as a piece of *(historical) data-retrieval*, rather than as a genuine finding, due to the present author. In retrospect, virtually *any mindful reader* of Prawitz, already familiar with the basics of λ-calculus, *could have came out with a similar proof-formalism*, even ignoring the Chrysippean antecedents.

§ 7.2 $\lambda\partial\Lambda$ is a subsystem of $\partial\lambda^*\Sigma$

It is easy to establish the fact that $\lambda\partial\Lambda$ *is a subsystem of* $\partial\lambda^*\Sigma$.[48] Indeed, define, in $\partial\lambda^*\Sigma$,

(df ▷) c ▷ a := ∂y.($\prec$a,y$\succ\star$c), y not free in a and c.

This yields $(\beta\lambda)$ and $(\eta\lambda)$, and, of course, monotony for the defined (▷)-operator. Set now, as before,

(df Λ) Λp.a[p] := Σ(p,z).(z$\star$a[p]), if z is not free in a[p], and
(df ▶) c▶α := ∂z.($\downarrow_\alpha$(z)$\star$c), if z is not free in c.

This yields $(\beta\Lambda)$ and $(\eta\Lambda)$, as well as the expected monotony conditions for the defined $[(\Lambda),(▶)]$-pair of operators.

The fact that the defined operators *inherit* the *intended* decoration ('typing') from the primitive decoration of the *definientia* is obvious.

§ 7.3 The $\partial\int$-calculus

Incidentally, the $(\Sigma$-$\downarrow)$-free fragment of ['type-free'] $\partial\lambda^*\Sigma$ – call it $\partial\lambda^*$, for convenience – admits of an alternative, more general formulation [at a decoration-free level].

Indeed, setting (x,y distinct free variables in e[x,y]),

(df $\int$) $\int$(x,y).e[x,y] := λx.∂y.e[x,y] (for the *split* operator),

we get, in $\partial\lambda^*$,

$(\beta\int)$ ⊢ $\prec$a,b$\succ$ $\star$ ($\int$(x,y).e[x,y]) = e[x:=a,y:=b],
$(\eta\int)$ ⊢ $\int$(x,y).($\prec$x,y$\succ$ $\star$ c) = c (x,y not free in c),

together with the expected monotony condition for $\int$, and it is obvious that we can trade $\int$ for λ in this context (at a decoration- / 'type-free' level), i.e., that one could have had, in the background, a calculus $\partial\int$, say, instead of $\partial\lambda^*$, in the above.

To see this, define, as before,

(df λ) λx.b[x] := $\int$(x,y).(y$\star$b[x]), y not free in b[x], as well as
(df ▷) c ▷ a := ∂y.($\prec$a,y$\succ\star$c), y not free in a and c,

in $\partial\int$. This yields the expected conditions $(\beta\lambda^*)$ and $(\eta\lambda^*)$, as well as $(\beta\lambda)$ and $(\eta\lambda)$, so that $\partial\int$ and $\partial\lambda^*$ are equationally equivalent, too.

As a bonus, for $\partial\int$ (Post-) consistency is straightforward. The latter is a (proper) subsystem of $\lambda\pi$: define the operator $\int$, in the $\lambda\pi$-calculus, by

---

[48] To show that $\lambda\partial\Lambda$ is a *proper* subsystem of $\partial\lambda^*\Sigma$ requires a more involved argument. I'd rather defer the details (too far from the subject of the present notes, anyway).

$\int[x,y].c[x,y] := \partial z.c[x:=\mathbf{1}(z),y:=\mathbf{2}(z)]$,

where $\mathbf{j}(c)$, j := 1, 2, are the usual $\lambda\pi$-projections, and $\partial \equiv \lambda$, for convenience.[49]

The latter definitional pattern of the $(\lambda,\triangleright)$-pair, in $\partial\int$ – actually, in $\lambda\pi$ – can be iterated in order to yield an infinite sequence of distinct $(\lambda,\triangleright)$-pairs satisfying conditions analogous to $(\beta\lambda)$ and $(\eta\lambda)$. Set

$\lambda_0 x.b := \partial x.b[x]$,
$c \triangleright_0 a := a \star c$, and, for n $\geq$ 0,

$\lambda_{n+1} x.b[x] := \lambda_n z.(b[x:=\mathbf{1}(z)] \triangleright_n \mathbf{2}(z))$, z fresh for b[x],
$c \triangleright_{n+1} a := \lambda_n y.(c \triangleright_n \prec a,y\succ)$,

whence, by induction, for n $\geq$ 0,

$(\beta\lambda_n) \vdash (\lambda_n x.b[x]) \triangleright_n a = b[x:=a]$,
$(\eta\lambda_n) \vdash \lambda_n x.(c \triangleright_n x) = c$ (x not free in c).

Some variations on this are also possible. For an alternative, see **§ 7.5**.

For $\partial\int$, the intended *art déco* would have been different, however. Actually, equational equivalence holds only in a decoration-free setting. In $\partial\int$, one should change the primitive (propositional) signature, by replacing the primitive C [implication] with D [the 'Sheffer-functor' *incompatibility*, or **nand**, i.e., semantically, negated classical conjunction], whereupon N [classical negation] becomes redundant, by setting Np := Dpp. The resulting ['typed'] calculus $\partial\int$[f,D], say – based on the witness primitives ∂, $\star$, $\int$ and the pair-construct π, as well as on the associated $\beta\eta$-conversion conditions $(\beta\partial)$, $(\eta\partial)$, and $(\beta\int)$, $(\eta\int)$, resp. – is, actually, an extension of $\partial\lambda^*$[f,N,C], with Cpq := DpNq, in $\partial\int$[f,D].[50]

§ 7.4 A few more extensions

Where T_1, T_2 are equational theories, let us write, for convenience, $T_1 \preceq T_2$, resp. $T_1 \simeq T_2$, for the fact that T_1 is a subsystem of T_2, resp. that T_1 and T_2 are equationally equivalent. From the above, we know already that

$$\lambda \preceq \lambda\partial \preceq \partial\lambda^* \simeq \partial\int \preceq \lambda\pi.$$

The reader can establish easily the fact that, in $\partial\lambda^*$, one can replace $(\beta\lambda^*)$ by the condition:

---

[49] The $\lambda\pi$-calculus is known to be consistent by a well-known lattice-theoretical (actually topological) construction due to Dana Scott (1969; cf. Scott 1973), as well as by constructive ('syntactical') means, as shown recently by Kristian Støvring (November 2005, rev. 2006).

[50] In his Warsaw lectures, Łukasiewicz alluded actually to the alternative – cf., e.g., Łukasiewicz 1929, Chapter II **§ 17** – but he was, apparently, distracted by *provability* details on Henry M. Sheffer 1913, 1913a and Jean Nicod 1917–1920, so that the idea was diluted. It is only in (very) recent times that the Peirce-Sheffer *nand* and *nor* connectives deserved a proper treatment in 'natural deduction' terms.

(β**p**) ⊢ ≺a,c≻ ⋆ f = c ⋆ (f ▷ a).

Let now [in the syntax of $\partial\lambda^*$], $\lambda\partial\mathbf{p}_\beta := \lambda\partial + (\beta\mathbf{p})$, and consider the (equational) extensions $\lambda\partial\mathbf{p} := \lambda\partial_\beta + (\eta\mathbf{p})$, $\partial\lambda^*\mathbf{p} := \partial\lambda^* + (\eta\mathbf{p})$ [in the syntax of $\partial\lambda^*$], and $\partial\!\!\int\!\mathbf{p} := \partial\!\!\int + (\partial\!\!\int)$ [in the syntax of $\partial\!\!\int$], with the additional conditions:

(η**p**) ⊢ ∂z.e[z] = λx.∂y.e[z:=≺x,y≻],
($\partial\!\!\int$) ⊢ ∂z.e[z] = $\int$(x,y).e[z:=≺x,y≻],

resp.[51] One can establish the fact that

$$\lambda \preceq \lambda\partial \preceq \lambda\partial\mathbf{p}_\beta \preceq \lambda\partial\mathbf{p} \simeq \partial\lambda^*\mathbf{p} \simeq \partial\!\!\int\!\mathbf{p} \preceq \lambda\pi^{52}.$$

The reader may also want to contemplate, in particular, the *separation properties* of the $\lambda\partial\mathbf{p}$-axiomatics, which consists of [1] pure β-η-λ-conditions, ($\beta\lambda$), ($\eta\lambda$), [2] pure β-η-∂-conditions, ($\beta\partial$), ($\eta\partial$), as well as (mixed) [3] β-η-π-conditions, (β**p**), (η**p**), on the primitive 'pairs' (π).

Summing up, the *intended* decoration ('typing') for $\lambda\partial\mathbf{p}$[f,N,C], based on the bare 'propositional' (quantifier-free) syntax:

formulas :: α, β := p | **f** | Nα | C$\alpha\beta$
w-variables :: x, y, z, ...
w-terms :: a,b,c,d,e,f := x | λx:α.b | f▷a | ∂x:Nα.e | c⋆a | ≺a,f≻,

is given by:

(λ) ⊢ λx:α.b[x] : C$\alpha\beta$, if [x:α] ⊢ b[x] : β,
(▷) ⊢ f▷a : β, if ⊢ f : C$\alpha\beta$ and ⊢ a : α,

(∂) ⊢ ∂x:Nα.e[x] : α, if [x:Nα] ⊢ e[x] : **f**,
(⋆) ⊢ c⋆a : **f**, if ⊢ c : Nα and ⊢ a : α,

(π) ⊢ ≺a,f≻ : NC$\alpha\beta$, if ⊢ a : α, and ⊢ f : Nβ,

i.e., in this version, the primitive rules of classical ['propositional'] logic, based on the primitive [**f**,N,C]-signature, are the Deduction Theorem, *modus ponens*, *reductio ad absurdum*, the 'law of (non-) contradiction', and the rule (π), a rule of 'NC-introduction', so to speak.

---

[51] We can also spot some obvious redundancy in $\partial\!\!\int\!\mathbf{p}$.
[52] For an explict way of inserting the ∂-segment – i.e., the [(∂)-(⋆)]-pair – in $\lambda\pi$, at decoration-free ['type-free'] level, see § **7.5**. All inclusions are *strict*. As for $\lambda\pi$, 'projections' – relative to a primitive π, where available – are definable in **p**-subsystems, but the resulting pairing is not 'surjective', i.e. ($\eta\pi$) ['surjectivity of pairing'] fails.

§ 7.5 More iterations

On a different route, one can define explicitly, in $\boldsymbol{\lambda\pi}$, a $(\partial,\star)$-pair satisfying $(\beta\partial)$ and $(\eta\partial)$, as well.

With a usual primitive pairing (i.e., pairs $\prec a,b \succ$ and projections $\mathbf{j}(c)$, $j:=1,2)$, as ever, let $[(\lambda),(\triangleright)]$ be an arbitrary (abs,cut)-pair satisfying $(\beta\lambda)$ and $(\eta\lambda)$, and define, successively:

(df Δ) $\Delta(c) := \lambda z.(c \triangleright \mathbf{1}(z) \triangleright \mathbf{2}(z))$,
(df ∇) $\nabla(f) := \lambda x.\lambda y.(f \triangleright \prec x,y \succ)$,

(df ∂_0) $\partial z.e[z] \equiv \partial_0 z.e[z] := \Delta(\lambda z.e[z])$,
(df $\star_0$) $c \star a \equiv c \star_0 a := \nabla(a) \triangleright c$,

(df λ_0) $\rho x.e[x] \equiv \lambda_0 x.e[x] := \partial x.e[x:=\Delta(x)]$,
(df $\triangleright_0$) $c \triangleright_0 a := \nabla(a) \star c$,

iterating, next, for any natural number n,

(df λ_{n+1}) $\lambda_{n+1} x.e[x] := \partial_n x.e[x:=\Delta(x)] \equiv \Delta(\lambda_n x.e[x:=\Delta(x)])$, with
(df ∂_n) $\partial_n z.e[z] := \Delta(\lambda_n z.e[z])$, and

(df $\triangleright_{n+1}$) $c \triangleright_{n+1} a := \nabla(a) \star_n c \equiv \nabla(c) \triangleright_n \nabla(a)$, with
(df $\star_n$) $c \star_n a := \nabla(a) \triangleright_n c$.

From this, we get the inversion

$(\beta\Delta) \vdash \nabla(\Delta(c)) = c$,
$(\eta\Delta) \vdash \Delta(\nabla(a)) = a$,

and, for any natural number n, $(\beta\text{-}\eta)$-conditions

$(\beta\lambda_n) \vdash (\lambda_n x.b[x]) \triangleright_n a = b[x:=a]$,
$(\eta\lambda_n) \vdash \lambda_n x.(c \triangleright_n x) = c$ (x not free in c),

$(\beta\partial_n) \vdash c \star_n (\partial_n z.e[z]) = e[z:=c]$,
$(\eta\partial_n) \vdash \partial_n z.(z \star_n a) = a$ (z not free in a),

as well as the expected monotony rules for the operations so defined.

In other words, unlike pure λ-calculus, $\boldsymbol{\lambda}$, the $\lambda\pi$-calculus, $\boldsymbol{\lambda\pi}$, contains infinitely many distinct (non-trivial) copies of itself, so to speak. In particular, the fragments $\boldsymbol{\lambda}^\Delta$, $\boldsymbol{\rho}^\Delta$, and $\boldsymbol{\lambda\rho}$, defined in the obvious way, by the corresponding $(\beta\text{-}\eta)$-conditions, are easily seen to be equationally equivalent.

The availability of the 'inversion' $[(\beta\Delta),(\eta\Delta)]$ ensures consistency for calculi containing, as a primitive, a $[(\Delta),(\nabla)]$-pair satisfying $(\beta\Delta)$ and $(\eta\Delta)$.

§ 7.6 Consistency matters

In the end, $\partial\lambda^*\Sigma$ – as well as $\lambda\partial p\Sigma$ (i.e., the extension of $\lambda\partial p$ with (Σ-$\downarrow$)-primitives) – is (*Post*) *consistent*, and so are the corresponding (DN)-extensions. For the (Σ-$\downarrow$)-free part of the proof, the result is already contained in the above. The genuine (Σ-$\downarrow$)-part consists of a trivial translation argument, collapsing the full system on its (Σ-$\downarrow$)-free fragment.

§ 7.7 Chrysippean 'polar' constructions

One might also notice that the analogous calculi $\partial\int(\mathbf{p})\Sigma[\mathbf{f},\mathrm{D},\Pi]$, based on $\{(\partial), (\star), (\int), (\pi), (\Sigma), (\downarrow)\}$, are 'polar' (Chrysippean) constructions, as well. The same remark applies to the corresponding (DN)-extensions. The basic equational conditions are $(\beta\partial)$, $(\eta\partial)$, $(\beta\int)$, $(\eta\int)$, while the **p**-extension has also $(\partial\int)$, whereupon either one of $(\eta\partial)$ or $(\eta\int)$ are redundant.

§ 7.8 (Lambek) C-monoids

As a final remark, *all consistency proofs* mentioned in this paper amount to an easy – even though oft slightly involved – exercise of (explicit) *definability in the* ('type-free') $\lambda\pi$-*calculus* ($\lambda\pi$). Algebraically speaking, *we are dealing with* (a rather specific class of) *monoids*.

Since (the intuitionistically decorated) $\lambda\pi$ is also known as 'the internal language of CCCs' [cartesian closed categories] among category theorists, most of the facts relevant here should also amount to *category-theoretic folklore*.[53]

§ 8 *Coda*

I hope my discussion above has made more or less clear *what* Jan Łukasiewicz – and his (very) young student Stanisław Jaśkowski, as well as his (equally) young colleague Alfred Tajtelbaum [Tarski][54] – *did actually know and/or could have known*, as regards 'natural deduction', during the mid- and late twenties. Why they did not invent something like [decorated / 'typed'] λ-calculus, in order to make things *conceptually clean*, evades me completely. It is up to my better informed – and more gifted – readers to speculate upon.

---

[53] The specific subject falling under the label *cartesian closed monoids* [CCMs, for short] – has been invented by Joachim Lambek sometime during the 1970's and has been vastly explored since, mainly in research on *categorical models of* λ-*calculus*. On the equational theory of CCM's, also known as *C-monoids* – equivalent with the ['type-free'] $\lambda\pi$-calculus – cf. Koymans 1982, 1984, and Lambek 1980, Lambek & Scott 1986, as well as the work of Adam Obtułowicz (Warsaw 1979 and later), Takanori Adachi (Tokyo 1982, 1983), Hirofumi Yokouchi (Tokyo 1983), and Pierre-Louis Curien (PhD Dissertation Paris 7, 1985), cited there.

[54] See Rezuş 1982, 1990 for relevant complements of information.

Acknowledgements. The present notes consist of a thorough revision of a draft dated Nijmegen, 9 October 2015, based on previous work. The bulk of the *historical* information on the *early Polish logic school* (Łukasiewicz, Jaśkowski, Tarski *et al.*) is based on research done as a graduate student at the University of Bucharest (1972–1977), while busy with a very different subject, abandoned since. My specific interest in 'Polish logic' has been first stimulated by several Romanian mathematicians and philosophers, including some of my teachers and friends. On the other hand, most of the *technical* details – on λ-calculus – appearing, mainly, in **§ 7**, go back to an equally old research project on *subsystems of λ-calculus*, submitted to the Chair of Dirk van Dalen [*Logic and Foundations of Mathematics*] at the University of Utrecht, in 1978. The original project, actually supervised by Henk Barendregt, was focused on the *strict* λ-calculi of Alonzo Church (especially Church's 'ordinal logic', appearing in his Princeton Lectures of 1935–1936), as well as on the proof theory of *relevance logics* – Anderson, Belnap *et al.* 1975, 1992 – and concerned only incidentally the proof theory of classical logic, as such. The subject has been revisited on several occasions since: first in the Chair of Nicolaas G. de Bruijn, at the University of Technology of Eindhoven (1982) – where I was involved in research on AUTOMATH – and next, after 1983, at the University of Nijmegen, with support from NWO [formerly ZWO], the Dutch National Science Research Foundation. The earliest *public* records (seminars, lectures, conferences, etc.), concerning *explicitly* what I was brought to call 'the witness theory of classical logic', are dated – as far as I can remember – *after the mid-eighties* (Canberra ACT 1986, Paris 1987, Nijmegen 1988, Karlsruhe 1990, etc.) Cf. Rezuş 1990, 1991. As to the the very last revisions (2016–2017), I am grateful to J. Roger Hindley (Swansea University, Wales, UK) for stylistical remarks and useful suggestions concerning previous drafts of these notes.

10 April 2017, revised 25 October 2019

12
The Curry-Howard Correspondence revisited (2019b)

The systematic ambiguity of the (loose, formularian way of understanding the) 'Law' of Non-Contradiction [LNC] has generated – amongst formularian thinkers – the *superstition* that classical logic is (unlike minimal resp. intuitionistic logic) 'badly behaved' proof-theoretically or else that it is 'computationally inconsistent' (sic!)[1].

As already suggested elsewhere, this kind of ambiguity is *not necessarily classically specific*, it might obtain already in the tiny [implication-negation]-fragment of the *Minimalkalkül*.

The fact is that the [LNC]-ambiguity spotted in Rezuş 2019a concerns only the traditional thinking on logic matters, viz., the elliptic, combinatorial, figurative, 'geometric', and (even) purely visual ruminations about formal proofs, while relying a rather vague idea about *what is actually an inference rule*.

If we decide to make the proof-information explicit, by stipulating equational conditions on classical proof operators, the claimed 'inconsistencies' are vanishing, as it becomes obvious that they are generated by an inadvertent, superficial way of thinking about logic, in terms of formulas, sets of formulas, and properties thereof, as, e.g., provability, consequence relations and the like, while ignoring proof properties, proof-behaviours, and proof operators.

In a way, the fact that a given tautology admits of equationally distinct (technically: non-convertible) witnesses is as expected, since the corresponding proof-calculi are (Post-) consistent. They are even *required* to be so, if we intend to manipulate proof *operators*, and not mere notational expedients, as, e.g. in axiomatics, in 'natural deduction' or in Gentzen-like sequent-systems.

---

[1] This kind of (global) objection – to classical logic – has no direct connection to the well-known, traditional Brouwerian objections, and stems mainly from former logic-folk recently requalified, mainly by hear-say, in 'theoretical' computer science [TCS] matters. Otherwise, a completely similar situation occurs, *mutatis mutandis*, in connection with the (allegedly pseudo-) concept of *classical existence* (viz. $\dot\exists \equiv \neg\forall\neg$), a [bibliographical] detail implicit already in Kolmogorov 1925 (!) that has been hardly noticed in the 'proof-theoretical' ruminations of the (post-modern) formularians, thus far.

The said confusions are *epistemic* in nature and have been equipped, gradually, with a un-welcome *ideological* component.

We *might* need a small *historical* détour in order to make the matter explicit.

Relying on Frege's specific choice of formularian primitives $[\rightarrow, \neg]$, the post-Fregean logical tradition has been brought to associate material implication with a kind of *inference carrier*.

The subsequent philosophical or technical (more or less) debates – during the early thirties – about a would-be 'definition' of the (intuitionistic, resp. minimal) logical connectives in terms of inference rules (sic) has blurred considerably the conceptual picture in as far classical logic is concerned.

On the technical side, such considerations have – beneficially, in a way – generated, later (during 1965–1969), a piece of elementary mathematics ('proof-mathematics', so to speak) associated to an extended decorated ('typed') λ-calculus, roughly a proof-system for the Kolmogorov-Glivenko-Heyting intuitionistic first-order logic (Kolmogorov meant to justify only the minimal logic, in fact, a subject discussed explicitly by Ingebrigt Johansson, 1937). The outcome, based, essentially, on Heyting's explanations of 'intuitionistic' logic (with some assistance from Kolmogorov 1925, 1932) is known nowadys as 'Curry-Howard Correspondence' (or 'Isomorphism') [CHC, for short] and the technicalities – in as far the first-order (logic) intuitionism is concerned – are due, essentially, to William A. Howard (MS 1969, published in Seldin & Hindley 1980), whence also the label.

> In retrospect, the reference to Curry is a piece of pious – although rather inadvertent – advertising, as Curry did not mean to associate his *'illative'* *logic* project – dating since the mid- and late thirties – to proof (resp. witness) theory in the sense meant today. On the other hand, Howard had plenty of predecessors (before 1969) on this line of thought, so, ultimately, the 'Curry-Howard' label is, historically speaking, a *misnomer*. In connection with classical logic, the first *technical* insights into the matter are due to Carew A. Meredith (cca 1950), Dag Prawitz (1964–1965) and Govert N. de Bruijn (1966–1967). Otherwise, 'witness theory' in the sense of the present notes has been already anticipated by Saunders MacLane (in his Göttingen PhD Dissertation of 1933). On this, see also McLarty 2007, and, specifically, the comments of Engeler 2015.

The least thing to say is that the so-called 'BHK interpretation' of the (Glivenko-Heyting) intuitionistic logic – with 'BHK' short for 'Brouwer-Heyting Kolmogorov' – whatever its epistemic value, has no direct connection with classical logic and with the classical proof-world. Mainly because, unlike in intuitionism (putting aside intuitionistic resp. minimal negation), the classical propositional logic connectives are *inter-definable*, in logical terms only.[2] In other words, they are linked-up *monolitically* together, so to speak, and there is no point in 'explaining' (conceptually, philosophically, epistemically and so on) either one of them in isolation, without reference to the other ones.

---

[2] In **HA** (the *Heyting Arithmetic*), one can also define explicitly the intuitionistic **or** by *extra-logical* means.

Roughly speaking, this makes a would-be Curry-Howard Correspondence for classical logic rather *unconventional*. There is, indeed, an essential ideological mis-match in such an attempt: we cannot actually 'explain' the proof-behaviour of material implication, for instance – that can be incidentally 'characterised', in superficial formularian terms, à la Peirce (1885), by the tautologies (K) [Simp], (S) [Frege], (P) [Peirce's Law] and (▷) [*modus (ponendo) ponens*] – without making appeal to the proof-behaviour of classical negation and *falsum* **f**, i.e., ultimately, to an expressively complete set of classical propositional logic primitives. Then what about a would-be Witness Theory for Classical Logic, understood as an equational theory of classical proofs?

We are supposed to make a *conceptual distinguo*: putting aside the ideological assumptions behind the BHK 'explanations' of the intuitionistic logic connectives, witnessing a provable formula is, by no means, *specific* to intuitionism.

> After all, the logicians used to 'witness' tautologies by exponential procedures (using valuations in 2-element Boolean algebras, i.e., via truth-tables) a while before formularian intuitionism was born. Practically, the truth-tables were already known – twenty-two centuries ago – to Chrysippus, who also invented the – more economical – modern refutation-techniques of witnessing, essentially the Beth-Hintikka tableaux (cf., e.g., van Ulsen 2000, Rezuş 2016 and the recent work of Susanne Bobzien, referred to there). On the other hand, the ancients used to think in terms of inference rules, not in terms of tautologies.

The idea of *formalising the rules of inference* occurred first in Frege's **GGA** (1:1893, 2:1903), although, as already observed earlier, defectively so. Even Gentzen (circa 1932–1933) – who was familiar with the early intuitionistic lore – did not manage to think properly, in terms of inference rules thought of as (mathematical) proof operators. This emerged later, in subsequent, more or less technical considerations on intuitionistic logic.

Let us recapitulate, once more, the *basic facts on file*.

(1) The main idea worth retaining from the Curry-Howard approach [CHC] is the association: *proof operator* ≃ *rule of inference*.

> As an aside, Howard's slogan 'formulas-as-types' echoes a typical piece of formularian thinking – and propaganda – in the context, as the general idea of a type in mathematics is rather vague, no matter what the (too many) 'type-theorists' that have emerged in the meantime would claim. In CHC, a 'type' is nothing but a *syntactic decoration*. Putting primitive 'data' (propositions, natural numbers, booleans, reals, etc.), while closing under function spaces, [set-theoretical] products and the like, amounts, more or less, to putting apples, potatoes, bananas, unicorns (and Santa Claus, for that matter) in the same bag: things would go awfully wrong as long as we intend to 'close' under negation, even minimally so. (Minimally – and intuitionistically – a 'negative type' would have – typically – a single 'inhabitant' [!], a rather weird idea of a proof [of an 'essentially negative property', to quote Brouwer].)[3]

---

[3] The higher-order – both impredicative and predicative – essentially 'positive', side of the story has been already vastly epitomised in Rezuş 1986, 1987, 1987a, 1987b

In minimal as well in intuitionistic logic, this kind of association was meant to be one-one. Inadvertently so, because, upon minute pondering, it turns out that this might not be exactly the case (cf. Rezuş 2019a, **§§ 3**–4).

For classical logic, the association is – definitely – many-one. In particular, the abrupt, equational proof DN-normalisation policy discussed in Rezuş 2019a is of no help, in this respect: it just makes the situation official, so to speak, by additional, explicit equational stipulations.

At this point, the most we can say is that: *a proof operator is a – technical, resp. formal – tool / way of disambiguating a rule of inference.*

So, incidentally, we are prone to error – to *conceptual errors*, namely – not only in classical, but also in intuitionistic (and, even, in minimal) logic, if we are going to manipulate ambiguous 'objects' – like those we are traditionally used to call 'inference rules' – by making appeal to visual, geometric and/or vague (pseudo-) mathematical ideas about sequences, labelled trees, block structures and the like, while ignoring their (characteristic, defining) equational behaviour qua proof operators.

(2) A second valuable idea to be retained – this time from Gentzen's *Inauguraldissertation* (1934–1935), although already incorporated in the CHC – is that of a *proof-détour* (*Beweisumweg*).

This addresses a slightly different logico-theoretical subject, viz. that of *proof-complexity*. (Cf. Rezuş 2017d, 2017e.)

Technically, in ('typed') λ-calculus terms, a proof-détour is a so-called *redex* (short for 'reducible expression', in jargon). This yields a 'concept of reduction' (by taking compatible reflexive-transitive closures).

> In recent times, the reduction-business has become a separate, rather ramified sub-discipline in theoretical computer science, going far beyond logic / proof theory. Cf. Baader & Nipkow 1998, TeReSe 2003, etc.

Theoretically speaking, proof-isomorphisms are to be generated from initial / 'primitive' proof-détours (by taking compatible reflexive-transitive-symmetric closures). In practice, this does not work smoothly, not even for the *Minimalkalkül*, let alone intuitionistic and classical logic.

> As an aside, since Gentzen 1934–1935 introduced classical logic on the very redundant (primitive) signature $[\rightarrow, \neg, \wedge, \vee, (\forall, \exists)]$ – as a proper extension of the Glivenko-Heyting logic – he could have chosen to have a *Double-Negation primitive* – ¬¬, say [sic] – as well, in order to advocate his int-elim strategy, perhaps. (Cf., e.g., his early considerations of 1932-1933 on the matter, available now in von Plato 2017.)

Here, the *technical* point is that – in order to make things 'work' *as intended* (in logic / proof theory) – we have to think, now and then, in formularian terms,

---

[Nijmegen Lectures, 1983–1987], as a sequel to Rezuş 1983 and Barendregt & Rezuş 1983. (Actually, the Nijmegen Lectures consisted of an extended gloss to N. G. de Bruijn's Eindhoven Lectures on proof theory and on the 'structure of the mathematical language' [1978 and later].)

too. This complicates considerably the matter, not only classically, but also intuitionistically, in a way.

> There are obvious proof-isomorphisms that are not generated by proper proof-détours. For intuitionism, cf., e.g., the pseudo-détours behind the ad hoc (ϖ)-manipulations mentioned in Rezuş 2019a, **§ 4.2.1**. There are many more such. For classical logic, an analogous situation can be encountered in $\lambda\gamma$-calculi. Cf. Rezuş 1990, 1991, and Rezuş 2019a, **§ 5.6**.

(3) The reader has noticed thus far the fact that, while pretending to talk about Frege's **BS**-booklet (1897), I have spent a considerable amount of time, in Rezuş 2019a, on the (Glivenko-Heyting) logic intuitionism and on its little brother called *Minimalkalkül* (Johansson 1937), a favourite subject – *avant la lettre* – of Kolmogorov 1925. This was, in fact, as intended, because I wanted to confront the badly tutored reader with a wrong way of thinking about logic, first.

Another basic idea worth mentioning here is that – whatever its conceptual, epistemic or philosophical merits – the minimal, as well as the intuitionistic proof theory is by no means (conceptually and – in the end – ideologically) consistent with its classical logic counterpart.

To re-assess the situation in plain terms: the fact that the intuitionistic (resp. the minimal) logic is a *proper subsystem* of classical logic refers only to bare provability matters, not to 'the proofs themselves', and, in particular, not to characteristic equational behaviours, as regards the corresponding proof operators.

In short, a would-be correct account of classical proofs (in proper equational, witness-theoretical terms) is not bound to recover, as 'classically correct', the would-be intuitionistically / minimally admissible ('reliable') proof-isomorphisms (or else the associated proof-détours, for that matter).

> This is technically feasible, anyway. I even attempted to argue at length for it, in Rezuş 1991, although, in retrospect, the outcome was not only unreadable for most of the mortals, but also unconvincing, conceptually. For the purpose, I had to invent a 'classical logic' – in witness-theoretical terms – suitable for the job (more or less in the spirit of Prawitz 1965). Roughly speaking, I intended to understand what did Brouwer exactly mean by 'unreliable' (*onbetrouwbaar*, in Dutch), as applied to genuinely classical rules of inference and/or to genuinely classical logical 'Laws'. In the end, there was not very much to understand, beyond what I knew already before: viz. that Brouwer meant to promote and defend a very different (and rather ad hoc) set of logical connectives and rules (why not also an intuitionistic **nand** or **nor**, indeed?) and took as granted the rather weird idea that logic is a time-bound, empirical affair: some time in the future (around 2518, say), we may eventually discover new logical connectives and proof-principles governing their mathematical behaviour (like, *mutatis mutandis*, for quarks in physics)! In the end, Brouwer could have better fared if he would have denied explicitly the *Principle of Omniscience*, on which classical logic is ultimately based. See also the useful remarks of van Atten 2017, on the *open-ended character* of 'Brouwer's logic'.

Why should we care, indeed, about a rather ad hoc fragment of classical logic while attempting to understand classical proof behaviours in proper mathematical terms? After all, the typical Brouwerian objections to classical logic concern only a traditional, obsolete, inadvertent, and conceptually misleading way of thinking about logic and its main concept, the concept of (a) proof (as used – on an everyday basis – in classical, 'practical' mathematics).

In retrospect, however, we have to be fair: the intuitionistic criticism of the traditional – essentially formularian – logic lore has *prompted us to think properly about logic as such*, namely *in proof-theoretical terms*.

(4) Putting aside the rather long, technical discussion in Rezuş 2019a, § 5, there is hardly a point in insisting, once more, on the merits and/or the shortcomings of the classical solution to the – intuitionistic – conceptual worries discussed at length in the above: I told the essentials of the story upon several occasions, already (cf., e.g., Rezuş 2017a, 2017b, and the comments appearing in Rezuş 2016, 2017, 2017d, 2017e etc.). Technically speaking, all proof-systems (equational witness theories) for classical logic – one can obtain, more or less combinatorially, a few dozens or so – are proper fragments of the $\lambda\pi$-calculus, appropriately decorated ('typed'). As suggested earlier, the basic idea is rather simple and consists of extending the overall picture, by taking, as an alternative – classically equivalent – starting point, a propositional primitive signature based on **f** (*falsum*), possibly negation ¬, and the genuinely classical △ (**nand**), for instance. Whence, among other things, the idea of *material implication qua inference carrier*, as well as a few other venerable, traditionally received views should vanish. – In a way, Frege was right to ignore the Deduction Theorem (λ): in classical logic, the inferential reading of the λ-operator – making material implication → into an inference carrier – is a mere accident, since, classically, A → B, means just 'A is incompatible with the negation of B', nothing else, while classical negation amounts to self-incompatibility [¬A means A △ A], etc.[4]

16 November 2018 – 30 September 2019

---

[4] On the closely related, original Chrysippean plan concerning the explanation of (classical) logic connectives and inference rules, see Rezuş 2016.

Part IV

Normal singleton bases

13

Tarski's Claim, thirty years later (2010)

voor YLA – *weer jarig*

Tarski's Claim (TC = Theorem 8 in Łukasiewicz & Tarski 1930) follows from simple considerations in (type-free) λ-calculus. The present note records essentially a proof of Lemma 1.1 in Rezuş 1980, i.e. TCL = the type-free λ-calculus variant of TC, as well as a few historical comments appearing there. Additional remarks are meant to ensure the fact that TCL can be transferred verbatim to typed λ-calculus [TCLT]. (TCLT is just a notational variant of the derivation of TC in ordinary Łukasiewicz / traditional style.)

§ 1 Tarski's Claim

Alfred Tarski claimed in 1925, without proof, the following (meta-) statement [hereafter TC]:

> Let $\mathcal{L}$ be a propositional logic in a propositional language containing at least implication (→). If L is finitely axiomatisable with *modus ponens* for → (and substitution) then $\mathcal{L}$ is also axiomatisable with a single axiom, and *modus ponens* (and substitution), provided it contains
> [K] p→(q→p) – [irrelevance]
> [D] p→(q→((p→(q→r))→r)) – [pairing]
> as theorems ('theses').

Apparently, the Claim above was first mentioned in print in 1929, by Stanisław Leśniewski, Tarski's PhD advisor[1]. Tarski published his Claim slightly later, still without a proof, as Theorem 8, in Łukasiewicz & Tarski 1930. Beyond

---

[1] Cf. Leśniewski 1929, § **10**, pp. 58–59, in the German original [= Leśniewski 2015, 2, pp. 546–547]. The information was repeated in Leśniewski 1938, § **10**, p. 14 [= Leśniewski 2015, 2, p. 583; see McCall 1967, Paper 6, p. 128, resp. Leśniewski 1992, 2, pp. 662—663, for an English translation], a paper scheduled for **Collectanea Logica**. [Added in proof (31 October 2019)]. For the fate of this journal, edited by Jan Łukasiewicz, see Sobociński 1949. The Leśniewski 1929 paper, supplemented by a

Leśniewski, the *original proof* of TC was known to other Polish logicians, during the early twenties (as, e.g., to Jan Łukasiewicz), as well as to some other people, mathematicians and/or philosophers, at a later time. Among them one could mention, for instance, Carew A. Meredith (who attended Łukasiewicz's lectures held at the Royal Irish Academy from 1947 on, cf. David Meredith's 1977, page 514) and his occasional collaborator, Arthur N. Prior (cf., e.g., Meredith & Prior 1963, § 10, page 181). In retrospect, one could have also included Bolesław Sobociński in the list. On the other hand, by the end of the seventies, any information as regards the 'original method' of proof seems to have been irretrievably lost. So, David Meredith (Carew's cousin), in correspondence:

> "The oft talked about but rarely if ever documented 'methods of Tarski' for finding single axioms have been a longstanding sore point with me: by the time I realized they were not in every book on logic, it was too late for me to ask either of the people – Łukasiewicz or Carew A. Meredith – who could have explained them to me. So I've been doubly irritated with myself." (Meredith 1979c, letter of 28 December 1979 to me, a reference from Rezuş 1980, fn. 12.)

[Added in proof (9 October 2019)]. On second thoughts, this is not exactly true. The 'methods of Tarski' were, at least implicitly – modulo Curry-Howard, so to speak – in print long ago, although only *in Polish*, as a matter of fact, viz. in Sobociński's Warsaw Master's Thesis of 1932 = *Z badań nad teorią dedukcji* [Investigations on the theory of deduction], Przegląd Filozoficzny 35, 1932, pp. 171–193. See the Appendix to this Part.

§ 2 A historical aside

Alfred Tarski was concerned with axiomatisability problems while working on his PhD dissertation, Tarski 1923, under Leśniewski[2]. Specifically, during the early twenties, he was involved, among other things, in studying the so-called 'protothetic', a 'logistic' system proposed by his *Doktovater*. In modern terms, Leśniewski's protothetic was meant to be the 'pure' logical segment of a larger, more ambitious foundational enterprise[3]. Tarski's doctoral thesis provided actually the starting point for Leśniewski's protothetic. Incidentally, this explains Leśniewski's frequent references to the work of his PhD student[4].

---

technical § 12, dated 1938 [i.e., Leśniewski 1938a, also meant for Collectanea Logica], has been reprinted in the recent Polish edition Leśniewski 2015, 2, pp. 489–569 [§§ 1–11], resp 630 713 [§ 12], and §§ 1–12 appear, in English translation, in Leśniewski 1992, 2, pp. 410–605.)

[2] Published first in Polish, in Tarski 1923a, English translation in Tarski 1956, pp. 1–23, etc. Much later, he came back to such problems (in algebra, mainly group theory), cf. e.g., Tarski 1938, 1968, in Tarski 1986, II, IV.

[3] A 'General Theory of Sets'; cf. the title of Leśniewski 1916, in Leśniewski 1992, 1, pp. 128–173, and Leśniewski 1927, in Leśniewski 1992, 1, pp. 174–382.

[4] For a complete list of references to Tarski, see, e.g., the Index of Leśniewski 1992, in *loc. cit.*, 2, page 794.

Technically, the main result of Tarski 1923 consists of the observation that the propositional connectives of the two-valued logic can be obtained from 'material' equivalence and the [propositional] universal quantifier. Whence, in the end, in view of the fact that the system contained appropriate type-distinctions, Leśniewski's 'General Theory of Sets' could have been based on three primitives, only – using equivalence, a general quantifier and something similar to the membership relation – by using a single axiom, together with several 'directives', i.e., in our terms, appropriate rules of inference.

§ 3 A personal aside (excerpted from Rezuş 1980, § 0, and fn. 12, etc.)

I came across TC – and the problem of reconstructing a would-be proof of it – some time around 1975, in Bucharest, when a Romanian mathematician interested in logic matters noticed, not without some irritation, that 'Tarski did not usually publish proofs' (this is only in part true). TC was mentioned, conversationally, as a case in point.[5] As a matter of fact, this 'practice' concerns other Polish logicians active within the Lwow-Warsaw Logic Seminar, as well, and it has a quite reasonable explanation: during the twenties and the early thirties, the members of the Seminar produced a large mass of results in a relatively new discipline, by then, so that they were forced to state even important results without insisting in ultimate details on the specific methods of proof therein involved, and this just in order to make available / understandable more involved ones. On the other hand, even though intriguing at a first look, TC is not a 'result' that could have been seen as an 'important' one, in itself. – As I was already familiar by then – i.e., during the mid-seventies – with the so-called 'formulas as types' approach (or else with the 'functionality theory' of H. B. Curry, popularized by Curry himself, as well as by Carew A. Meredith, during the sixties[6], I strongly suspected one could eventually prove TC in (typed) λ-calculus. As nothing came out after a one moment's reflection (and I was, anyway, involved in some other – more interesting – things), a would-be proof of TC along such lines was duly post-poned. However, I came back later to it, in Geneva, around 1978–1979, while working for a PhD on (subsystems of) λ-calculus under Dirk van Dalen (Utrecht). The actual thesis supervisor was Henk Barendregt, so, among other things, I was supposed to become familiar with what came to be known, later on, as 'The Bible' of the discipline ('type-free λ-calculus': Barendregt 1981, first edition, then still in manuscript). Barendregt had plenty of entertaining Exercises in the Bible, and many more in

---

[5] Another example could have been even the Deduction Theorem [DT]. Although usually credited to Jacques Herbrand (circa 1930), the Frenchman comes much later into the picture on this: Tarski was aware of DT already around 1921. (On Herbrand's version of DT for classical first-order logic, see Herbrand 1930, Chapter III, 2.4, Herbrand 1930a, Porte 1982, Grattan-Guinness 2000, p. 550, etc.)

[6] The label comes actually from Howard 1980, a manuscript circulated from 1969 on, whence the current terminology in the literature on λ-calculus: the 'Curry-Howard Correspondence', cf. Sørensen & Urzyczyn 2006, etc. The first book-length documentation of 'Curry-Howard' in print appears in Stenlund 1972.

his trans-finite personal archive. In particular, the problem of the 'single point bases' for the (pure) 'type-free' λ-calculus appears in Barendregt 1981, Chapter 8, pp. 161–162, as well as in Exercise 8.5.1 (containing two examples from J. B. Rosser: correspondence with Barendregt of 1971) and 8.5.15 (with four more examples from Carew A. Meredith, Henk Barendregt, Corrado Böhm and J. B. Rosser; other examples of the kind – due to Carew A. Meredith, Ivo Thomas, J. B. Rosser, Corrado Böhm, W. L. van der Poel, J. W. van Briemen, etc., most of them unpublished – are also mentioned in Rezuş 1980). As Barendregt's book concerned mainly the 'type-free' λ-calculus, there was no special reason to mention the 'typed' case separately there (although Meredith's construction in Exercise 8.5.15, based on Meredith & Prior 1963, involved implicitly a 'typed' case, motivated by logic reasons alone). A fortiori, there was no special reason to mention TC, in the Bible, either. Nevertheless, I 'translated' TC in Biblical [type-free λ-calculus] terms and came, after a (short) while, with a rather simple solution (fitting on less than a page in print). This is the main object of the present note and comes next.

§ 4 The Singleton Basis Claim

[TCL, *idest* Tarski's Claim 'translated' in type-free λ-calculus (Lemma 1.1. in Rezuş 1980).

> Let $\mathcal{A}$ be a set of (closed) λ-terms [combinators], such that [1] $\mathcal{A}$ is closed under application and β – reduction, and [2] $\mathcal{A}$ has a finite basis. Then $\mathcal{A}$ has also a singleton basis, provided it contains the combinators $\mathsf{K} := \lambda xy.x$ and $\mathsf{D} := \lambda xyz.zxy$.

Proof. Some convenient notation first. Set $<X,Y> := \lambda z.zXY$ (x not free in X, Y). So $\vdash \mathsf{D}XY = <X,Y>$.[7] Iterate this '[by accumulating] to the left', in the obvious way, taking care of the limit case (and counting from 1):

$F_1 = <X_1] := X_1,$
$F_2 = <X_1,X_2] := <X_1,X_2>,$
$F_3 = <X_1,X_2,X_3] := <F_2,X_3> = <<X_1,X_2>, X_3>,$

---

[7] This simulation of (ordered) pairs comes from Frege's **Grundgesetze der Arithmetik** (**GGA** 1:1893). In this respect, Church's initial paper on λ-calculus (Church 1932–1933) contains an independent rediscovery. [Added in proof (23 October 2019). In fact, Frege anticipated, in his **GGA**, 1.1893, the 'pure' λβ-calculus, $\boldsymbol{\lambda}_\beta$, while Russell even formulated, in passing, around 1903, the 'pure' λβη-calculus $\boldsymbol{\lambda}_{\beta\eta}$ (cf. Russell 1903a, first paragraph of **3a**, p. 50, *et sqq.*), although he could not find any use for it (see also Klement 2003 and Rezuş 2015a, fn 2). According to H. B. Curry (Curry 1980, p. 88), Church's earliest concerns with λ-calculus can be traced back to Göttingen, circa 1928. From private correspondence with Alonzo Church [1980, 1983], it follows that he managed to read Frege's **Grundgesetze** – as well as Schönfinkel 1924 – only later, after writing Church 1932–1933. Cf. also Scott 1980, and Cardone & Hindley 2009.] Whence also the terminology used here: 'Frege-Church pairs'.

and so on. In other words, the inductive step is:

$F_{k+1} = <X_1,\ldots, X_{k+1}] := <F_k, X_{k+1}>$ (up to k < n-1).

That is to say, for n>0, ⊢ D...DX$_1$...X$_n$ = F$_n$ = $<X_1,\ldots, X_n]$ (n-1 times D). (Here one can read β-equality = β-conversion as β-reduction, as well.) Let $\{X_1,\ldots, X_n\}$ be the finite basis of the hypothesis. For n = 1 there is nothing to show. If n > 1 set

$F := F_n = <X_1, \ldots, X_n] = $ D...DX$_1$...X$_n$, as above, and
$G := <K,KK,F] = <K,KK,<X_1, \ldots, X_n]].$

Obviously, G is in the set $\mathcal{A}$ (since so are K, D and therefore F, as $\mathcal{A}$ is supposed to be closed under β-reduction). On the other hand, one checks easily (just a matter of milliseconds with a λ-calculus reduction machine) that

⊢ GG = K,
⊢ G(KK)K = F = $<X_1, \ldots, X_n]$,

whereas extracting the X_i's (0<i<n+1) from F := F$_n$ is straightforward, even by hand:

⊢ FK...K = $<X_1, \ldots, X_n]$K...K = $<X_1, \ldots, X_k]$
(n-k times K postponed, k ∈ {1,...,n-1}, so, in particular,
⊢ FK = $<X_1, \ldots, X_n]$K = X_1 (for k = n-1, i.e., n-k=1), and
⊢ F$_k$(KK)K = $<X_1,\ldots,X_k](KK)K = X_k$, for all k ∈ {2,...,n}. QED.

[As promised, the argument fits on less than a page in print.]

§ 5 Remarks

(1) Certainly, one could have 'accumulated' Frege-Church pairs 'to the right' in F, as well (Barendregt's own preferred way of gathering in his book of 1981), but the latter choice would have required different 'projections' (and so 'extraction patterns'), beyond the mere pretty old K. [In practice, the underlying 'extraction patterns' were meant to be short and 'easy', first of all.]

(2) Obviously, the assumption that $\mathcal{A}$ contains D can be weakened [down] to: $\mathcal{A}$ is closed under the 'rule' (D): X,Y ∈ A ⇒ <X,Y> ∈ $\mathcal{A}$. (Yet, Tarski, as well as his former Polish colleagues – Stanisław Jaśkowski and Mordchaj Wajsberg excepted, perhaps – was less concerned with rules, however; in any case, he vastly preferred a *Satzlogik* instead of a *Regellogik*, so to speak: an algebraist's state of mind, more or less.)

(3) [TCLT = the 'typed' variant of TCL]. TCL exemplified in the above can be repeated in typed λ-calculus. (This step is particularly boring[8], but it is, ultimately, trivial for the case in point. In general, such a step is *necessary*,

---

[8] Especially so, if you don't already have at hand a convenient Robinson-like soft-engine, to do 'unifications', and so to find 'most general unifiers', for you, in milliseconds.

however, because we can have singleton bases for sets of closed terms in type-free calculus for which the 'extraction pattern' has no 'typed' counterpart. As reported by Arthur N. Prior in 1963, an example of the latter kind was produced first by Carew A. Meredith, some time around 1956. Cf. Meredith & Prior 1963, § 9 [*A Combinatory Base Without C-Positive Analogue*] and/or the Appendix of Rezuş 1982 [*On a Singleton Basis for the Set of Closed Lambda-terms*]. Meredith's 1956-example was $G := \lambda xyz.y(\lambda u.z)(xz)$, with $GGG := \lambda xy.y(xx)$ [untypeable, of course], the latter term being used in the 'extraction' of $C_* [= CI] = G(GGG)G$, as well as I, $K' [= CK]$ and, ultimately, K. – Actually, Carew A. Meredith claimed the latter fact without a proof in print, and it took us quite a while – several people on two continents, not just me, appropriate software included – before I was able to realise how it was meant to be done!)

(4) The proof of TC appearing in Rezuş 1982 is a generalisation of the above (in a variant of the 'typed' λ-calculus, using [Carew A.] Meredith's 'condensed detachment' and the like). As the paper is available in print, there is no point in pausing on it again, once more.

§ 6 A moral

As a matter of fact, there is no need to repeat the previous construction in 'typed' λ-calculus, because, *in this case*, TCLT ['typed'] follows from TCL ['type-free'] and the observation that F and G are head normal forms (hnf's) of *a very special kind*, since we have also the following rather obvious Metatheorem:

> Let $H := \lambda x_1 \ldots x_m.x_j X_1 \ldots X_n$ be a closed term in hnf, such that the X_i's are closed terms. If all the X_i's are stratifiable (= typeable = have a principal type [scheme]) then so is H (and a principal type [scheme] of H can be derived explicitly / effectively from the principal types of its 'components' X_i).

Indeed, our G was a Frege-Church pair

$$G := <K,KK,<X_1, \ldots, X_n]] = <<K,KK>,F> = \lambda x.x(\lambda y.yK(KK))F,$$

i.e., of the form H above, as well as <K,KK>, with K and KK typeable; so G must be typeable if so is F. But F is a Frege-Church pair, too, and by (repeating inside-out) the same argument, F is typeable because so were supposed to be its 'ultimate' subterms X_i ($0<i<n+1$). As an[other] aside, a similar argument applies – *mutatis mutandis* – to Church n-tuples in general (n ≻ ?), i.e., terms of the form $\lambda x.xX_1 \ldots X_n$, with all X_i's typeable closed terms ($0<i<n+1$). Since most examples of singleton bases for the full (pure) λ-calculus mentioned in passing earlier (as, e.g., those provided by J. B. Rosser, W. L. van der Poel, J. W. van Briemen, etc.) were just Church n-tuples of typeable combinators, with n > 2, there was no need to check the typeability of the derivations (the 'extraction patterns') in detail in order to establish the fact that their (principal) types were also single axioms for Heyting's pure implication. (An example, the shortest of the kind, is Rosser's $G = <K,S,K> = \lambda x.xKSK$, with $\vdash GGG = K$

and $\vdash G(GG) = S$; cf. Barendregt 1981, Proposition 8.1.4.) The Moral is that – in such (rather specific) conditions – we can always forget about *what we prove* (i.e. the formulas / types themselves) and pay attention only to the *form of the proofs* (i.e. to the corresponding λ-terms).

§ 7 Addendum

Based, apparently, on an approach due to Dolph Ulrich (2004–2005), John Halleck provided by e-mail (Halleck 2010, dated 23 September 2010), the following variant of the proof above (a fully 'typed' version of it, with detailed 'traditional'/ Łukasiewicz-style derivations in propositional logic: a remarkable achievement, as well as a unusual example of patience, perhaps, quite rare by post-modern standards, these days). On notational reasons, we start, this time, counting from 0. So the finite basis of the hypothesis is going to be now $\{X_0, \ldots, X_n\}$, n natural (with, for n = 0, nothing to show, as above). Set also $K(X) := \lambda x.X$ (x not free in X), for arbitrary X. We iterate the Frege-Church pairs as before, but with $K(X_j)$ – instead of X_j (parentheses are just to improve readability here) – after the first one:

$F_0 = X_0$,
$F_{k+1} := <F_k, K(X_{k+1})>$,

so that

$F := F_n = <X_0, K(X_1), \ldots, K(X_n)]$

(equally compact and pretty readable). Now, once we have K, we can get X_0 and $K(X_j) := \lambda x.X_j$ (0<j<n+1), as before, whence also the X_j's (0<j<n+1), as well, since $\vdash KXY = X$, for arbitrary X, Y. (Halleck used actually a different 'extraction pattern', a trifle more involved.) Of course, in view of the above, the use of K's in F is not necessary (but, as already announced, I'm just recording the content of Halleck's note in type-free λ-terms). Finally, set

H := <F,KK> and
G := K(<H,KK>) = K(<<F,KK>,KK>).

Halleck first noticed that

$\vdash GGG = K(KK)$,

and obtained K as

$\vdash GG(GGG) = K$.

Actually, we have, in detail,

$\vdash GX = <H,KK>$, for arbitrary X (so, in particular, choose X := G), while
$\vdash GXK = H = <F,KK>$, for arbitrary X (choose X := G, again),

wherefrom one can get also F := F_n, by

⊢ HK = F = F_n.

In particular, Halleck's G is a special hnf of the kind discussed earlier (a term of the form KX is typeable if so is X, G is typeable if so is <H,KK>, the latter pair is typeable if so is H, while H is a Frege-Church pair with typeable 'ultimate' components, etc.). So, in view of the Metatheorem referred to in the above, there is no need to check in detail the 'typed' version of the argument, either.

§ 8 A final comment

(On the non-technical topic of 'conceptual means'.) To be fair, Halleck's point in his note was meant to show that Tarski *could have* produced a proof of TC 'with the conceptual means of his time', that is: without using (typed) λ-calculus (the type-free λ-calculus appeared in 1932–1933 in print while the 'typed' variant came out even later, around 1937–1938. Halleck's argument – 'full-spelling' in Old Polish – was meant to invalidate a previous (non-technical) claim of mine, which, roughly, consisted of saying that Tarski must have known / anticipated some λ-calculs and/or combinatory logic (at least an 'applied' form of it, as Henk Barendregt used to think about such things in his early 'type-free' life), and that, as (I said) 'there is, essentially, no other way of proving TC' (an approximate self-quote). Halleck's argument showing TCLT in the spirit of the early twenties (Łukasiewicz-style derivations in propositional logic) does not 'prove', however, the fact that my non-technical claim is wrong. After all, I could have provided myself a full, 'typed' variant of the proof of TC above, as well (by writing down due derivations in propositional logic à la Łukasiewicz, with substitutions displayed explicitly, and so on): the outcome would have been even somewhat simpler / shorter than the alternative one, as displayed in Halleck 2010. (As already noted above, there was no need to use cancellators, K's, in Halleck's F.) Whereby, my proof of TC (TCLT, in fact) would have used 'the conceptual means of the early twenties', as well! Both arguments (Halleck's, as well as mine, in its TCLT variant) are ultimately 'based on typed λ-calculus' – actually on 'typed combinatory logic', because we can also rephrase the whole story, once more, in terms of the Schönfinkel-Curry 'combinatory logic', anyway. Yet, in order to produce the argument for TC, one must be aware of the abstract (reduction / conversion) behavior of K and D, at least (resp. of K and <X,Y>), or else, one must be aware of 'equivalences' like KXY = X, DXYZ = <X,Y>Z = ZXY, whatever the actual notation, and the 'intended meanings'. Historically speaking, such things *could have been* available to Tarski at some time-point before 1925, I suppose. In fact, Moses Schönfinkel did read his paper on combinators, Schönfinkel 1924, in Göttingen, at a local mathematical conference, on 7 December 1920, and this pioneering paper (actually prepared for publication – from the author's talk – by Heinrich Behmann) was already in print by 1924. Moreover, Leśniewski referred to Schönfinkel 1924 explicitly (as well as to John von Neumann [1927], who used a related form of 'combinatory logic' in his foundational papers), although the Pole claimed somewhat later he was not 'acquainted with' Schönfinkel's paper while preparing Leśniewski

1929. (See the Index to Leśniewski 1992, for exact references.) In view of this, whatever the *actual historical detail*, it is sensible to suppose that Tarski *could have* used, around 1925, some form of the so-called 'typed combinatory logic' and / or some form of the 'formulas as types' approach (kind of a crude variant of the Curry-Howard Correspondence), at least for the purpose of proving TC. (A minor result, in the end, as already noticed in the above.) The fact that neither Tarski (nor Leśniewski, Łukasiewicz and, later, Jaśkowski, for that matter), did exploit the [Curry-Howard] 'correspondence' in order to recover a would-be meta-theory (of proofs / derivations) can be explained in various ways, and is irrelevant in this discussion. We can speculate, at best. Yet, I'd rather defer such speculations, for a would-be less technical talk.

27 September 2010 – 31 October 2019

14

Tarski singleton bases: 1925–1932 (2019)

Foreword

Tarski's Claim of 1925, cf. Rezuş 2010 – published first as Theorem 8 in Łukasiewicz & Tarski 1930, *without* a proof in print – says, roughly, that:

> If a propositional calculus $\mathcal{L}$ with at least implication [C] as a primitive is finitely axiomatisable with *modus ponens* [for C] and substitution as only rules of inference, and contains, moreover, the tautologies [K] := CpCqp and [D] := CpCqCCpCqrr, then $\mathcal{L}$ can be also axiomatised with a single axiom and the same rules.[1]

Sometime around 1978–1979, I managed to prove Tarski's Claim – as well as some variants thereof for proper subsystems of the classical propositional logic without derivable [K] and the like – using (typed) λ-calculus.[2]

As an intriguing historical detail, the 'type-free' λ-calculus has first appeared in print around 1932, while the '[simply] typed' λ-calculus goes back to 1940 at best.[3] This makes Tarski's Claim – and his original 'methods of proof' – somewhat mysterious, since Tarski could not have had the right kind of (abstract) tools at hand around 1925.[4]

---

[1] There was also a variant of the Claim, with [D_1] := CpCqCrCCpCqCrss in place of [D] ($\equiv$ [D_0]). See the Łukasiewicz example **2.4** [= Ł$_{38,1927}$], mentioned under **§ 2**, below. For a more general case, see Rezuş 1980, 1982.

[2] As a matter of fact, the actual proof of Tarski's Claim amounts to an easy λ-calculus exercise. See Rezuş 1980, 1982, the historical comments appearing in Rezuş 2010, and, possibly, Rezuş 1990, 2017, 2017a, 2017b, 2017d for the general theory behind this kind of results.

[3] As is well-known, both achievements are due, essentially, to Alonzo Church.

[4] In the end, 'my' method of proof (i.e., the λ-argument of Rezuş 1980, 1982, 2010) is, essentially, the *only* way of proving Tarski's Claim. Formally, the original Tarski argument (1925), as recovered below (**§ 3**) – or else as in Rezuş 1980, 1982, 2010 – is *isomorphic* to my λ-calculus argument of 1979. The fact that Tarski did not have, in 1925, a λ-calculus notation – or a would-be general theory – at hand is immaterial.

Now, according to some authors[5], the information regarding the original 'methods of Tarski' (1925) of finding single axioms ('singleton bases') for logics with implication [C] and *modus ponens* [for C], plus substitution, as only rules of inference was already lost, at least by the late 1970's, and, definitely so, after the death of Alfred Tarski (26 October 1983).[6]

So, for instance, the late David Meredith, the American cousin of Carew A. Meredith[7], in a letter of 28 December 1979 (Meredith 1979c), to me:

> "The oft talked about but rarely if ever documented 'methods of Tarski' for finding single axioms have been a longstanding sore point with me: by the time I realized they were not in every book on logic, it was too late for me to ask either of the people – Łukasiewicz or Carew A. Meredith – who could have explained them to me. So I've been doubly irritated with myself." (a reference from Rezuş 1980, footnote 12, resp. Rezuş 2010.)

In retrospect, it appears that this is not exactly the case: the statement that Tarski's original 'methods of proof' were lost (by the late 1970's or so) is only half-true.

The said 'methods' of Tarski were actually mentioned and exemplified explicitly in an early *Polish* paper by Bolesław Sobociński, viz. in Sobociński 1932, otherwise quoted in Rezuş 1980, 1982, as well.[8]

Moreover, according to V. Frederick Rickey (an American PhD student of Sobociński at the University of Notre Dame), Sobociński sketched a proof of Tarski's Claim in a class on 'Deductive Theories', during the fall semester of 1968 at the University of Notre Dame.

> "He [Sobociński (AR)] knew about this work [of Tarski (AR)] from his time as a student at the University of Warsaw. He entered [the Warsaw University (AR)] in 1926 and soon became active in the research program which searched for single axioms of various systems." (Rickey 2019).

As I could gather, the 1968-proof mentioned by Fred Rickey (actually preserved in the Sobociński *Nachlass* deposited at the University of Notre Dame) is pretty much the same as the one we can derive from Sobociński's Warsaw Master's Thesis, i.e., Sobociński 1932.

---

[5] People rather well-informed in early Polish-logic matters, even better than me.

[6] Cf. Rezuş 1980, 1982, 2010. It did not occur to me to ask – in 1979 – Tarski himself about, but the oversight was a good idea, after all, since, by doing so, I would only have managed to make myself ridiculous: the story was already in print, at least implicitly. See below.

[7] Carew A. Meredith (1904–1976) was a student of Jan Łukasiewicz in Dublin, Ireland. See the obituary by David Meredith 1977.

[8] This was, by the way, Sobociński's Warsaw Master's Thesis supervised by Łukasiewicz with, likely, some assistance from Tarski. See also Sobociński 1955–1956.

Incidentally, Tarski's genuine example of 1925 is *not* the most general one possible.[9]

I will record, next (**§ 2**), the bulk of Sobociński 1932 on so-called 'non-organic[10] singleton bases' for the classical [logic] 'CN-calculus', exemplifying Tarski's *original* 'methods of proof', with Tarski's first finding (1925) of the kind, and some variants thereof, concocted somewhat later (1927, resp. 1932) by Jan Łukasiewicz and Bolesław Sobociński, as mentioned in Łukasiewicz & Tarski 1930 and in Sobociński 1932.

§ 1 Prerequisites

Throughout the remaining of these notes, the term 'λ-calculus' refers to the 'extended λ-calculus' **λπ**, with primitive pairs, projections and so-called 'surjectivity of pairing'. The corresponding sub-system (without pairs and projections), **λ**, will be referred to as 'pure λ-calculus'.[11]

The reader is supposed to be familiar with, at least, Rezuş 2017b[12], including the appropriate rudiments of λ-calculus, proof- and category theory, referred to *ad loc*.

The 'type-free' equational theory of main concern here is a (syntactic) variant of **λπ** – or else, equivalently, of the theory of the *Lambek C-monoids* (Lambek & Scott 1986).[13]

The *art deco* ['typing'] is, essentially, as in Rezuş 2017b[14].

1.0 Notation and terminology. (1) As we allow multi-letter combinator-names, the usual λ-cuts (the so-called 'applications') are abbreviated with round parentheses F(X) := F▷X. (However, if both F and X are variables, there is no danger of confusion, so we may also write xy := x(y) ≡ x▷y, as ever.)

---

[9] For details, and more examples of the kind, see **§ 2**.

[10] A tautology is said to be *organic* – terminology from Stanisław Leśniewski, Mordchaj Wajsberg, and Bolesław Sobociński – if no proper subformula of it is a tautology, otherwise it is said to be *non-organic*. *Mutatis mutandis*, this way of speaking can be transferred to combinators (closed λ-calculus terms). Obviously ('by construction', so to speak), Tarski's 'methods' of finding single axioms yield only non-organic bases.

[11] For a survey of pure **λ**, cf., e.g., Barendregt 1981, 1984², Rezuş 1981 (pure *strict* λ-calculus and extensions), and the text-books Hindley & Seldin 1986, 2008.

[12] On λγ-calculi, see also Rezuş 1990, 2017a, 2017c. On the historical side, see Rezuş 2017 (Łukasiewicz, Jaśkowski, 'natural deduction'), and Rezuş 2017d, 2017e ('natural deduction' in Gentzen's *Nachlass*).

[13] For all *practical purposes*, the λ∂-calculus **λ∂p** of Rezuş 2017b – a *proper* subsystem of **λπ** – should largely suffice, however.

[14] As regards the 'typed' λ-calculus, familiarity with the recent monograph Barendregt *et al.* 2013 is not required, but is a bonus. (See also the reviews Hindley 2014, and Rezuş 2015.) Although based on slightly different assumptions (Curry 'typing'), an excellent introduction into the pure 'typed' λ-calculus (and the main sub-systems thereof) is Hindley 1997. Cf. also Hindley & Seldin 1986, 2008.

(2) The names of the pure combinators I, K, B, C, S, W are the standard ones (as in Curry's publications, for instance). Multi-letter combinator names are, in general, mnemonic (so, e.g., ⊢ CB = C(B), ⊢ CC = C(C), ⊢ CS = C(S), ⊢ CK = C(K) = KI = K(I), ⊢ 1 := BI = B(I) etc.)

(3) The 'exponential' notation $X^n(Y)$, n>0, amounts to an implicit use of the *Church numerals* (Barendregt 1981, 1984[2], Rezuş 1981), possibly modulo λ-conversion, $n_{ch} := \lambda x.\lambda y.x(...(x(y)...)$, n times x, n ≥ 0, where the limit case (n = 0) is 0 ≡ KI := λx.I. (So ⊢ $X^n(Y) = n_{ch}(X)(Y)$, as expected.)

(4) The Frege-Church pairs (Frege 1893 [1], Church 1932–1933, 1935–1936, 1941, Barendregt 1981, 1984[2], Rezuş 1981) are iterated – resp. 'accumulated' – inductively, either to the left, as in:

<$X_1, ..., X_n, X$] := <<$X_1, ..., X_n$],X>

or to the right, as in:

[X, $X_1, ..., X_n$> := <X,[$X_1, ..., X_n$>>

(n ≥ 1), in the expected way, where the limit cases (n = 1) are <X] ≡ [X> ≡ <X> [sic]. So <X,Y,Z] ≡ <<X,Y>,Z>, [X,Y,Z> ≡ <X,<Y,Z>>, and so on.

On the other hand, the Church n-tuples, n ≥ 1,

<$X_1, ..., X_n$> := λz.z(X_1) ... (X_n), z fresh for $X_1, ..., X_n$,

are as usual (the limit case, n = 1, is <X> := λz.z(X), with z fresh for X).

(5) For the corresponding 'projections', we set $K_i^n := \lambda x_1. \ldots .\lambda x_n.x_i$, and $P_i^n := <K_i^n>$, for 0<i<n, with n>1. So $K^n(K) \equiv K_{n+1}^{n+2}$, for n>1, say.

(6) If X is a combinator with 'type' α, we write usually ⊢ X : [X] ≡ α, for this, while the 'type-free' version of X is denoted by |X| (sic).

(7) Further, for any (quantifier-free) formula ('type') α, lh(α) is *the length of α*, counting bare 'letters', i.e., both propositional variables and connective-occurrences, in Łukasiewicz notation. (Example: lh(CpCqp) = 5.)[15]

(8) Finally, in a decorated ('typed') context, the 'applicative' notation F(X) := F▷X stands for the [C. A.] Meredith 'condensed detachment' (cf. Rezuş 1982, Kalman 1983, Hindley & Meredith 1990, Wos & Veroff 2001, Wos 2007, etc.).

(9) With such conventions, we have, for instance:

⊢ <X, Y, Z] : CCCC[X]C[Y]rrC[Z]ss,
⊢ [X, Y, Z> : CC[X]CCC[Y]C[Z]rrss,
⊢ <X, Y, Z> : CC[X]C[Y]C[Z]rr,

where r, s are fresh for [X], [Y], [Z], so, in general, for ⊢ X . [X], we can easily retrieve (a normal form of) X from (the form of) its '(principal) type' [X].[16]

---

[15] With an apology to otherwisely minded readers, the spelling of formulas (decorations or 'types') in the present notes is going to be Łukasiewicz-like (prefix 'Polish' notation). Otherwise, consider writing down (and printing) – with infix notation and parentheses etc. – a CN-formula α, with lh(α) = 139, say (an actual example of the kind was, in fact, provided in Sobociński 1932).

[16] If [X] is not a 'most general' (or else a 'principal') 'type' of X, some additional tinkering might be, of course, also required.

This being said, the Sobociński 1932-list of examples is already transparent.

§ 2 Non-organic singleton bases (1925–1932)

Recall, first, the Sobociński 1932-list [Sobociński 1932, § 1, Endnote 5] of *non-organic* singleton bases (1925–1932).

2.0 Notation. Hereafter, the subscript on a combinator ('generator', 'singleton basis') label stands for the length of its 'type' (actually, 'principal type scheme' or 'principal types', in Curry's terms), followed by the year the result was found. E.g., [$T_{53,1925}$] is **T**arski's original singleton-basis, and [$S_{36,1927}$] is a singleton-basis of length 36, found by **S**obociński in 1927. The remaining Ł-combinators – here: singleton bases – are due to Łukasiewicz (found in 1927 and 1933, resp.).[17]

2.1 Tarski (1925), $T_{53,1925}$. Tarski's original example (1925) of a singleton basis for the classical, two-valued CN-calculus was

⊢ $T_{53,1925}$:= <F,<S,T>,K] : [$T_{53,1925}$] ≡
CCCCCsCtCtCuCuuCCCCCpCqrCCpqCprCCCNpNqCqpxxyyCCpCqpzz,

where

⊢ K : CpCqCp,
⊢ S : CCpCqrCCpqCpr, as ever, and
⊢ F := λf:s.λg:t.λh:t.λx:u.λy:u.y : CsCtCtCCuCuu,
⊢ T [≡ C^∂] := λf.CNpNq.λy:q.∂z:Np.(f▷z)⋆y : CCNpNqCqp.

This is a special case of a more general pattern discused below.[18] Tarski's original solution works for the 'type-free' combinator | $T_{53,1925}$ |, as well.

Subsequently (1927–1932), Łukasiewicz and Sobociński found more such examples, by [1] modifying slightly Tarski's original packaging method and by [2] using appropriate extraction recipes (or patterns).

The main interest, at the time, consisted of finding witnesses G, with ⊢ G : [G], such that the tautology [G] has a shorter length, i.e., here, lh([G]) < 53 (counting the number of symbols – propositional variables and C-N-letters – occurring actually in [G]). The results – nine more examples – were duly reported in Sobociński 1932, § 1, Endnote 5. The shortest one of the kind – i.e., obtained by using 'Tarski methods' – was found by Łukasiewicz, in 1932.[19]

---

[17] Recall that, in λ∂-terms, εz.e[z] := ∂z.(z⋆e[z]), and ϖ(e) := ∂z.e, with z not free in e (Rezuş 2017a, 2017b). Here, ε stands (after 'typing') for the proof operator corresponding to Cardano's *consequentia mirabilis*, and ϖ for the proof operator corresponding to the *ex falso* rule.

[18] Obviously, one could have had the most general F, with ⊢ F ≡ K^4(I) : CsCtCwC-CvCuu, instead, choosing propositionally distinct variables in all components.

[19] This example – **2.10** = $Ł_{29,1932}$, below – or, rather, the λ-calculus version of the original Łukasiewicz proof, as recorded by Sobociński 1932, is discussed at length in **§ 4**.

In view of the above, the transcription of the corresponding witnesses in ('typed') λ-calculus terms is, more or less, straightforward.

The extraction steps required by the remaining examples (**2.2**–**2.9**) are left as exercises to the reader.[20]

2.2 Łukasiewicz (1927), Ł$_{43,1927}$ ≡ [K^3(K), S, T> ≡ <K^3(K),<S,T>>

[Ł$_{43,1927}$] ≡ CCCsCtCtCpCqpCCCCCpCqrCCpqCprCCCNpNqCqpvvee ≡ CC [K(K(KK))] CCC [S] C [T] vvee ≡ CC [K^3(K)] CCC [S] C [T] vvee

2.3 Łukasiewicz (1927), Ł$_{39,1927}$ ≡ [K$^3_1$, S, T>

[Ł$_{39,1927}$] ≡ CCCaCbCdaCCCCCpCqrCCpqCprCCCNpNqCqpvvee ≡ CC [K$^3_1$] CCC [S] C [T] vvee

2.4 Łukasiewicz (1927), Ł$_{38,1927}$ ≡ <K^2(K), R, CB> (sic), where we have set ⊢ R : CCNstCCNsNts, and ⊢ CB : CCpqCCqrCpr.[21]

[Ł$_{38,1927}$] ≡ CCCaCbCdCedCCCNstCCNsNtsCCCpqCCqrCprzz ≡ CC [K^2(K)] C [R] C [CB] zz

2.5 Sobociński (1927), S$_{36,1927}$ ≡ <K$^3_1$, F>, | F | := λf.λg.λh.λx.εy.(g(X)(hx)), with X := ∂z.((fz(hx)y)⋆x), where, in F, the most general decorations ('principal typing') on bound variables are to be restored as in [f:CNmCqCNrNp] [g:CmCqr] [h:Cpq] [x:p] [y:Nr] [z:Nm].

[S$_{36,1927}$] ≡ CCCaCbCdaCCCNmCqCNrNpCCmCqrCCpqCpree ≡ CC [K$^3_1$] C [F] ee, where [F] ≡ CCNmCqCNrNpCCmCqrCCpqCpr

2.6 Łukasiewicz (1927), Ł$_{35,1927}$≡ <K$^3_1$, F>, | F | := λf.λg.λh.λx.εy.(g(X)(hx)), as above, but with X := ∂z.((fz(hx)(hx))⋆x), where the most general decorations on bound variables, in F, are to be restored, now, as in [f:CNmCqCqNp] [g:CmCqr] [h:Cpq] [x:p] [y:Nr] [z:Nm]. This is, actually, a variation on Sobociński's S$_{36,1927}$, i.e., example **2.5**.

[Ł$_{35,1927}$] ≡ CCCaCbCdaCCCNmCqCqNpCCmCqrCCpqCpree ≡ CC [K$^3_1$] C [F] ee, where [F] := CCNmCqCqNpCCmCqrCCpqCpr

2.7 Łukasiewicz (1927), Ł$_{33,1927}$

[Ł$_{33,1927}$] ≡ CCCpCqpCCCNrCsNtCCrCsuCCtsCtuvCwu.[22]

---

[20] The industrious reader can, eventually, find the proper derivations by using appropriate condensed detachment [CD] software, or else a would-be implementation of the λ∂-calculus, say. For the record, John Halleck managed to obtain (Halleck 2018) the derivations mentioned in **§ 3**, with a CD soft-machine of his own, in about two seconds CPU-time. On a typical λ-calculus 'reduction machine', the same kind of work could have been achieved in milliseconds.

[21] For R – a version of the genuinely classical *reductio ad absurdum*, after 'typing' – one can take R := λf:CNst.λg:CNsNt.∂z:Ns.((g⊳z)⋆(f⊳z)).

[22] Cf. Łukasiewicz & Tarski 1930, **§ 2**, Satz 9, page 8 [= Paper IV, in Tarski 1956, **§ 2**, Theorem 9, page 44].

This example is slightly involved. It is appropriate to use some convenient shorthand. Set ⊢ G := Ł$_{33,1927}$: [Ł$_{33,1927}$] ≡ [G]. With

[K] := CpCqp
α := CNrCsNt
β := CrCsu
γ := CCtsCtu
η := CαC$\beta\gamma$

we have

[G] := CC[K]CCαC$\beta\gamma$vCwv ≡ CC[K]CηvCwv

(Note that η is a tautology.) Where K := λx:pλy:q.x, ⊢ K : [K], as ever, consider the following *assumption-list* (or else 'proof-context', written down vertically, and indenting as appropriate in order to ease readability):

[F : C[K]Cηv ≡ C[K]CCαC$\beta\gamma$v]
　[x : w]
　　[f : α ≡ CNrCsNt]
　　　[g : β ≡ CrCsu]
　　　　[h : Cts]
　　　　　[y : t]
　　　　　　[z : Nr]

(This is, more or less, as in a so-called 'natural deduction' display.[23] Note also that [F : C[K]Cηv] ⊢ F(K) : Cηv ≡ CCαC$\beta\gamma$v.)

Omitting decorations on free witness variables – appearing in proof-contexts – we have, successively, in outermost proof-context [F,x]:

[f,g,h,y,z] ⊢ fz : CsNt
[f,g,h,y,z] ⊢ hy : s
[f,g,h,y,z] ⊢ fz(hy) : Nt
[f,g,h,y,z] ⊢ fz(hy)⋆y : **f**
[f,g,h,y] ⊢ ∂z:Nr.fz(hy)⋆y : r
[f,g,h,y] ⊢ g(∂k:Nz.fz(hy)⋆y) : Csu
[f,g,h,y] ⊢ X[f,g,h,y] := g(∂z:Nr.fz(hy)⋆y)(hy) : u
[f,g,h] ⊢ λy:t.X[f,g,h,y] : Ctu
[f,g] ⊢ λh:Cts.λy:t.X[f,g,h,y]: CCtsCtu = γ
[f] ⊢ H := λg:β.λh:Cts.λy:t.X[f,g,h,y] : C$\beta\gamma$
⊢ H := λf:α.λg:β.λh:Cts.λy:t.X[f,g,h,y] : CαC$\beta\gamma$ = η.

So H is a closed term (i.e., here, a $\lambda\partial$-combinator). Furthemore,

---

[23] See, e.g., Rezuş 2017 and, *mutatis mutandis*, Rezuş 1990. Otherwise, the display is also closely matching an AUTOMATH-like notation. Cf. Rezuş 1983.

$$[F,x] \vdash F(K)(H) : v$$
$$[F] \vdash \lambda x{:}w.F(K)(H) : Cwv, \text{ and, in the end}$$
$$\vdash G := \lambda F{:}C[K]C\eta v.\lambda x{:}w.F(K)(H) : CC[K]C\eta vCwv$$
$$\equiv CC[K]CC\alpha C\beta\gamma vCwv \equiv [G] \equiv [Ł_{33,1927}].$$

The corresponding 'type-free' combinator is thus:

$$|\,G\,| \equiv \lambda F.\lambda x.F(|K|)(|H|), \text{ where}$$
$$|\,K\,| \equiv_\alpha \lambda u.\lambda v.u, \text{ and}$$
$$|\,H\,| \equiv \lambda f.\lambda g.\lambda h.\lambda y.X[f,g,h,y] \equiv \lambda f.\lambda g.\lambda h.\lambda y.g(\partial z.fz(hy){\star}y)(hy).$$

The remaining examples (**2.8–2.10**) – due to Łukasiewicz (1932) – are, obviously, Frege-Church pairs. The first two (**2.8–2.9**) are left as exercises. The last example (**2.10**) is discussed in § 4.

2.8 Łukasiewicz (1932), $Ł_{33,1932}$

$$[Ł_{33,1932}] \equiv CCCaCCNpNuCCCpqCrsCtCCsuCrpCCbbdd$$

2.9 Łukasiewicz (1932), $Ł_{31,1932}$

$$[Ł_{31,1932}] \equiv CCCpCCNqCrNsCCtrCCrsCtqCCtCutvv.$$

2.10 Łukasiewicz (1932), $Ł_{29,1932}$

$$[Ł_{29,1932}] \equiv CCCpCCNqCrsCCsqCtCrqCCNuCuvww.$$

§ 3 Tarski's first singleton basis (1925)

According to Sobociński 1932, Tarski's first singleton generator ['single-axiom basis'] (1925) for the two-valued propositional logic, based on C and N [material implication and classical negation] as primitives, with *modus ponens* and uniform substitution as only rules of inference, was:

$$\vdash T_{53,1925} : [T_{53,1925}]$$
$$\equiv CCCCCsCtCtCvCvvCCCCCpCqrCCpqCprCCCNpNqCqpxxyyCCpCqpzz,$$

where $T_{53,1925}$ stands for a witness of the corresponding 53-letter CN-tautology $[T_{53,1925}]$.

By the time Tarski was pondering on the problem – sometime during the early twenties – it was known that the classical (two valued) CN-calculus has a finite (axiomatic) basis, containing at least $[K] \equiv CpCqp$, $[S] \equiv CCpCqrCCpqCpr$, and finitely many 'N-axioms' [tautologies containing negation] in a set $\mathcal{N}$, say (cf., e.g., Frege's *Begriffsschrift*, 1879, and Rezuş 2019a). In particular, Łukasiewicz established the fact that one can take $\mathcal{N} = \{\,[T]\,\}$, where $[T] := CCNpNqCqp$ [= a variant of a genuinely classical *contraposition* 'law'].

As I noticed about forty years ago (1979), this very specific *logic* problem amounts to a special case of a rather trivial exercise in the pure λ-calculus. The ['type-free'] λ-calculus equivalent is a typical packaging / unpackaging problem in pure λ-calculus, combinatorial in nature, and concerns the so-called *singleton bases* for sets of combinators (closed λ-terms), closed under conversion (resp. reduction), that can be generated explicitly from finitely many combinators. (Cf. Rezuş 1980, 1982, 2010, 2017c.)

A one moment's reflection shows that a would-be *normal form* of the witness $T_{53,1925}$ must be of the form <F, X, Y] [≡ <<F,X>,Y>], for appropriate decorated ['typed'] *combinators* F, X, and Y. That much for the packaging step.

As to unpackaging, let's work out, by way of example, a generic *extraction pattern* from triples of the form <F, X, Y], in 'type-free' terms, first.

In this case (i.e., for Frege-Church pairs), we need, essentially, projections $\mathsf{P}_1^2 := \lambda z.z(\mathsf{K})$ and $\mathsf{P}_2^2 := \lambda z.z(\mathsf{KI})$, or else just the usual combinators, $\mathsf{K} := \lambda x.\lambda y.x$, and $\mathsf{KI} := \lambda x.\lambda y.y = \mathsf{K}(\mathsf{I})$, used in a postponement-régime, since we have ⊢ <X,Y>(K) = X, and ⊢ <X,Y>(KI) = Y.

Set $\mathsf{F} := \mathsf{K}^4(\mathsf{I}) \equiv \mathsf{K}(\mathsf{K}(\mathsf{K}(\mathsf{K}(\mathsf{I}))))$ and $\mathsf{G} := \mathsf{G}_{\mathsf{X},\mathsf{Y}} \equiv$ <F,X,Y], for arbitrary X, Y. This yields:

(1) ⊢ F(X₁)(X₂)(X₃)(X₄)(Y) = I(Y) = Y, and
(2) ⊢ G(G) ≡ <F,X,Y](G) ≡ <<F,X>,Y>(G)
   = G<F,X>(Y)
   ≡ <<F,X>,Y>(< F,X>)(Y)
   = <F,X>(<F,X>)(Y)(Y)
   = <F,X>(F)(X)(Y)(Y)
   = F(F)(X)(X)(Y)(Y) = Y [from (1)].

So, for arbitrary X, Y, Z, and with $\mathsf{KK} := \mathsf{K}^1(\mathsf{K}) \equiv \mathsf{K}(\mathsf{K})$, one has:

(3) ⊢ G(K)(KK)(Z) ≡ <F,X,Y](K)(KK)(Z) ≡ <<F,X>,Y>(K)(KK)(Z) = K<F,X>(Y)(KK)(Z) = <F,X>(KK)(Z) = KK(F)(X)Z = K(X)(Z) = X.

Set Y := K and X := <S,T>, for arbitrary T.

That is, this time, one has G := <F,<S,T>,K] ≡ <K⁴(I),<S,T>,K], with arbitrary T. So, the above would yield

(2a) ⊢ G(G) = K, and
(3a) ⊢ G(K)(KK)(Z) = <S,T>, i.e., a Frege-Church pair.

Now, the first (postponement) projection (K) – from (2a) – gives

(4) ⊢ <S,T>(K) [= G(K)(KK)(Z)(K)] = K(S)(T) = S,

while the second (postponement) projection (KI) can be obtained as ever (from S and K), by,

(5) ⊢ I = S(K)(K), and
(6) ⊢ KI := K(I) ≡ K(S(K)(K)),

so that we can also extract T:

(7) ⊢ <S,T>(KI) [= G(K)(KK)(Z)(KI)] = KI(S)(T)= I(T) = T.

Altogether, G yields K, S, and T, for arbitrary T.

Here, the internal structure of T is immaterial, so, in particular, it can be a Frege-Church pair or, even, an n-tuple generated from such pairs, i.e, something of the form <X₁, ..., Xₙ], say.

The 'type-free' argument above proves the following meta-theorem:

3.1 Theorem (Tarski 1925). Let λ^+ be an extension of the pure λ-calculus λ, whose (closed) terms are generated by K, S and $X_1, \ldots, X_n$ (n>0). Then, where T := $<X_1, \ldots, X_n>$, G := $<K^4(I),<S,T>,K]$ is a singleton basis for λ^+.

3.2 Remark. The preceding ('type-free') argument can be repeated, *verbatim*, in 'typed' terms, since a Frege-Church pair is always 'typeable', if so are its ultimate components, here: I, K, S, and T (cf. Rezuş 2010, for a discussion of the matter in a more general setting).

In particular, putting T := T [sic], in the above[24], with $\vdash$ T : [T] ≡ CNpNqC-qCp, yields *the most general form* of Tarski's singleton basis for the (classical propositional) CN-calculus, on the original 1925-pattern, viz.:

$$\vdash\ <K^4(I),<S,T>,K] : CCCC[K^4(I)]CCC[S]CC[T]xxyyC[K]zz,$$

where the indended ('most general') deco's / 'typings' of the components are, as expected:

$\vdash K^4(I) : [K^4(I)] \equiv CsCtCwCvCuu,$
$\vdash K : [K] \equiv Cp_1Cq_1p_1,$
$\vdash S : [S] \equiv CCp_2Cq_2rCCp_2q_2Cp_2r,$
$\vdash T : [T] \equiv CNpNqCqp,$

(with all propositional variables distinct), while Tarski's original basis $T_{53,1925}$ identified (unnecessarily) $t \equiv w$, $v \equiv u$, $p \equiv p_1 \equiv p_2$, and $q \equiv q_1 \equiv q_2$.

3.3 Remarks. (1) The ('type-free') result above is a special case of a more general (meta-) theorem, viz. *Tarski's Claim* (1925), in Rezuş 2010. In the latter paper, I had G := $<K,KK,F]$, where F := $<X_1, \ldots, X_n]$, and KK := K(K), with extraction pattern given by:

$$\vdash G(G) = K, \text{ and } \vdash G(KK)(K) = F,$$

so the extraction of the components of F could be done using only K.

(2) Otherwise, there are many possible variations on this. Worth noting is a solution due to John Halleck 2010 (cf. Rezuş 2010 [*Addendum*]).

Set, for emphasis, $K(X) := \lambda x.X$ (x not free in X), for arbitrary X, and

$F_0 := X_0,$
$F_{k+1} := <F_k, K(X_{k+1})>$, i.e.,
$F := F_n \equiv [X_0, K(X_1), \ldots, K(X_n)>,$

with, further,

$H := <F, KK>$ and
$G := K<H, KK>) \equiv K(<<F,KK>,KK>).$

---

[24] A (normal) witness for [T] in (extensions of) $\lambda\partial[f,N,C]$ could be the (genuinely classical) combinator C^∂, with $\vdash C^\partial := \lambda x{:}CNpNq.\lambda y{:}q.\partial z{:}Np.(x(z)\star y) : CNpN\text{-}qCpCq.$

Then

$\vdash G_3 := G(G)(G) = \mathsf{K}(\mathsf{KK}) \equiv \mathsf{K}^2(\mathsf{K})$, and finally,
$\vdash G(G)(G_3) = \mathsf{K}$.

In detail,

$\vdash G(X) = \langle H,\mathsf{KK}\rangle$, for arbitrary X (in particular, choose X := G), while
$\vdash G(X)(\mathsf{K}) = H = \langle F,\mathsf{KK}\rangle$, for arbitrary X (choose X := G, again),

so we get

$\vdash H(\mathsf{K}) = F = F_n$.

(3) In view of a side-remark of Rezuş 2019, **§ 3**, there are *infinitely many* solutions of the kind (countably many, though), pairwise distinct.

§ 4 A Łukasiewicz singleton basis (1932)

From Sobociński 1932, we have also

$[Ł_{29,1932}] \equiv \mathsf{CCCpCCNqCrsCCsqCtCrqCCNuCuvww}$

as a singleton basis for the (classical) CN-calculus.

This was, apparently, by 1932, the *shortest* singleton basis of the kind, that could be found by using Tarski's original 'methods' of 1925.

Let $\vdash \mathsf{K}_p(X) := \lambda x{:}p.X : \mathsf{C}p\alpha$ (where x, p are fresh for $\vdash X : \alpha$).

Obviously, $Ł_{29,1932}$ is a Frege-Church pair, i.e., of the form G := $\langle X,Y\rangle$, where X := $\mathsf{K}_p(F)$, and $\vdash Y := \mathsf{CO}_{p,q} \equiv \lambda z{:}\mathsf{N}p.\lambda x{:}p.\varpi_q(z\star x) : \mathsf{CN}p\mathsf{C}pq$, with

$\vdash F := \lambda f{:}\mathsf{CNqCrs}.\lambda g{:}\mathsf{Csq}.\lambda h{:}t.\lambda z{:}r.\epsilon y{:}\mathsf{Nq}.g(fyz) : \mathsf{CCNqCrsCCsqCtCrq}$.

So, in the end,

$\vdash Ł_{29,1932} := \langle \mathsf{K}_p(F),\mathsf{CO}_{u,v}\rangle : \mathsf{CCCpCCNqCrsCCsqCtCrqCCNuCuvww}$;

in other words, Łukasiewicz still followed the original Tarski idea of packaging.

Igoring uniform substitutions and recording deco's (here: 'types' = formulas) only *as needed*, the original Łukasiewicz derivations, reported by Sobociński 1932, **§ 1**, translate into to the following *charabia*[25]:

---

[25] This is just Meredith 'condensed detachment' [CD] notation (due to Carew A. Meredith). As expected, the low positive integers (here: 1 to 33) are *local* combinator-labels. Further, = stands for *conversion* in an appropriate extension of the pure λ-calculus – actually, in **λ∂p[f,C,N]** – so that each line has to be established separately, by using, possibly, a CD theorem prover or else an appropriate λ-calculus 'reduction-machine'. In view of a well-known theorem (the 'unification theorem' of John Alan Robinson 1965; cf., e.g., Rezuş 1982), 'most general' substitutions (in CN-formulas) can be uniquely restored, up to uniform reletterings, so there is no need to mention them explicitly.

1 := G ≡ Ł$_{29,1932}$
2 := 1(1) ≡ G(G) = K(1) = K(BI) : CtCCsqCsq
3 := 1(2) ≡ G(G(G)) = CO = C(O) : CNpCpq
4 := 2(3) = BI ≡ 1 = I : CCsqCsq ≡ [1$_{s,q}$]
[Here, ⊢ 1 = I is ($\lambda\eta$)-conversion, but we have [1$_{s,q}$] ≡ [I$_{Csq}$], anyway.]
5 := 1(4) = G(BI)
6 := 5(3)
7 := 6(1) ≡ 5(3)(1) = KI = K(I) : CtCpp ≡ [KI$_{t,p}$]
8 := 5(7) : CCpqCrCpq ≡ [K$_{Cpq,r}$]
9 := 8(8) : CsCCpqCrCpq ≡ [K$_s$(K$_{Cpq,r}$)]
10 := 5(9)
11 := (10)(5)
12 := (11)(7)
13 := (12)(3)
14 := (13)(8)
15 := (14)(7)
16 := 5(15)
17 := (12)(16)
18 := (17)(1) = KP = K(P) : CsCCCpqpp ≡ [K$_s$(P$_{p,q}$)]
19 := (18)(1) = P : CCCpqpp ≡ [P$_{p,q}$] [= 'Peirce's Law']
20 := (12)(2)
21 := (20)(19)
22 := (21)(7) = B(W) ∘ C : CCpCqCprCqCpr
23 := (22)(12) = B : CCqrCCpqCpr ≡ [B$_{q,r,p}$]
24 := (23)(19) = B(P)
25 := (23)(24) = B = B(B(P)) = B²(P)
26 := (25)(12) = CB = C(B) : CCpqCCqrCpr ≡ [CB$_{p,q,r}$]
27 := (22)(3) = B(W)(O) = O : CpCNpq ≡ [O$_{p,q}$] [= 'Duns Scotus']
[Here, ⊢ (B(W)∘C)(C(O)) = B(W)(C(C(O))) = B(W)(O) = W∘O = O, since we have ⊢ C(C(F)) = F, for arbitrary combinators F, and $\hat{\varGamma}$ ⊢ ϖ_{Cpq}(e)X = ϖ_q(e), for $\hat{\varGamma}$ ⊢ e : f and $\hat{\varGamma}$ ⊢ X : p, for any proof-context $\hat{\varGamma}$.]
28 := (16)(16)
29 := (28)(7)
30 := (29)(7)
31 := 5(30)
32 := (22)(31) = KE = K(E) : CtCCNqqq ≡ [K$_t$(E$_q$)]
33 := (32)(7) = E : CCNppp ≡ [E$_p$] [= the 'Law of Clavius'],

and { [CB$_{p,q,r}$], [O$_{p,q}$], [E$_p$] } is the basis of Łukasiewicz 1929, for the classical, two-valued CN-calculus.

The corresponding 'type-free' combinator is thus:

⊢ | Ł$_{29,1932}$ | := < λx.λf.λg.λh.λz.ϵy.g(fyz), λz.λx.ϖ(z⋆x) >.

4.1 Remark. Note that – unlike for the Tarski original example of 1925, which required only pure λ-calculus-derivations, i.e., the implicational fragment of intuitionistic logic – the $\lambda\partial$-derivations from $Ł_{29,1932}$ use also specific properties of genuinely classical proof operators.

Acknowledgements. For obvious reasons, mentioned also in the main text and in the references, I am indebted to the late David Meredith (Merrimack NH), as well as to John Halleck (Salt Lake City UT), and V. Frederick Rickey (Cornwall NY). Rickey made also useful historical and stylistical comments on a previous version of this note.

9 August – 10 October 2019

Old folklore on λ-calculus generators (2015b)

This note, historical in character, concerns a few venerable λ-calculus generators [singleton bases] that are not obtained directly via the Tarski Theorem (1925) of Rezuş 1982.[1] Otherwise, what follows below is old λ-calculus folklore of the 1970's, extracted from Rezuş 1980. Apparently, except for Rosser's G_5, given as an exercise in Henk Barendregt's monograph (Barendregt 1981), the examples listed here are not in print.

Unless otherwise stated, the *notation* is as in Barendregt 1981. In particular, we have:

Combinators. I, K, S are the usual combinators [here: closed λ-terms] and so is B := λxyz.(x(yz)), with also X∘Y = BXY := λz.X(Yz) [z not free in X,Y]. However, KX := λv.X [v not free in X], etc.

Exponents (actually, the [n]'s are the *Church numerals*, for n > 0):

[1]XY := XY, [n+1]XY = X([n]XY).
Example: [3]XY := X(X(XY)).

Frege-Church sequences.

$<X_1, ..., X_n>$:= λx.xX_1...X_n [x not free in X_i; i := 1,...,n].
Example: $<K>$:= λx.xK.

Singleton bases (*λ-generators*) Rezuş 1980. All generators G_i, i := 1,...,9, listed below are not *organic*[2] (i.e., here, they contain closed λ-terms in normal form).

G_1 := λx.(xS(KK))K [H. P. Barendregt]
G_2 := λx.(xS([3]KI))K = λx.x(xS(K(K(KI))))K [C. Böhm]

---

[1] See also Rezuş 2010. In particular, the discussion (appearing in Rezuş 1980) of generators found by other people was not included in Rezuş 1982, so that most of these findings remained unpublished thus far.
[2] Terminology as in Sobociński 1932, Rezuş 1980, 1982.

$G_3 := \langle [4]KI, S, K\rangle = \langle K(K(K(KI))), S, K\rangle$ [J. Barkley Rosser]
$G_4 := \langle K(KK), K, K, S\rangle$ [J. Barkley Rosser]
$G_5 := \langle K, S, K\rangle$ [J. Barkley Rosser]
$G_6 := \langle I, [3]KS, KK\rangle$ [C. Böhm, W. L. van der Poel]
$G_7 := \langle I, K(K\circ(KK))\rangle$ [C. Böhm, W. L. van der Poel]
$G_8 := \langle K, K, S, K\rangle$ [W. L. van der Poel]
$G_9 := \langle S, S, K, K, K, S\rangle$ [J. W. van Briemen].

Extraction patterns. Let $G := G_1$ [Barendregt]. Then

$\vdash K = GGG, \vdash S = GK = G(GGG)$.

Let $G := G_5$ [Rosser]. Then

$\vdash K = GGG, \vdash S = G(GG)$.

Let $G := G_i$, $i := 2, 3, 7, 8$ ('Böhm-like' extraction pattern). Then

$\vdash K = GG, \vdash S = GK = G(GG)$.

Let $G := G_i$, $i := 4, 6, 9$ ('Schönfinkel-like' extraction pattern, Schönfinkel 1924). Then

$\vdash S = GG, \vdash K = GS = G(GG)$.

Sources. (1) G_1, G_2: From a note of H. P. Barendregt, Stanford University, (October 1971). (2) G_3, G_4, G_5: From a letter of J. B. Rosser to H. P. Barendregt, Rosser 1971 (December 20, 1971). (3) G_6–G_9: From a note of W. L. van der Poel (Delft University of Technology, The Netherlands) to A. Rezuş (September 1979).[3]

Remarks (May 23, 2015).

(1) The shortest λ-generator of the kind (i.e., non-organic) known in the seventies was Rosser's G_5 [= $\langle K, S, K\rangle$]. (Remark **1.3** in Rezuş 1980.)

The shortest one I know of so far was found recently (2015) by John Halleck (University of Utah, Salt Lake City, UT). This is $H := K(\langle S, K\rangle) = \lambda xy.ySK$. For Halleck's H, the most economic SK-extraction-pattern seems to be

---

[3] G_6 and G_7 have been 'found after a hint of Corrado Böhm with the help of a computer' (W. L. van der Poel, 1979). J. W. van Briemen was a student and a collaborator of W. L. van der Poel at the Delft University of Technology, during the 1970's. Putting aside his pioneering contributions to computer science, both practical and theoretical (cf. van der Poel 1988, 2003), W. L. van der Poel was also interested in logic (as e.g., in axiomatisability problems, cf. van der Poel 1969) and in λ-calculus (van der Poel et al. 1980). Otherwise – *pace* Corrado Böhm, Wolf Gross et al. (Universities of Pisa and Rome, Italy) with their CUCH-*machine*; cf., e.g., Böhm & Gross 1965, Böhm 1966, Ausiello 1971, Böhm & Dezani 1972, Böhm et al. 2017, etc. – van der Poel and his collaborators and students of the Delft University of Technology (mainly Gerrit van der Mey, a blind computer scientist) were among the first in Europe to implement λ-calculus and combinatory 'reduction machines', as well as Curry's 'bracket abstraction algorithm(s)', in various Lisp dialects and/or in C.

⊢ S = HHH, ⊢ K = HH(SH) = HH(HHHH).

H is typeable, of course (as H is a head normal form, and K, S are typeable), viz. its most general type [scheme] is (in Łukasiewicz notation):

⊢ H : CpCCCCqCrsCCqrCqsCCtCutvv.

(2) The shortest *organic* λ-generator known to date is due to Carew A. Meredith (circa 1956; cf. Meredith & Prior 1963, Rezuş 1982a), viz.

$G_0 := \lambda xyz.y(Kz)(xz)$, where $Kz := \lambda v.z$, as above.

Notably, although G_0 is typeable, its most general type [scheme] is *not* an axiom for minimal, resp. intuitionistic implication [with *modus ponens*], but $M_0 := K(G_0)$ [$= \lambda uxyz.y(Kz)(xz)$] is. As regards G_0, see the Appendix [Rezuş 1982a] to Rezuş 1982 for an easy SK-extraction pattern and other details. For M_0 and some variations on it, see, *e.g.*, the notes of Dolph Ulrich 2007 and, possibly, Ulrich 1999.

(3) The [typeable] λ-generators dating of the late 1920's and the early 1930's (due to Alfred Tarski, Jan Łukasiewicz, Mordchaj Wajsberg and Bolesław Sobociński, and implicit in Sobociński's survey Sobociński 1932, already referred to obliquely in Rezuş 1980, 1982, will be discussed in detail elsewhere.

Acknowledgements. I am grateful to Henk Barendregt (Radboud University, Nijmegen), Corrado Böhm (University of Rome, *La Sapienza*) and Willem L. van der Poel (University of Technology, Delft) – for reasons already stated elsewhere – as well as to John Halleck (University of Utah, Salt Lake City, UT) for reviving my interest in the matter of this note, via Ulrich's and his own recent findings. J. Roger Hindley (Swansea University, Wales, UK) spotted an editorial inadvertence in a previous draft.

23 May 2015 – 3 May 2017

16
Implications and generators (2019c)

There is little interest in talking about *arbitrary* 'fragments' of classical logic; we should better qualify the matter, as appropriate.

(1) *Bases and generators.* A set $\mathcal{L}$ of witness-terms closed under conversion (equality) is said to have *basis* $\mathcal{B}$ *relative to a set* $\mathcal{R}$ *of witness operators* if $\mathcal{L}$ can be generated from $\mathcal{B}$ (and witness-variables) with the operators of $\mathcal{R}$ alone. This terminology applies to undecorated ('type-free') as well as to decorated ('typed') terms. Of course, in the limit case, a basis can be also empty relative to a given set $\mathcal{R}$ of operators, in which case we can speak about *closure under* $\mathcal{R}$ (*by explicit definability*)[1]. In applications, we are usually interested in sets with a *finite* basis relative to a *finite* set of operators.

Mutatis mutandis, the same way of speaking can be applied to sets of tautologies (understood as propositional schemes[2]) and rules of inferences.

The formularian logic tradition privileged logics with a primitive implication ($\to$) and a primitive rule ($\triangleright$) [*modus ponens*] (also known as *detachment*) for ($\to$), so that bases were understood as *axiomatic bases*, i.e., finite sets of tautologies (axiom schemes), while a logic – containing a primitive ($\to$) – was supposed to contain specific axiom schemes and to be closed under a given set $\mathcal{R}$ of rules, where $\mathcal{R}$ contained ($\triangleright$). Axiomatic bases relative to the singleton set { ($\triangleright$) } were usually called – mainly in the Polish Lvov-Warsaw 'School' (cf. Rezuş 2018a, etc.) – *normal bases*, the rule of uniform substitution being therein included.[3] Singleton normal bases are also called *generators*. This terminology, applies, *mutatis mutandis*, to the corresponding witness combinators.

(2) *Abstraction-strictness and linearity* (Rezuş 1977). A monadic abstraction operator ♮ (here: λ, ρ, ∂, ϵ) – term-form ♮x.a[x] – is said to be *strict* if the free

---

[1] More generally, we may think of a combinator as being a specific witness-operator so that a basis is going to be just a set of witness-operators in the extended sense.

[2] Ignoring thus the 'rule' of uniform substitution.

[3] 'Deductively closed' sets of tautologies admitting of an infinite (normal) basis alone – or with no basis at all – have been studied only incidentally (as, e.g., by Karl Schröter, around 1937; cf. his 1952).

variable x occurs (actually) at least once in a[x], and *linear* if x occurs at most once in a[x]. Similarly, an abstraction operator is *strictly linear* if it is both strict and linear. Witness-terms containing only strict (linear, resp. strictly linear) abstractions are said to be strict (linear, strictly linear resp.).

> Non-strict witness-terms that are not in normal form may reduce – resp. be convertible (equal) – to strict witness-terms and analogously as regards non-linearity *vs* linearity. Cases in point, using the examples below: ⊢ K(S) □ K = B ≡ λf.λg.λx.f(g(x)) resp. ⊢ S(K)(K) = I ≡ λx.x, and so on. There is no ambiguity involved, since the terminology is supposed to define *fields* of witness-terms in as far their *well-formedness* is concerned: a strict (linear, strictly linear resp.) equational theory (typically, a λ-calculus) is strict (linear, strictly linear resp.) if its well-formed terms are so, etc. (Rezuş 1981, 1982; in λ-calculus, this way of speaking goes back to the late seventies).

Examples. (1) Frege's axiomatic **BS**-basis contains a single non-strict linear witness combinator (K). The remaining witness combinators taken as primitive in **BS** are strict, while only (S) is not linear.

(2) The following ('pure') combinators are strictly linear:

(I) ⊢ I[A] := λx.x = S(K)(K) ≡ K □ K = C(K)(K)
: A → A [Identity].
where F □ G := S(F)(G),
(B) ⊢ B[A,B,C] := λf.λg.λx.f(g(x)) = S(K(S))(K) ≡ K(S) □ K
: (A → B) → ((C → A) → (C → B)) [Prefixing]
(C) ⊢ C[A,B,C] := λf.λx.λy.f(y)(x) = (B □ K(K)) ∘ S
: (A → (B → C)) → (B → (A → C)) [Commutation],
where F ∘ G := B(F)(G),
(CI) ⊢ CI[A,B] := λx.λy.λy.y(x) = C(I)
: A → ((A→ B) → B) [Assertion]
(CB) ⊢ CB[A,B,C] := λf.λg.λx.g(f(x)) = C(B)
: (A → B) → ((B → C) → (A → C)) [Suffixing]
(CC) ⊢ CC[A,B,C] := λx.λy.λz.y(z)(x) = C(C) = B ∘ CI
: A → ((B → (A → C)) → (B → C)) [Selfcommutation]
(D) ⊢ D[A,B,C] := λx.λy.λz.z(x)(y) = C ∘ CI
: A → (B → ((A → (B → C)) → C)) [Pair]

(3) Other examples of ('pure') strict, non-linear combinators:

(CS) ⊢ CS[A,B,C] := λf.λg.λx.gx(fx) = C(S)
: (A → B) → ((A → (B → C)) → (A → C)) [Commuted Frege],
(W) ⊢ W[A,D] := λf.λx.f(x)(x) = CS(I)
: (A → (A → B)) → (A → B) [Hilbert], and
(S_M) ⊢ S_M[A,B,C] := λf.λg.λx.f(gx)x = B(W) ∘ B
: (A → (B → C)) → ((B → A) → (B → C)) [Meredith].

In particular, *strictness* and *strict linearity* admit of extensions to λ-calculi with (primitive) pairs – as, e.g., **λπ** – resp. to 'logic' systems with (a primitive) conjunction (a free variable is *strict in* a pair ≺ a,b ≻, if it occurs 'evenly' in a and b, i.e., if it occurs actually in both components). Although awfully 'syntactic', this kind of restriction yields a proper proof-theoretical account of so-called *relevance logics with* (a primitive) *conjunction* (relevance-preserving and), lacking and-or-distributiviy, typically the 'Lattice $\mathcal{R}$', referred to in passing below.

§ 1 *'Implications'.* A while ago, the founders of *relevance logic* – Alan Ross Anderson and Nuel Belnap Jr. – have noticed the fact that '[M]*aterial implication is no more a kind of implication than a blunderbuss is a kind of bus.*' (Anderson & Belnap Jr. 1962, and 1975, **1**, **§ 1.1**)[4]

Putting aside matters of logical baptism, virtually, all known 'concepts' of (an) implication worth looking into are just truncations – or modifications – of a blunderbuss that does not shoot.

More specifically, the idea of (an) implication qua *inference carrier* is a formularian invention and rests on a *conceptual blunder*.

What follows next is a cocktail of ad hoc examples, popular in the current (formularian) logic literature, that can be described, in a straightforward way, in witness-theoretical terms.[5]

§ 1.1 *Pure minimal implication and pure* $\boldsymbol{\lambda}$. The pure λ-calculus $\boldsymbol{\lambda}$ agrees, at bottom, with the implicational fragment of Johansson's 'minimal' logic. This might be a mere *accident*, however.

(1) Note that { (K), (C), (S) } and { (K), (C), (CS) } are (deductively resp. definitionally) equivalent in the presence of ($\triangleright$), since $\vdash$ S(F)(G) = CS(G)(F) and $\vdash$ C(C(F)) = F. So we can replace (S) by (CS) in the original, *redundant* **BS**-axiomatics with (C) primitive.

(2) Recall that { (K), (S) } is a normal basis – with ($\triangleright$), thus – for the pure λ-calculus $\boldsymbol{\lambda}$.

> In the form presented there, the argument – also referred to as 'abstraction algorithm' in the λ-calculus literature – is due to Paul Rosenbloom (1950). H. B. Curry used originally a different basis – with primitive B, C, and W, in place of S – in his Göttingen *Inaugurardissertation* (1930), inspired apparently by Hilbert's early work on axiomatics. In logic proper, the result – although

---

[4] 'The term *blunderbuss* is of Dutch origin, from the Dutch word *donderbus*, which is a combination of *donder*, meaning 'thunder', and *bus*, meaning 'pipe' (Middle Dutch: *busse*, box, tube, from Late Latin, *buxis*, box, from Ancient Greek *pyxís*, box: esp. from boxwood).' (Wikipedia) Actually, '[t]he *blunderbuss* is a firearm with a short, large caliber barrel which is flared at the muzzle and frequently throughout the entire bore, and used with shot and other projectiles of relevant quantity and/or caliber. The blunderbuss is commonly considered to be an early predecessor of the modern shotgun, with similar military and defensive use. It was effective only at short range, lacking accuracy at long range. A blunderbuss in handgun form was called a dragon, and it is from this that the term dragoon evolved.' (*ibid.*)

[5] For so-called 'intermediate' systems resp. 'implications' – living in the no-man's land between classical and intuitionistic logic – of minor interest here, see, e.g., Dummett 1959, Bull 1962, 1964, Thomas 1962, Meredith 1966, etc.

credited sometimes to Jacques Herbrand (PhD Dissertation Paris 1930, cf. also his 1930a) – is, in fact, due to Alfred Tarski (circa 1921).[6]

So { (K), (C), (CS) } is a normal basis for λ, too. In fact, (C) is redundant, so { (K), (CS) } is a normal basis for λ, as well (Carew A. Meredith, *apud* Prior 1962, *Appendix I*, § **12.1**). Indeed, set F □ G := CS(G)(F). Then one has ⊢ S = (K(CS) □ K) □ K(CS) and also ⊢ S(F) = CS □ K(F).

Alternatively, one can modify as appropriate the *abstraction algorithm* of Rezuş 2019a, § **1.8**, by using (CS) as a primitive combinator instead of (S) [sic].

(3) (Carew A. Meredith; Prior *ibid.*, § **1.1**) The set { (K), (C), (S_M) } is a *non-redundant* normal basis for λ. Indeed, we have, successively: ⊢ B = S_M ∘ C ∘ K, ⊢ CB = C(B), ⊢ F ∘ G = S_M(C(K(F)))(G), whence, finally, ⊢ S = C(S_M ∘ CB), while ⊢ S_M ≡ λf.λg.λx.f(gx)x = B(W) ∘ B, with ⊢ B = K(S) □ K, and ⊢ W = C(S)(I).

A vast amount of variations on this can be found in Wajsberg 1937, 1939, Prior 1962, *Appendix I*, Ulrich 1999, 2007, Rezuş 2018a, etc. Such equivalences can be also established nowadays with appropriate software, in matter of milliseconds, by using either 'traditional' implementations of the [C.A.] Meredith *condensed detachment* [CD] or else a λ-calculus 'reduction machine'. Notably, the first software implementation of the CD-operator is due to David Meredith (Carew's American cousin), during the early fifties. (Cf. e.g., Hindley & Meredith 1990.) On Carew A. Meredith, see also Prior 1956, 1962[2] [the Appendices], Meredith & Prior 1963, Kalman 1964, 1983, Meredith 1977 [obituary], Rezuş 1982, 1982a, Copeland 1996 [*passim*], Ulrich 2001, Wos & Veroff 2001, Humberstone 2011 [*passim*], etc.

(4) *Mutatis mutandis*, (1)–(3) above hold for λ[→], as well. Note, however, that a singleton basis for the 'type-free' λ need not be a basis for the 'typed' λ[→]. Cf. § **2** below.

§ **1.2** *Pure material implication.* The following remarks answer the somewhat ad hoc question: 'What witness-terms can be decorated with → alone?'

The second well-known axiomatisation of classical (propositional) logic is due to Charles S. Peirce (1885). Peirce considered also propositional quantifiers, relying on the primitive signature [→,∀^{p}], where ∀^{p} stands for the universal propositional quantifier, whereupon **f** [*falsum*] could be defined as **f** := ∀^{p}p.p. Incidentally, the (minimal) quantifier rules for ∀^{p} guarantee also the provability of the genuinely intuitionistic *ex falso*-tautology ($\dot{\omega}$) **f** → A. Leaving ($\dot{\omega}$) aside, we get an axiom system for pure classical implication (→), with (▷) [PP] for (→).

(1) (Charles S. Peirce 1885, Jan Łukasiewicz, cca 1929). Let

(P) ⊢ P[A,B] : ((A → B) → A) → A [Peirce's Law].

---

[6] See Porte 1982, Grattan-Guinness 2000, p. 550, etc.

The tautologies witnessed by (I), (CB), (C), (P) axiomatise, with ($\triangleright$) [PP], the classical tautologies containing material implication ($\rightarrow$) alone.

In $\boldsymbol{\lambda\gamma}[\mathbf{f},\rightarrow]$, one has $\sim A := A \rightarrow \mathbf{f}$, as ever, and we can set:

(P^γ) $\vdash P^\gamma[A,B] := \lambda f{:}(A \rightarrow B){\rightarrow} A).\varepsilon z{:}{\sim}A.f(\lambda x{:}A.\varpi[B](zx))$
: $((A \rightarrow B) \rightarrow A) \rightarrow A$.

Analogously, in $\boldsymbol{\lambda\partial}[\mathbf{f},\rightarrow,\neg]$ (where $\neg$ is a primitive), one has:

(P^∂) $\vdash P^\partial[A,B] := \lambda f{:}(A \rightarrow B){\rightarrow} A).\varepsilon z{:}\neg A.f(\lambda x{:}A.\varpi[B](z\star x))$
: $((A \rightarrow B) \rightarrow A) \rightarrow A$.

Here, (I) is redundant, because we have (Arthur N. Prior 1958):

(W$_=$) $\vdash W = P \circ CB \equiv P \circ C(B)$, and
(I$_=$) $\vdash I = P \circ C(W)$,

for both (P)'s, i.e., (P^γ) in $\boldsymbol{\lambda\gamma}$, and ($P^\partial$) in $\boldsymbol{\lambda\partial}$ (and extensions).[7]

There are many variations on this, since the Peirce axiomatics above amounts, roughly, to (BCI) + (P) + ($\triangleright$), i.e., to *strictly linear* (λ) + ($\triangleright$) + (P), or else to $\mathcal{BCI}$ + (P).

> See the discussion of the $\mathcal{BCI}$-axiomatics, following below. Here, one can replace the combinator (P) by the corresponding rule – an abstraction (proof) operator – of course. There are some – rather weird – equivalent alternatives to the Peirce set, in terms of rules / proof operators alone, I shall go into later on (cf. (4) below).

(2) In 1921, while working on his Warsaw PhD Dissertation, Alfred Tarski used the tautologies witnessed by (K), (CB) and

(T) $\vdash T[A,B,C] : ((A \rightarrow B) \rightarrow C) \rightarrow ((A \rightarrow C) \rightarrow C))$,
where, for instance, T[A,B,C] stands, in $\boldsymbol{\lambda\partial}[\rightarrow]$, for
(T^∂) $\vdash T^\partial[A,B,C] := \lambda f{:}(A{\rightarrow}B){\rightarrow}C.\lambda g{:}A \rightarrow C.\varepsilon z{:}\neg C.f(\lambda x{:}A.\varpi[B](z\star(gx)))$
: $((A \rightarrow B) \rightarrow C) \rightarrow ((A \rightarrow C) \rightarrow C))$,

with ($\triangleright$) and propositional quantifiers in order to axiomatise the full classical propositional logic (Prior 1962, *Appendix I*).[8]

Still without (K), it is easy to see that one can replace [formula] (P) by [formula] (T), in the Peirce axiomatics (1).

Indeed, we have:

(P$_=$) $\vdash P = C(T)(I)$, as well as
(T$_=$) $\vdash T = P^* \circ P^*$, where
($P^*_=$) $\vdash P^* = B(P) \circ CB = \lambda x.\lambda y.P(y \circ x)$,

---

[7] Deriving (W$_=$) and (I$_=$), in $\boldsymbol{\lambda\partial}$, say, should be an entertaining exercise.
[8] One can obtain the corresponding $\lambda\gamma$-witness term (T^γ) in the obvious way.

i.e., only *strictly linear transformations* T ⇌ P, whereas the equations – resp. derivations – above can be established for both (P)'s and (T)'s, i.e., for (P$^\gamma$), (T$^\gamma$) in $\boldsymbol{\lambda\gamma}$, and (P$^\partial$), (T$^\partial$), resp. in $\boldsymbol{\lambda\partial}$ (and extensions).

See, e.g., Wajsberg 1937, § 6 I, for a hint: Tarski's T can be obtained as T := F(I), where F is a witness for Wajsberg's A$_1$ [≡ CCsrCCCpqrCCpsr, in Łukasiewicz notation], *ad loc.*

(3) *The Tarski-Bernays axioms for material implication.* Subsequently, Tarski axiomatised (circa 1926) the pure classical implication by (K), (CB) and (T), with (▷). Paul Bernays noticed slightly later (1928) that (T) can be replaced by (P) in Tarski's 1926-set. The latter axiomatics { (K), (CB), (P), (▷) } has been taken, later on, as a reference set by most workers in the area. Cf., e.g., Wajsberg 1937, 1939, the survey of Prior 1962, *Appendix I*, etc.

(4) *Variations on Peirce: the Kanger Rule, Wajsberg etc.* Stig Kanger (1924–1988) noticed, around 1960, that – granted (I) – both (P) and (▷) [*modus (ponendo) ponens*] (sic) can be replaced, in this context, by the single inference rule:

(▷$^+$) ⊢ f : A → B, ⊢ g : (A → C) → A ⇒ ⊢ f ▷$^+$ g : B [the Kanger Rule] [KR]

(cf. Beth & Leblanc 1960, resp. Leblanc 1979, pp. 382–384, and, possibly, Ulsen 2000, pp. 247–249). As regards provability (resp. rule-derivability), we can set, indeed,

(P$^\pm_=$) ⊢ P$^+$[A,B] := λf:(A→B)→A.(I[A] ▷$^+$ f) : ((A → B) → A) → A, and

(▷$_=$) ⊢ f : A → B, ⊢ a : A ⇒ ⊢ f ▷ a := f ▷$^+$ (λx:A→B.a), x not free in a, resp.

(▷$^\pm_=$) ⊢ f : A → B, ⊢ g : (A → C) → A ⇒ ⊢ f ▷$^+$ g := f ▷ P[A,C](g) : B, for the appropriate combinator P[A,C].

So, in particular, on the signature [**f**,→], with inferential ∼A := A → **f**, as ever, the set { (λ), (▷$^+$), (ϖ) } is a basis for the (classical) propositional logic, while, on [**f**,→,¬] (with primitive negation ¬), we need also (ρ) and a would-be (primitive) rule

(·$^+$) ⊢ f : ¬A, ⊢ g : (A → B) → A ⇒ ⊢ f ·$^+$ g : **f**,

for the purpose, in order to be able to retrieve (·), i.e., the Law of Non-Contradiction, on the Kanger-pattern above.

More variations on Peirce's 'Law' (resp. on the 'Rule of Peirce') can be found in the survey Wajsberg 1937, 1939.

§ 1.3 *Pure 'weak' implication* (Ivan Efimovič Orlov 1928, Moh Shaw Kwei 1950, Alonzo Church 1951). The strict terms of pure $\boldsymbol{\lambda}$ can be generated from the (non-redundant) normal basis { (S), (B), (C), (I) }. In logic, this corresponds to the implicational fragment $\mathcal{R}$[→] of the relevance logic $\mathcal{R}$ (Alan Ross Anderson and Nuel Belnap Jr. 1957–1958).

For copious formularian details on $\mathcal{R}$ and relevance logics in general, see the treatise Anderson & Belnap Jr. 1975, 1992. A monographical study of the ('type-free') strict λ-calculus, also known as λI-*calculus* in the literature, can be found in Rezuş 1981. See also Church's Princeton Lectures Notes, Church 1935–1936, and Church 1941. For Orlov's anticipation (1928) of $\mathcal{R}$, see Alves 1992, Došen 1992, 1993, Bazhanov 2001, 2003, Stelzner 2002, etc.

§ 1.4 *Pure linear and strictly linear implication* (Carew A. Meredith, cca 1950).

(1) The linear terms of the pure λ-calculus $\boldsymbol{\lambda}$ can be generated from the (non-redundant) normal basis { (B), (C), (K) }. For instance, I is linearly definable as I := C(K)(X), for arbitrary (linear) X.

(2) Analogously, the strictly linear terms of pure $\boldsymbol{\lambda}$ can be generated from the (non-redundant) normal basis { (B), (C), (I) }.

> This information can be obtained by refining the argument [the *abstraction algorithm*] supposed to 'define' the corresponding λ-abstractors. For strictness – and strict linearity – see Rezuş 1981. The original argument for strict 'type-free' terms, is due to J. Barkley Rosser (Princeton PhD Dissertation, 1935), but it cannot be transferred as such to the 'typed' case, i.e., to the purely implicational relevance logic $\mathcal{R}[\rightarrow]$. See also Rosser 1942, 1953. In logic, the Deduction Theorem for $\mathcal{R}[\rightarrow]$ is due to Alonzo Church 1951, 1951a. The purely implicative logics $\mathcal{BCK}[\rightarrow]$ and $\mathcal{BCI}[\rightarrow]$ are due to Carew A. Meredith (circa 1950). (In fact, $\mathcal{BCK}[\rightarrow]$ is already implicit in Fitch's Yale PhD Dissertation 1934; cf. Fitch 1936. See also Jaśkowski 1963.) For $\mathcal{BCI}[\rightarrow]$ see also Smiley 1955, Urquhart 1972, 1973, Rezuş 1982, 1987 etc.

> Incidentally, $\mathcal{BCI}[\rightarrow]$ is the purely implicational fragment of Girard's 'linear logic' (1987).[9]

> There is a huge amount of formularian lore in print on the strictly linear extensions of $\mathcal{BCI}[\rightarrow]$ (a special case of 'relevance logics'), as well as on the (linear) extensions of $\mathcal{BCK}[\rightarrow]$ (also known as $\mathcal{BCK}$-logics[10].).

> In formularian terms, 'linearity' in logic is usually referred to by the colloquial label 'logics without contraction' (cf. Ono & Komori 1985, Komori 1987, Girard 1987, 2011, Restall 1994, Bull 1996, Bunder 1996, etc.), and counts as a species of 'sub-structurality' – because, in Gentzenese, such 'logics' lack the 'structural' rule (ctc) – whereas their (formularian) algebraic counteparts are referred to as BCK- resp. BCI-'algebras' (see, e.g., Iséki 1966, Imai & Iséki 1966, Iséki & Tanaka 1978, Komori 1987, Meng & Jun 1994, Iorgulescu 2008 [BCK], resp. Meng 1987, Huang 2006 [BCI], etc).

(3) All ('type free') linear terms of $\boldsymbol{\lambda}$ can be decorated ('typed') on the signature $[\rightarrow]$, with the same 'typing' rules as for minimal (resp. intuitionistic)

---

[9] As an aside, Girard's 'linear logic' (1987, 2011, etc.) is nothing but a modal (Lewis-like, $\mathcal{S}4$-ish) extension of a (strictly linear) subsystem of $\mathcal{R}$, viz. of its *non-distributive* variant, $\mathcal{LR}$ [*Lattice* $\mathcal{R}$] (cf. Thistlewaite *et al.* 1985, 1987, Rezuş 1987).

[10] Notably, the Łukasiewicz many-valued logics are very specific $\mathcal{BCK}$-logics.

implication. (For a proof, see Hindley 1989. Of course, this implies normalisability for the linear terms of $\boldsymbol{\lambda}$.)

§ 1.5 *'Abelian' implication* (Carew A. Meredith, cca 1957; Ettore Casari, cca 1979; Robert K. Meyer & John K. Slaney, cca 1980). Adding the axiom scheme

[A] $((A{\to}B){\to}B){\to}A$ ['relativity']

to $\mathcal{BCI}[{\to}]$ yields the purely implicational fragment $\mathcal{A}[{\to}]$ of the 'Abelian Logic' $\mathcal{A}$ of Robert K. Meyer and John K. Slaney 1980, 1981, 1989, 2000[11] Obviously, (I) is redundant in the extension (it can be obtained from (A) by a single application of *modus ponens*), and it is an easy exercise (rather tricky, though, if done by hand) to show that (C) is redundant as well, so that $\mathcal{A}[{\to}]$ has a non-redundant normal basis containing (A) and (B) alone. Alternatively, { (A), (CB) } is a normal basis for $\mathcal{A}[{\to}]$, too.[12] (These remarks are due to Carew A. Meredith, cca 1957.)

Notably, in view of the 'relativity' axiom A, $\mathcal{A}[{\to}]$ is *not* a fragment of the implicational fragment of classical logic. Consistency is straightforward, though: one can interpret $\mathcal{A}[{\to}]$ in the pure equivalential calculus of Łukasiewicz 1939. So, $\mathcal{A}[{\to}]$ is a fragment (subsystem) of classical logic, anyway.

In particular, $\mathcal{A}[{\to}]$ is the *logic of Abelian groups*, a fact already known to Henry George Forder (1889–1981), Carew A. Meredith *et al.* since the mid- or late nineteen fifties.[13]

As an amusing *aside*, on the primitive signature [**f**,$\to$], one can also extend $\mathcal{BCI}[{\to}]$ consistently, in the same way as for $\mathcal{A}[{\to}]$, by replacing (A) with its (proper) substitution instance

(Δ^γ) $((A{\to}\mathbf{f}){\to}\mathbf{f}){\to}A$,

while defining the corresponding inferential negation, $\sim A := A{\to}\mathbf{f}$, as ever.

In this case, one obtains $\mathcal{BCI}\Delta^\gamma[\mathbf{f},{\to}]$, say, which is the same thing as the so-called 'multiplicative' fragment of Girard's 'linear logic' $\mathcal{LL}$ (ignoring the other two $\mathcal{LL}$-constants), otherwise, a detail which Girard 1987a missed (see also Rezuş 1987c).

---

[11] As advertised by its proponents, $\mathcal{A}$ is the *logic of lattice-ordered Abelian groups*. A similar logic system and variants ('comparative logics') have been proposed by Ettore Casari around 1979 (cf. Casari 1981, 1984, 1985, 1987, 1989, 1991 etc.).

[12] In view of the fact that ⊢ CC = L(L) = B ∘ CI = λx.λy.λz.(yzx), ⊢ C = CC(CC)(CC) = λx.λy.λz.(xzy), and ⊢ CI = C(I) = λx.λy.(yx), it is enough to obtain CI or CC. So, for instance, Branden Fitelson obtained, quite economically [correspondence: October 2012] CI as ⊢ CI = F(F), with F := A ∘ B(B) ∘ B, using OTTER software, while John Halleck got [correspondence, ditto] CC as ⊢ H(G), with G := B(B) ∘ B ∘ B, and H := B(B) ∘ B ∘ B(A), using specific condensed detachment [CD] software of his own.

[13] Cf., e.g., Forder 1968 and Meredith & Prior 1968 [§ 4 *Abelian groups*], reporting conversations on the subject dated 1957. See also Tarski 1938.

Worth noting is the fact that both $\mathcal{A}[\to]$ and $\mathcal{BCIA}^\gamma[\mathsf{f},\to]$ admit of elegant witness theories that have the same 'type-free' cognate, viz. the *strictly linear* fragment of the 'type-free' $\lambda\gamma$-calculus (cf. Rezuş 1990, 1991, 1993, 2019a, 2019b).

Of course, in ('type-free') $\lambda\gamma$, one can have $\mathsf{A} \equiv \lambda x.\gamma y.(x(y))$, while $\vdash \gamma z.e[z] = \mathsf{A}(\lambda z.e[z])$, as expected.[14]

§ 2 *Singleton bases.* Among other things, the pioneers – mainly Polish, although not only – were obsessed with the *economy of* (expressive) *means* in logic, and used to pay special attention to finite *non-redundant* normal bases, and to *singleton normal bases*, in particular.

> This is a subject derived, essentially, from Stanisław Leśniewski, with his PhD student Alfred Tarski (by then Tajtelbaum), Jan Łukasiewicz and his own students – Stanisław Jaśkowski, Mordchaj Wajsberg, Bolesław Sobociński, and, later, Carew A. Meredith – as main contributors to the subject. There is a rather vast – now nearly forgotten – literature on this kind of trifles. Cf. Rezuş 2018a, for a survey, and Rezuş 1982, 2010, 2019 for Tarski's pioneering work (cca 1921–1926) on the matter.

§ 2.1 *Singleton bases for pure minimal implication and pure* λ. As the pure ('typed') λ-calculus agrees with the proof theory of the *minimal* – resp. *intuitionistic* – implication, one might suspect that the problem of finding a normal singleton basis for the pure minimal implication amounts to the same thing as to that of finding a singleton basis for ('typed') pure λ. This is, indeed, the case.

Sufficient conditions have been first found by Alfred Tarski around 1925[15] Notably, Tarski's original 'method' applied to *arbitrary* subsystems of classical logic, with at least implication ($\to$) as a primitive, and admitting of a (finite) normal axiomatisation (viz. with *modus ponens* for $\to$ and substitution, as only primitive rules of inference), and containing the (minimal) tautology $\mathsf{A} \to (\mathsf{B} \to \mathsf{A})$, i.e. *mutatis mutandis*, in λ-calculus terms, the ('typed') combinator $\mathsf{K}[\mathsf{A},\mathsf{B}]$.[16]

Moreover, Tarski's idea can be also used in a 'type-free' setting, and applies to subsystems of pure λ and/or extensions, as well. (Cf. Rezuş 1982, 2010, for details.)

Note, however, that a (normal) basis for the 'type-free' λ need not be a basis for the 'typed' $\lambda[\to]$. The following (counter-) example is due to Carew A.

---

[14] For more details on the witness theory of Abelian groups see also **§ 2.5** below, and Rezuş 2019d.

[15] No proof in print. Explicit solutions, based on λ-calculus, have been first obtained by the present author sometime before 1979; cf., e.g., Rezuş 1982, and the subsequent comments appearing in Rezuş 2010.

[16] The latter condition can be dispensed with in favour of 'local' K's, by using the concept of *restricted* – resp. *local* – *solvability*, an idea going back to Kleene 1934, 1935. Cf. Rezuş 1978, 1981, and the explicit constructions of Rezuş 1982.

Meredith (circa 1956, as reported by Arthur N. Prior, in 1963, without a proof in print):

$$\mathsf{G} := \lambda x.\lambda y.\lambda z.y(\lambda u.z)(xz)$$

is a generator ({ (G) } is singleton basis) for 'type-free' λ, but the singleton containing:

$$\mathsf{G}[A,B,C,D] := \lambda x{:}A{\to}B.\lambda y{:}(C{\to}A){\to}(B{\to}D).\lambda z{:}A.y(\lambda u{:}C.z)(xz)$$
$$: (A \to B) \to (((C \to A) \to (B \to D)) \to (A \to D))$$

is not a (normal singleton) basis for $\lambda[\to]$.

> A proof can be found in Rezuş 1982a [i.e., Rezuş 1982, *Appendix*]. See also Rezuş 2010, Remark **3**. Surprisingly, the '(most generally) typed' K(G) *is* a generator for $\lambda[\to]$. Anecdotally, the proof of the Meredith claim has been obtained *by hand*, in October 1981, after a huge amount of unsuccessful ruminations using specific λ-calculus software ('reduction machines') implemented by Willem L. van der Poel and his students and colleagues from the Delft University of Technology. (See also Rezuş 2015b.)

An inconvenience of the 'Tarski method' of obtaining 'single axioms' consists of the fact the explicit constructions yield only so-called *non-organic* singleton bases.[17] Organic singleton bases for minimal / intuitionistic implication have been found later on by Mordchaj Wajsberg, Carew A. Meredith *et al.* Cf. Sobociński 1932, 1955–1956, Ulrich 1999, 2007, and Rezuş 2015c, 2018a, 2019, etc. for details.

§ 2.2 *Singleton bases for pure material implication.* As pure material implication admits of a finite normal axiomatisation, results similar in nature to those obtained by Tarski (1925) – applying to the implicational fragment of classical logic – have been subsequently found by Jan Łukasiewicz, Mordchaj Wajsberg and Bolesław Sobociński (cf., again, Sobociński 1932, 1955–1956, and Rezuş 2018a, 2019). Worth noting is the following.

The 'shortest' singleton basis for material implication (Jan Łukasiewicz 1948; cf. Rezuş 1990, **3.20**, 2017a, 2017b). Let, in the ('type-free') $\lambda\gamma$-calculus,

($\mathsf{Ł}^\gamma$) $\mathsf{Ł}^\gamma := \lambda x.\lambda y.\lambda z.\varepsilon u.y(x(\lambda v.\varpi(uv)))$.

Of course, { ($\mathsf{Ł}^\gamma$) } is *not* a singleton basis for the ('type-free') $\lambda\gamma$-combinators. (One cannot obtain, $\varpi := \lambda x.\gamma v.x$, for instance.) On the other hand, for $\lambda\gamma[\mathsf{f},\to]$, with inferential $\sim A := A \to \mathsf{f}$, one can generate, with ($\triangleright$) [PP], the witness combinators for the classical tautologies containing material implication alone from the $\lambda\gamma$-witness-term:

($\mathsf{Ł}^\gamma$) $\mathsf{Ł}^\gamma[A,B,C,D] := \lambda x{:}(A{\to}B){\to}C.\lambda y{:}C{\to}A.\lambda z{:}D.\varepsilon u{:}{\sim}A.y(x(\lambda v{:}A.\varpi[B](uv)))$
$: ((A \to B) \to C) \to ((C \to A) \to (D \to A))$,

---

[17] Recall that a tautology is *organic* for a logic $\mathcal{L}$ if no proper sub-formula of it is a theorem of $\mathcal{L}$, and, *mutatis mutandis*, similarly for combinators / closed λ-terms.

by using the *non-redundant* classical signature, viz. inside the full (quantifier-free) classical logic based on [**f**,→], where the corresponding 'type-free' $\lambda\gamma$-combinator is: $\mathsf{t}^\gamma := \lambda\mathrm{x}.\lambda\mathrm{y}.\lambda\mathrm{z}.\epsilon\mathrm{u}.\mathrm{y}(\mathrm{x}(\lambda\mathrm{v}.\varpi(\mathrm{uv})))$, above.

Alternatively, one can do the same thing inside the full (quantifier-free) classical logic based on [**f**,→,¬], in $\boldsymbol{\lambda}\boldsymbol{\partial}$[**f**,→,¬] (and extensions), with (▷) and a $\lambda\partial$-witness combinator:

(t^∂) t^∂[A,B,C,D] $:= \lambda\mathrm{x}{:}(A{\to}B){\to}C.\lambda\mathrm{y}{:}C{\to}A.\lambda\mathrm{z}{:}D.\epsilon\mathrm{u}{:}\neg A.\mathrm{y}(\mathrm{x}(\lambda\mathrm{v}{:}A.\varpi[B](\mathrm{u}{\star}\mathrm{v})))$
    $: ((A \to B) \to C) \to ((C \to A) \to (D \to A))$,

with 'type-free' $\lambda\partial$-cognate: $\mathsf{t}^\partial := \lambda\mathrm{x}.\lambda\mathrm{y}.\lambda\mathrm{z}.\epsilon\mathrm{u}.\mathrm{y}(\mathrm{x}(\lambda\mathrm{v}.\varpi(\mathrm{u}{\star}\mathrm{v})))$.

Notably, while estimating *proof-complexity* in bare formularian terms, by counting symbols in formulas, the Łukasiewicz single axiom (1948) for material implication (→), appears to be the shortest one possible.

> Otherwise, the ('type-free') combinators (t^γ), resp. (t^∂) have no specific personality, so to speak. Actually, their form – and complexity – depends heavily on the choice of the primitive proof operators to work with, even in a 'typed', proper witness-theoretical setting. Roughly speaking, *proof-theoretical complexity* in classical logic is – prima facie, and among other things – a matter of *notational relativity*.

Longer normal singleton bases for material implication – counting lengths by formularian standards – have been found previously by Mordchaj Wajsberg, Jan Łukasiewicz, and Carew A. Meredith (cf. Rezuş 2018a, 2019).

§ 2.3 *Singleton bases for pure 'weak' implication.* According to Rezuş 1982, the strict fragment of $\boldsymbol{\lambda}$ admits of a singleton normal basis and analogously for $\mathcal{R}$[→].

> As to the 'Rezuş $\mathcal{R}$[→]-axiom' R$_\to$[18] this is, actually, obtained as a straighforward application of a modified theorem of Tarski (1925), discussed in Rezuş 1982. For a general method of constructing singleton bases in strict cases (as, e.g., in the absence of K, say), cf. Rezuş 1978, 1982 ('local solvability', based on an idea of S. C. Kleene, PhD Dissertation, Princeton 1934). For a discussion and explicit Meredith-style condensed detachment derivations (using Automated Reasoning tools), see, e.g., Fitelson 2001, Wos & Veroff 2001, Ulrich 2007a, Wos 2007, 2007a, etc.

§ 2.4 *Singleton bases for pure linear and strictly linear implications* (Carew A. Meredith, cca 1950). The linear (resp. strictly linear) fragment of $\boldsymbol{\lambda}$ admits of a singleton normal basis (cf. Rezuş 1982, Ulrich 2007b, 2007c).

> For linearity, the result is due to Alfred Tarski 1925 (unpublished; proofs in Rezuş 1982, 2010, 2018a, 2019). For strict linearity, see the refinements of Tarski's result in Rezuş 1982. A list of singleton bases for the strictly linear combinators (due to Carew A. Meredith) – resp. single-axioms for the corresponding, purely implicative logic $\mathcal{BCI}$[→] can be found in Prior 1962, *Appendix I*. See also Ulrich 2007b (for pure $\mathcal{BCK}$), 2007c (for pure $\mathcal{BCI}$).

---

[18] CCCpCCCqqCCrrCCssCCttCpuuCCCCCvwCCwxCvxCCCCCyCyzCyzCCCde-CCefCdfgghhCCiCCCjjCCkkCCllCCmmCinnoo [Łukasiewicz-notation].

§ 2.5 *Singleton bases for 'Abelian' implication* (Carew A. Meredith cca 1957, Rezuş 2012). The strictly linear fragment of $\lambda\gamma$ – and so, *mutatis mutandis*, the pure 'Abelian' implication $\mathcal{A}[\to]$ – admit(s) of singleton normal bases. The shortest example (counting C-symbols and propositional variables, in formularian terms)

[$M_\mathcal{A}$] $(A \to B) \to (((C \to B) \to A) \to C)$

is due, as ever, to Carew A. Meredith[19], with, as possible witnesses,

$(M_\mathcal{A})$ ⊢ $M_\mathcal{A} := \lambda x{:}A{\to}B.\lambda y{:}(C{\to}B){\to}A.\gamma z{:}C{\to}B.(x \triangleright (y \triangleright z))$
: $[M_\mathcal{A}] \equiv (A \to B) \to (((C \to B) \to A) \to C)$,

$(M'_\mathcal{A})$ ⊢ $M'_\mathcal{A} := \lambda x{:}A{\to}B.\lambda y{:}(C{\to}B){\to}A.\gamma z{:}C{\to}(C{\to}B).(x \circ y \circ z)$
: $[M_\mathcal{A}] \equiv (A \to B) \to (((C \to B) \to A) \to C)$, etc.[20]

§ 2.6 For additional information on singleton bases for various λ-calculi resp. fragments of classical (propositional) logic, see Sobociński 1932, 1955–1956, Rezuş 1982, 1982a, 2010, 2015c, Fitelson 2001, Ulrich 1999, 2001, 2007, 2007a, 2007b, 2007c, Wos & Veroff 2001, and the survey Rezuş 2018a.

16 November 2018 – 11 November 2019

---

[19] Cf. Meredith & Prior 1968 [**§ 4** *Abelian groups*], already mentioned before, containing condensed detachment [CD] derivations from (B) – resp. (CB) – and (A).
[20] $M'_\mathcal{A}$ has been obtained *by accident*.

17
Appendix: Bolesław Sobociński 1932, § 1

We found useful to insert here an English translation of the relevant section(s) of Sobociński's Warsaw Master's Thesis (1932), concerning the singleton bases for the classical propositional logic based on material implication [C], classical negation [N], and *modus ponens* [for C], as found by Alfred Tarski, Jan Łukasiewicz, Mordchaj Wajsberg and Bolesław Sobociński himself during 1925–1932. Otherwise, the attempt to find witnesses for the *organic* normal generators [singleton bases] mentioned by Sobociński in the text (examples to be analysed in detail elsewhere, in a more general setting) should make up a useful exercise in witness theory (or else in the appropriate 'natural deduction' system). Conversely, the proofs – in ('typed') λ-calculus terms – to the effect that the corresponding witness combinators are actually normal bases for the classical CN-calculus of propositions should be somewhat easier to obtain than during the mid-twenties and the early thirties, for the reader who is already equipped with an appropriate proof-engine, able to perform Meredith *condensed detachment* [CD] derivations, for instance. The reader may want to note, however, the fact that the early Polish workers in the area managed to do such proofs *by hand*, including the required 'most general' substitutions. (Sobociński's *first example* in point, $S_{139,1927}$, obtained in 1927, and mentioned in **§ 1.3**, below, contained 139 letters: a real *technical challenge* even for an experienced OTTER user.)

Bolesław Sobociński, *Z badań nad teorją dedukcji* [Investigations on the theory of deduction][0].

This Appendix contains a translation of the *Introduction* [here: § 0] and § 1, of the Master's Thesis of Bolesław Sobociński (Warsaw 1932), *Z badań nad teorją dedukcji* [Investigations on the theory of deduction], originally published in *Przegląd Filozoficzny 35*, 1932, pp. 171–193. On editorial reasons, the layout of the English version is slightly different from that of the Polish original. For instance, the author's Endnotes are printed here as usual footnotes, while some details (proofs etc.) appearing in the original Endnotes has been, conveniently, inserted into the main text. In order to ease readability and cross-referencing, the text has been segmented as appropriate. The editorial additions / comments appear in square brackets. I am indebted to Andrzej Bułecka (Kraków) for checking the translation.[1]

[**§0 Introduction**]

In the work following below we shall give a few results concerning investigations on the theory of deduction. All results mentioned here that are not already obtained by the author will be given by specifying the names of the persons concerned and, in as far they are not yet in print, by acknowledging their gracious permission.

By the expression 'theory of deduction' [*teorja dedukcji*] or 'ordinary theory of deduction' [*zwykła teorja dedukcj*] or 'implication-negation [resp. CN-] theory of deduction' [*implikacyjno-negacyjna teorja dedukcj*] we shall understand here a system of propositional calculus inferentially equivalent to the quantifier-free theory of deduction given by Jan Łukasiewicz in his lecture notes **Elements of Mathematical Logic**[2], which relies on the primitive terms implication [denoted by C] and negation [denoted by N] alone.[3] By the expression '[equipped] with the normal rules' [*o normalnych dyrektywach*] or, if the rules are not mentioned, we shall always understand that the system under focus has rules equiform to the rules set up [*przyjętemi*] in that system by J. Łukasiewicz in the above-mentioned work[4]. It should be also noted that, whenever we speak about a non-bivalent quantifier-free system of the theory of deduction, it will always understood a

---

[0] Introduction and § 1, Przegląd Filozoficzny 35, 1932, pp. 171–177, & Endnotes, pp. 188–190.

[1] [*Translator's Note*. Everywhere here, the term 'single-axiom' is to be understood as meaning 'normal singleton [one-element] (axiomatic) basis', in the terminology of the early Polish Logic 'school', or else – *mutatis mutandis* – as meaning 'generator' in λ-calculus. In other words, 'single', in such contexts, does not mean 'isolated', 'unique', etc. See also the author's [Endnote 3] below.]

[2] [Endnote 1] [**188**] Łukasiewicz (1929), pp. 36 ff.

[3] [Footnote (*)] For typographic and technical reasons [*wzgędów*], in the original paper all the footnotes appeared – as Endnotes – at the end of the [main] text.

[4] [Endnote 2] [**188**] Łukasiewicz (1929), pp. 63–64.

partial system of the ordinary theory of deduction. By the expression 'thesis' [*teza*] it will be always understood – unless there will be a more detailed explanation – a true expression of the theory of deduction. It should be also noted that we use here the symbolism of the theory of deduction invented by J. Łukasiewicz, whose rules are set out in Łukasiewicz (1929).

§1 Some organic theses that are single-axioms for the CN-theory of deduction[5].

[**1.0 The first single-axiom of Alfred Tarski (1925),** $T_{53,1925}$] The first single axiom for the [CN-] theory of deduction has been found by Alfred Tarski [in 1925]. Tarski's single-axiom [cf. $T_{53,1925}$, i.e., I., in **§ 1.2**, below] was, however, not organic [*nieorganiczne*].

[**1.1 The 29-letter single-axiom of Jan Łukasiewicz,** $Ł_{29,1932}$] Subsequently, Jan Łukasiewicz and the author of the present work have discovered shorter single axioms for the CN-theory of deduction.[6]

The shortest single-axiom for the [CN-] theory of deduction [known so far] has been obtained by J. Łukasiewicz.[7] This thesis, mentioned here with [the author's] permission, has 29 letters, it is not organic, and looks as follows:

CCCpCCNqCrsCCsqCtCrqCCNuCuvww. [$Ł_{29,1932}$]

[We show first, following Jan Łukasiewicz, that this thesis is, indeed, a single axiom for the CN-theory of deduction. [Endnote 5] [**188**]]

Proof:
1. CCCpCCNqCrsCCsqCtCrqCCNuCuvww.
    1 p/CpCCNqCrsCCsqCtCrq, q/Csq, r/Csq, s/Ns, u/s, v/q, w/CtCCsqCsq * C1u/Csq, v/Ns w/CCNsCsqCtCCsqCsq - 2.
2. CtCCsqCsq. [K1 = K(BI)]
    1 u/p, v/q, w/CNpCpq * C 2 t/CpCCNqCrsCCsqCtCrq, s/Np, q/Cpq - 3.
3. CNpCpq. [O]
    2 t/CNpCpq * C3 - 4.
4. CCsqCsq. [1 = BI]

---

[5] [Endnote 3] [**188**] For 'single-axiom', see Łukasiewicz-Tarski (1930), § 2 *Satz 8* (page 7). For 'organic thesis', see *ibid.*, § 2 *Satz 9* (page 8).

[6] [Endnote 4] [**188**] Cf. Łukasiewicz-Tarski (1930), § 2 *Satz 8*. This applies to the CN-theory of deduction. The single-axiom based on Sheffer's [negation of] 'disjunction' [**nor**] is due to the French logician Jean Nicod (1917–1920). The phrase 'the thesis A is shorter than the thesis B' means that [the formula] A has less symbols (Latin upper- and lower-case letters) than [the formula] B, in Łukasiewicz notation.

[7] [Endnote 5] [**188**][**189**] [Obtained in] August 17, 1932. Since this is one of the most important results in the field of research on single axioms for the theory of deduction, and given the fact that an accurate knowledge of it might be useful for different people in their research, I shall give here, in agreement with the author and with his permission, his own proof to the effect that the given thesis is actually a single axiom for the theory under consideration.

17 Appendix: Bolesław Sobociński 1932, § 1

1 p/CNuCuv, w/CCNqCrsCCsqCtCrq *
 C 4 s/CNuCuv, q/CCNqCrsCCsqCtCrq - 5.
5. CCNqCrsCCsqCtCrq.
 5 q/p, r/p, s/q * C 3 - 6.
6. CCqpCtCpp.
 6 q/CCpCCNqCrsCCsqCtCrqCCNuCuvp * C 1 w/p - 7.
7. CtCpp. [KI = CK]
 5 r/p, s/p, t/y * C 7 t/N q - 8.
8. CCpqCrCpq.
 8 p/Cpq, q/CrCpq, r/s * C 8 - 9.
9. CsCCpqCrCpq.
 5 r/Cpq, s/CrCpq, q/s * C 9 s/Ns - 10.
10. CCCrCpqsCtCCpqs.
 10 y/Nr, s/CCqrCtCpr, t/u * C 5 q/y, r/p, s/q - 11.
11. CuCCpqCCqrCtCpr.
 11 u/CtCpp * C 7 - 12.
12. CCpqCCqrCtCpr.
 12 p/Np, q/Cpq * C 3 - 13
13. CCCpqrCtCNpr.
 13 y/CrCpq ° C 8 - 14.
14. CtCNpCrCpq.
 14 t/CtCpp * C 7 - 15.
15. CNpCrCpq.
 5 q/p, s/Cpq * C 15 - 16.
16. CCCpqpCtCrp.
 12 p/CCpqp, q/CtCrp, r/p, t/s * C 16 - 17.
17. CCCtCrppCsCCCpqpp.
 17 t/CpCCNqCrsCCsqCtCrq y/CNuCuv * C 1 w/p - 18.
18. CsCCCpqpp. [KP]
 18 s/CtCpp * C 7 - 19.
19. CCCpqpp. [P]
 12 p/Cpq, q/CCqrCtCpr, r/s, t/u * C 12 - 20.
20. CCCCqrCtCprsCuCCpqs.
 20 q/CqCpr, s/CqCpr, t/q * C 19 p/CqCpr, q/r - 21.
21. CuCCpCqCprCqCpr.
 21 u/CtCpp * C 7 - 22.
22. CCpCqCprCqCpr.
 22 p/Cpq, q/Cqr, r/Cpr * C12t/Cpq - 23. [189]
23. CCqrCCpqCpr. [B]
 23 q/CCrqr * C 19 p/r - 24.
24. CCpCCrqrCpr.
 23 q/CpCCrqr, r/Cpr p/s * C 24 - 25.
25. CCsCpCCrqrCsCpr.
 25 s/Cpq, p/CqR, r/Cpr * C12t/CCprq - 26.
26. CCpqCCqrCpr. [CB]
 22 p/Np, q/p, r/q * C 3 q/CNpq - 27.
27. CpCNpq. [O]
 16 p/CpCqp, q/p * C 16 q/Cqp, t/p, r/q - 28.
28. CtCrCpCqp. [K(KK) = K²(K)]

28 t/CtCpp * C 7 - 29.
29. CrCpCqp. [KK]
    29 y/CtCpp * C 7 - 30.
30. CpCqp. [K]
    5 s/N q * C 30 p/Nq, q/r - 31.
31. CCNqqCtCrq.
    22 p/CNqq, q/t, r/q * C 31 r/CNqq - 32.
32. CtCCNqqq. [KE]
    32 t/CtCpp, q/p * C 7 - 33.
33. CCNppp. [E]

Here, the theses 26 [CB], 27 [O], and 33 [E] are the axioms of Łukasiewicz (1929)[8].

[1.2 Non-organic single axioms for the CN-theory of deduction (Tarski & Łukasiewicz 1925–1932)] Since, except for the 33-letter axiom of Łukasiewicz (1927) [VII, below], the results are not in print, we mention here, with the permission of the authors, the most important (non-organic) single-axioms for the CN-theory of deduction known so far.

I. The 53-letter axiom of Alfred Tarski (1925) [$T_{53,1925}$]:
    CCCCCsCtCtCuCuuCCCCCpCqrCCpqCprCCCNpNqCqpxxyyCCpCqpzz.
II. The 43-letter axiom of Jan Łukasiewicz (1927) [$Ł_{43,1927}$]:
    CCCsCtCtCpCqpCCCCCpCqrCCpqCprCCCNpNqCqpvvee.
III. The 39-letter axiom of Jan Łukasiewicz (1927) [$Ł_{39,1927}$]:
    CCCaCbCdaCCCCCpCqrCCpqCprCCCNpNqCqpvvee.
IV. The 38-letter axiom of Jan Łukasiewicz (1927) [$Ł_{38,1927}$]:
    CCCaCbCdCedCCCNstCCNsNtsCCCpqCCqrCprzz.
V. The 36-letter axiom of the present work (Bolesław Sobociński 1927) [$S_{36,1927}$]:
    CCCaCbCdaCCCNmCqCNrNpCCmCqrCCpqCpree.
VI. The 35-letter axiom of Jan Łukasiewicz (1927) [$Ł_{35,1927}$]:
    CCCaCbCdaCCCNmCqCqNpCCmCqrCCpqCpree.
VII. The 33-letter axiom of Jan Łukasiewicz (1927) [$Ł_{33,1927}$][9]:
    CCCpCqpCCCNrCsNtCCrCsuCCtsCtuvCwu.
VIII. The 33-letter axiom of Jan Łukasiewicz (1932) [$Ł_{33,1932}$]:
    CCCaCCNpNuCCCpqCrsCtCCsuCrpCCbbdd.
IX. The 31-letter axiom of Jan Łukasiewicz (1932) [$Ł_{31,1932}$]:
    CCCpCCNqCrNsCCtrCCrsCtqCCtCutvv. [190]
X The 29-letter axiom of Jan Łukasiewicz [$Ł_{29,1932}$] mentioned in the above.
    CCCpCCNqCrsCCsqCtCrqCCNuCuvww.

[1.3 Organic single-axioms for the CN-theory of deduction (Sobociński 1927–1932, Łukasiewicz 1930)] Jan Łukasiewicz has also first raised the issue of finding organic single axioms for the CN-theory of deduction.

---
[8] [Note (*)] [190] See *loc. cit.*, page 121, and Łukasiewicz-Tarski (1930), § 2, *Satz 6*, page 6.
[9] [Note (**)] [190] Cf. Łukasiewicz-Tarski (1930), § 2, *Satz 9*, page 8.

262   17 Appendix: Bolesław Sobociński 1932, § 1

As far as we know from the contemporary logical literature, the author of the present work[10] has first solved the problem, indicating a number of such (organic) theses, ever shorter. They are presented here in the order in which they have been found.

Evidence to the effect that the theses cited below are single-axioms of the [CN-] theory deduction, are given only for the 139-letter, the 60-letter and the 47-letter theses.

As to the remaining theses, mentioned below, the arguments meant to show that they are actually single-axioms [for the CN-theory of deduction] are much similar and will be skipped. The evidence consists of showing that one can obtain, with the usual inference rules, the following theses:

CpCqp [K],
CCpCqrCCpqCpr [S], and
CCNpNqCqp [C^{∂}]],

which make up, as shown by J. Łukasiewicz, a complete axiomatic presentation of the CN-theory of deduction.[11]

The fact that the formulas mentioned here are actually (organic) theses can be checked by truth-tables.[12]

Before going into the list of organic single axioms, recall also that the thesis

CCCpqCCrstCCuCCrstCCpuCst [$\overrightarrow{W_{1931}}$]

has been shown by Mordchaj Wajsberg to be a single axiom for the (pure) implicational theory of deduction[13]. This means that all the implicational theses of the usual [CN-] theory of deduction can be obtained from Wajsberg's thesis, with the usual inference rules, and, in particular, the theses:

α CpCqp [K]
β Cpp [I]
γ CCCpqrCqr [CBK] [**173**]
δ CCpCqrCqCpr [C]
ϵ CCpCqrCCpqCpr [S]
ξ CCCCqCqrsCqrCqr [Ξ]

---

[10] [Endnote 6] [**190**] Cf. Łukasiewicz-Tarski (1930), § 2, Satz 9, page 8.
[11] [Endnote 7] [**190**] Cf. Łukasiewicz-Tarski (1930), page 6, note 9.
[12] [Endnote 8] [**190**] In this way, one can see that no [well-formed] part of any of the theses given below would evaluate to 1, for any combination of 0 and 1, inserted in place of variables. On the method of testing by truth value tables [0's and 1's], see also Kotarbiński (1929), pp. 192–196. [*Translator's Note*. This was, likely, a piece of typical Warsaw logical humour, by the late twenties, as we have exponential growth: Sobociński's $S_{139,1927}$, for instance, has 12 distinct propositional variables [abekmpqrstuv], so its brute-force 'witness' would be a truth-table with $2^{12} = 4096$ entries ('valuations')!]
[13] [Endnote 9] [**190**] For the 'implicational theory of deduction', see Łukasiewicz (1929), pp. 76–77, Łukasiewicz-Tarski (1930), **§ 4**, as well as the dissertation of Morchaj Wajsberg (1931), *Satz 30*. On this dissertation see also Wajsberg (1932), page 261.

17 Appendix: Bolesław Sobociński 1932, § 1

I. [Sobociński] [$S_{139,1927}$], 139-letter axiom[14]:

1. CCCaCCCbCCCvCNmNsCNkNsCCCpqCCrsmCCuCCrst
CCpuCsteeCCCbCCCvCNmNsCNtNsCCCpqCCrsmCCuCCrst
CCpuCskeeCCCbCCCvCNmNsCNtNsCCCpqCCrsmCCuCCrst
CCpuCskee.

Proof:
 1 a/CaCCCbCCCvCNmNsCNtNsCCCpqCCrsmCCuCCrstCCuCstee, k/t *
 C1k/t - 2.
2. CCCbCCCvCNmNsCNtNsCCCpqCCrsmCCuCCrstCCpuCstee.[15]
 2 b/CbCCCvCNmNsCNtNsCCCpqCCrsmCCuCCrstCCpuCst,
 e/CCCvCNmNsCNtNsCCCpqCClrsmCCuCCrstCCpuCst *
 C2e/CCCvCNmNsCNtNsCCCpqCCrsmCCuCCrstCCpuCst - 3.
3. CCCvCNmNsCNtNsCCCpqCCrsmCCuCCrstCCpuCst.
 3 v/CbCCCvCNmNsCNtNsCCCpqCCrsmCCuCCrstCCpuCst, m/t *
 C2e/CNtNs - 4,
4. CCCpqCCCrstCCuCCrstCCpuCst. [$W_{1931}^{\rightarrow}$]

As noted above, with thesis 4, one can obtain the theses α [K], β [I], γ [CBK], δ [C], ϵ [S], and ξ [Ξ]. So we can add them here without proof.

 γ p/CvCNmNs, q/CNtNs, r/CCCpqCCrsmCCuCCrstCCpuCst * C3 - 5.
5. CCNtNsCCCpqCCrsmCCuCCrstCCpuCst.[16]
 δ p/CNtNs, q/CCpqCCrsm, r/CCuCCrstCCpuCst * C 5 - 6.
6. CCCpqCCrsmCCNtNsCCuCCrstCCpuCst.
 6 m/Cpq * Cαp/Cpq, q/Crs - 7.
7. CCNtNsCCuCCrstCCpuCst.
 δ p/CNtNs, q/CuCCrst, r/CCpuCst * C 7 - 8.
8. CCuCCrstCCNtNsCCpuCst.
 8 u/t * Cαp/t, q/Crs - 9.
9. CCNtNsCCptCst.
 δ p/CNtNs, q/Cpt, r/Cst * C 9 - 10.
10. CCptCCNtNsCst.
 10 t/p, s/q * Cβ - 11.
11. CCNpNqCqp [C^{∂}].

Once we have obtained the theses α [K], ϵ [S], and 11 [C^{∂}], we have a proof of the fact that, in agreement with what has been said above, thesis 1 [$S_{139,1927}$] can be accepted as a single-axiom for the [CN-] theory of deduction. [174]

II. [Sobociński] [$S_{101,1927}$], 101-letter axiom[17]:

---

[14] [Endnote 10] [190] Obtained in December 1927.
[15] [Endnote 11] [190] [Note that] thesis 2 appearing in this proof is itself a single-axiom for the [CN-] theory of deduction. It is not organic. It has been set [*wyszukało*] especially for the purpose of finding an organic axiom.
[16] [Cf. thesis 4, under III., and thesis 6, under IV., below.]
[17] [Endnote 10] [190] Obtained in December 1927.

CCCaCCCbCgCdCnCxCsteeCCCbCCyCNtNsCCCpqCzmCCuCzt
CCpuCskeeCCCbCCCvCNmNsCNtNsCCCpqCCrsmCCuCCrst
CCpuCskee.

III. [Sobociński] [$S_{60,1928}$], 60-letter axiom[18]:

1. CCCvCCNtNsmCCNxaCCbCexCCugCCpuCstCdCCCpq
CCrsxCCuCCrstCCpuCst.

Proof:
    1 v/CvCCNtNsm, m/CCCpqCCrstCCuCCrstCCpuCst, x/t, a/Ns,
        b/Cpq, e/Crs, g/CCrst * C1x/t, a/Ns, b/Cpq, e/Crs, g/CCrst, d/CNtNs - 2.
2. CdCCCpqCCrstCCUCCrstCCpuCst.
    2 d/CdCCCpqCCrstCCuCCrstCCpuCst * C 2 - 3.
3. CCCpqCCrstCCuCCrstCCpuCst. [$W^{\rightarrow}_{1931}$]

As noted in the above, with the thesis 3, one can obtain the theses α [K], β [I], γ [CBK], δ [C], ϵ [S], and ξ [Ξ]. So we add them here without proof.
    ξ q/CNtNs, r/CCCpqCCrsmCCuCCrstCCpuCst,
        s/CCNmaCCbCemCCugCCpuCst *
    C 1 v/CNtNs, m/CCCpqCCrsmCCuCCrstCCpuCst, x/m, d/CNtNs - 4.
4. CCNtNsCCCpqCCrsmCCuCCrstCCpuCst.[19]

From thesis 4 one can obtain the theses α [K], β [I], γ [CBK], and δ [C], in the same way as it was done earlier, in the case of [$S_{139,1927}$] [the 139-letter single-axiom mentioned under I.] above, thereby obtaining also the thesis:

CCNpNqCqp [C^∂],

and, since one has already the theses α [K] and ϵ [S], we have provided the required evidence.

IV. [Sobociński] [$S_{55,1928}$], 55-letter axiom[20]:

CCCvCumCzCCCpqaCCuaCCpzCNtNsCdCCCpqCCrstCCuCCrstCCpzCst.

V. [Sobociński] [$S_{47.1,1928}$], first 47-letter axiom[21]:

1. CCCvCCCpqCCrstmCzCCuaCCpuCNtNsCzCCuCCrstCCpucst.

Proof:
    1 v/CvCCCpqCCrNtNsm, m/CCuCCrNtNsCCpuCNNsNNt,
        z/CCpqCCrst, a/CCrNtNs*C1z/CCpqCCrst, a/CCrNtNs, s/Nt, t/Ns - 2.
2. CCCpqCCrstCCuCCrstCCpuCst. [$W^{\rightarrow}_{1931}$]

As noted above, from thesis 2 one can obtain the theses α [K], β [I], γ [CBK], δ [C], ϵ [S], and ξ [Ξ]. So we add them here without proof.
    γ p/CvCCCpqCCrstm, q/CzCCuaCCpuCNtNs, r/CzCCuCCrstCCpuCst *

---

[18] [Endnote 12] [190] Obtained in the spring of 1928.
[19] [Cf. thesis 5, under I. above.]
[20] [Endnote 13] [190] Obtained in the autumn of 1928.
[21] [Endnote 13] [190] Obtained in the autumn of 1928.

17 Appendix: Bolesław Sobociński 1932, § 1

C 1 - 3. [**175**]

3. CCzCCuaCCpuCNtNsCzCCuCCrstCCpuCst.
   γ p/z, q/CCuaCCpuCNtNs, r/CzCCuCCrstCCpuCst * C 3 - 4.
4. CCCuaCCpuCNtNsCzCCuCCrstCCpuCst.
   γ p/Cua, q/CCpuCNtNs, r/CzCCuCCrstCCpuCst * C 4 - 5.
5. CCCpuCNtNsCzCCuCCrstCCpuCst.
   γ p/Cpu, q/CNtNs, r/CCCpqCCrsmCCuCCrstCCpuCst * C5z/CCpqCCrsm

- 6.

6. CCNtNsCCCpqCCrsmCCuCCrstCCpuCst.[22]

From thesis 6 one can obtain the theses α [**K**], β [**I**], γ [**CBK**], and δ [**C**], in the same way as it was done in the case of [$S_{139,1927}$] [the 139-letters single-axiom, mentioned under I.] above, thereby obtaining also the thesis

CCNpNqCqp [C^∂],

and, since one has already the theses α [**K**] and ϵ [**S**], we have provided the required evidence.

VI. [Sobociński] [$S_{47.2,1928}$], second 47-letter axiom[23]:

CCCvCCCpqCCrstCCuCCrstmCdCeCCpuCNtNsCdCeCCpuCst.

VII. [Sobociński] [$S_{47.3,1928}$], third 47-letter axiom[24]

CCCvCCCpqCCrstCCuCCrstCCpumCdCeCkCNtNsCdCeCkCst[25].

Each of the above-mentioned organic theses (I–VII) – that are single-axioms of the theory of deduction – is so constructed that it involves, in some way, the organic thesis discoverd by M. Wajsberg. Wajsberg's thesis is a single-axiom of the implication theory of deduction, and its structure yields a somewhat general scheme, that can involve not only Wajsberg's thesis, but also any other one with the same property[26], such that the resulting organic thesis is a single-axiom of the theory under focus.

So, when Łukasiewicz found, in 1930, a thesis that is a single-axiom of the implicational theory of deduction, and, at the same time, significantly shorter than the 17-letter thesis of Wajsberg [$W^{\rightarrow}_{1931}$], viz.

CCCpqCrsCtCCspCrp [$Ł^{\rightarrow}_{17.1,1930}$][27],

---

[22] [Cf. thesis 5, under I. above.]
[23] [Endnote 13] [**190**] Obtained in the autumn of 1928.
[24] [Endnote 13] [**190**] Obtained in the autumn of 1928.
[25] [Endnote 14] [**190**] The third 47-letter axiom is much worse than the first resp. the second 47-letter axioms, in that it contains, at the same length, more pairwise distinct [*różnokształtny*] variables than the other two.
[26] [Endnote 15] [**190**] Namely an organic single-axiom for the implicational theory of deduction.
[27] Cf. [Endnote 16] [**190**]. See Łukasiewicz (1931a), page 17.

he applied it immediately to one of the schemes given here, thereby obtaining a 37-letter thesis that is a single-axiom of the theory of deduction.

VIII. [Łukasiewicz] [Ł$_{37.1,1930}$], first 37-letter axiom:

CCCvCCCpqCrsmCzCtCCsuCNpNrCzCtCCspCrp.

Łukasiewicz also observed that such an axiom could be a thesis similar to thesis VIII and based on the same scheme, but with a different distribution of negation(s) [o innem rozmieszczeniu negacji], namely:

IX. [Łukasiewicz] [Ł$_{37.2,1930}$], second 37-letter axiom:

CCCvCCCpqCrsmCzCtCCNrpCNspCzCtCCspCrp. [176]

In the summer of this year [1930], Łukasiewicz found yet another 17-letter organic thesis, which is a single-axiom of the implicational theory of deduction. It is presented here with his kind permission:

CCCpqCrsCCspCtCrp [Ł$_{17.2,1930}^{\rightarrow}$].

This thesis can be used with the help of one of these schemes in such a way that one could obtain:

X. [Łukasiewicz] [Ł$_{37.3,1930}$], third 37-letter axiom:

CCCvCCCpqCrsmCzCCsuCtCNpNrCzCCspCtCrp.

One should note that the results (obtained by different authors and reported here) concerning non-organic and organic theses, that are also single-axioms of the implication-negation, as well as of the implicational theory of deduction, are not final. Namely, the question:

> which thesis is the shortest single-axiom of the implication-negation, resp. implicational theory of deduction?

has not been answered yet. Settling down [rozstrzygnięcie] this issue must be one of the most immediate and most important tasks of the researchers in the theory of deduction. [177]

Bolesław Sobociński's references for the Appendix

1 Tadeusz Kotarbiński 1929 Elementy teorji poznania, logiki formalnej i metodologji nauk [Elements theory of knowledge, formal logic and methodology of science], Wydawnictwo Zakładu Narodowogo Imienia Ossolińskich [Ossolineum Publishers], Lwów 1929. (English translation: [374].)
2 Stanisław Leśniewski 1928 O podstawach matematyki [On the foundation of mathematics] [II = second paper], [Chapter IV] Przegląd Filozoficzny 31 (3), 1928, pp. 261–291. (Digitized online @ [6]. Reprinted in [426]. English translation in [425], **1**, pp. 227–263.)

3 — 1929 *Grundzüge eines neuen Systems der Grundlagen der Mathematik. Einleitung und §1-11*, Sonderabdruck (mit unveränderter Pagination) aus dem XIV Bande der Fundamenta Mathematicae, Warsaw 1929. (Digitized online © [3]. Reprinted in [426]. English translation in [425], [2] pp. 410–605.)

4 Jan Łukasiewicz 1929 Elementy logiki matematycznej [Elements of mathematical logic], Skrypt autoryzowany opracował M. Presburger. Z częściowej subwencji Senatu Akademickiego Uniw. Warsz. Nakładem Komisji Wydawniczej Koła Matematyczno-Fizycznego Słuchaczów Uniwersytetu Warsz., Warszawa 1929 (Wydawnictw Koła tom XVIII). [Authorized typo-script (of lecture notes) prepared by Mojzesz Presburger. Subsidized in part by the Senate of the Warsaw University. Financed by the Publishing Committee of the Association of Students of Mathematics and Physics of the Warsaw University, Publications Series of the Association of Students of Mathematics and Physics of the Warsaw University, Volume XVIII, Warsaw 1929.] (Second edition: Państwowe Wydawnictwo Naukowe [PWN: Polish Scientific Publishers], Warsaw 1958. English translation [441].) [= Lukasiewicz 1]

5 — 1931a *Uwagi o aksjomacie Nicoda i o 'dedukcji uogólniajacej'* [Notes on the axiom of Nicod and on 'generalising deduction'], Odbitka z Księgi Pamietkowej polskiego Towarzystwa Filozoficznego we Lwowie, 12.II 1904 – 12.II 1929, Lwów 1931. Skład główny w Księgarni S. A. Książnica-Atlas. [*Separatum* from the Memorial Book of the Polish Philosophical Society in Lwów, 12 II 1904 – 12 II 1929, Lwów 1931. Printed by Książnica-Atlas SA, etc.] (Reprinted in [440]. English translation in [442] pp. 179–196.)

6 — 1931b *Dowód zupełności dwuwartościowego rachunku zdań* [*Ein Vollständigkeitsbeweis des zweiwertigen Aussagenkalküls*], Odbitka ze sprawozdań z posiedzeń Towarzystwa Naukowego Warszawskiego XXIV 1931, Wydział III. [*Separatum* from the Comptes Rendus des séances de la Société des Sciences et des Lettres de Varsovie (Classe III) 24, 1931, Warsaw 1932]

7 Jan Łukasiewicz & Alfred Tarski 1930 *Badania nad rachunkiem zdań* [*Untersuchungen über den Aussagenkalkül*], Comptes Rendus des séances de la Société des Sciences et des Lettres de Varsovie 23 (Classe III) 1930, Warsaw 1930, [pp. 30–50]. (Reprinted in [737], **1**, pp. 323–343. Polish translation in [440], pp. 129-143. English translation (revised text), in: [734], pp. 38–59, reprinted in [442], pp. 131–152.) [= Łukasiewicz-Tarski]

8 Jean Nicod 1917–1920 *A reduction in the number of primitive propositions of logic*, Proceedings of the Cambridge Philosophical Society 19, 1917–1920, pp. 32–41 (i.e., [501]).

9 Alfred Tarski 1930 *O niektórych podstawowych pojęciach metamatematyki* [= *Über einige fundamentalen Begriffe der Metamathematik*]. Odbitka ze sprawozdań z posiedzcen Towarzystwa Naukowego Warszawskiego XXIII. 1930. Wydział III. [*Separatum* from Comptes Rendus des séances de la Société des Sciences et des Lettres de Varsovie 23, 1930, Classe III, Warszawa 1930, pp. 22–29] (Reprinted in [737], **1**, pp. 313–320. English translation in [734], pp. 30–37.) [= Tarski 1]

10 Mordchaj Wajsberg 1931*Aksjomatyzacja trójwartościowego rachunku zdan* [Axiomatization of the trivalent propositional calculus]. Praca przedstawiona Uniwersytetowi Warszawskiego celem uzyskania dyplomu doktora filozofji i przyjęta przez referentów Prof. Prof. J. Łukasiewicza i St. Mazurkiewicza. Odbitka ze sprawozdań z posiedzeń Towarzystwa Naukowego Warszawskiego XXIV, 1931, Wydział III. [Work presented to the University of Warsaw in order to obtain the diplom of the doctor in philosophy, with, as referents, Prof. J. Łukasiewicz

and St. Mazurkiewicz. *Separatum* from the reports of the meetings of the Warsaw Scientific Society XXIV, 1931, Section III = Comptes Rendus des séances de la Sociéte des Sciences et des Lettres de Varsovie 24, 1931, Classe III, Warsaw 1931)] (English translation in [450], pp. 264–284, reprinted in [791], pp. 12–29.)

11 — 1932 *Ein neues Axiom des Aussagenkalküls in der Symbolik von Sheffer*. Aus den Monatsheften für Mathematik und Physik, XXXIX. Band, 2 Heft, Leipzig 1932. (English translation in [791], pp. 37–39.) [= Wajsberg 1]

Part V

Guidelines

18

Catalogue of concepts and notations

§ 0 Witness grammar: a synopsis

We survey the notational conventions concerning (1) the witness operators and the witness combinators, (2) the equational conditions supposed to characterise the witness operators, and the witness theories thereby defined, as occurring most frequently in the book.

Accordingly, the reader should be able to use this synopsis as a kind of (multi-lingual) dictionary, without being supposed to chase would-be cross-references in the main text or to make appeal to the cumulative *Index of subjects*.

Recall that each witness operator has an associated *term-form* – essentially, a grammatical ('syntactic') stipulation, together with a stratification / decoration ('typing') constraint – and characteristic *equational conditions*, defining it: as in abstract algebra, such conditions concern oft two or more operators, simultaneously. Typically, we have (β)- and (η)-conditions, as in, *mutatis mutandis*, the usual λ-calculi, as well as specific (ζ)-conditions, slightly more involved.

As a rule, once the syntax and the ('typical') decoration of a term-form is known, such conditions can be spelled out in 'type-free' terms, without danger of confusion, so, on economy reasons, we use a 'type-free' spelling while stating them.[1]

For witness combinators X, with $\vdash X : [X]$, where $[X]$ is the 'type' of X, the internal stratification ('typing') of X can be recovered from its 'type' $[X]$, so that the 'type-free' spelling is sufficient. In consequence, where $| X |$ is the 'type-free' cognate of a witness combinator X, we write, more economically, $\vdash | X | : [X]$, instead of the redundant $\vdash X : [X]$, somewhat à la Curry.[2] Here, the use of the turnstile indicates the fact that the combinator is *correctly constructed*.

---

[1] Note that the (ζ)-conditions make sense in a 'type-free' régime, as well: they are automatically interpreted in models of the ('type-free') extensional $\lambda\pi$-calculus, e.g., in a Scott λ-model.

[2] Recall that, everywhere here, the 'typing' is supposed to be *rigid*, as in a Church / de Bruijn type-assignment 'discipline', more or less. Specifically, unlike in a Curry-style 'typing' – in 'functionality theory' or else in 'illative' logic, for instance – we

18 Catalogue of concepts and notations

On economy reasons, again, we suppress the *parametric* context information, as well (specifically, whenever an assumption context $\hat{\varGamma}$, say, occurs on both sides of a meta-conditional), and ignore the so-called 'structural' rules, meant to manipulate such contexts, as they are, obviously, redundant witness-theoretically.

Whenever the typography allows it, the combinators are set in serif type.

Cummulative syntax. Unless explicitly specified otherwise, the régime of metavariables is as usual, viz.:

individual variables: u, v, ...[3]
individual terms: s, t, ...
propositional variables: p, q, r, ...
fomulas: A, B, C, D, ...
witness variables: x, y, z, ...
witness terms: a, b, c, d, e, f, g, ...

With **f** [*falsum*], ¬ [negation], and → [implication] as primitives, one has **v** := ¬**f** [*verum*] and ∼A := A → **f** [inferential negation], as defined notions.[4]

The other connectives – mainly the genuinely classical ones– come in polar / opposite pairs – additive *vs* multiplicative, on the disjunction- *vs* conjunction-like pattern (in this order) – viz. (△, ∧), (→, ↛), (←, ↚), (∨, ▽), and the self-duals (↔, ↮), as in Chapter 8 (i.e., Rezuş 2016), for instance.

---

have always (UT), 'unicity of typing'. See the review Rezuş 2015a, for the larger 'type-theoretical' context.

[3] In the ad hoc λ-theory **λγ̄**, u, v, ... range over atomic (pseudo-) scalars, whereas the (pseudo-scalar) abstraction terms of the form γ̄u.e[u] are well-formed, provided e[u] contains only subterms of the form u ⋉ a, whenever u occurs free in e[u]. [*Aside.* In Parigot's λμ-calculi, the atomic [pseudo-] scalars are ranged over by lower-case Greek letters. Cf. Parigot 1997, and, *mutatis mutandis*, Sørensen & Urzyczyn 2006. As this is conflicting with our usual notational practice concerning the Greek types, we have used the notation for *atomic* scalars instead. (If uncomfortable with this exception, the reader may want to imagine coloured [meta-] letters for the purpose.) The λμ-syntax is *conceptually confused*, in fact: the μ-variables are supposed to be witness-variables enjoying a special régime as regards substitution etc. The confusion goes back to Gentzen's ad hoc usage of L-sequents 'multiple on the right', and amounts to a characteristic mis-understanding of the classical logic *proof-grammar*, as endemically present in *formularian* logic (mainly in so called 'Gentzenisations', and in proofs of 'cut-elimination'). This is not very important, in the end, since Parigot's **λμ** (here **λγ̄**) occurs only locally in the text (Chapter 3 ≡ Rezuş 2017b), as a mere digression. As regards Gentzen, for a clean version of the same story (essentially *proof-normalisation* in classical logic), see, e.g., Rezuş 2017d, 2017e, and, possibly, our discussion of **λγ**, and **λγṫ**, in Rezuş 1990, 2017b, 2017a.]

[4] Casually – mainly in classical logic contexts – one has **f** ≡ ⊥ and **v** ≡ ⊤. (Note, however, that, in some 'weaker' logics – as e.g., in so-called 'substructural' logics – these constats may be distinct.) As a rule, ¬ stands for a primitive negation.

Analogously, the (first-order) quantifiers are grouped in polar pairs ($\bar{\forall}$, $\forall$), and ($\exists$, $\bar{\exists}$), resp.[5]. Propositional quantifiers are denoted similarly, by superscripting a 'p'.

We have used, now and then, the Łukasiewicz frontal notation (N, C, K, A, etc.), for connectives, in some Chapters – those concerning 'Polish' logic, mainly – in which cases, on readability reasons, the formulas are ranged over by α, β, ..., instead, while the 'Polish' quantifiers are denoted by $\Pi \equiv \forall$ and $\Sigma \equiv \exists$, resp., with, if necessary, $N\Pi \equiv \bar{\Pi} \equiv \bar{\forall}$ and $N\Sigma \equiv \bar{\Sigma} \equiv \bar{\exists}$, resp.

Otherwise, the synopsis records only scalar notions as instantiated to first-order logics: the instantiations to logics with popositional and/or second-order quantifiers are completely similar.

In general, the colloquial nomenclature – applying to inference rules and to tautologies – comes from the traditional (formularian) logic lore and is supposed to bear a mnemonic character.

§ 1 Witness operators and witness combinators

§ 1.1 Witness operators as rules of inference

(a) singulary (one-place) operators $\simeq$ single-premiss rules
  ($Cl_\vdash$) [Assertion qua rule]
     $\vdash <a> := \lambda z.z(a) = Cl[A,B](a) : (A \to B) \to B$,
     if $\vdash a : A$, provided z is not free in a
  (ϖ) [ex falso quodlibet]
     $\vdash \varpi_A(e) = \partial x{:}\neg A.e = \gamma x{:}{\sim}A.e : A$, if $\vdash e : \mathbf{f}$,
     provided x is not free in e
  (∇) [DN-i $\equiv$ double negation introduction]
     $\vdash \nabla(a) : \neg\neg A$, if $\vdash a : A$
  (Δ) [DN-e $\equiv$ double negation elimination]
     $\vdash \Delta(c) : A$, if $\vdash c : \neg\neg A$
  $(\pi_1)_1$ [Left projection associated to $(\pi)_1$-adjunction]
     $\vdash \mathbf{1}(c) \equiv \pi_1(c) : A$, if $\vdash c : C \equiv A \wedge B$
  $(\pi_2)_1$ [Right projection associated to $(\pi)_1$-adjunction]
     $\vdash \mathbf{2}(c) \equiv \pi_2(c) : B$, if $\vdash c : C \equiv A \wedge B$
     where either $\wedge$ is primitive or Δ is primitive
  $(\pi_1)_2$ [Left projection associated to $(\pi)_2$-adjunction]
     $\vdash \mathbf{1}(c) \equiv \pi_1(c) : A$, if $\vdash c : C \equiv \neg(A \to B)$,
  $(\pi_2)_2$ [Right projection associated to $(\pi)_2$-adjunction]
     $\vdash \mathbf{2}(c) \equiv \pi_1(c) : \neg B$, if $\vdash c : C \equiv \neg(A \to B)$,
     where both $\to$ and $\neg$ are primitive

---

[5] Otherwise, by our standards, $\forall$ and $\bar{\exists}$ should be both 'multiplicative', while their 'polar' cognates – $\bar{\forall}$ and $\exists$, resp.– should be 'additive', respecting thus duality, as well as a formularian tradition that takes $\forall$-propositions to be infinite conjunctions and $\exists$-propositions to be infinite disjunctions.

274   18 Catalogue of concepts and notations

(τ) [¬-~]-transform] (in inversion [$\bar{\tau},\tau$])
$\vdash \tau[A](c) : {\sim}A$, if $\vdash f : \neg A$

($\bar{\tau}$) [~-¬]-transform] (in inversion [$\bar{\tau},\tau$])
$\vdash \bar{\tau}[A](f) : \neg A$, if $\vdash f : {\sim}A$

(b) binary operators $\simeq$ two-premiss rules

($\triangleright$) [PP ≡ modus (ponendo) ponens, cut operator associated to $\lambda \equiv \lambda^{\triangleright}$]
$\vdash f(a) \equiv f \triangleright a : B \equiv f(a)$, if $\vdash f : A \to B$, $\vdash a : A^6$

($\triangleright^+$) [KR ≡ 'Kanger's Rule' for material implication]
$\vdash f \triangleright^+ g \equiv f \triangleright (P[A,C] \triangleright g) : B$, if $\vdash f : A \to B$, $\vdash g : (A \to C) \to A$,
where one has **either**
(1) $P \equiv P^{\partial}$ and $\triangleright \equiv \triangleright_{\partial}$, with $\triangleright^+ \equiv \triangleright^+_{\partial}$, **or**
(2) $P \equiv P^{\gamma}$ and $\triangleright \equiv \triangleright_{\gamma}$, with $\triangleright_K \equiv \triangleright^+_{\gamma}$, say[7]

($\triangleleft$) [reversed modus (ponendo) ponens, cut operator associated to $\lambda^{\triangleleft}$]
$\vdash a \triangleleft f : B$, if $\vdash f : A \to B$, $\vdash a : A$

($\star$) [LNC ≡ 'law' of non-contradiction, cut operator associated to ∂]
$\vdash c \star a : \mathbf{f}$, if $\vdash c : \neg A$, $\vdash a : A$

($\cdot$) [LNC, cut operator associated to ρ]
$\vdash c \cdot a : \mathbf{f}$, if $\vdash c : \neg A$, $\vdash a : A$

($\bar{\triangleright}$) [TT ≡ modus (tollendo) tollens]
$\vdash f \bar{\triangleright} c : \neg A$, if $\vdash f : A \to B$, $\vdash c : \neg B$

($\ltimes$) [scalar restricted LNC, cut operator associated to $\bar{\gamma}$ (in $\lambda\bar{\gamma}$ only)]
$[u{:}{\sim}A] \vdash u \ltimes a : \mathbf{f}$, if $\vdash a : A$ (u atomic (pseudo-) scalar)

(D$_\vdash$) [Pairing qua rule]
$<a,b> := \lambda z.x(a)(b) = D[A,B](a)(b) = <b> \circ <a>$
$: (A \to (B \to C)) \to C$,
if $\vdash a : A$, $\vdash b : B$, provided z is not free in a, b

---

[6] [Notation. We write oft fa := f(a), fab := f(a)(b), etc., if confusions are unlikely. On the other hand, we allow multi-letter combinators [set in serif type], although one has, in general, $\vdash XY = X(Y)$, where = stands for conversion.] Recall that, in classical logic (i.e., if $\to$ stands for *material* implication), without appropriate equational specifications concerning the behaviour of the witness operators associated to the classical negation, for instance, this notation is *systematically ambiguous*. If negation (here ¬) is primitive, then, as a default, one takes $\triangleright := \triangleright_{\partial}$ – in (extensions of) decorated ('typed') $\lambda\partial$, for instance – and one has also a *definable* cut operator $\triangleright_{\gamma}$ (corresponding to a derivable *modus ponens* rule), whereas if negation (here ~) is defined 'inferentially [${\sim}A := A \to \mathbf{f}$], one has a *primitive* cut operator $\triangleright := \triangleright_{\gamma}$.

[7] Conversely, $\vdash f \triangleright a = f \triangleright_{\partial} a = f \triangleright^+_{\partial}$ (λm.A)D.a), If $\vdash f : A \to B$, $\vdash a : A$, with x not free in a, and $\vdash P^{\partial}[A,B] = \lambda x{:}(A \to B) \to A.(I[A] \triangleright^+_{\partial} x)$, and analogously for the inferential versions ($\triangleright_{\gamma}$ resp. P^{γ}). Incidentally, this points out to the fact that, in classical logic, there are *infinitely many equationally distinct* witness operators corresponding to *modus ponens*, as well (classical logic is *infinitistic at bottom*, so to speak). One must, in fact, take our (colloquial) use of $\simeq$ in the 'correspondence' *witness operator $\simeq$ rule of inference* with a grain of salt, since, in classical logic, the association is, in general, many-one: a witness operator is just a way of analysing an *intuitive idea* in exact terms, so to speak.

18 Catalogue of concepts and notations   275

- $(\pi)_1$ [Adjunction rule, pair-forming operator]
  $\vdash \prec a,b \succ \equiv \pi(a,b) : A \wedge B$, if $\vdash a : A, \vdash b : B$
  where either $\wedge$ is primitive or $\triangle$ is primitive
- $(\pi)_2$ [Adjunction rule, pair-forming operator]
  $\vdash \prec a,b \succ \equiv \pi(a,b) : \neg(A \to B)$, if $\vdash a : A, \vdash b : \neg B$
  where both $\to$ and $\neg$ are primitive
- $(\circ)$ [$\to$-Transitivity rule, Composition]
  $\vdash f \circ g := \lambda x{:}A.f \triangleright (g \triangleright x) = B[C,B,A](f)(g) : A \to B$,
  if $\vdash f : C \to B, \vdash g : A \to C$ [x not free in f, g]
- $(\square)$ [Self-distribution qua rule]
  $\vdash f \square g := \lambda x{:}A.(f \triangleright x) \triangleright (g \triangleright x) = S[A,C,B](f)(g) : A \to B$,
  if $\vdash f : A \to (C \to B), \vdash g : A \to C$ [x not free in f, g]

(c) abstraction operators and abstraction terms

- (c1) monadic abstractors ∥ term-forms
  - $(\lambda^\triangleright) \equiv (\lambda)$ [DT $\equiv$ 'deduction theorem' for $\to$]
    $\vdash \lambda x{:}A.b[x] : A \to B$, if $[x{:}A] \vdash b[x] : B$
  - $(\lambda^\triangleleft)$ [DT $\equiv$ 'deduction theorem' for $\leftarrow$]
    $\vdash \lambda^\triangleleft y{:}B.a[y] : A \leftarrow B$, if $[y{:}B] \vdash a[y] : A$
  - (ρ) [minimal / intuitionistic $\to$-introduction]
    $\vdash \rho x.e[x] : \neg A$, if $[x{:}A] \vdash e[x] : \mathbf{f}$
  - (∂) [reductio ad absurdum]
    $\vdash \partial z{:}\neg A.e[z] = \epsilon z{:}\neg A.\varpi_A(e[z]) : A$, if $[z{:}\neg A] \vdash e[z] : \mathbf{f}$
  - (ϵ) [consequentia mirabilis]
    $\vdash \epsilon z{:}\neg A.a[z] = \partial z{:}\neg A.z \star a[z] : A$, if $[z{:}\neg A] \vdash a[z] : A$
  - $(\wp^\partial)$ ['Peirce's Rule']
    $\vdash \wp^\partial z{:}(A{\to}B).a[z] = \epsilon z{:}\neg A.a[z{:=}\lambda x{:}A.\varpi_B(z \star x)] : A$,
    if $[z{:}A{\to}B] \vdash a[z] : A$
  - (γ) [Inferential reductio ad absurdum]
    $\vdash \gamma z{:}{\sim}A.e[z] = \varepsilon z{:}{\sim}A.\varpi_A(e[z]) : A$, if $[z{:}{\sim}A] \vdash \mathbf{f}$
  - (ε) [Inferential consequentia mirabilis]
    $\vdash \varepsilon z{:}{\sim}A.a[z] = \gamma z{:}{\sim}A.z \triangleright a[z] : A$, if $[z{:}{\sim}A] \vdash a[z] : A$
  - $(\wp^\gamma)$ [Inferential 'Peirce's Rule']
    $\vdash \wp^\gamma z{:}(A{\to}B).a[z] = \varepsilon z{:}{\sim}A.a[z{:=}\lambda x{:}A.\varpi_B(z \triangleright x)] : A$,
    if $[z{:}A{\to}B] \vdash a[z] : A$
  - $(\bar\gamma)$ [Scalar restricted reductio ad absurdum (in $\lambda\bar\gamma$ only)]
    $\vdash \bar\gamma u{:}{\sim}A.e[u] : A$, if $[u{:}{\sim}A] \vdash e[u] : \mathbf{f}$ [for atomic (pseudo-) scalar u]
- (c2) dyadic abstractors ∥ term-forms
  - $(\int)$ ['Split'-rule]
    $\vdash \int(x{:}A,y{:}B).e[x,y] : A \triangle B$, if $[x{:}A,y{:}B] \vdash e[x,y] : \mathbf{f}$ (x, y distinct)
- (c3) mixed (complex) abstractors ∥ term-forms
  - $(\chi)_1$ [Jaśkowski reductio ad absurdum]
    $\vdash \chi(x,y{:}\neg A).(c[x],a[y]) : A$, if $[x{:}\neg A] \vdash c[x] : \neg B, [y{:}\neg A] \vdash a[y] : B$
  - $(\chi)_2$ [Jaśkowski reductio ad absurdum, α-variant]
    $\vdash \chi(z{:}\neg A).(c[z],a[z]) : A$, if $[z{:}\neg A] \vdash c[z] : \neg B, [z{:}\neg A] \vdash a[z] : B$

(d) scalar forms (first-order)

276   18 Catalogue of concepts and notations

- (▶) [(Direct) Instantiation, associated to Λ]
  [binary scalar operator $\simeq$ single-premiss first-order rule (sic)]
  $\vdash f \blacktriangleright t : A[t]$, if $\vdash f : \forall u.A[u]$,
  where t is a scalar term substitutable for u
- (Λ) [(direct) Generalisation, monadic scalar abstractor]
  $\vdash \Lambda u.a[u] : \forall u.A[u]$, if $[u] \vdash a[u] : A[u]$
- (Π) [Counterexample, scalar pair-forming operator]
  $\vdash \downarrow_t(a) \equiv \Pi(t,a) : \exists u.\neg A[u] \equiv \neg \forall u.A[u]$, if $\vdash a : \neg A[u:=t]$
- (Π_1) [Counterexample-scalar, Choice, 'Hilbert's epsilon']
  $\vdash \backslash_t(c) \equiv \Pi_1^t(c) \doteq t$, if $\vdash c : \exists u.\neg A[u]$
  ['some u such that $\neg A[u]$', where $\doteq$ is scalar equality]
- (Π_2) [Counterexample-witness: a witness for $\neg A[v]$, for some $v \doteq \backslash(c)$]
  $[v] \vdash \theta(c) \equiv \Pi_2(c) : \neg A[v]$, if $\vdash c : \exists u.\neg A[u]$, with v fresh for c, etc.
  [both (Π_1) and (Π_2) occur in ad hoc $\boldsymbol{\Pi}$-systems only]
- (Σ) [Indirect Generalisation from no counterexamples, scalar 'Split'-rule]
  $\vdash \Sigma(u,x:\neg A[u]).e[u,x] : \forall u.A[u]$, if $[u][x:\neg A \vdash e[u,x] : \mathbf{f}$

§ 1 2 Witness combinators

- (A) $\vdash \mathsf{A}[A,B] := \lambda x.\gamma y.y(x) : ((A \to B) \to B) \to A$
  ['Relativity', only in $\mathcal{A}[\to] \equiv$ pure abelian logic]
- (B) $\vdash \mathsf{B}[A,B,C] := \lambda x.\lambda y.\lambda z.x(y(z)) = \mathsf{S}(\mathsf{K}(\mathsf{S}))(\mathsf{K}) \equiv \mathsf{K}(\mathsf{S}) \square \mathsf{K}$
  $: (A \to B) \to ((C \to A) \to (C \to B))$
  [Prefixing]
- (BI) $\vdash \mathsf{BI}[A,B] \equiv \mathbf{1}[A,B] := \lambda x.\lambda y.x(y) \equiv \mathsf{B}(\mathsf{I}) =_\eta \mathsf{I}[A \to B]$
  $: (A \to B) \to (A \to B)$
  [Identity]
- (C) $\vdash \mathsf{C}[A,B,C] := \lambda x.\lambda y.\lambda z.x(z)(y) = (\mathsf{B} \square \mathsf{K}(\mathsf{K})) \circ \mathsf{S}$
  $: (A \to (B \to C)) \to (B \to (A \to C))$
  [Commutation],
- (C^λ) $\vdash \mathsf{C}^\lambda[A,B] \equiv \mathsf{C}[A,B,\mathbf{f}] \equiv \lambda x.\lambda y.\lambda z.x(z)(y)$
  $: (A \to \sim B) \to (B \to \sim A)$
  [Inferential (Minimal) C-Contraposition]
- (C^M) $\vdash \mathsf{C}^\mathsf{M}[A,B] := \lambda x.\lambda y.\rho z.(xz) \cdot y$
  $: (A \to \neg B) \to (B \to \neg A)$
  [Minimal C-Contraposition]
- (C^ρ) $\vdash \mathsf{C}^\rho[A,B] := \lambda x.\lambda y.\rho z.(xz) \star y$
  $: (A \to \neg B) \to (B \to \neg A)$
  [(Classically Witnessed) Minimal C-Contraposition]
- (C^∂) $\vdash \mathsf{C}^\partial[A,B] \equiv \mathsf{T}[A,B] := \lambda x.\lambda y.\partial z.(xz) \star y$
  $: (\neg A \to \neg B) \to (B \to A)$
  [Classical C^∂-Contraposition]
- (CB) $\vdash \mathsf{CB}[A,B,C] := \lambda x.\lambda y.\lambda z.y(x(z)) = \mathsf{C}(\mathsf{B})$
  $: (A \to B) \to ((B \to C) \to (A \to C))$
  [Suffixing]

18 Catalogue of concepts and notations   277

(CB$^\lambda$) $\vdash$ CB$^\lambda$[A,B] := λx.λy.λz.y(xz)
: (A $\to$ B) $\to$ ($\sim$B $\to$ $\sim$A)
[Inferential (Minimal) CB-Contraposition]

(CB^M) $\vdash$ CB^M[A,B] := λx.λy.ρz.y.(xz)
: (A $\to$ B) $\to$ ($\neg$B $\to$ $\neg$A)
[Minimal CB-Contraposition]

(CB$^\rho$) $\vdash$ CB$^\rho$[A,B] := λx.λy.ρz.y$\star$(xz)
: (A $\to$ B) $\to$ ($\neg$B $\to$ $\neg$A)
[(Classically Witnessed) Minimal CB$^\rho$-Contraposition]

(CB$^\partial$) $\vdash$ CB$^\partial$[A,B] := λx.λy.∂z.y$\star$(xz)
: ($\neg$A $\to$ B) $\to$ ($\neg$B $\to$ A)
[Classical CB$^\partial$-Contraposition]

(CC) $\vdash$ CC[A,B,C] := λx.λy.λz.y(z)(x) = C(C) = B $\circ$ CI
: A $\to$ ((B $\to$ (A $\to$ C)) $\to$ (B $\to$ C))
[Selfcommutation]

(CI) $\vdash$ CI[A,B] := λx.λy.λy.y(x) = C(I) : A $\to$ ((A$\to$ B) $\to$ B)
[Assertion]

(CO$^\partial$) $\vdash$ CO$^\partial$[A,B] := λx.λy.ϖ(x$\star$y) : $\neg$A $\to$ (A $\to$ B)
[ex contradictione quodlibet, Commuted Scotus]

(CO$^\gamma$) $\vdash$ CO$^\gamma$[A,B] := λx.λy.ϖ(xy) : $\sim$A $\to$ (A $\to$ B)
[inferential Commuted Scotus]

(CS) $\vdash$ CS[A,B,C] := λx.λy.λz.yz(xz) = C(S)
: (A $\to$ B) $\to$ ((A $\to$ (B $\to$ C)) $\to$ (A $\to$ C))
[Commuted Frege]

(D) $\vdash$ D[A,B,C] := λx.λy.λz.z(x)(y) = λx.λy.(<x,y>) = C $\circ$ CI
: A $\to$ (B $\to$ ((A $\to$ (B $\to$ C)) $\to$ C))
[Pairing qua tautology]

(E$^\partial$) $\vdash$ E$^\partial$[A] := λx.ϵy.(x$\star$y) = λx.γy.y(x$\star$y) : ($\neg$A $\to$ A) $\to$ A
[consequentia mirabilis qua tautology, the 'Law of Clavius']

(E$^\gamma$) $\vdash$ E$^\gamma$[A] := λx.εy.(xy) = λx.γy.y(xy) = P$^\gamma$[A,f] : ($\sim$A $\to$ A) $\to$ A
[inferential consequentia mirabilis qua tautology]

(I) $\vdash$ I[A] := λx.x = S(K)(K) = K $\square$ K = C(K)(K) : A $\to$ A
[Identity]

(K) $\vdash$ K[A,B] := λx.λy.x : A $\to$ (B $\to$ A)
[Simplification, Irrelevance]

(KI) $\vdash$ KI[A,B] := λx.λy.y = CK = C(K) : A $\to$ (B $\to$ B)
[Commuted Simplification]

(Ł^∂) Ł^∂[A,B,C,D] := λx.λy.λz.ϵu.y(x(λv.ϖ(u$\star$v)))
: ((A $\to$ B) $\to$ C) $\to$ ((C $\to$ A) $\to$ (D $\to$ A)) $\equiv$ [Ł^∂] $\equiv$ [Ł^γ]
[Łukasiewicz generator for $\mathcal{C}$[$\to$], pure material implication]

(Ł^γ) Ł^γ[A,B,C,D] := λx.λy.λz.εu.y(x(λv.ϖ(uv))) : [Ł^γ] $\equiv$ [Ł^∂]
[Inferential Łukasiewicz generator for pure material implication)]

(M$_\mathcal{A}$) $\vdash$ M$_\mathcal{A}$:= λx.λy.γz.x$\triangleright$(y$\triangleright$z)
: (A$\to$B)$\to$(((C$\to$B)$\to$A)$\to$C) $\equiv$ [M$_\mathcal{A}$]
[C. A. Meredith's generator for $\mathcal{A}$[$\to$], pure abelian logic]

278   18 Catalogue of concepts and notations

$(\mathsf{M}'_{\mathcal{A}})$ ⊢ $\mathsf{M}'_{\mathcal{A}} := \lambda x.\lambda y.\gamma z.(x \circ y \circ z) : [\mathsf{M}_{\mathcal{A}}]$ (sic) [ditto]
(O^{∂}) ⊢ $\mathsf{O}^{\partial}[A,B] \equiv \mathsf{O}[A,B] := \lambda x.\lambda y.\varpi(y \star x) : A \to (\neg A \to B)$
   [ex contradictione quodlibet, 'Scotus Law', Explosion]
(O^{γ}) ⊢ $\mathsf{O}^{\gamma}[A,B] := \lambda x.\lambda y.\varpi(yx) : A \to (\sim A \to B)$
   [Inferential 'Scotus Law']
(P^{∂}) ⊢ $\mathsf{P}^{\partial}[A,B] := \lambda x.\epsilon y.x \triangleright (\lambda z.\varpi(y \star z)) : [\mathsf{P}^{\partial}] \equiv [\mathsf{P}^{\gamma}]$
   ['Peirce's Law']
(P^{γ}) ⊢ $\mathsf{P}^{\gamma}[A,B] := \lambda x.\varepsilon y.x \triangleright (\lambda z.\varpi(y \triangleright z)) : ((A \to B) \to A) \to A \equiv [\mathsf{P}^{\gamma}]$
   [Inferential 'Peirce's Law']
(S) ⊢ $\mathsf{S}[A,B,C] := \lambda x.\lambda y.\lambda z.x(z)(y(z))$
   $: (A \to (B \to C)) \to ((A \to B)) \to (A \to C))$
   [Frege, Selfdistribution on the major]
$(\mathsf{S_M})$ ⊢ $\mathsf{S_M}[A,B,C] := \lambda x.\lambda y.\lambda z.x(yz)(z) = B(W) \circ B$
   $: (A \to (B \to C)) \to ((B \to A) \to (B \to C))$
   [C. A. Meredith]
(T) ⊢ $\mathsf{T}[A,B] \equiv \mathsf{C}^{\partial}[A,B] := \lambda x.\lambda y.\partial z.(xz) \star y$
   $: (\neg A \to \neg B) \to (B \to A)$
   [Classical C^{∂}-Contraposition, local notation]
(W) ⊢ $\mathsf{W}[A,B] := \lambda x.\lambda y.x(y)(y) = \mathsf{CS}(\mathsf{I}) = \mathsf{C}(\mathsf{S})(\mathsf{I})$
   $: (A \to (A \to B)) \to (A \to B)$
   [Hilbert, Contraction]
$(\dot{\varpi})$ ⊢ $\dot{\varpi}[A] := \lambda x.\varpi(x) = \lambda x.\gamma y.x = \lambda x.\partial y.x : \mathbf{f} \to A$
   [ex falso quodlibet qua tautology]
(∇^{∂}) ⊢ $\nabla^{\partial}[A] \equiv \nabla[A] := \lambda x.\partial y.y \star x : A \to \neg\neg A$
   [DN-i qua tautology]
(∇^{γ}) ⊢ $\nabla^{\gamma}[A] \equiv \mathsf{CI}[A,\mathbf{f}] := \lambda x.\lambda y.y(x) : A \to \sim\sim A$
   [Inferential DN-i qua tautology]
(Δ^{∂}) ⊢ $\Delta^{\partial}[A] \equiv \Delta[A] := \lambda x.\partial y.x \star y : \neg\neg A \to A$
   [DN-e qua tautology]
(Δ^{γ}) ⊢ $\Delta^{\gamma}[A] := \lambda x.\gamma y.x(y) : \sim\sim A \to A$
   [Inferential DN-e qua tautology]
(τ) ⊢ $\tau[A] := \lambda x.\lambda y.x \cdot y : \neg A \to \sim A$
   [LNC ≡ (the minimal) 'law' of non-contradiction qua tautology]
$(\bar{\tau})$ ⊢ $\bar{\tau}[A] := \lambda x.\rho y.x(y) : \sim A \to \neg A$ [8]
(Ω) ⊢ $\Omega : \mathbf{v} \equiv \neg\mathbf{f}$

§ 2 Equational theories

All equational theories surveyd below are consistent and make sense at 'type-free' level. As expected, they become witness theories after 'typing'. The abstract

---

[8] This yields minimal ¬-introduction, because ⊢ $\rho x{:}A.e[x] = \bar{\tau}[A](\lambda x{:}A.e[x]) : \neg A$, if [x: A] ⊢ e[x] : **f**.

('type-free') syntax is as usual (cf. Rezuș 2017b, 2017a), and, unless otherwise specified explicitly, all operators are supposed to be *algebraic* (i.e., monotone, 'respecting', thus, equality; here: conversion). Further, for two equational theories T_i (i := 1, 2), $T_1 \preceq T_2$ means that T_1 is a subsystem of T_2, while $T_1 \simeq T_2$ means that T_1 and T_2 are equationally equivalent.

§ 2.1 Generic folds

Any extensional mono-fold ♮ – based on [♮,≀] – satisfies

$(\beta♮)$ ⊢ (♮x.b[x]) ≀ a = b[x:=a],
$(\eta♮)$ ⊢ ♮x.(c≀x) = c (x not free in c),

where ♮ is a *monadic abstractor*, and ≀ is its *associated cut*. In general, with

(df ♯) ♯z.a[z] := ♮z.(a[z]≀z), as the *iterator* of the decomposition, and
(df ♭) ♭(c) := ♮z.c (z not free in c), as its associated *cancellator*,

♮ admits of the *linear decomposition*:

(♮) ⊢ ♮z.c[z] = ♯z.♭(c[z]),

$(\beta♯)$ ⊢ (♯z.a[z]) ≀ c = a[z:=c] ≀ c,
$(\eta♯)$ ⊢ ♯z.a = a, if z is not free in a,
$(♯♯)$ ⊢ ♯x.♯y.a[x,y] = ♯z.a[x,y:=z],

$(\beta♭)$ ⊢ ♭(e)≀ c = e,
$(♯♭)$ ⊢ ♯z.(♭(a[z]) ≀ z) = ♯z.a[z],

In particular, this obtains for the (extensional) mono-fold (≡ 'pure' λ-calculus) **λ** ≡ $\lambda_{\beta\eta}$, based on [λ,▷], where ♮ ≡ λ, and ≀ ≡ ▷. For **λ**, the iterator is $\bar{W}$x.a[x] := W▷(λx.a[x]) and its cancellator is $\bar{K}$(c) := K▷c, where W and K are the usual combinators of the 'pure' λ-calculus.

Analogously, for the mono-fold ∂, based on [∂,⋆], where ♮ ≡ ∂, and ≀ ≡ ⋆, satisfying

$(\beta\partial)$ ⊢ c ⋆ (∂z.e[z]) = e[z:=c],
$(\eta\partial)$ ⊢ ∂z.(z⋆a) = a (z not free in a),

(with cuts, ⋆, written down in reversed order), the iterator is εz.a[z] := ∂z.(z⋆a[z]) and its cancellator is ϖ(e) := ∂z.e, with z not free in e, and one has a typographical variant of the λ-decomposition above, viz.:

(∂) ⊢ ∂z.c[z] = εz.ϖ(c[z]),

$(\beta\epsilon)$ ⊢ c ⋆ (εz.a[z]) = c ⋆ a[z:=c],
$(\eta\epsilon)$ ⊢ εz.a = a, if z is not free in a,
$(\epsilon\epsilon)$ ⊢ εx.εy.a[x,y] = εz.a[x,y:=z],

$(\beta\varpi) \vdash c \star \varpi(e) = e,$
$(\epsilon\varpi) \vdash \epsilon z.(z\star\varpi(a[z])) = \epsilon z.a[z].$

The original mono-fold ♮ can be retrieved back equationally from its linear decomposition ♮♭, by setting:

$(df\ \natural) \vdash \natural x.a[x] := \natural x.\flat(a[x]).$

In particular, with notational conventions as above, the linear decomposition of ∂ is $\epsilon\varpi := (\beta\epsilon) + (\eta\epsilon) + (\epsilon\epsilon) + (\beta\varpi) + (\epsilon\varpi)$, and one has, equationally, $\partial \simeq \epsilon\varpi$.

§ 2.2 Generic mono-folds: basic properties

Extensional mono-folds are known to be monoids. They are also *sub-cartesian* as folds and, somewhat unexpectedly, in generic mono-folds, ♮ – based on [♮,⥱] – the cut (⥱) is explicitly definable in *strictly linear* terms. Defining *identity, composition, Frege-Church sequences*, and the associated projections as usual, by:

(df I) $I := \natural x.x,$
(df o) $f \circ g := \natural z.(f⥱(g⥱z))$ [z fresh for f, g],

(df p) $<a_1, ..., a_n> := \natural z.(z⥱a_1⥱ ... ⥱a_n)$, z not free in a_i $(0 < i < n+1, n > 0)$

(df K_i^n) $K_i^n := \natural x_1. ... .\natural x_n.x_i$ $(0 < i < n+1),$
(df P_i^n) $P_i^n := <K_i^n>$ $(0 < i < n+1)$, with, in particular,

(df K, KI) $K := K_1^2, KI := K_2^2,$

(df $\tilde{1}, \tilde{2}$) $\tilde{1} := <K>, \tilde{2} := <KI>$

we have

$\vdash I \circ f = f = f \circ I,$
$\vdash (f \circ g) \circ h = f \circ (g \circ h),$

$\vdash P_i^n(<a_1, ..., a_n>) = a_i$ $(0 < i < n+1),$

$\vdash \tilde{1}\ (<a, b>) = a, \vdash \tilde{2}\ (<a, b>) = b,$

$\vdash <a_n> \circ\ ...\ \circ <a_1> = <a_1, ..., a_n>.$

With the generic 'type-free' notation, the usual *strictly linear* combinators are:

(df B) $B := \natural x.\natural y.(x \circ y),$
(df CB) $CB := \natural x.\natural y.(y \circ x),$

18 Catalogue of concepts and notations

(df C) $C := ƛx.ƛy.ƛz.(x\wr z\wr y)$,
(df CC) $CC := ƛx.ƛy.ƛz.(y\wr z\wr x)$,
(df CI) $CI := ƛx.ƛy.(y\wr x)$,
(df D) $D := ƛx.ƛy.(<x,y>)$,

so that:

$\vdash C \circ C = I$,
$\vdash C(C(a)) = a$,
$\vdash CI = C(I) = ƛx.<x>$,
$\vdash D = C \circ CI = ƛx.ƛy.(<y>\circ<x>)$,
$\vdash C = CC(CC)(CC) = ƛx.ƛy.(<y>\circ x)$,
$\vdash C(a) = CB(a) \circ CI = ƛx.(<x>\circ a)$,
$\vdash C(a)(b) = <b>\circ a$,
$\vdash CB = C(B), \vdash B = C(CB)$,
$\vdash CB(a) = <a> \circ B$,
$\vdash B(a) = <a> \circ CB$,
$\vdash CC = C(C) = B \circ CI = ƛx.ƛy.(<x>\circ y)$,
$\vdash CC(a)(b) = <a>\circ b$,
$\vdash CC(a) = ƛx.<a>\circ x$.

In the end, with

(df ∋) $f \ni a := <f,a> \circ CC$,

one has

(∈=) $\vdash f \ni a = <a> \circ C(f) = f(a) \ [\equiv f\wr a]$.[9]

Note. In witness-theoretical applications, we use a mono-fold $\rho := [\rho,\cdot]$, where ρ is a new monadic abstractor, and $\cdot$ is its associated cut, so that, at 'type-free' level, one has di-folds $\lambda\rho := \lambda + \rho$, and $\rho\partial := \rho + \partial$, resp. In general, any ('type-free') di-fold contains an inversion (in fact, infinitely many such). In particular, for $\rho\partial$, for instance, one can obtain this as $\Delta := [\Delta, \nabla]$, with

(df Δ) $\Delta(c) := \partial z.(c \cdot z)$ [z fresh for c],
(df ∇) $\nabla(a) := \rho x.(x \star a)$ [x fresh for a],

whence:

($\beta\Delta$) $\vdash \nabla(\Delta(c)) = c$,
($\eta\Delta$) $\vdash \Delta(\nabla(a)) = a$.

---

[9] This is implicit in Church 1937. Extensionality is required here.

282   18 Catalogue of concepts and notations

§ 2.3 Cartesian folds

Further, given an *extensional pairing*, i.e., a triple $[\pi,\pi_1,\pi_2]$ with term-forms
$\prec a,b \succ := \pi(a,b)$, $\mathbf{j}(c) := \pi_j(c)$, j :=1, 2, resp., satisfying:

$(\beta\pi_1) \vdash \mathbf{1}(\prec a,b \succ) = a$,
$(\beta\pi_2) \vdash \mathbf{2}(\prec a,b \succ) = b$,
$(\eta\pi) \vdash \prec \mathbf{1}(c),\mathbf{2}(c) \succ = c$,

the generic (extensional) *cartesian mono-fold* is $\natural\pi := \natural + (\beta\pi_1) + (\beta\pi_2) + (\eta\pi)$.

In particular, one has the (extensional) cartesian folds $\boldsymbol{\lambda}\pi := \boldsymbol{\lambda} + (\beta\pi_1) + (\beta\pi_2) + (\eta\pi)$ and $\boldsymbol{\partial}\pi := \boldsymbol{\partial} + (\beta\pi_1) + (\beta\pi_2) + (\eta\pi)$.

Setting, in $\boldsymbol{\partial}\pi$,

(df λ) $\lambda x.b[x] := \partial z.(\mathbf{2}(z) \star b[x:=\mathbf{1}(z)])$, z fresh for b[x],
(df $\triangleright$) $c \triangleright a := \partial y.(\prec a,y \succ \star c)$, y not free in a and c,

one has $(\beta\lambda)$ and $(\eta\lambda)$, as well as

$(\beta p) \vdash \prec a,c \succ \star f = c \star (f \triangleright a)$,

already in $\boldsymbol{\partial}\pi$, so that, equationally, $\boldsymbol{\lambda} \preceq \boldsymbol{\lambda}\pi \preceq \boldsymbol{\partial}\pi$, where the inclusions are strict. In other words, $\boldsymbol{\partial}\pi$ is a di-fold, since it contains (equationally) $\boldsymbol{\lambda}\boldsymbol{\partial} := \boldsymbol{\lambda} + \boldsymbol{\partial}$. Incidentally, from the above, one has also $\boldsymbol{\lambda}\boldsymbol{\partial} \simeq \boldsymbol{\lambda}\epsilon\varpi := \boldsymbol{\lambda} + \epsilon\varpi$.

Moreover, every (extensional) cartesian mono-fold $\natural\pi$ is also a *cyclic equational theory*, as it contains an *inversion* $[\uparrow,\downarrow]$, with

(df $\uparrow$) $\uparrow(c) := \natural x.\natural y.(c \wr \prec x,y \succ)$ [x, y fresh for c],
(df $\downarrow$) $\downarrow(a) := \natural z.((a \wr \mathbf{1}(z)) \wr \mathbf{2}(z))$ [z fresh for a],

satisfying thus

$(\beta\uparrow) \vdash \downarrow(\uparrow(a)) = a$, and
$(\eta\uparrow) \vdash \uparrow(\downarrow(c)) = c$.

For (the 'type-free') $\boldsymbol{\lambda}\pi$, the operators $\uparrow$ and $\downarrow$ correspond to the usual 'curry-ing' / 'un-curry-ing' operators, of course.

In general, however, there is no need for a 'surjective' ($\equiv$ extensional) pairing, in order to get this, since the result obtains already for any di-fold (as shown above, di-folds are, at 'type-free' level, cyclic equational theories). In fact, any di-fold — and so any (extensional) cartesian mono-fold — contains a n-fold, for each n $\geq$ 2, and, in particular, for any (generic) inversion $\Uparrow := [\uparrow,\downarrow]$, one has, equationally (and redundantly so), at 'type-free' level, $\natural\pi \simeq \natural\pi\Uparrow := \natural\pi + \Uparrow$.

Note. For witness-theoretical applications, we use, however, a decorated ('typed') inversion $\Delta := [\Delta,\nabla]$, instead, with

$(\beta\Delta) \vdash \nabla(\Delta(c)) = c$,
$(\eta\Delta) \vdash \Delta(\nabla(a)) = a$.

18 Catalogue of concepts and notations

and witness theories $T^\Delta := T + (\beta\Delta) + (\eta\Delta)$, where T contains at least one (decorated) mono-fold, and the inversion $^\Delta := [\Delta,\nabla]$, represents a double-negation isomorphism, relative to a primitive negation $\neg$, i.e., we have also

$(\Delta) \vdash \Delta(c) : A$, if $\vdash c : \neg\neg A$,
$(\nabla) \vdash \nabla(a) : \neg\neg A$, if $\vdash a : A$,

so that, after 'typing', $\partial\pi$ and $\partial\pi^\Delta$, for instance, are equationally distinct theories, and similarly for $\lambda\pi$ and $\lambda\pi^\Delta$.

§ 2.4 Calculi of $\lambda\partial$-conversion

Furthermore, in $\partial\pi$, one has also

$(\beta\lambda^*) \vdash \prec a,c \succ \star (\lambda x.b[x]) = c\star(b[x:=a])$,
$*) \equiv (\eta p) \vdash \lambda x.\partial y.(\prec x,y\succ \star c) = c$ (x, y not free in c),
$*) \equiv (\zeta p) \vdash \partial z.e[z] = \lambda x.\partial y.e[z:=\prec x,y\succ]$ (x,y fresh for e[z]),

while, with the (pure) dyadic *split* abstraction operator $\int$, defined as:

$(\text{df } \int) \int[x,y].c[x,y] := \partial z.c[x:=\mathbf{1}(z),y:=\mathbf{2}(z)]$ (z fresh for c[x,y], x$\neq$y),

we get:

$(\lambda) \vdash \lambda x.b[x] = \int(x,y).(y\star b[x])$, y not free in b[x],
$(\int \lambda\partial) \vdash \int(x,y).e[x,y] = \lambda x.\partial y.e[x,y]$, and

$(\beta \int) \vdash \prec a,b\succ \star (\int(x,y).c[x,y]) = c[x:=a,y:=b]$,
$(\eta \int) \vdash \int(x,y).(\prec x,y\succ \star c) = c$ (x, y not free in c),
$(\zeta \int) \vdash \partial z.e[z] = \int(x,y).e[z:=\prec x,y\succ]$ (x, y fresh for e[z]).

Setting

$\lambda\partial p_\beta := \lambda\partial + (\beta p)$,
$\lambda\partial p_{\beta\eta} [\equiv \lambda\partial p] := \lambda\partial + (\beta p) + (\eta p)$,
$\lambda\partial p_\zeta := \lambda\partial + (\beta p) + (\zeta p)$,

$\partial\lambda^* := \partial + (\beta\lambda^*) + (\eta\lambda^*)$,
$\partial\lambda^*_\zeta := \partial + (\beta\lambda^*) + (\eta\lambda^*) + (\zeta\lambda^*)$,

$\partial\int := \partial + (\beta \int) + (\eta \int)$,
$\partial\int_\zeta := \partial + (\beta \int) + (\eta \int) + (\zeta \int)$,

we get the (strict) inclusions and equivalences:

$(\preceq) \lambda, \partial \preceq \lambda\partial \preceq \lambda\partial p_\beta \preceq \lambda\partial p_{\beta\eta} \simeq \partial\lambda^* \simeq \partial\int \preceq \lambda\partial p_\zeta \simeq \partial\lambda^*_\zeta \simeq \partial\int_\zeta \preceq \partial\pi$.

§ 2.5 Calculi of $\lambda\gamma$-conversion and neighbours

Setting in $\lambda\partial p_\zeta$, say,

(df $\dot\tau$) $\dot\tau(c) := \lambda x.(c\star x)$ (x not free in c),
(df γ) $\gamma z.e[z] := \partial z.e[z:=\dot\tau(z)]$,

(df ε) $\varepsilon z.a[z] := \gamma z.(z\triangleright a[z])$,
(df ϖ) $\varpi(e) := \gamma z.e \; [\equiv \partial z.e]$ (z not free in e), and

(df μ) $\mu(a,b) := \lambda z.(b \triangleright (z \triangleright a))$ (z not free in a, b),

yields, successively,

($\beta\dot\tau$) ⊢ $\dot\tau(c) \triangleright a = c \star a$,
($\mu\dot\tau$) ⊢ $\mu(a,\dot\tau(b)) = \dot\tau(\prec a,b\succ)$,

($\bar\beta\gamma$) ⊢ $\gamma x.(x\triangleright(\gamma y.e[x,y])) = \gamma z.e[x,y:=z]$,
($\eta\gamma$) ⊢ $\gamma z.(z\triangleright a) = a$ (z not free in a),
($\zeta\gamma$) ⊢ $\gamma z.e[z] = \lambda x.\gamma y.e[z:=\mu(x,y)]$,
($\zeta_\beta\gamma$) ⊢ $\gamma z.e[z] \triangleright a = \gamma z.e[z:=\mu(a,z)]$,

($\dot\tau\gamma$) ⊢ $\dot\tau(c) \triangleright \gamma z.e[z] = e[z:=\dot\tau(c)]$,
($\xi\gamma$) ⊢ $\gamma z.e_1 = \gamma z.e_2 \Rightarrow\; \vdash \gamma z.(f\triangleright e_1) = \gamma z.(f\triangleright e_2)$,

($\hat\beta\gamma$) ⊢ $\gamma x.c[x\triangleright(\gamma y.e[x,y])] = \gamma z.c[e[x,y:=z]]$[10],
($\gamma\varepsilon\varpi$) ⊢ $\gamma z.e[z] = \varepsilon z.\varpi(e[z])$,

($\hat\beta\varepsilon\varepsilon$) ⊢ $\varepsilon x.c[x\triangleright\varepsilon y.a[x,y]] = \varepsilon z.c[z\triangleright a[x,y:=z]]$,
($\eta\varepsilon$) ⊢ $\varepsilon z.a = a$ (z not free in a),
($\varepsilon\varepsilon$) ⊢ $\varepsilon x.\varepsilon y.a[x,y] = \varepsilon z.a[x,y:=z]$
($\zeta\varepsilon$) ⊢ $\varepsilon z.a[z] = \lambda x.\varepsilon y.(a[z:=\mu(x,y)]\triangleright x)$, and

($\hat\beta\varepsilon\varpi$) ⊢ $\varepsilon z.c[z\triangleright\varpi(e[z])] = \varepsilon z.c[e[z]]$,
($\zeta\varpi$) ⊢ $\varpi(e) = \lambda x.\varpi(e)$ (x not free in e),
($\varepsilon\varpi$) ⊢ $\varepsilon z.\varpi(z\triangleright a[z]) = \varepsilon z.a[z]$.

Let, in the appropriate syntax (and assumming, in each case, monotony for the primitives):

$$\lambda\gamma := \lambda + (\hat\beta\gamma) + (\eta\gamma) + (\zeta\gamma),$$
$$\lambda\gamma\dot\tau := \lambda\gamma + (\dot\tau\gamma),$$
$$\lambda\varepsilon := \lambda + (\hat\beta\varepsilon\varepsilon) + (\eta\varepsilon) + (\zeta\varepsilon) + (\varepsilon\varepsilon),$$
$$\lambda\varepsilon\varpi := \lambda\varepsilon + (\hat\beta\varepsilon\varpi) + (\zeta\varpi) + (\varepsilon\varpi).$$

We get

---

[10] As ever, c[a] ≡ c[x:=a].

($\preceq_\gamma$) $\boldsymbol{\lambda} \preceq \boldsymbol{\lambda\gamma} \preceq \boldsymbol{\lambda\gamma\dot{\tau}} \preceq \boldsymbol{\lambda\partial p_\zeta} \preceq \boldsymbol{\lambda\pi}$, and
($\preceq_\varepsilon$) $\boldsymbol{\lambda} \preceq \boldsymbol{\lambda\varepsilon} \preceq \boldsymbol{\lambda\varepsilon\varpi} \simeq \boldsymbol{\lambda\gamma}$.

Some remarks are in order.

(1) In $\boldsymbol{\lambda\gamma}$, one can trade $(\hat{\beta}\gamma)$ for $(\bar{\beta}\gamma) + (\xi\gamma)$, while

(2) for $\boldsymbol{\lambda\varepsilon\varpi}$, one must turn $(\gamma\varepsilon\varpi)$ into a definition, and one can make some economy, by realising that $(\varepsilon\varepsilon)$ follows from the $(\varepsilon\text{-}\varpi)$ conditions.

(3) *Mutatis mutandis*, without $(\eta\lambda)$ one obtains analogous 'intensional' λ-calculi, as, for instance, $\boldsymbol{\lambda_\beta\gamma}$, say.

(4) Further, leaving out $(\zeta\gamma)$ from $\boldsymbol{\lambda\gamma}$ yields the 'Glivenko' $\lambda\gamma$-calculus $\boldsymbol{\lambda\gamma_G}$, whereas replacing $(\zeta\gamma)$ by its β-version $(\zeta_\beta\gamma)$, in $\boldsymbol{\lambda_\beta\gamma}$, yields the 'Kolmogorov' $\lambda\gamma$-calculus $\boldsymbol{\lambda_\beta\gamma_K}$. (So, $\boldsymbol{\lambda\gamma} \simeq \boldsymbol{\lambda_\beta\gamma_K} + (\eta\lambda)$, in fact.) Notably, at 'type-free' level, the latter two λ-calculi are contained in the 'pure' λ-calculus $\boldsymbol{\lambda} \equiv \boldsymbol{\lambda_{\beta\eta}}$.

§ 2.6 C-monoids

Putting things in algebraic terms, the generic (extensional) cartesian mono-fold $\natural\pi$ is a *C-monoid* in the sense of Lambek (Lambek & Scott 1986), where the monoid structure is guaranteed by (df I) and (df ∘), we have cartesian pairs, and 'categorical' combinators $\bar{1}, \bar{2}$, given by:

(df P) $[f,g] := \natural z.\prec f\wr z, g\wr z\succ$ [z fresh for f, g],
(df $\bar{1}, \bar{2}$) $\bar{1} := \natural z.\mathbf{1}(z)$, $\bar{2} := \natural z.\mathbf{2}(z)$,

whereas the generic inversion is given by $\uparrow := [\uparrow,\downarrow]$ above, whence the cartesian closed monoid (C-monoid, for short) $\mathbf{C}^\circ$, characterised equationally by:

(id) $\vdash \mathrm{I} \circ f = f = f \circ \mathrm{I}$,
(ass) $\vdash (f \circ g) \circ h = f \circ (g \circ h)$,

($\beta \mathrm{P}_1$) $\vdash \bar{1} \circ [f,g] = f$,
($\beta \mathrm{P}_2$) $\vdash \bar{2} \circ [f,g] = g$,
($\eta \mathrm{P}$) $\vdash [\bar{1} \circ f, \bar{2} \circ f] = f$,

($\beta\uparrow$) $\vdash \downarrow(\uparrow(c)) = c$,
($\eta\uparrow$) $\vdash \uparrow(\downarrow(a)) = a$.

In the end, since, formally, $\boldsymbol{\lambda\pi}$ is the same thing as $\boldsymbol{\partial\pi}$, qua 'type-free' equational theories, one has $\boldsymbol{\partial\pi} \simeq \mathbf{C}^\circ$, equationally (cf. Lambek & Scott 1986, for the other half of the argument that simulates the extensional [cartesian] mono-fold structure in terms of 'categorical' combinators, i.e., in $\mathbf{C}^\circ$).

§ 2.7 Scalar extensions

Let $\partial\Pi$ be a scalar (extensional) mono-fold based on $[\partial,\star]$, i.e., with grammar including:

scalars :: s, t := u | \(a) | ...
w-terms:: a, b, c, d, e := x | ∂z.e | c $\star$ a | $\downarrow_t$(c) | θ(c) | ...

and satisfying at least $(\beta\partial)$, $(\eta\partial)$, and:

$(\beta\Pi_1)$ ⊢ \($\downarrow_t$(c)) $\doteq$ t (where $\doteq$ stands for scalar equality),
$(\beta\Pi_2)$ ⊢ $\theta(\downarrow_t(c))$ = c,
$(\eta\Pi)$ ⊢ $\downarrow_t(\theta(c))$ = c, if \(c) $\doteq$ t,

Define, in $\partial\Pi$,

(df Σ) Σ(u,x).e[u,x] := ∂z.e[u:=\(z),x:=θ(z)], z fresh for a[u,x],
(df Λ) Λu.a[u] := Σ(u,z).(z$\star$a[u]), z fresh for a[u],
(df ▶) c ▶ t := ∂z.($\downarrow_t$(z)$\star$c), z not free in c.

This yields:

$(\beta\Lambda)$ ⊢ (Λu.a[u]) ▶ t = a[u:=t],
$(\eta\Lambda)$ ⊢ Λu.(c▶u) = c, (u not free in c),

$(\Sigma\Lambda\partial)$ ⊢ Σ(u,x).e[z] = Λu.∂x.e[u,x],

$(\beta\Lambda^*)$ ⊢ $\downarrow_t$(c) $\star$ (Λu.a[u]) = c$\star$(a[u:=t]),
$(\eta\Lambda^*)$ ⊢ Λu.∂z.($\downarrow_u$(z) $\star$ c) = c (u, z not free in c),
$(\zeta\Lambda^*)$ ⊢ ∂z.e[z] = Λu.∂x.e[z:=$\downarrow_u$(x)] (u, x fresh for e[z]),

$(\beta\Sigma)$ ⊢ $\downarrow_t$(a) $\star$ (Σ(u,x).c[u,x]) = c[u:=t,x:=a]),
$(\eta\Sigma)$ ⊢ Σ(u,x).($\downarrow_u$(x) $\star$ c) = c (u, x not free in c),
$(\zeta\Sigma)$ ⊢ ∂z.e[z] = Σ(u,x).e[z:=$\downarrow_u$(x)] (u, x fresh for e[z]).

Let T be any one of the λ-calculi mentioned previously, containing at least $\lambda\partial$, and define scalar extensions:

$T\Lambda := T + (\beta\Lambda) + (\eta\Lambda)$,
$T\Lambda^* := T + (\beta\Lambda^*) + (\eta\Lambda^*)$,
$T\Lambda^*_\zeta := T + (\beta\Lambda^*) + (\eta\Lambda^*) + (\zeta\Lambda^*)$,

$T\Sigma := T + (\beta\Sigma) + (\eta\Sigma)$,
$T\Sigma_\zeta := T + (\beta\Sigma) + (\eta\Sigma) + (\zeta\Sigma)$.

One has the following (strict) inclusions and equivalences:

$(\preceq_{\partial\Pi})$ $T \preceq T\Lambda \preceq T\Lambda^* \simeq T\Sigma \preceq T\Lambda^*_\zeta \simeq T\Sigma_\zeta \preceq \partial\pi\Pi$.

One may also add systems containing an inversion $[\Delta,\nabla]$, taken as a primitive. The latter are contained in $\partial\pi\Pi := \partial\pi + (\beta\Pi_1) + (\beta\Pi_2) + (\eta\Pi)$.

29 January – 13 February 2020

19

Vademecum: a guide to the bibliography

§ 1 In retrospect, it turned out that our historical and conceptual data are in need for *revision*.

(1) As regards *intuitionism* proper, the need to distinguish between Brouwer's genuine views on *logic* (actually: proof theory) and *formalisation*, on the one hand, and the *formularian* understanding of the issues in the work of his immediate followers (as, e.g., his PhD student Arend Heyting, and, possibly, A. N. Kolmogorov and V. I. Glivenko), on the other, has re-emerged, now and then, in private discussions with Brouwer *connoisseurs*, mainly Mark van Atten. The least thing to say, in this respect, is that the *formularian*, Kolmogorov-[Glivenko]-Heyting account of Brouwer's ideas along the so-called *BHK-interpretation* [short for 'Brouwer-Heyting-Kolmogorov'] does not match Brouwer's original views on 'intuitionistic reasoning'. Whether the latter – a would-be *genuine intuitionistic logic* – could be actually *formalised*, by our current standards of *formalisation*, is a matter open for debate.

(2) On the other hand, during the last decades, it has become clear that, on the *technical* side, the formal counterpart of the *BHK*-reading of the intuitionistic connectives, in terms of proof-operations – currently kwown as *Curry-Howard Correspondence* (or *Isomorphism*) [CHC] – is by no means *intuitionistically specific*.

Classical logic – as well as many non-classical logics sharing genuinely classical features (as, e.g., various modal [à la Lewis] and relevance logics [1]) – admits of elegant *witness-theoretical* descriptions – based on the same principles (essentially λ-calculus) – that are, in general, conceptually neater and simpler than it is the case for the Kolmogorov-[Glivenko]-Heyting variety of (formalised) intuitionism.

(3) Extensions of the CHC to classical logic and to 'intensional' logics with classical features and negation defined inferentially ($\lambda\gamma$-*calculi*, based essentially

---

[1] Including the so-called *linear* logics, that are, actually, very specific and baroque combinations of both.

on Kolmogorov [363], Glivenko [237], [238], and Prawitz [560]) have been first presented publicly in Nijmegen lectures (1986–1988), and in some other places ([600], [601], [602], etc.). See also the subsequent detailed records [604], [605], [607], [618], [620].[2]

(4) For extensions to logics with a *primitive* (classical) *negation*, and some attempts to *data recovery* from traditional formularian logic (Chrysippus and Stoic logic, Łukasiewicz, Jaśkowski, Gentzen), see [616], [617], [619], [622], in THIS VOLUME, and, possibly, [621].

(5) On the *conceptual* and *technical* side, the *alternative approach* has been summarised – under the general heading *witness theory* – in a series of *Notes on λ-calculus and logic*, most of them collected in THIS VOLUME.

(6) '*Begriffsschrift*' *beweistheoretisch abgeleitet*. As to our great-grandfather, Frege, I meant to examine in detail the effect of the *Brouwerian epistemic shift of paradigm* on Frege's *Begriffsschrift* [216] (**BS** 1879), [218] (**GGA** 1:1893), analysing the basics of his *concept*[ual] *script* in mathematical, proper *witness-theoretical* terms.[3]

(7) *The λ-calculus background*. Since the proof-syntax of concern in THIS VOLUME involves abstraction operators, all relevant concepts (free vs bound variables, open vs closed terms, subterms, substitution, α-conversion, etc.) are supposed to be understood, *mutatis mutandis*, as in λ-calculus. Cf. [39] (the main survey), [584] ('strict' λ-calculi and 'ordinal logic', à la Church, Kleene, Rosser and Turing), [318], [319], and [404], [405] For 'typed' λ-calculi, see [316], and [43], as well as the reviews of the latter monograph mentioned *ad loc*.[4]

> Since Frege's **GGA**, 1:1893, and Russell's ruminations of 1903 that remained in MS – now recovered as [648] – the λ-calculus has been (re-) invented three times, independently, by three different mathematicians (Moses Schönfinkel [December 1920, in print 1924], Haskell B. Curry [around 1927–1928, in print 1930], and Alonzo Church [1927–1928, in print 1931-1932]), and it took several

---

[2] Alternative, rather ad hoc variants of λγ-calculi have been discussed, under different labels, in the TCS-literature during the early 1990's, by Timothy G. Griffin, Chetan R. Murthy, Michel Parigot, Morten Heine Sørensen, and Niels Jakob Rehof, mainly in connection with the call-by-value λ-CS-calculi of Matthias Felleisen, [198]. See [605], [607], [618], and [700], for the main references.

[3] Specifically, in [626], resp. [608], I was only concerned with the *first-order* segment of BS, resp. GGA. So, save for casual remarks, I did not go into the neverending debates of the most recent Frege-experts, as how to 'repair' or to 'reconstruct' a would-be higher-order 'Frege logic'. Otherwise, witness-theoretically, the matter is, more or less, trivial, and amounts to an appropriate *indexing* of the higher-order proof operators (cf., e.g., [126]). So, in particular, since the bulk of my ('technical') talk in the above concerned Frege and his followers, an appropriate *subtitle* of this book could have been '**Frege** *versus* **Brouwer**'.

[4] For historical details see [357], [358], [643], and the surveys [109], [671]. In this guise, [130] came too late to my attention to be accounted for, but see [609].

decades – about four, counting with a generous approximation – until the corresponding formalisms were *properly* understood in *mathematical* terms (Dana Scott, December 1969). Factual history is not necessarily logical, but Dana Scott has also explained to us, in a perspicuous way (cf. his [668]), how things *could have happened*.

§ 2 As regards the literature extant in print, a *shortlist* is as follows.

(0) For the general context (*rappel des faits* regarding the earlier fomularian trends in 'mathematical' logic, mainly during the first decades of the XX-th century), see, e.g., [490] (the 'emergence' of first-order logic), [251] ('mathematical roots', 'logicism' and 'foundations of mathematics'), [446] ('nine itineraries from Russell to Tarski', roughly, the period 1900–1935), and the references mentioned by these authors, *ad loc*.

The specifics of main concern in this book – to be taken as prerequisites – are recorded *selectively* next.

(1) *The 'Minimalkalkül' and the 'intuitionistic' logic.* The relevant *historical* and *technical* pointers to this debate, including A. N. Kolmogorov (1925–1932), V. I. Glivenko (1928–1929), A. Heyting (1928–1930 and later), G. Gentzen (1932–1934), H. Freudenthal, and I. Johansson (1934–1936) can be found in the original papers by these authors (for Gentzen, see now [534], pp. 190–192), as well as in the contributions of Dirk van Dalen, Anne S. Troelstra, Walter P. van Stigt, Mark van Atten, and Johannes J. C. Kuiper, listed in the references.

For 'intuitionistic' logic and the *Minimalkalkül*, the main sources are Kolmogorov [363], [364], [365], [367], Glivenko [237], [238], Heyting [289], [290], [292], [293], [294], [295], [296], [297], [298], [299], Freudenthal [219], Johansson [341], and Gödel [244].

Heyting's own ('historical') debt to the early *formularian establishment* (Russell & co) has been meticulously documented in print by Anne S. Troelstra, one of his PhD students: one can hardly add something new on the subject.

On historical and closely related matters, the relevant secondary literature include [753], [754], [755], [756], [757], [758], [760], [761], [226], [454], [637], [149], [150], [764], [535], [536], [605] (BHK-bibliography updated in 2000), [646], [261], [155], [286], [23], [24], [25], [135], [487], and [213].

For Gödel's casual concerns with intuitionism, see his [241], [242], [243], [244], [246], [the *Dialectica* interpretation, 1958], [247], [ditto, revised in 1972], as well as Kreisel's [396], [399], and Tait's [716], [719], [720], [721].

On Wittgenstein and intuitionism, see, e.g., [249].

On intuitionistic logic, in general, see also the introductory texts [172], [149], [155], [182], [480], the textbooks [356], [158], [663], the monograph [763], and the surveys [383], [385], [387], [477], [478], and [234].

For rudiments of (classical and intuitionistic) proof theory taken in the sense of the formularian tradition, see the lectures (resp. textbooks) [724], [231] (Part I), [235] (Parts I–II), [233], [548], [549], [662], and [663].

For *Brouweriana*, see the bibliographical entries under Brouwer – mainly [79], [81] – resp. (secondary sources) under Mark van Atten (*et al.*), Dirk van

Dalen, Michael Detlefsen, Miriam Franchella, Johannes J. C. Kuiper, Enrico Martino, Walther van Stigt, Anne S. Troelstra, and Wim Veldman, appearing in the references.

The Brouwer bibliography can be found in [157]. For Heyting's bibliography and the *Nachlass*, see [300], [502], and [753], [759].

The Index of the Troelstra Archive [*Rijksarchief*, North Holland] is now available online [775].

Finally, on the *Curry-Howard Correspondence* [CHC] (for minimal and intuitionistic logic), see [229], [230], [233], [235], [704] (the first book-length monograph of the subject in print), [763], [591], [594], [595], [596], [598], and the survey [700].[5]

The relevant AUTOMATH literature can be now found online @ [1]. See also the entries under de Bruijn, appearing in the references, the PhD Dissertation [147], the monograph [590], and the collection [498].[6]

(2) *Gentzen sequent systems*. There is a huge amount of (formularian) literature in print on Gentzen sequent (L-) systems (the 'cut-elimination' industry). Of interest are, mainly, [139], [141], [143], [356], [346], [484], [485], [763], [418], [231], [233], [234], [235], [479], [663], [494], [66] [534], etc.[7]

(3) *'Natural deduction'*. For 'natural deduction' à la Jaśkowski (1926–1927), and for the Gentzen N-systems (1932–1933), see [139], [143], [201], [560], [16], [17], [496], the historical notes of [531], [532], [263], and the comments and references of [617].

> Again, factual history moves slowly: it has taken other forty years (1927 to 1967 or so) to the formularian logical establishment to understand that the so-called 'natural deduction' is nothing but an *elliptic* graphical resp. visual representation of (appropriately decorated / 'typed') combinators (closed λ-terms), resp. open λ-terms), and that the Gentzen L-systems are just special cases of 'natural deduction'.

(3) *Axiomatics*. On axiomatics for intuitionistic and other fragments of classical logic, worth mentioning are [790] (survey paper), [345] (on separation matters), [571] [*Appendix 1*] (survey of formal systems), the monograph [326] (the most recent philosophical bible of propositional logics), and, possibly, the historical references of [625], [626], [629].

---

[5] For related philosophical comments see, for instance, [709] [on proofs as 'constructions'], [748] [on 'what is a proof?' (sic)], and [573] [on Wittgenstein as a 'precursor' of CHC (sic)].

[6] The vast literature on Martin-Löf's 'constructive type theory' and its very recent international HoTT-prolongations [https://homotopytypetheory.org/book/ etc.] – not of direct concern here – have been left out, since the relevant data can be recovered from the bibliography of the 'HoTT-book'. Ditto for the more recent – rather ramified – research on proof-checking and / or automated-proof emulations of AUTOMATH, a separate subject, in fact.

[7] The newest trends have been conveniently left aside.

(4) *Classical logic.* As for *classical logic*, on the *historical* side, the Chysippean way of understanding it is documented in my earlier notes (2009), available now as [616].⁸ It follows that *Chrysippus' method of constructing* classical logic – as retrieved and recorded in [616] – admits of several *witness-theoretical formulations* [sic].⁹

> Between Chrysippus (280 BC – cca. 207 BC) and Frege (1848–1925) we had a vacuum of about twenty-two centuries, where nearly nothing of substance happend in *logical theory*. Surprisingly, however, the 'Chrysippean method' can be encountered – at least partly, now and then – in the moderns (in Frege, in Hilbert and, even, in the work of the *recentiores*, as, e.g., in the Beth-Hintikka 'tableaux' techniques, in resolution, etc.) On this, see, again, my notes [616] (2016), and the recent work of Susanne Bobzien [71] (2019).

Frege's *Regellogik* of **GGA** has been described in [608], while his *Satzlogik* of **BS** has been anaysed witness-theoretically in [627].

On (my way of understanding) Kolmogorov and Glivenko, see, e.g., [604], [607], and the recent notes [619], [620].

The Polish (Łukasiewicz-Jaśkowski) 'natural deduction' [ND] episode has been discussed at length, in witness-theoretic terms, in [617]. On Gentzen's ND, see [621], and [622].

(5) *Early 'Polish logic' and its colonies.* The 'early' Poles (i.e., the Lvov-Warsaw 'school' [LWS]) are relatively well-documented in print. We have, for instance, good editions – collections of reprints – of papers by Stanisław Leśniewski, Jan Łukasiewicz, Alfred Tarski, and Mordchaj Wajsberg. See also [450]. On Bolesław Sobociński – who exported, next to Tarski, after the WWII, formularian logic overseas – see [634], [624] (survey of didactic, scientific and editorial activities), [625] (work on axiomatics), [652] (larger context, at the Notre Dame University), and [634] (description of the *Nachlass*, currently deposited at the Notre Dame University). For secondary literature on the LWS, see the references of [617] and the collection [222].

As for the hinterlands, the 'Polish logic' emulation in Romania was *synchronic* – it dates from the early nineteen thirties – and is best illustrated by the publications of Grigore C. Moisil, [482], [484], [485], and Eugen Gh. Mihăilescu, [475], [476], who defected, both, rather early in their mathematical career to Polish-styled 'mathematical logic'. See also [749] for the typical output of a rather industrious (post-war) Romanian epigone of Jan Łukasiewicz (although, basically, a mathematician [geometry and analysis]). One can likely say that the work of Łukasiewicz, for instance – as well as of some of his students and

---

⁸ This was actually meant as an introduction to a longer talk about the 'modern readers' of Chrysippus, deferred for a separate publication.

⁹ We can, actually, proceed *systematically* – and combinatorially, by 'syntactic fibrations', so to speak – cf. [616], **§ 3** – in order to retrieve the possible variants.

collaborators – was better known, at least for a while, in Romania than in Poland or elsewhere in Western Europe and overseas.[10]

Otherwise, the early 'Polish logic' survived, mainly after WWII, in emigration, via Jan Łukasiewicz (in Ireland), Alfred Tarski, Bolesław Sobociński et al. (in the United States), etc. So Łukasiewicz influenced directly several oustanding logicians – as, e.g., Carew A. Meredith and Arthur N. Prior – while Tarski – the founder the 'Berkeley school' of model theory – and Sobociński – through his teaching and editorial activities at the University of Notre Dame, Notre Dame IN – configured a significant part of the mainstream post-war logical research in the United States.

(6) *Hilbert und die Göttinger.* Hilbert and his fellow *Göttinger* – before Gentzen – would *certainly* deserve a separate treatment. Relevant (historical) pointers can be found in, e.g., [518], [257], and [258], as well as in the editorial notes (by William Ewald, Wilfried Sieg et al.) to Hilbert's Lectures on the Foundations of Mathematics and Physics, 1891–1933, appearing in [312] (vol. 2, 1894–1917, [forthcoming] 2020), and in [311] (vol. 3, 1917–1933, in print 2013).

For Hilbert and his immediate descendance, see the entries under David Hilbert, Paul Bernays, John von Neumann, and Wilhelm Ackermann, in the references.

Further specific Hilbertian issues – mainly *finitism* and the fate of the so-called *Hilbert Programme* after Gödel's [240] – have been discussed at length – among many others – by Paul Bernays (see [54], [55], [56], [59], [60], [62], the lectures [58] and the collection [64]), Georg Kreisel[11], William Tait (cf. [715], [717] [finitism], [716], [719], [720] [Gödel], [721] [Gödel and finitsm], and, possibly, the collection [718]), Wilfried Sieg (cf. [681], [682], [684], [685], [686], [687], [688]), and Michael Detlefsen ([161], [164], etc.).

As regards the Eastern emulation, Hilbert's *Beweistheorie* and – more recently – Heyting and Gentzen were *en vogue* in Soviet Union, as well. See, e.g., the entries under Al'bert G. Dragalin [172] (cf. also the obituary [22]) and Grigori E. Minc [= Mints] in the references, mainly [479] and the surveys [477], [478].[12] This explains, in a way, why the Soviet logic research has remained – as

---

[10] Actually, Moisil – although essentially an algebrist – was also among the earliest *connoisseurs* of Heyting and Gentzen, in Europe, as regards formularian intuitionism. More or less in the footsteps of Moisil, nearly all relevant Romanian workers on mathematical logic were professional algebraists and focused their research mainly on algebraic logic and (Tarskian) model theory. (As an example, a thorough survey of the axiomatics for lattices and Boolean algebra can be found in [645], now revised as [516]).

[11] If possible, the reader should have a look at *all* Kreisel-items listed in the references. Otherwise, the Kreisel *Nachlass* is now publicly available for consultation at the Department of Special Collections and University Archives, Stanford University Libraries, Stanford CA.

[12] The hinterlands were less privileged, in this respect, although some Soviet mathematicians involved in mathematical logic research (including Hilbert's *Beweistheorie*, intuitionism, and a local trend of 'constructivism', as well as model theory), like,

in the West – *essentially formularian*. Curiously, Kolmogorov's (and Glivenko's) work on intuitionism deserved limited attention in the USSR.

(7) *The 'historical' Frege versus Frege'*. As noted, in passing, in Rezuș 2019a, Frege's debt to the 'tradition' (logic and philosophy) is a rather mysterious subject: he did not mention his sources, and we can, at best, make educated guesses in this respect. See, however, [19], [742], [106], [786], and [402], [403].[13]

On Frege's 'new paradigm for logic' – otherwise a slogan on which the Fregeans, both old and new, do not agree, in content – see, e.g., [741], [287], [784], [785], [786], [580], [747], [409], [180], [181], [228] (pp. 272–277), [160], [353], [654], [403], [453], [67]. etc. Worth mentioning is also the recent 'neo-Fregean' literature, concernd with consistency problems in **GGA** and fragments, with as main contributors, George Boolos (cf. [72], Part II *Frege studies*), Chrispin Wright, Bob Hale, [804], [252], Richard G. Heck Jr., [264], [265], [266], *et al.*

As surprising as it might appear, in spite of the vast Frege literature available so far – the *Frege' industry*, disserting *ad nauseam* about *Frege'*, the alleged precursor and founder of the Anglo-Saxon analytic philosophy'[14] – I was unable to find – save for minor exceptions – a *relevant* treatment of Frege's *proof theory* in print.

24 October 2015 – 31 December 2019

---

e.g., A. A. Markov, P. S. Novikov, and A. I. Mal'cev, have made their name in Eastern European countries, too. A case in point was Pyotr Sergeevič Novikov (1901–1975), whose elementary textbook [504], first published in 1959 – otherwise the first original textbook on the subject in the country, translated in about five foreign languages during 1964–1975, – was used as a standard course of mathematical logic – undergraduate level – in some Romanian and East German universities.

[13] Historically, we are on quicksands, here. As an aside, a few years ago, Susanne Bobzien gave a lecture, at the The Keeling Center for Ancient Philosophy (The S. V. Keeling Memorial Lecture 2015–2016, London, 18 March 2016), bearing the intriguing title *Did Frege plagiarize the Stoics?* [sic]. I could not attend the lecture, but its title was, certainly, more than a *bon mot*. It is unlikely Frege was familiar with Chrysippus and the Stoic logic, in general, although the specific choice of the axioms in **Begriffsschrift** (1879), as well as his later 'semantical' considerations (on the distinction *Sinn / Bedeutung* and on the objective status of the propositions, for him *Gedanken*) are reminiscent of typical issues going back to Stoic lore. As for the would-be association with the medieval *logica modernorum* (cf. [19], [106], etc.), such connections are interesting intellectual exercises, perhaps, but the fact is that we cannot establish a direct (or a mediate) connection of Frege to the early *terministae* either. *Pace* Frege's casual allusions to Leibniz, the same kind of remarks applies to the would-be links to Bernard Bolzano and / or to Hermann Grassmann, for instance. Cf., e.g., [710], and [251], [403].

[14] Cf. [178], [179], [180], [181], [353], [453], etc. For *Frege'*, see the comments of the late Ivor Grattan-Guinness, in his [251], **4.5.1**. Characteristically, in Romania, for instance, Frege counts as a *'classic of philosophy'* – cf. [783], [784], [785], [786] – and one of my former colleagues (Călin Candiescu, [106]) has even made him into a full-fledged Kantian.

References

1. *** The AUTOMATH Archive (2004), online @ the University of Technology, Eindhoven, www.win.tue.nl/automath/.

2. *** Euclides (*De Nederlandse Vereniging van Wiskundeleraren & KNAW* [The Royal Dutch Academy of Sciences], Amsterdam) = the journal of the Dutch Association of the Mathematics Teachers [NVvWL], online archive @ https://archief.vakbladeuclides.nl/.

3. *** EUDML, The European Digital Mathematics Library (*Institut Fourier*, Université de Grenoble-Alpes, Grenoble, and *Interdisciplinary Centre for Mathematical and Computational Modelling*, University of Warsaw), online @ https://eudml.org/. [Fundamenta Mathematicae, etc.].

4. *** Congress [PPZM] Proceedings (1929) Księga Pamiątkowa Pierwszego Polskiego Zjazdu Matematycznego [Commemorative Book of the First Polish Mathematical Congress], Lwów, 7–10 IX 1927 [Dotatek do Annales de la Société Polonaise de Mathématique], Kraków [Czonkami Drukarni Universytetu Jagiellonskiego] 1929, online @ www.ptm.org.pl/zawartosc/i-zjazd-ptm-1927-e-book.

5. *** MacTutor History of Mathematics Archive, School of Mathematics & Statistics, University of St Andrews, @ http://www-history.mcs.st-andrews.ac.uk/

6. *** Przegląd Filozoficzny [Philosophical Review] (Warsaw, 1897–1939), digitised @ *Wielkopolska Biblioteka Cyfrowa* [The Greater Polish Digital Library], Poznań = http://www.wbc.poznan.pl/publication/105589.

7. *** Dolph Ulrich Home Page, *Sentential Calculi*, Purdue University, West Lafayette IN, @ https://web.ics.purdue.edu/~dulrich/Home-page.htm.

8. Vito Michele Abrusci (1982) *Paul Hertz's logical works: contents and relevance*, in Atti del Convegno Internazionale di Storia della Logica (San Gimignano, 4–8 December 1982), Cooperativa Libreria Universitaria Editrice, Bologna 1982, pp. 369–374.

9. Wilhelm Ackermann (1925) *Begründung des 'tertium non datur' mittels der Hilbertschen Theorie der Widerspruchsfreiheit*, Mathematische Annalen 93, 1925, pp. 1–36. (Originally, *Inauguraldissertation* Göttingen 1924.)
10. — (1940) *Zur Widerspruchsfreiheit der reinen Zahlentheorie*, Mathematische Annalen 117, 1940, pp. 162–194.
11. Sergeĭ Ivanovič Adĭan (2001) *K stoletiĭu so dnĭa roždeniĭa Pyotra Sergeeviča Novikova* [On the centenary anniversary of Pyotr Sergeevič Novikov] [Russian], Uspexi Matematičeskix Nauk 56, 4 (340), 2001, pp. 177–184 = Russian Mathematical Surveys 56 (4), 2001, pp. 793–802. (See also *Pyotr Sergeevič Novikov, s semidesĭatiletĭu so dnĭa roždeniĭa* [Pyotr Sergeevič Novikov, on his 70th anniversary] [Russian], Uspexi Matematičeskix Nauk 26, 5 (161), 1971, pp. 231–241.)
12. Marcello D'Agostino (1999) *Tableau methods for classical propositional logic*, in Marcello D'Agostino et al. (eds.) Handbook of Tableau Methods, Kluwer Academic Publishers, Dordrecht 1999, pp. 45–123.
13. Alexander of Aphrodisias [Maximilian Wallies] (1883) In Analyticorum Librum Priorum Commentarium (edited by Maximilian Wallies), G. Reimer, Berlin 1883 [CAG 2.1]
14. Elias H. Alves (1992) *The first axiomatization of paraconsistent logic*, Bulletin of the Section of Logic (University of Lódź, Polish Academy of Sciences) 21 (1), 1992, pp. 19–20.
15. Alan Ross Anderson, and Nuel D. Belnap Jr. (1962) *The pure calculus of entailment*, The Journal of Symbolic Logic 27 (1), 1962, pp. 19–52.
16. Alan Ross Anderson, Nuel D. Belnap Jr. et al. (1975) Entailment, *The Logic of Relevance and Necessity*, 1, Princeton University Press, Princeton NJ 1975. (Second printing, with corrections 1990^2.)
17. Alan Ross Anderson, Nuel D. Belnap Jr., J. Michael Dunn et al. (1992) Entailment, *The Logic of Relevance and Necessity*, 2, Princeton University Press, Princeton NJ 1992.
18. Irving H. Anellis (1995) *Peirce rustled, Russell pierced: How Charles Peirce and Bertrand Russell viewed each other's work in logic, and an assessment of Russell's accuracy and rôle in the historiography of logic*, Modern Logic 5 (3), 1995, pp. 270–328.
19. Ignacio Angelelli (1967) Studies on Gottlob Frege and Traditional Philosophy, D. Reidel Publishing Company, Dordrecht 1967.
20. — (1975) *On Saccheri's use of the 'consequentia mirabilis'*, in Akten des II. Internationalen Leibniz-Kongresses, 1972, Steiner Verlag, Wiesbaden 1975, 4, pp. 19–26.
21. Hans Friedrich August von Arnim, and Maximilian Adler (eds.) (1903–1924) Stoicorum Veterum Fragmenta, In Aedibus B. G. Teubneri, Lipsiae [B. G. Teubner, Leipzig] 1903–1924 [1:1905. *Zeno et Zenonis discipuli*; 2:1903. *Chrysippi fragmenta logica et physica*; 3:1903. *Chrysippi fragmenta moralia & Fragmenta successorum Chrysippi*; 4:1924. *Indices*, by Maximilian Adler]. Reprint [*editio stereotypa*]: In Aedibus B. G. Tuebneri, Stutgardiae [B. G. Teubner, Stuttgart] 1964. [SVF] (For translations in modern languages, see [176] [French], [36], [575] [Italian], and the collection [325] [German]. The Romanian translation, currently in progress [= HANS VON ARNIM (ed.) Fragmentele stoicilor vechi 1, *Zenon și discipolii lui Zenon*, Humanitas, Bucharest 2016 [*Surse clasice*], tr. by Filotheia Bogoiu, and Cristian Bejan], reached only the first von Arnim volume to date, irrelevant for the history of logic.)

22. Sergei N. Artemov, Boris A. Kušner [Kushner], Grigori Minc [Mints], Elena Nogina, and Anne S. Troelstra (1999) *In Memoriam: Albert G. Dragalin, 1941–1998*, Bulletin of Symbolic Logic 5 (3), 1999, pp. 389–391.
23. Mark van Atten (2005) *The correspondence between Oskar Becker and Arend Heyting*, in Volker Peckhaus (ed.) Oskar Becker und die Philosophie der Mathematik, Wilhelm Fink Verlag, München 2005 [*Neuzeit und Gegenwart*], pp. 119–142.
24. — (2009) *The hypothetical judgement in the history of intuitionistic logic*, in Clark Glymour, Wang Wei, and Dag Westerståhl (eds.) Logic, Methodology, and Philosophy of Science 13, *Proceedings of the 2007 International Congress in Beijing*, King's College Publications, London 2009, pp. 122–136.
25. — (2017) *The development of intuitionistic logic*, The Stanford Encycpopedia of Philosophy, 10 July 2008, revised 8 November 2017.
26. — (2017a) *Predicativity and parametric polymorphism of Brouwerian implication* [preprint], Paris, 19 October 2017.
27. Mark van Atten, Pascal Boldini, Michel Bourdeau, and Gerhard Heinzmann (eds.) (2008) One Hundred Years of Intuitionism (1907–2007), *The Cerisy Conference*, Birkhäuser, Basel, etc. 2008 [*Publications des Archives Henri-Poincaré.*]
28. Mark van Atten, and Dirk van Dalen (2002) *Intuitionism*, in Dale Jaquette (ed.) A Companion to Philosophical Logic, Blackwell, Oxford 2002, pp. 513–530.
29. Mark van Atten, and Göran Sundholm (2008) *The proper explanation of intuitionistic logic: on Brouwer's proof of the Bar Theorem*, in [27], pp. 60–77.
30. — (2014) *Intuïtionistische logica en het scheppend subject* [Intutionistic logic and the creative subject] [Dutch], Nieuw Archief voor Wiskunde (5) 15 (2), [June] 2014, pp. 124–130. (Paper for Dirk van Dalen's 80th anniversary. The section on *Het scheppend subject* [The creative subject], pp. 127–130, was written by Mark van Atten alone.)
31. — (eds., trs.) (2017) L. E. J. BROUWER's *'Unreliability of the logical principles': A new translation, with an introduction*, History and Philosophy of Logic 38 (1), 2017, pp. 24–47.
32. Mark van Atten, Göran Sundholm, Michel Bourdeau, and Vanessa van Atten (eds., trs.) (2014) [L. E. J. BROUWER] *'Que les principes de la logique ne sont pas fiables.' Nouvelle traduction française annotée et commentée de l'article de 1908 de L. E. J. Brouwer*, Revue d'Histoire des Sciences 67 (2), 2014, pp. 257–281.
33. Catherine Atherton (1993) The Stoics on Ambiguity, Cambridge University Press, Cambridge UK 1993 [*Cambridge Classical Studies*].
34. Giorgio Ausiello (1971) *Automatic reduction of* CUCH *expressions by means of the value method*, Atti del I Congresso Nazionale dell'AICA (Naples, 26–29 September1968), Rome, 1971, pp. 174–184.
35. Franz Baader, and Tobias Nipkow (1998) Term Rewriting and All That, Cambridge University Press, Cambridge 1998 [first edition], 1999 [paperback].
36. Mariano Baldassari (1984) La logica stoica. *Testimonianze e frammenti*, Litotipografia Malinverno, Como 1984 [not seen].
37. Henk Barendregt (1971) [*Two λ-calculus generators*] [MS], Stanford University, October 1971.
38. — (1974) *Pairing without conventional constraints*, Zeitschrift für mathematischen Logik und Grundlagen der Mathematik 20, 1974, pp. 289–306.
39. — (1981, 1984^2, 2012^R) The Lambda Calculus, *Its Syntax and Semantics*, North-Holland Publishing Company, Amsterdam, etc. 1981 (second, revised edition, ibid. 1984^2, reprinted by College Publications, London 2012^R) [*Studies in Logic and the Foundations of Mathematics 103*]. (The first edition has been reviewed

by Erwin Engeler in The Journal of Symbolic Logic 49 (1), [March] 1984, pp. 301–303. Russian translation [Lambda-isčislenie, *Ego sintaksis i semantika*] of the first [updated] edition, by G. E. Minc [= Gregory Mints], edited by Alexander S. Kuzičev, 'Mir' Publishers, Moskow 1984.)

40. — (1996) *Kreisel, lambda calculus, a windmill and a castle*, in Piergiorgio Odifreddi (ed.) Kreiseliana, *About and around Georg Kreisel*, A. K. Peters, Wellesley MA 1996, pp. 3–14.

41. Henk Barendregt, and Adrian Rezuş (1983) *Semantics for Classical* AUTOMATH *and related systems*, Information and Control [currently: Information and Computation] 59 (1–3), 1983, pp. 127–147.

42. Henk Barendregt, and Silvia Ghilezan (2000) *Lambda terms for natural deduction, sequent calculus and cut-elimination*, Journal of Functional Programming 10, 2000, pp. 121–134.

43. Henk Barendregt, Wil Dekkers, and Richard Statman *et al.* (2013) Lambda Calculus with Types, Cambridge University Press, Cambridge UK etc. & ASL, Urbana IL 2013 [*Perspectives in Logic*]. (Reviewed by J. Roger Hindley in Bulletin of the London Mathematical Society 46 (5), 2014, pp. 1110–1112, and by Adrian Rezuş in [611].)

44. Jonathan Barnes (1980) *Proof destroyed*, in Jonathan Barnes *et al.* (eds.) Doubt and Dogmatism, Clarendon, Oxford 1980, pp. 161–181.

45. — (1985) *Theophrastus and hypothetical syllogistic*, in William W. Fortenbaugh *et al.* (eds.) Theophrastus of Eresus, *On His Life and Work*, New Brunswick 1985. (Also in Jürgen Wiesner (ed.) Aristoteles Werk und Wirkung. *Paul Moraux gewidmet* 1, Walter de Gruyter, Berlin 1985, pp. 557–576.)

46. — (1996) *The catalogue of Chysippus' logical works*, in Keimpe A. Algra, Peter W. van der Horst, and David T. Runia (eds.) Polyhistor. *Studies in the History and Historiography of Ancient Philosophy*, E. J. Brill, Leiden etc. 1996, pp. 169–184. (*Festschrift* Jaap Mansfeld.)

47. — (1999) *Aristotle and Stoic logic*, in Katerina Ierodiakonou (ed.) Topics in Stoic Philosophy, Clarendon Press, Oxford 1999, pp. 230–153.

48. — (2007) Truth, Etc. *Six Lectures on Ancient Logic*, Clarendon Press, Oxford 2007.

49. Valentin A. Bažanov [V. A. Bazhanov] (2001) *The origins and becoming of non-classical logic in Russia (XIX – the turn of XX century)*, in Werner Stelzner, and Manfred Stöckler (eds.) Zwischen traditioneller und moderner Logik. *Nichtklassische Ansätze*, MENTIS, Paderborn 2001 [*Perspektiven der analytischen Philosophie, Neue Folge*], pp. 205–217.

50. — (2003) *The scholar and the 'Wolfhound Era': The fate of Ivan E. Orlov's ideas in logic, philosophy, and science*, Science in Context 16 (4), 2003, pp. 535–550.

51. Oskar Becker (1957) *Über die vier Themata der stoischen Logik*, in OSKAR BECKER Zwei Untersuchungen zur antiken Logik, Harrassowitz, Wiesbaden 1957, pp. 27–10.

52. Fabio Bellissima, and Paolo Pagli (1996) Consequentia mirabilis, *Una regola logica tra matematica e filosofia*, Leo Olschki Editore, Florence 1996 [*Biblioteca di storia della scienza 38*].

53. Lambertus [Bert] Salomon van Benthem Jutting (1977) Checking Landau's 'Grundlagen' in the AUTOMATH System, PhD Dissertation, University of Technology, Eindhoven 1977, under N. G. de Bruijn [TH Eindhoven Report AUT 79-46]. (Reprinted, without the AUTOMATH proof-text of Landau [413], by Mathematisch Centrum [*CWI*], Amsterdam 1979 [*Mathematical Centre Tracts 83*].)

54. Paul Bernays (1930) *Die Philosophie der Mathematik und die Hilbertsche Beweistheorie*, Blätter für deutsche Philosophie 4, 1930, pp. 326–367. (Reprinted in [64], pp. 17–61.)
55. — (1935) *Hilberts Untersuchungen über die Grundlagen der Arithmetik*, in [310], pp. 196–216.
56. — (1935-1936) *Sur le platonisme dans les mathématiques*, L'Enseignement Mathématique 34, 1935–1936, pp. 52–69. (German version [*Über den Platonismus in der Mathematik*], in [64], pp. 62–78.)
57. — (1935–1936a) *Quelques points essentiels de la metamathématique*, L'Enseignement Mathématique 34, 1935–1936, pp. 70–95.
58. — (1935–1936b) Logical Calculus, Lectures delivered at the Institute for Advanced Study, Princeton 1935–1936, mimeographed Lectures Notes, Princeton NJ 1936 [125 pp.] [not seen].
59. — (1950) *Mathematische Existenz und Widerspruchsfreiheit*, in Études de philosophie des sciences en hommage à Ferdinand Gonseth, Éditions du Griffon, Neuchâtel 1950, pp. 11–25. (Reprinted in [64], pp. 92–106.)
60. — (1954) *Zur Beurteilung der Situation in der beweistheoretischen Forschung*, Revue Internationale de Philosophie 8, 1954, pp. 9–13. (See also the *Discussion*, ibid., pp. 15–21.)
61. — (1965) *Betrachtungen zum Sequenzen-Kalkül*, in Anna-Teresa Tymieniecka, and Charles Parsons (eds.) Contributions to Logic and Methodology, in Honor of J. M. Bocheński, North-Holland Publishing Company, Amsterdam 1965, pp. 1–44.
62. — (1967) *Hilbert, David*, in Paul Edwards (ed.) Encyclopedia of Philosophy 3 MacMillan & Free Press, New York 1967, pp. 496–504.
63. — (1969) *Paul Hertz*, in Neue Deutsche Biographie 8, Duncker & Humblot, Berlin 1969, s.v. [= pp. 711 et sq.].
64. — (1976) Abhandlungen zur Philosophie der Mathematik, Wissenschaftliche Buchgesellschaft, Darmstadt 1976.
65. Evert Willem Beth, and Hugues Leblanc (1960) *A note on the intuitionistic and the classical propositional calculus*, Logique et analyse 3, 1960, pp. 174–176, reprinted in [418], pp. 382–384. (On Kanger's Rule.)
66. Katalin Bimbó (2014) Proof Theory, *Sequent Calculi and Related Formalisms*, Chapman and Hall / CRC Press, Boca Raton FL 2014 [*Discrete Mathematics and Its Applications*].
67. Patricia A. Blanchette (2012) Frege's Conception of Logic, Oxford University Press, New York 2012.
68. Susanne Bobzien (1996) *Stoic syllogistic*, Oxford Studies in Ancient Philosophy 14, 1996, pp. 133–192.
69. — (1999) *Logic: The Stoics*, in K. A. Algra, J. Barnes, J. Mansfeld, and M. Schofield (eds.) The Cambridge History of Hellenistic Philosophy, *Part 2: Logic and Language*, Cambridge University Press, Cambridge 1999, pp. 92–157.
70. — (2003) *Stoic Logic*, in Brad Inwood (ed.) Cambridge Companion to Stoic Philosophy, Cambridge University Press, Cambridge UK 2003, 2005^R, pp. 85–123.
71. — (2019) *Stoic sequent logic and proof theory*, History and Philosophy of Logic 40 (3), 2019, pp. 234–265.
72. George Boolos (1998) Logic, Logic, and Logic (edited by Richard Jeffrey, with Introductions and Afterword by John P. Burgess), Harvard University Press, Cambridge MA and London 1998. (Collected papers. Cf. Part II, *Frege studies*.)

73. Ludwik Borkowski, and Jerzy Słupecki (1963) Elementy logiki matematycznej i teorii mnogości [Elements of Mathematical Logic and Set Theory] [Polish], PWN [Polish Scientific Publishers], Warsaw 1963, 1966^2, 1969^3, etc. (Russian translation [Elementy matematičeskoĭ logiki i teoriĭa množestv], by Oleg Fedorovič Serebrĭanikov], Izdatel'stvo 'Progress', Moskow 1965. English translation [Elements of Mathematical Logic and Set Theory], by Olgierd Adrian Wojtasiewicz, Pergamon Press, Oxford 1967 [International Series of Monographs in Pure and Applied Mathematics 96].)
74. Corrado Böhm (1966) The CuCh as a formal and description language, in T. B. Steele (ed.) Formal Language Description Languages for Computer Programming, North-Holland Publishing Company, Amsterdam, 1966, pp. 179–197.
75. Corrado Böhm, and Wolf Gross (1965) Introduction to the CuCh, in E. R. Caianiello (ed.) Automata Theory, Academic Press, New York 1966), pp. 35–65.
76. Corrado Böhm, and Mariangiola Dezani (1972) A CuCh-machine: The automatic treatment of bound variables, International Journal of Computer and Information Sciences 1, 1972, pp. 171–191.
77. Corrado Böhm et al. (2017) The CuCh Machine, Curry-Church machine for functional programming, University of Rome 1, 2017 [web-site of Luigi Mazzucchelli].
78. Luitzen Egbertus Jan Brouwer (1905) Leven, kunst en mystiek [Life, art, and mysticism] [Dutch], J. Waltman Jr., Delft 1905 (English translation [Life, art, and mysticism], in Notre Dame Journal of Formal Logic 37 (3), 1996, pp. 389–429, with an introduction by Walter P. van Stigt, ibid., pp. 381–387.)
79. — (1907) Over de grondslagen der wiskunde [On the foundations of mathematics] [Dutch], Maas & van Suchtelen, Amsterdam 1907. (PhD Dissertation, University of Amsterdam 1907. Reprint, with additions, edited by Dirk van Dalen, as Over de grondslagen der wiskunde, Mathematisch Centrum [CWI], Amsterdam 1981. An expanded version of the latter edition [L.E.J. Brouwer en de grondslagen van de wiskunde], also edited by Dirk van Dalen, has been issued by Epsilon, Utrecht 2001 [reprint 2005]. English translation [On the Foundations of Mathematics], in [92], 1, pp. 11–101.)
80. — (1908) Over de grondslagen der wiskunde [On the foundations of mathematics] [Dutch], Nieuw Archief voor Wiskunde 8, 1908, pp. 326–328. (English translation [On the foundations of mathematics] in [92], pp. 105–106.)
81. — (1908a) De onbetrouwbaarheid der logische principes [The unreliability of the logical principles] [Dutch], Tijdschrift voor Wijsbegeerte 2, 1908, pp. 152–158. (Reprinted in [84]. French translations: by [1] Jean Largeault [Qu'on ne peut pas se fier aux principes logiques], in Jean Largeault (ed., tr.), Intuitionisme et théorie de la démonstration, Librairie Philosophique Vrin, Paris 1992 (reprint 2002) [Mathesis], pp. 15–23, by [2] Jacques Bouveresse [Les principes logiques ne sont pas sûrs], in François Rivenc, and Philippe de Rouilhan (eds.) Logique et fondements des mathématiques, Anthologie (1850–1914), Éditions Payot, Paris 1992, pp. 379–392, and by [3] Mark van Atten, and Michel Bourdeau [Que les principes de la logique ne sont pas fiables], in [32]. English translations, by [1] Arend Heyting [Unreliability of the logical principles] in [92], pp. 107–111, and by [2] Mark van Atten, and Göran Sundholm, in [31].)
82. — (1912) Intuïtionisme en formalisme [Intuitionism and formalism] [Dutch], Clausen, Amsterdam 1912. (Inaugural Lecture, University of Amsterdam, October 14, 1912. Commercial edition issued by Noordhoff, Groningen 1912. Reprinted in Wiskundig Tijdschrift 9, 1919, pp. 180–211, resp. in [84]. English translation [Intuitionism and formalism], by Arnold Dresden, in Bulletin of the American

Mathematical Society 20 (2), [November] 1913, pp. 81–96, reprinted in [92], pp. 123–138 [editorial notes, pp. 570–571], and in Paul Benacerraf, and Hilary Putnam (eds.) Philosophy of Mathematics, *Selected Readings*, Prentice Hall, Englewood Cliffs NJ 1964^1, pp. 66–77; Cambridge University Press, New York 1984^2, pp. 77–89.)

83. — (1917) *Addenda en corrigenda: Over de grondslagen der wiskunde* [On the foundations of mathematics: addenda and corrigenda] [Dutch], Nieuw Archief voor Wiskunde 12, 1917, pp. 439–445. (English translation in [92], pp. 145–149.)

84. — (1919) Wiskunde, waarheid, werkelijkheid [Mathematics, truth, reality] [Dutch], Noordhoff, Groningen 1919.

85. — (1928) *Intuitionistische Betrachtungen über den Formalismus*, Proceedings of the Royal Dutch Academy of Sciences (KNAW, Amsterdam) 31, 1928, pp. 374–379. (English translation in [445], pp. 40–44.)

86. — (1929) *Mathematik, Wissenschaft und Sprache*, Monatshefte für Mathematik und Physik 36, 1929, pp. 153–164. (English translation in [445], pp. 45–53.)

87. — (1933) *Willen, weten, spreken* [Willing, knowing, and speaking] [Dutch], in Luitzen Egbertus Jan Brouwer, Jacob Clay, Arnold de Hartog, Gerrit Mannoury, Hugo Pos, Géza Révész, Jan Tinbergen, and Johannes van der Waals Jr. De uitdrukkingswijze der wetenschap. *Kennistheoretische openbare voordrachten gehouden aan de Universiteit van Amsterdam gedurende de kursus 1932–1933* [The way of expressing science. Epistemological public lectures held at the University of Amsterdam during the academic year 1932–1933], Noordhoff, Groningen 1933, pp. 45–63. (English translation in [706], pp. 418–431.)

88. — (1948) *Essentieel negatieve eigenschappen* [Essentially negative properties] [Dutch], Proceedings of the Royal Dutch Academy of Sciences (KNAW, Amsterdam) 51, pp. 963–964 [= Indagationes Mathematicae 10, 1948, pp. 322–323]. (English translation [*Essentially negative properties*] in [92], pp. 478–479.)

89. — (1949) *Consciousness, philosophy and mathematics*, in Evert Beth, Hugo Pos, and Jan Hollak (eds.) Proceedings of the 10th International Congress of Philosophy (Amsterdam, August 1948), vol. 2, North-Holland Publishing Company, Amsterdam 1949, pp. 1235–1249. (Reprinted in [92], pp. 480–494, and in Paul Benacerraf, and Hilary Putnam (eds.) Philosophy of Mathematics, *Selected Readings*, Prentice Hall, Englewood Cliffs NJ 1964^1, pp. 78–84, Cambridge University Press, New York 1984^2, pp. 90–96.)

90. — (1952) *Historical background, principles and methods of intuitionism*, South African Journal of Science 49, 1952, pp. 139–146. (Reprinted in [92], pp. 508–515.)

91. — (1955) *The effect of intuitionism on classical algebra of logic*, Proceedings of the Royal Irish Academy 57, 1955, pp. 113–116. (Reprinted in [92], pp. 551–554.)

92. — (1975) Collected Works, Vol. 1 *Philosophy and Foundations of Mathematics* (edited by Arend Heyting), North-Holland Publishing Company, Amsterdam / Oxford, and American Elsevier Publishing Company Inc., New York 1975. (Reviewed by Georg Kreisel, in Bulletin of the American Mathematical Society 1 (1), 1977, pp. 86–93.)

93. Jacques Brunschwig (1980) *Proof defined* (translated by Jennifer Barnes), in Jonathan Barnes et al. (eds.) Doubt and Dogmatism, Clarendon, Oxford 1980, pp. 125–160.

94. Nicolaas Govert de Bruijn (1967) *Verificatie van wiskundige bewijzen door een computer. Een voorstudie ten behoeve van een projekt* AUTOMATH [Verification of mathematical proofs by a computer. A preliminary study for a project AUTOMATH]

[Dutch]. Text of a talk, University of Technology, Eindhoven, 9 January 1967 [typescript: 11 + 5 pp.] [= AUT 67–16]. (The first public AUT-report.)

95. — (1968) AUTOMATH, a language for mathematics, TH[E] Report 68-WSK-05, November 1968 [47 pp.] [= AUT-68-01] (Reviewed by H. J. Schneider, Zfm 174, 484. Revised version with a Commentary by the author [December 1981], in Jörg Sieckmann, and Graham Wrightson (eds.) Automation of Reasoning 2, *Classical Papers on Computational Logic 1967–1970*, Springer Verlag, Berlin & Heidelberg 1983 [*Symbolic Computation*], pp. 159–200.)

96. — (1970) *The mathematical language* AUTOMATH, *its usage, and some of its extensions*, M. Laudet, D. Lacombe, L. Nolin, and M. Schützenberger (eds.) Symposium on Automatic Demonstration (held at Versailles / France, December 1968), Springer Verlag, Berlin etc. 1970 [*Lecture Notes in Mathematics 125*], pp. 29–61.

97. — (1978) Taal en structuur van de wiskunde [Language and structure of mathematics] [Dutch], Lecture Notes, Department of Mathematics and Computing Science, Eindhoven University of Technology, Spring Semester 1978. (Unpublished. See, however, the separate installments published [in Dutch] in Euclides 55, 1979–1980, pp. 7–12, 66–72, 262–268, and 429–435, @ [2] and [497], [227].)

98. — (1980) *A survey of the* AUTOMATH *project*, in [672], pp. 579–606.

99. — (1981) Formalizing the Mathematical Vernacular, Preprint University of Eindhoven, Department of Mathematics and Computing Science, March 1981 [48 pp.]. (Revised as *The Mathematical Vernacular, a language for mathematics with typed sets*, in [498], pp. 865–935. See also [97].)

100. — (1987) *Generalizing* AUTOMATH *by means of a lambda-typed lambda-calculus*, in D. W. Kueker et al. (eds.) Mathematical Logic and Theoretical Computer Science, Marcel Dekker, New York 1987, pp. 71–92 [*Lecture Notes in Pure and Applied Mathematics 106*]. (Reprinted in [498], pp. 313–337.)

101. — (1990) Gedachten rondom AUTOMATH [Reflections on AUTOMATH] [Dutch], Preprint: University of Eindhoven, Department of Mathematics and Computing Science, March 1990 [24 pp.]. (For an English version of this report, cf. [498], pp. 201–228. See also The AUTOMATH Archive [1].)

102. Robert A. Bull (1962) *The implicational fragment of Dummett's LC*, The Journal of Symbolic Logic 27 (2), 1962, pp. 189–194. (Cf. [177], [745], [103], [459].)

103. — (1964) *Some results for implicational calculi*, The Journal of Symbolic Logic 29 (1), 1964, pp. 33–39. (Cf. [177], [745], [102], [459].)

104. — (1996) *Logics without contraction I*, in [133], pp. 317–336.

105. Martin Bunder (1996) *Logics without contraction II*, in [133], pp. 337–336.

106. Călin Candiescu (1980) Gottlob Frege și filosofia analitică a limbajului [Gottlob Frege and the analytical philosophy of language] [Romanian] PhD Dissertation, University of Bucharest, Bucharest 1980 [unpublished].

107. Girolamo Cardano (1570) De proportionibus, Sebastian Henricpetri, Basle 1570. (*Praefatio*, pp. [i–ii], *Tabula propositionum de proportionibus* [Analytic Table of Contents], pp. [iii–xiii], *De proportionibus*, pp. 1–270 [i.e., *Hieronymi Cardani Mediolanensis, civisque Bononiensis, medici, De proportionibus, seu operis perfecti, liber quintus*], Officina Henricpetrina [Sebastian Henricpetri = Sebastian Petri], Basileae anno salutis M.D.LXX, mense martio (= March 1570)]. The Basle edition contains also Ars magna and De aliza regula and is available online @ www.archive.org and at the Archimedes Project [The Digital Research Library, Deutsche Forschungsgemeinschaft, Max Planck Institute for the History

of Science, Berlin, and the National Science Foundation (USA), Harvard University, Department of the Classics]. Reprinted in [108], 4, [= *Hieronymi Cardani Mediolanensis, civisque Bononiensis, philosophi, medici et mathematici clarissimi,* Opus novum de proportionibus *numerorum, motuum, ponderum, sonorum, aliarumque rerum mensurandarum, non solum geometrico more stabilitum, sed etiam variis experimentis et observationibus rerum in natura, solerti demonstratione ilustratum, ad multiplices usus accommandatum*], pp. 463–603.)

108. — (1663) Opera omnia [*Hieronymi Cardani Mediolanensis*], edited by Karl Spon, 10 vols. in f°, Lugduni [Lyons, France] 1663. (Reprographic reprints: (1) Fromann Verlag, Stuttgart / Bad Cannstatt 1966 and (2) Johnson Reprint Corporation, New York & London 1967. For *consequentia mirabilis* in the Opus novum De proportionibus, cf., especially, the *Scholium ad Propositionem 201 [ducentesimaprimam]*, 4.580 = [107], p. 231. Full context: Prop. 201, Lemmas 1–2 and the *Scholium*, pp. 4.579–4.581 = [107], pp. 230–232.)

109. Felice Cardone, and J. Roger Hindley 2009 *History of lambda-calculus and combinatory logic*, in Dov M. Gabbay, and John Woods (eds.) Handbook of the History of Logic 5, *Logic from Rusell to Church*, North-Holland Publishing Company, Amsterdam, etc. 2009, pp. 723–817.

110. Ettore Casari (1979) *Remarks on comparison and superlation*, in Sergio Bernini (ed.) Atti del Congresso Nazionale di Logica 1979, Bibliopolis, Naples 1981, pp. 261–271.

111. — (1984) *Note sulla logica aristotelica della comparazione*, Sileno 10, 1984, pp. 131–146.

112. — (1985) *Logica e comparativi*, in Corrado Mangione (ed.), Scienza e filosofia. *Saggi in onore di Ludovico Geymonat*, Garzanti, Milan 1985, pp. 392–418.

113. — (1987) *Comparative logics*, Synthese 73, 1987, pp. 421–449.

114. — (1989) *Comparative logics and Abelian l-groups*, in Ruggero Ferro, Cinzia Bonotto, Silvio Valentini, and Alberto Zanardo (eds.) Logic Colloquium '88, North-Holland Publishing Company, Amsterdam 1989 [*Studies in Logic and the Foundations of Mathematics 127*], pp. 161–190.

115. — (1991) *Logics on pregroups*, in Giovanni Corsi, and Giovanni Sambin (eds.) Atti del Congresso 'Nuovi Problemi della Logica e della Filosofia della Scienza' (Viareggio, January 8–13, 1990), 2 *Logica*, Cooperativa Libreria Universitaria Editrice, Bologna 1991, pp. 39–58.

116. Alonzo Church (1932–1933) *A set of postulates for the foundation of logic*, Annals of Mathematics (2) 33, 1932, pp. 346–366, and (second paper) 34, 1933, pp. 839–864. (Reprinted in [130], pp. 51–67, 68–87.

117. — (1935) *A proof of freedom from contradiction*, Proceedings of the National Academy of Sciences of the United States of America 21, 1935, pp. 275–281. ('Communicated March 26, 1935.' Reprinted in [130], pp. 99–104.)

118. — (1935–1936) Mathematical Logic, Lecture Notes [by F. A. Ficken, H. G. Landau, H. Ruja, R. R. Singleton, N. E. Steenrod, J. H. Sweer, and F. J. Weyl], Princeton University, October 1935 – January 1936, Princeton NJ 1936 [typescript, 114 pp.]. (Revised in [121]. This is a course on $\lambda\delta$I-calculus containing the first example of ordinal logic in the sense of Alan Turing [766]. No relation to [125]. See also [117], [584], Chapters V–VII, pp. 91–181, and the Princeton University [Firestone] Library record, in [806] [item in Box 10, Folder 2].)

119. — (1937) *Combinatory logic as a semigroup* (abstract), Bulletin of the American Mathematical Society 43, 1937, p. 333.

120. — (1940) *A formulation of the simple theory of types*, The Journal of Symbolic Logic 5 (2), 1940, pp. 56–68. (Reprinted in [130], pp. 171–183.)
121. — (1941) The Calculi of Lambda-conversion, Princeton University Press, Princeton NJ 1941 [*Annals of Mathematical Studies 6*] (Second printing 1951^2. A condensed, revised version of the Princeton Lecture Notes [118]. Reprinted in [130], pp. 201–255. See also [39], Chapter 9, and [584], Chapters I–IV.)
122. — (1951) *The weak positive implicational calculus* (abstract), The Journal of Symbolic Logic 16 (3), 1951, p. 238. (On the logic counterpart, modulo Curry-Howard [CHC] – i.e., the pure implicational fragment of relevance logic $\mathcal{R}[\rightarrow]$, cf. [16] – of the strict λ-calculus [also known as λI-calculus] of [118], [121]. See also [123] [for relevance logic with propositional quantifiers], [268] [for 'positive' relevance logics with (relevant) conjunction], and [39], Chapter 9, [584], Chapters I–IV, [316], 6C1, etc.)
123. — (1951a) *A weak theory of implication*, in A. Menne, A. Wilhelmy, and H. Angstl (eds.) Kontrolliertes Denken, *Untersuchungen zum Logikkalkül und der Logik der Einzelnwissenschaften (Festgabe zum 60. Geburtstag von Prof. W. Britzelmayr)*, Kommisions Verlag Albert Meiner, Munich 1951, pp. 22–37. (Reprinted in [130], pp. 307–318. On the implicational fragment of relevance logic with propositional quantifiers $\mathcal{R}[\rightarrow,\forall]$. Deduction Theorem for $\mathcal{R}[\rightarrow]$, etc. Cf. [16], [17]. For neighbours and extensions, see Russell's [649] ['extended (classical) propositional logic' (with propositional quantifiers)], resp. Girard's [230] ['System $\mathcal{F}$', i.e. intuitionistic propositional logic with propositional quantifiers]. Cf. [591], [594], [595], [596].)
124. — (1951b) *Minimal logic* (abstract) The Journal of Symbolic Logic 16 (3), 1951, p. 239. (See also [142], [463], [465], and [466].)
125. — (1956) Introduction to Mathematical Logic, 1, Princeton University Press, Princeton NJ 1956. (Many reprints, as, e.g., 1970^6. No other volume in print. See, however, the recent edition [130].)
126. — (1972) *Axioms for functional calculi of higher order*, in Richard S. Rudner, and Israel Scheffler (eds.) Logic and Art, *Essays in Honor of Nelson Goodman*, Bobbs-Merrill Company, Indianapolis IN, and New York 1972, pp. 197–213.
127. — (1980) Letter to Adrian Rezuş, dated Freeport, Bahamas, 8 September 1980.
128. — (1983) Letter to John W. Dawson [25 July 1983], in [683], *Appendix D*, pp. 177–178.
129. — (1984) *Russell's theory of identity of propositions*, Philosophia Naturalis [*Archiv für Naturphilosophie und die philosophischen Grenzgebiete der exakten Wissenschaften und Wissenschaftsgeschichte*] 21 (2/4), 1984, pp. 513–522. (Reprinted in [130], pp. 815–821.)
130. — (2019) Collected Works (edited by Tyler Burge and Herbert B. Enderton), The MIT Press, Cambridge MA, and London & ASL Urbana IL, [April] 2019.
131. Alonzo Church, and J. Barkley Rosser *Some properties of conversion*, Transactions of the American Mathematical Society 39, 1936, pp. 472–482. (Reprinted in [130], pp. 130–138.)
132. Christophorus Clavius [likely, Christoph Klau SJ] (1611) [*Christophori Clavii Bambergensis e Societate Jesu*] Opera mathematica [*V tomis distributa ab auctore nunc denuo correcta et plurimis locis aucta...*], 5 vols., in f°, edited by Reinhard Eltz and published by Anton Hierat, Moguntiae [i.e., Mainz] 1611–[1612]. (Microfilm-edition, issued by the Pius XII Memorial Library, St. Louis University, St. Louis, Missouri. Paper copy consulted in 'The L. E. J. Brouwer Collection',

Provinciale Bibliotheek van Friesland, Leeuwarden NL [formerly, Brouwer's personal library]. Vol. 1: COMMENTARIA In Euclidis Elementa geometrica [ET] In Sphaerica Theodosii..., in two parts: 1.1, pp. 1–638, and 1.2, pp. 1–248, with separate page-numbering.)

133. B. J. [Jack] Copeland (ed.) (1996) **Logic and Reality**, *Essays on the Legacy of Arthur Prior*, Clarendon Press, Oxford 1996.

134. John Corcoran (1974) *Remarks on Stoic deduction*, in John Corcoran (ed.) **Ancient Logic and Its Modern Interpretations**, D. Reidel Publishing Company, Dordrecht 1974, pp. 169–181.

135. Thierry Coquand (2007) *Kolmogorov's contribution to intuitionistic logic* [translated, from French, by Emmanuel Kowalski], in Éric Charpentier, Annick Lesne, and Nikolaï K. Nikolski (eds.) **Kolmogorov's Heritage in Mathematics**, Springer, Berlin etc., 2007, pp. 19–40. (The collection is a translation of *L'Héritage de Kolmogorov en mathématiques*, Éditions Belin, Paris 2004.)

136. Pierre-Louis Curien (1985) **Categorical Combinators, Sequential Algorithms, and Functional Programming**, PhD Dissertation [*Thèse de Doctorat d'État*], Paris 7 [= LITP Report 85-27], Paris 1985. (Also published by Pitman Publishing Ltd., London & Wiley, New York 1986 [*Research Notes in Theoretical Computer Science*]. Second, revised edition, Birkhäuser, Basle etc. 1993.)

137. Haskell B. Curry (1930) *Grundlagen der kombinatorischen Logik*, American Journal of Mathematics 52, 1930, pp. 509–535, 789–834. (Reprinted with an English translation by Fairouz Kamareddine, and Jonathan P. Seldin, as: **Foundations of Combinatory Logic / Grundlagen der kombinatorischen Logik**, College Publications, London 2016 [*Logic PhDs 1*]. This edition contains also the *Errata in Curry's thesis*, pp. 167–169. Originally, *Inauguraldissertation*, Göttingen 1930, under David Hilbert.)

138. — (1948-1949) *A simplification of the theory of combinators*, Synthese 7, 1948-1949, pp. 391–399. (Reviewed by S. C. Kleene, in The Journal of Symbolic Logic 17 (1) [March] 1952, p. 76.)

139. — (1950) **A Theory of Formal Deducibility**, University of Notre Dame, Notre Dame IN 1950 [*Notre Dame Mathematical Lectures 6*]. (Originally, Lectures given at Notre Dame University.)

140. — (1952) *The system* L𝒟, The Journal of Symbolic Logic 17 (1), 1952, pp. 35–42. (The logic of 'strict negation'. The paper contains results reported at the *International Congress of Mathematicians*, Cambridge MA, in 1950.)

141. — (1952a) *On the definition of negation by a fixed proposition in inferential calculus*, The Journal of Symbolic Logic 17 (2), 1952, pp. 98–104.

142. — (1953) *Deduction theorem based on* PB *and* PI, Manuscript, dated 15 July 1953 [Curry *Nachlass*, T 53.07.15 A], 6 pp. ('[T]here is no minimal logic in the sense of Church.' Cf. Church's [124] and the Curry-Church correspondence, cca 1953. See also [464], [466], [751], [752].)

143. — (1963) **Foundations of Mathematical Logic**, McGraw Hill, New York 1963. (Reprinted by Dover Publicatins, Inc., New York 1977.)

144. — (1980) *Some philosophical aspects of combinatory logic*, in Jon Barwise, H. Jerome Keisler, and Kenneth Kunen (eds.) **The Kleene Symposium**, North-Holland Publishing Company, Amsterdam etc. 1980 [*Studies in Logic and the Foundations of Mathematics 101*], pp. 85–101.

145. Haskell B. Curry, and Robert Feys (1959) **Combinatory Logic 1**, North-Holland Publishing Company, Amsterdam 1958 [*Studies in Logic and the Foundations of Mathematics*].

146. Haskell B. Curry, J. Roger Hindley, and Jonathan P. Seldin (1972) **Combinatory Logic 2**, North-Holland Publishing Company, Amsterdam 1972. [*Studies in Logic and the Foundations of Mathematics*].
147. Diederik Ton van Daalen (1980) **The Language Theory of** AUTOMATH, Eindhoven 1980. (PhD Dissertation, University of Technology Eindhoven, under the supervision of N. G. de Bruijn and W. Peremans.)
148. Dirk van Dalen (1980, 2013^5) **Logic and Structure**, Springer-Verlag, Berlin & Heidelberg 1980^1, 1983^2, 1994^3, 2004^4, 2013^5 [*Universitext*].
149. — (1985, 2002) *Intuitionistic logic*, in Dov M. Gabbay, and Franz Guenthner (eds.) **Handbook of Philosophical Logic 3** [first edition], D. Reidel Publishing Company, Dordrecht 1985. pp. 225–340. (Revised version: *Intuitionistic logic*, in Dov M. Gabbay, and Franz Guenthner (eds.) **Handbook of Philosophical Logic 5** [second edition], Kluwer Academic Publishers, Dordrecht 2002, pp. 1–114.)
150. — (1988) *Kolmogorov and Brouwer on constructive implication and the 'Ex Falso' rule* [Russian], Uspexi Matematičeskix Nauk 59 (2), 1988, pp. 53–64. (English original in **Russian Mathematical Surveys** 59 (2), 2004, pp. 247–257.)
151. — (1990) *The war of the frogs and the mice, or the crisis of the 'Mathematische Annalen'*, **The Mathematical Intelligencer** 12 (4), 1990, pp. 17–31.
152. — (1999, 2005) **Mystic, Geometer, and Intuitionist**, *The Life of L. E. J. Brouwer*, 1 *The Dawning Revolution*, 2 *Hope and Disillusion*, Oxford University Press / Clarendon Press, Oxford 1:1999, 2:2005.
153. — (2000) *Brouwer and Fraenkel on intuitionism*, **The Bulletin of Symbolic Logic** 6 (3), [September] 2000, pp. 284–310.
154. — (2000a) *The development of Brouwer's intuitionism*, in [269], pp. 117–152.
155. — (2001) **L. E. J. Brouwer, Een Biografie**, *Het heldere licht van de wiskunde* [L. E. J. Brouwer, A Biography – The Shiny Light of Mathematics] [Dutch], Prometheus – Bert Bakker, Amsterdam 2001 (reprint 2002, etc.)
156. — (2001a) *Intuitionistic logic*, in Lou Goble (ed.) **Philosophical Logic**, Blackwell, Oxford 2001, pp. 224–257.
157. — (2008) *A bibliography of L. E. J. Brouwer*. in [27], pp. 343–390. (A revised version of **Utrecht Logic Group Preprint Series 175**, 1997, 'updated April 14, 2008'.)
158. — (2013) **L. E. J. Brouwer – Topologist, Intuitionist, Philosopher**, *How Mathematics is Rooted in Life*, Springer, London 2013. (A revised version of [152].)
159. Martin Davis (ed.) (1965) **The Undecidable**, *Basic Papers on Undecidable Propositions, Unsolvable Problems and Computable Functions*, Raven Press, Hewlett, New York 1965 (reprinted by Dover Publications, Inc., Mineola NY 2004^R [*Dover Books on Mathematics*]).
160. William Demopoulos (ed.) (1995) **Frege's Philosophy of Mathematics**, Harvard University Press, Cambridge MA 1995.
161. Michael Detlefsen (1986) **Hilbert's Program**, *An Essay on Mathematical Instrumentalism*, D. Reidel Publishing Company, Dordrecht, etc. 1986 [*Synthese Library 182*]. (Reviewed by David A. Auerbach, in **The Journal of Symbolic Logic** 54 (2) [June] 1989, pp. 620–622.)
162. — (1990) *Brouwerian intuitionism*, **Mind** 99 (396), 1990, pp. 501–534. (Reprinted in [167], pp. 208–250.)
163. — (1992) *Poincaré against the logicians*, **Synthese** 90 (3), 1992, pp. 349–378.
164. — (1992a) *On an alleged refutation of Hilbert's Program using Gödel's first incompleteness theorem*, in [168] pp. 199–235.
165. — (2011–2012) *Poincaré versus Russell sur le rôle de la logique dans les mathématiques* (traslated by Sébastien Gandon), **Les Études Philosophiques** 2 (97), 2011–2012, pp. 153–178.

166. Michael Detlefsen, and Andrew Arana (2011) *Purity of methods*, Philosopher's Imprint 11 (2) [January] 2011, pp. [1–39].
167. Michael Detlefsen (ed.) (1992) Proof and Knowledge in Mathematics, Routledge, London & New York 1992.
168. — (1992a) Proof, Logic and Formalization, Routledge, London & New York 1992.
169. Charles Lutwidge Dodgson (1879) Euclid and His Modern Rivals, Macmillan & Co, London 1879, 1885^2. (Reprints: Dover Publications, Mineola NY 1973^R, 2004^R, Cambridge University Press, New York etc. 2009^R, 2016^R [digital] etc.)
170. Kosta Došen (1992) *The first axiomatization of relevant logic*, Journal of Philosophical Logic 21 (4), 1992, pp. 339–356.
171. — (1993) *A historical introduction to substructural logics*, in Peter Schroeder-Heister, and Kosta Došen (eds.) Substructural Logics, Clarendon Press, Oxford 1993, pp. 1–36.
172. Al'bert Grigor'evič Dragalin (1979) Matematičeskiy intuicionizm, Vvedenie v teoriyu dokazatel'stv [Mathematical Intuitionism, Introduction to Proof Theory] [Russian], Izdatel'stvo 'Nauka' [Science Publishers], Moskow 1979 [*Matematičeskaĭa logika i osnovaniĭa matematiki*]. (Translated into English, by Eliott Mendelson, as Mathematical Intuitionism, Introduction to Proof Theory, AMS, Providence RI 1988 [*Translations of Matematical Monographs 67*]. Reviewed by Ieke Moerdijk, Bulletin of the American Mathematical Society (n.s.) 22 (2) [April] 1990, pp. 301–304.)
173. Lech Dubikajtis (1967) *Stanisław Jaśkowski*, Wiadomości Matematyczne 10, 1967, pp. 15–28.
174. — (1967a) *Stanisław Jaśkowski (1906–1965)*, Ruch Filozoficzny 25 (3–4), 1967, pp. 187–198.
175. — (1975) *The life and works of Stanisław Jaśkowski*, Studia Logica 34, (2), 1975, pp. 109–116.
176. Richard Dufour (ed., tr.) (2004) CHRYSIPPE Œuvre philosophique 1–2 [texte grec et français] (Textes traduits et commentés par Richard Dufour), Les Belles Lettres, Paris 2004 [*Fragments*].
177. Michael [M.A.E.] Dummett (1959) *A propositional calculus with denumerable matrix*, The Journal of Symbolic Logic 24 (2), 1959, pp. 97–107.
178. — (1973) Frege, Philosophy of Language, Duckworth, London 1973. (Also: Harper & Row Publishers, New York 1973.)
179. — (1981) The Interpretation of Frege's Philosophy, Duckworth, London 1981.
180. — (1991) Frege and Other Philosophers, Oford University Press, Oxford 1991.
181. — (1991a) Frege, Philosophy of Mathematics, Duckworth, London 1991.
182. — (1997) Elements of Intuitionism, Clarendon Press, Oxford 1997, 2000^2 [*Oxford Logic Guides 39*].
183. Glenn Durfee (1997) A Model for a List-oriented Extension of the Lambda-calculus. Master's Thesis (supervised by Dana Scott), School of Computer Science, Carnegie Mellon University, Pittsburgh PA, May 1997 [= CMU-CS-97-151].
184. Umberto Eco (2019) On the Shoulders of Giants (translated by Alastair McEwen), Harvard University Press [Belknap Press], Cambridge MA 2019 [ISBN 978-0-674-24089-6]. (See also: On the Shoulders of Giants, *The Milan Lectures*, Penguin Random House, London [November] 2019.)
185. Urs Egli (1967) Zur stoischen Dialektik, Sandoz, Basle 1967. (*Inauguraldissertation Phil.*, University of Bern 1967.)
186. — (1978) *Stoic syntax and semantics*, in Jacques Brunschwig (ed.) Les Stoïciens et leur logique, Librairie Philosophique J. Vrin, Paris 1978, pp. 135–154. (Also 2006^2, *deuxième édition révue, augmentée et mise à jour*, ibid., pp. 131–148.)

308   References

187. — (1979) *The Stoic concept of anaphora*, in Rainer Bäuerle, Urs Egli, and Arnim von Stechow (eds.) **Semantics from Different Points of View**, Springer, Berlin, etc. 1979 [*Springer Series in Language and Communication 6*], pp. 266–283.
188. — (1993) *Neue Züge im Bild der stoischen Logik*, in Klaus Döring, and Theodor Ebert (eds.) **Dialektiker und Stoiker**. *Zur Logik der Stoa und ihrer Vorläufer*, Franz Steiner Verlag, Stuttgart 1993, pp. 129–139.
189. — (2000) *Anaphora from Athens to Amsterdam*, in Klaus von Heusinger, and Urs Egli (eds.) **Reference and Anaphoric Relations**, Kluwer [Springer], Dordrecht 2000 [*Studies in Linguistics and Philosophy 27*], pp. 17–29.
190. Mihai Eminescu (1879) [*Noi amândoi avem același dascăl*] [We – both of us – have gotten the same teacher] [Romanian] in Dumitru Panaitescu-Perpessicius (ed.) MIHAI EMINESCU **Opere** [Works] 4, *Poezii postume* [Posthumous poems] [Romanian] (A critical edition), [MS 2.279, 74–75], Editura Academiei Republicii Populare Romîne [The Publishing House of the Romanian Academy], Bucharest 1952.
191. Erwin Engeler (2015) *A forgotten theory of proofs?* ETH Zürich, 13 February 2015 [forthcoming]. (On MacLane's Göttingen PhD Dissertation of 1933, [444].)
192. John Etchemendy (1990) **The Concept of Logical Consequence**, Harvard University Press, Cambridge MA 1990.
193. William Ewald (2005) *Hilbert's wide program*, in René Cori, Alexander Razborov, Stevo Todorčević, and Carol Wood (eds.) **Logic Colloquium 2000**, ASL, Urbana IL & A. K. Peters Ltd., Wellesley MA 2005 [*Lecture Notes in Logic 19*], pp. 228–251.
194. William Ewald (ed., tr.) (1996) **From Kant to Hilbert**, *A Sourcebook in the Foundations of Mathematics*, 1–2, Clarendon Press, Oxford 1996 (reprinted 2005).
195. Solomon Feferman (1979) *What does logic have to tell us about mathematical proofs?*, **Mathematical Intelligencer 2**, 1979, pp. 20–24. (Revised in [197], Chapter 9, pp. 177–186.)
196. — (1996) *Kreisel's 'unwinding' program*, in Piergiorgio Odifreddi (ed.), **Kreiseliana**, *About and Around Georg Kreisel*, A. K. Peters, Wellesley MA 1996, pp. 247–273.
197. — (1998) **In the Light of Logic**, Oxford University Press, Oxford etc. 1998 [*Logic and Computation in Philosophy*].
198. Matthias Felleisen (1987) **The Calculi of λ_v-CS Conversion**, *A Syntactic Theory of Control and State in Imperative Higher Order Programming Languages*, PhD Dissertation, Indiana University, Bloomington IN 1987 [= Technical Report 226 Indiana University, Department of Computer Science, Bloomington IN, August 1987, xvii + 230 pp.]. (See also: *A syntactic theory of sequential control* [with Daniel P. Friedman, Eugene Kohlbecker, and Bruce Duba], in **Theoretical Computer Science 52**, 1987, pp. 205–237, and *The revised report on the syntactic theories of sequential control and state* [with Robert Hieb], Rice University, Department of Computer Science, Houston TX = Technical Report 89-100, December 1989, 30 pp.)
199. Frederic Brenton Fitch (1934) **A System of Symbolic Logic that Avoids the Paradoxes without a Theory of Types**, PhD Dissertation, Yale University 1934.
200. — (1936) *A system of formal logic without an analogue to the Curry W operator*, **The Journal of Symbolic Logic 1** (3), 1936, pp. 92–100.
201. — (1952) **Symbolic Logic**, *An Introduction*, Ronald Press, New York 1952.

202. Branden Fitelson (2001) *New elegant axiomatizations of some sentential logics*, Department of Philosophy, Stanford University 2001.
203. — (2012) [OTTER-derivations in pure Abelian logic], Letter [*e-mail*] to Adrian Rezuş, 13 October 2012.
204. Henry George Forder (1968) *Groups from one axiom*, The Mathematical Gazette 52 (381), [October] 1968, pp. 263–266
205. Miriam Franchella (1994) *Heyting's contribution to the change in research into the foundations of mathematics*, History and Philosophy of Logic 15 (2), 1994, pp. 149–172.
206. — (1994a) *Brouwer and Griss on intuitionistic negation*, Modern Logic 4 (3), 1994, pp. 256–265.
207. — (1994b) L. E. J. Brouwer pensatore eterodosso, *L'intuizionismo tra matematica e filosofia*, Guerini e Associati, Milan 1994.
208. — (1995) *L. E. J. Brouwer towards intuitionistic logic*, Historia Mathematica 22 (3), 1995, pp. 304–322.
209. — (1995a) *Like a bee on a windowpane: Heyting's reflections on solipsism*, Synthese 105 (2) [November] 1995, pp. 207–251.
210. — (1995b) Il terzo escluso: genesi e critiche di un principio, Edizioni dell'Arco, Milan 1995.
211. — (2007) *Philosophies of intuitionism: why we need them*, Teorema 26 (1), 2007, pp. 73–82.
212. — (2008) Con gli occhi negli occhi di Brouwer, *Filosofie della matematica a confronto con l'intuizionismo*, Polimetrica [International Scientific Publisher], Monza - Milan 2008 [*Ratio – Studi e testi di filosofia contemporanea*].
213. — (2018) *Shaping the enemy: foundational labelling by L. E. J. Brouwer and A. Heyting*, History and Philosophy of Logic 40 (1), 2018, pp. 152–181.
214. Michael Frede (1974) Die stoische Logik, Vandenhoeck & Ruprecht, Göttingen 1974 [*Abhandlungen der Akademie der Wissenschaften zu Göttingen, 3. Folge, 88, Philologisch-Historische Klasse*] (*Habilitationsschrift*, Göttingen 1972).
215. — (1974a) *Stoic vs. Aristotelian syllogistic*, Archiv für Geschichte der Philosophie 56, 1974, 1–32. Reprinted in MICHAEL FREDE Essays in Ancient Philosophy, Oxford University Press, Oxford 1987, pp. 99–124.
216. Gottlob Frege (1879) Begriffsschrift, *Eine der arithmetischen nachgebildete Formelsprache des reinen Denkens*, Verlag Louis Nebert, Halle a/S 1879. [**BS**] (There are two English translations of **BS**: *Concept Script* (translated by Stefan Bauer-Mengelberg), in [288], and: Conceptual Notation and Related Articles, with a biography and introduction (translated by Terrell Ward Bynum), Clarendon Press, Oxford & Oxford University Press, New York NY 1972 [*Oxford Scholarly Classics*].)
217. — (1884) Die Grundlagen der Arithmetik, *Eine logisch-mathematische Untersuchung über den Begriff der Zahl*, Wilhelm Koebner, Breslau 1884. (English translation: [The Foundations of Arithmetic, *A Logico-mathematical Enquiry into the Concept of Number*], by J. L. Austin, Blackwell & Mott. Ltd, London 1950. Revised edition Blackwell, Oxford 1952, 1959^R, reprinted by Harper & Brothers, New York 1960, Northwestern University Press, Evanston IL 1980, etc. Romanian translation [Fundamentele aritmeticii, *o cercetare logico-matematică asupra conceptului de număr*], by Sorin Vieru, Humanitas, Bucharest 2000.)
218. — (1893–1903) Grundgesetze der Arithmetik, *begriffsschriftlich abgeleitet*, I. Band, Verlag von Hermann Pohle, Jena 1893; II. Band, *ibid.* 1903. [**GGA:1**], [**GGA:2**] (Anastatic reprint by Olms Verlag, Hildesheim 1962. Reset in modern notation

as Grundgesetze der Arithmetik, *begriffsschriftlich abgeleitet*, Band I und II (In moderne Formelnotation transkribiert und mit einem ausführlichen Sachregister versehen von Thomas Müller, Bernhard Schröder und Rainer Stuhlmann-Laeiz), MENTIS, Paderborn 2009. **GGA** has been partially translated into English as The Basic Laws of Arithmetic (tr. by Montgomery Furth), University of California Press, Berkeley and Los Angeles CA 1964. A complete English translation of **GGA** has been issued as Basic Laws of Arithmetic, *Derived using Concept-script*, Volumes I & II [in one] (translated and edited by Philip A. Ebert & Marcus Rossberg, with Crispin Wright), Oxford University Press, Oxford 2013.)

219. Hans Freudenthal (1937) *Zur intuitionistischen Deutung logischer Formeln*, Compositio Mathematica 4, 1937, pp. 112–116. (Received December 27, 1934. Reprinted in [300] pp. 331–335, followed by the *Remarks* of Heyting, [293], dated October 16, 1935, and a last word [*Nachwort*] by the author [six lines], dated April, 29, 1936 [= Compositio Mathematica 4, 1937, pp. 117–118, reprinted in [300], pp. 336–337]. The issue containing the paper and the exchange with Heyting has appeared in 1936. See also the review by A. N. Kolmogorov in Zentralblatt für Mathematik und ihre Grenzgebiete, 0015.24201. For a previous [1930] debate on the subject, as reflected in the Freudenthal-Heyting correspondence, see [758], pp. 203–208.)

220. Laurence S. Gagnon (1976) NOR *logic: a system of natural deduction*, Notre Dame Journal of Formal Logic 17 (2), 1976, pp. 293–294.

221. Galenus [John Spangler Kieffer] (1964) GALEN's Institutio Logica (Translation by John Spangler Kieffer), Johns Hopkins, Baltimore 1964.

222. Ángel Garrido, and Urszula Wybraniec-Skardowska (eds.) (2018) **The Lvov-Warsaw School, Past and Present**, Birkhäuser [Springer International Publishing], Basle etc. [June] 2018 [*Studies in Universal Logic*].

223. Gerhard Gentzen (1933) *Über die Existenz unabhängiger Axiomensysteme zu unendlichen Satzsystemen*, Mathematische Annalen 107, 1933, pp. 329–350. (English translation [*On the existence of independent axioms systems for infinite sentence systems*], by Manfred E. Szabo, in [225] pp. 29–52.)

224. — (1934–1935) *Untersuchungen über das logische Schliessen*, Mathematische Zeitschrift 39, 1934–1935, pp. 176–210, and 405–431. (French translation [*Recherches sur la déduction logique*], with comments, by Robert Feys and Jean Ladrière, Presses Universitaires de France, Paris 1955. Russian translation [*Issledovaniĭa logičeskix vyvodov*], by A. V. Idel'son, in A. V. Idel'son, and G. E. Minc (eds., trs.) Matematičeskaĭa teoriĭa logičeskogo vyvoda [The Mathematical Theory of Logical Deduction], Izdatel'stvo 'Nauka' [Science Publishers], Moskow 1967 [*Matematičeskaĭa logika i osnovaniĭa matematiki*], pp. 9–76. English translation [*Investigation into logical deduction*], by Manfred E. Szabo, in [225] pp. 68–131. Originally, *Inauguraldissertation*, University of Göttingen 1933, under David Hilbert [the actual thesis supervisor was Paul Bernays] See also Gentzen's *Nachlass*, in [534], and the review [621].)

225. — (1969) Collected Papers [= *The Collected Papers of Gerhard Gentzen*], edited and translated by Manfred E. Szabo, North-Holland Publishing Company, Amsterdam, etc. 1969 [*Studies in Logic and the Foundations of Mathematics 55*]. (Reviewed by Georg Kreisel, in The Journal of Philosophy 68 (8), 1971, pp. 238–265.)

226. G. N. Georgacarakos (1982) *The semantics of minimal intuitionism*, Logique et Analyse 25 (100), 1982, pp. 383–397.

227. Jan Herman Geuvers, and Rob P. Nederpelt (2014) **Type Theory and Formal Proof, An** *Introduction* (with a Foreword by Henk Barendregt), Cambridge University Press, Cambridge UK 2014.
228. Donald Gillies (1992) *The Fregean revolution in logic*, in Donald Gillies (ed.) **Revolutions in Mathematics**, Clarendon Press, Oxford & Oxford University Press, New York 1992, pp. 265–305.
229. Jean-Yves Girard (1971) *Une extension de l'interprétation de Gödel à l'analyse, et son application à l'élimination des coupures dans l'analyse et la théorie des types*, in Jens-Erik Fenstad (ed.) **Proceedings of the Second Scandinavian Logic Symposium**, North-Holland Publishing Company, Amsterdam 1971 [*Studies in Logic and the Foundations of Mathematics 63*] pp. 63–92.
230. — (1972) **Interprétation fonctionelle et élimination des coupures de l'arithmétique d'ordre supérieur**, PhD Dissertation [*Thèse de Doctorat d'État*], University of Paris 7, Paris 1972.
231. — (1987) **Proof Theory and Logical Complexity, 1**, Bibliopolis, Naples 1987 [*Studies in Proof Theory 1*]. (No other volume in print.)
232. — (1987a) *Linear logic*, **Theoretical Computer Science 50** (1), 1987, pp. 1–101. (Cf. the comments of [597].)
233. — (1989) **Proofs and Types**, Cambridge University Press, Cambridge UK 1989. (Revised Lecture Notes for a *Cours de DEA* [*Diplôme d'Études Approfondies*], given at the University of Paris 7, 1986–1987, translated into English by Paul Taylor and Yves Lafont.)
234. — (2000) *Du pourquoi au comment : la théorie de la démonstration de 1950 à nos jours*, in Jean-Paul Pier (ed.) **Development of Mathematics 1950–2000**, Birkhäuser / Springer Verlag, Basle etc. 2000. (Preprint, Marseille 1997 [34 pp.].)
235. — (2011) **The Blind Spot**, *Lectures on Logic*, European Mathematical Society, ETH Zürich 2011. (A revised set of Lectures on Proof Theory given, in 2004, at the University of Rome 3, issued previously in French, as **Le Point aveugle**, *Cours de logique* [1: *Vers la perfection*, 2: *Vers l'imperfection*], by Hermann, Paris 1:2006, 2:2007 [*Visions des sciences*]. Reviewed by Christian Retoré, and Thomas Seiller, in **La Gazette des Mathématiciens – Société Mathématique de France** (Paris) **146**, October 2014, pp. 137–144.)
236. — (2016) **Le Fantôme de la transparence**, Éditions Allia, Paris 2016. (A popular, colloquial summary of the author's discomfort with and allergies to the Tarskian model theory and to the Anglo-Saxon analytic philosophy.)
237. Valeriĭ Ivanovič Glivenko (1928) *Sur la logique de M. Brouwer*, **Académie Royale de Belgique**, *Bulletin de la Classe des Sciences* (5) **14**, 1928, pp. 225–228.
238. — (1929) *Sur quelques points de la logique de M. Brouwer*, **Académie Royale de Belgique**, *Bulletin de la Classe des Sciences* (5) **15**, 1929, pp. 183–188. (English translation by A. Rocha, in [445], pp. 301–305.)
239. — (1938) **Théorie générale des structures**, Hermann, Paris 1938 [*Actualités scientifiques et industrielles 652* (= *Exposés d'analyse générale IX*)].
240. Kurt Gödel (1931) *Über formal unentscheidbare Sätze der Principia Mathematica und verwandter Systeme I*, **Monatshefte für Mathematik und Physik 38**, 1931, pp. 173–198. (English translation [*On formally undecidable propositions of 'Principia Mathematica' and related systems I*], by Jean van Heijenoort, in [288], and in [248], 1, *Publications 1929–1936*, 1986, pp. 144–195. The first English translation, by Bernard Meltzer, with an introduction by R. B. Braithwaite, was published by Basic Books Inc., New York 1962 [reprint Dover Publications Inc., Mineola NY 1992]. A second English traslation, by Elliott Mendelson, appeared in [159].)

241. — (1932) *Zum intuitionistischen Aussagenkalkül*, Anzeiger der Akademie der Wissenschaften in Wien 69, 1932, pp. 65–66. (Reprinted, with an additional comment, in Ergebnisse eines mathematischen Kolloquiums 4, 1933, p. 40, and in [248], 1, *Publications 1929–1936*, 1986, pp. 222–224. English translation [*On the intuitionistic propositional calculus*], in *ibid.*, pp. 223–225. *Introductory Note* by Anne S. Troelstra, *ibid.*, pp. 222–223.)

242. — (1933e) *Zur intuitionistischen Logik und Zahlentheorie*, Ergebnisse eines mathematischen Kolloquiums 4, 1933, pp. 34–38. (Reprinted in [248], 1, *Publications 1929–1936*, 1986, pp. 286–294. English translation [*On the intuitionistic logic and arithmetic*], by Stefan Bauer-Mengelberg and Jean van Heijenoort, in *ibid.*, pp. 287–295. *Introductory Note* by Anne S. Troelstra, *ibid.*, pp. 282–287.)

243. — (1933f) *Eine Interpretation des intuitionistischen Aussagenkalkül*, Ergebnisse eines mathematischen Kolloquiums 4, 1933, pp. 39–40. (Reprinted in [248], 1, *Publications 1929–1936*, 1986, pp. 300–302. English translation [*An interpretation of the intuitionistic propositional calculus*], in *ibid.*, p. 301. *Introductory Note* by Anne S. Troelstra, *ibid.*, pp. 296–299.)

244. — (1941) *In what sense is intuitionistic logic constructive?*, in [248], 3, *Unpublished Essays and Lectures*, 1995, pp. 189–200. (With an *Introductory Note*, by Anne S. Troelstra, *ibid.*, pp. 186–189. Lecture given at Yale University, April 14, 1981; an early account of the 'Dialectica interpretation'.)

245. — (1944) *Russell's mathematical logic*, in P. A. Schilpp (ed.) The Philosophy of Bertrand Russell, Open Court Publishing Company, La Salle IL & Northwestern University Press, Evanston IL 1944 [*Library of Living Philosophers* 5], pp. 123–153. (Reprinted in Paul Benacerraf and Hilary Putnam (eds.), Philosophy of Mathematics, *Selected Readings*, Prentice Hall, Englewood Cliffs NJ 1964 (= [second edition], Cambridge University Press, Cambridge UK 1984^2, pp. 447–469) and in [248], 2, *Publications 1938–1974*, 1990, pp. 119–141, with an *Introductory Note*, by Charles Parsons, *ibid.*, pp. 102–118.)

246. — (1958) *Über eine bisher noch nicht benützte Erweiterung des finiten Standpunktes*, Dialectica 12, 1958, pp. 280–287. (Reprinted in [248], 2, *Publications 1938–1974*, 1990, pp. 241–251. English translation [*On a hitherto unutilized extension of the finitary standpoint*] in *ibid.*, pp. 241–251. *Introductory Note* to [246], by Anne S. Troelstra, *ibid.*, pp. 217–241.)

247. — (1972) *On an extension of finitary mathematics which has not yet been used*, in [248], 2, *Publications 1938–1974*, 1990, pp. 271–280. (A revised and expanded English translation of [246]. See also the *Introductory Note* to [246], by Anne S. Troelstra, in [248], 2, *Publications 1938–1974*, 1990, pp. 217–241.)

248. — (1986–2003) Collected Works, 1–5, edited by Solomon Feferman et al., Oxford University Press, Oxford 1986–2003.

249. Wenceslao J. Gonzalez (1991) *Intuitionistic mathematics and Wittgenstein*, History and Philosophy of Logic 12 (?), 1991, pp. 167–183.

250. Josiah B. Gould (1974) *Deduction in Stoic logic*, in John Corcoran (ed.) Ancient Logic and Its Modern Interpretations, D. Reidel Publishing Company, Dordrecht 1974, pp. 151–168.

251. Ivor Grattan-Guinness (2000) The Search for Mathematical Roots, 1870–1940, *Logics, Set Theories and the Foundations of Mathematics from Cantor through Russell to Gödel*, Princeton University Press, Princeton NJ and Oxford 2000. (Reviewed by William Ewald, in Bulletin of the American Mathematical Society (n.s.) 40 (1), 2002, pp.125–129.)

252. Bob Hale, and Crispin Wright (2001) **The Reason's Proper Study**, *Essays Towards a Neo-Fregean Philosophy of Mathematics*, Clarendon Press, Oxford & Oxford University Press, New York 2001. (See also [804].)
253. John Halleck (2010) Letter [*e-mail*] to Adrian Rezuş et al., 23 September 2010.
254. — (2012) [CD-derivations in pure Abelian logic], Letter [*e-mail*] to Adrian Rezuş, 15 October 2012.
255. — (2015) *Replacement G2, checked with an evaluator*, Letter [*e-mail*] to Adrian Rezuş, 12 May 2015.
256. Rudolf Haller (1993) **Neopositivismus**, *Eine historische Einführung in die Philosophie des Wiener Kreises*, Wissenschaftliche Buchgesellschaft, Darmstadt 1993.
257. Michael Hallett (1994) *Hilbert's axiomatic method and the laws of thought*, in Alexander George (ed.) **Mathematics and Mind**, Oxford University Press, New York 1994, pp. 158–200.
258. — (1995) *Hilbert and logic*, in Mathieu Marion and Robert Cohen (eds.) **Québec Studies in the Philosophy of Science, 1: Logic, Mathematics, Physics and the History of Science**, Kluwer Academic Publishers, Dordrecht 1995 [*Boston Studies in the Philosophy of Science 177*], pp. 135–187.
259. Daniel Hartwig (2010) **Guide to the Georg Kreisel Papers** (SC0136), Stanford University Libraries, Deptartment of Special Collections and University Archives, Stanford CA, October 2010.
260. — (2010a) **Georg Kreisel: Correspondence with Jean van Heijenoort** (SC0233), Stanford University Libraries, Department of Special Collections and University Archives, Stanford CA, October 2010.
261. Allen P. Hazen (1995) *Is even minimal negation constructive?*, **Analysis 55** (2), 1995, pp. 105–107.
262. Allen P. Hazen, and Francis Jeffry Pelletier (2012) *Natural deduction*, in Dov M. Gabbay, Francis Jeffry Pelletier, and John Woods (eds.) **Handbook of the History of Logic 11** (*A History of Its Central Concepts*), Elsevier / North-Holland Publishing Company, Amsterdam etc. 2012, pp. 341–414.
263. — (2014) *Gentzen and Jaśkowski natural deduction: fundamentally similar but importantly different*, **Studia Logica 102** (6), 2014, pp. 1103–1142.
264. Richard G. Heck Jr. (1993) *The development of arithmetic in Frege's 'Grundgesetze der Arithmetic'*, **The Journal of Symbolic Logic 58** (2), [June] 1993, pp. 579–601. (Revised as Chapter 2 in [265].)
265. — (2011) **Frege's Theorem**, Clarendon Press, Oxford & Oxford University Press, New York 2011
266. — (2013) **Reading Frege's Grundgesetze**, Clarendon Press, Oxford & Oxford University Press, New York 2013.
267. Georg Wilhelm Friedrich Hegel (1812–1816) **Wissenschaft der Logik**, Schrag, Nürnberg 1.1:1812, 1.2:1813, 2:1816. (Modern 'textkritische Edition' by Friedrich Hogemann and Walter Jaeschke, Meiner Verlag, Hamburg 1978–1981.)
268. Glen H. Helman (1977) **Restricted Lambda-abstraction and the Interpretation of Some Non-classical Logics**, University Microfilms, Ann Arbor MI 1977. (PhD Dissertation, University of Pittsburgh, under the supervision of Nuel Belnap Jr. Typescript [iv + 217 pp.], dated Pittsburgh PA 1977. Curry-Howard Correspondence for 'positive' relevance logics, containing relevant implication and conjunction alone. For a summary, see [17], Chapter XI, § 71 [*Relevant implication and relevant functions*], pp. 402–423.)
269. Vincent F. Hendricks, Stig Andur Pedersen, and Klaus Frovin Jørgensen (eds.) (2000) **Proof Theory: History and Philosophical Significance**, Kluwer Academic Publishers, Dordrecht 2000 [*Synthese Library 292*].

270. Jacques Herbrand (1930) *Recherches sur la théorie de la démonstration*, PhD Dissertation, University of Paris (Sorbonne) 1930 [dated 14 April 1929, defended 11 June 1930], 128 pp. © [3]. (Also: **Prace Towarzystwa Naukowego Warszawskiego, Wydział III, 33** [= Travaux de la société des Sciences et des Lettres de Varsovie, Classe III, Sciences Mathématiques et Physiques, 33, Varsovie 1930. Reprinted in [272], pp. 35–153. Partial English translation [Chapter V: *Investigations in proof theory: The properties of true propositions*], with comments and notes, by Burton Dreben and Jean van Heijenoort, in [288], pp. 525–581. Full English translation [*Investigations in proof theory*], by Warren Goldfarb, in [273], pp. 44–202. On Herbrand's version of the Deduction Theorem [for first-order classical logic], see Chapter III, 2.4 [= Dissertation, pp. 61–62, resp. [272], pp. 90–91, [273], pp. 107–108]. Cf. also [271], [558], and [251], p. 550.)

271. — (1930a) *Les bases de la logique hilbertienne*, **Revue de Métaphysique et de Morale 37**, 1930, pp. 243–255. (Reprinted in [272], pp. 155–166. English translation [*The principles of Hilbert's Logic*], by Warren Goldfarb, in [273], pp. 203–214.)

272. — **Écrits logiques** (edited by Jean van Heijenoort), Presses Universitaires de France, Paris 1968. (Reviewed by Paul Bernays, in **The Journal of Symbolic Logic 36** (3), 1971, pp. 523–524.)

273. — **Logical Writings** (edited [and translated] by Warren Goldfarb), D. Reidel Publishing Company, Dordrecht 1971. (An English translation of [272].)

274. Paul Hertz (1922) *Über die Axiomensysteme kleinster Satzzahl für ein gewisses System von Sätzen und den Begriff des idealen Elementes*, **Jahresbericht der deutschen Mathematikervereinigung** (2. Abteilung), **31**, 1922, pp. 154–157.

275. — (1922a) *Über Axiomensysteme für beliebige Satzsysteme, I. Teil. Sätze ersten Grades*, (*Über die Axiomensysteme von der kleinsten Satzzahl und den Begriff des idealen Elementes*), **Mathematische Annalen 87**, 1922, pp. 246–269. (Dated Göttingen, 15 September 1921, received: 25 October 1921. Translation [*On axiomatic systems for arbitrary systems of sentences*], by Javier Legris, in Jean-Yves Béziau (ed.) **Universal Logic,** *An Anthology. From Paul Hertz to Dov Gabbay*, Birkäuser, Basle 2012, pp. 11–29, with an Introduction by the translator, i.e., [419], pp. 3–10.)

276. — (1923) *Über Axiomensysteme für beliebige Satzsysteme, II. Teil. Sätze höheren Grades*, **Mathematische Annalen 89**, 1923, pp. 76–100, and 246–269. (Dated Göttingen 20 May 1922, received: 2 June 1922.)

277. — (1923a) *Über das Denken und seine Beziehung zur Anschauung, Erster Teil: Über den funktionalen Zusammenhang zwischen auslösendem Erlebnis und Enderlebnis bei elementaren Prozessen*, Julius Springer, Berlin 1923. (*Vorwort und Einleitung*, pp. i–x, dated Göttingen, July 1923. A book on congnitive science, somewhat avant la lettre, otherwise with no echo in the epoch. No 'Part II' in print.)

278. — (1928) *Reichen die üblichen syllogistischen Regeln für das Schließen in der positiven Logik elementarer Sätze aus?*, **Annalen der Philosophie und philosophischen Kritik (Leipzig)** 7, 1928, pp. 272–277. (Dated Göttingen, 16 July 1928.)

279. — (1929) *Über Axiomensysteme beliebiger Satzsysteme*, **Annalen der Philosophie und philosophischen Kritik (Leipzig)**, **8**, 1929, pp. 178–204. (Text of a conference given at a meeting of the *Deutsche Matematikervereinigung* held at the University of Hamburg in September 1928, with small additions.)

280. — (1929a) *Über Axiomensysteme für beliebige Satzsysteme* [III], **Mathematische Annalen 101**, 1929, pp. 457–514. (Received: 30 May 1928.)

281. — (1929b) *Über Axiomensysteme vor Satzsysteme*, **Jahresbericht der deutschen Mathematikervereinigung** (2. Abteilung), **38**, 1929, pp. 45–46.

282. — (1931–1932) *Vom Wesen des Logischen, insbesondere der Bedeutung des 'modus barbara'*, Erkenntnis 2, 1931–1932, pp. 369–392.

283. — (1935) *Sur la nature de la logique, de ses catégories et de ses vérités*, L'Enseignement Mathématique 34, 1935, pp. 95–97. (Summary of a conference presented at the University of Geneva, on 22 June 1934, in a series of *Conférences internationales des sciences mathématiques : Colloque sur la logique mathématique*, 18–23 June 1934. The first part of the conference is published as [284], the second part develops ideas of [282].)

284. — (1935a) *Über das Wesen der Logik und der logischen Urteilsformen*, Abhandlungen der Fries'schen Schule, Neue Folge (Göttingen), 6 (part 2), 1935, pp. 225–272. (New Series published by *Verlag Öffentliches Leben*, Inh. Erich Irmer, Berlin. The reference contains an 'extended presentation' of the first part of the Geneva 1934 conference, [283].)

285. — (1937–1938) *Sprache und Logik*, Erkenntnis 7, 1937–1938, pp. 309-324. (Communication to the Fourth International Congress for the Unity of Science, Cambridge MA 1938.)

286. Dennis E. Hesseling (2003) **Gnomes in the Fog**, *The Reception of Brouwer's Intuitionism in the 1920s*, Birkhäuser, Basel 2003 [*Science Networks, Historical Studies 28*]. (A revised version of the author's PhD Dissertation, Utrecht 1999. Reviewed by Mark van Atten, in The Bulletin of Symbolic Logic 10 (3), 2004, pp. 423–427, and by Jeremy Avigad, in The Mathematical Intelligencer 28 (4), 2006, pp. 71–74. See also the comments of [614].)

287. Jean van Heijenoort (1967) *Logic as calculus and logic as language*, in Robert Sonné Cohen, and Marx W. Wartofsky (eds.) **Proceedings of the Boston Colloquium for the Philosophy of Science 1964–1966**, Springer, Dordrecht 1967 [*Boston Studies in the Philosophy of Science 3*], pp. 440–446.

288. — Jean van Heijenoort (ed.) (1967) **From Frege to Gödel**, *A Source Book in Mathematical Logic, 1879–1931*, Harvard University Press, Cambridge MA 1967 [reprinted 1971].

289. Arend Heyting (1930) *Die formalen Regeln der intuitionistischen Mathematik*, I–III, Sitzungsberichte der Preussischen Akademie der Wissenschaften [*Physikalische-Mathematische Klasse*], 1930, I: pp. 52–56, II: pp. 57–71, III: pp. 158–169. (Reprinted in [300], pp. 191–205. Excerpts in [755], pp. 153–162. English translation of I, in [445], pp. 311–327.)

290. — (1930a) *Sur la logique intuitionniste*, Académie Royale de Belgique, *Bulletin de la Classe des Sciences* (5) 16, 1930, pp. 957–963. (Reprinted in [300]. English translation in [445], pp. 306–310.)

291. — (1931) *Die intuitionistische Grundlegung der Mathematik*, Erkenntnis 2, 1931, pp. 106–115; reprinted in [300], pp. 240–250.

292. — (1934) **Mathematische Grundlagenforschung**, *Intuitionismus, Beweistheorie*, Springer Verlag, Berlin, etc. 1934 (reprographic reprint 1974^{R}) [*Ergebnisse der Mathematik und ihre Grenzgebiete 3.4*].

293. — (1937) *Bemerkung zu dem Aufsatz von Herrn Freudenthal 'Zur intuitionistischen Deutung logischer Formeln'*, Compositio Mathematica 4, 1936, pp. 117–118. (The issue containing the paper has appeared in 1936. Cf. [219]. See also the review by A. N. Kolmogorov, in Zentralblatt für Mathematik und ihre Grenzgebiete, 0015.24201.)

294. — (1955) **Les Fondements des mathématiques**, *Intuitionnisme, Théorie de la démonstration*, Gauthier Villars, Paris-Louvain 1955. (A translation of [292], with additions.)

295. — (1956) Intuitionism, *An Introduction*, North-Holland Publishing Company, Amsterdam 1956 [*Studies in Logic and the Foundations of Mathematics*]. (Revised: 1966^2, 1971^3. Russian translation [Intuicionizm, Vvedenie], by V. A. Yankov, edited with comments by Andreĭ Andreevič Markov, Izdatel'stvo 'Mir', Moskow 1965 [*Biblioteka sbornika 'Matematika'*].)
296. — (1958) *Blick von der intuitionistischen Warte*, Dialectica 12, 1958, pp. 332–345; reprinted in [300], pp. 560–573.
297. — (1962) *After thirty years*, in Ernst Nagel, Patrick Suppes, and Alfred Tarski (eds.) Logic, Methodology and Philosophy of Science, Stanford University Press, Stanford CA, 1962, pp. 194–197; reprinted in [300], pp. 640–643. ([Anonymous] Romanian translation [*După treizeci de ani*], in [750], pp. 353–367.)
298. — (1968) *Wijsbegeerte van de wiskunde* [Philosophy of mathematics] [Dutch], Algemeen Nederlandse Tijdschrift voor Wijsbegeerte en Psychologie 60, 1968, pp. 140–153; reprinted in [300], pp. 711–724.
299. — (1974) *Intuitionistic views on the nature of mathematics*, Bollettino dell'Unione Matematica Italiana 9, 1974, pp. 122–134; reprinted in Synthese 27 (1974), pp. 79–91, and in [300], pp. 743–755.
300. — (1980) Collected Papers (edited by A. S. Troelstra, J. Niekus, and H. van Riemsdijk), Mathematical Institute, University of Amsterdam, Amsterdam 1980 [xvii + 821 pp.] (Reprographic reprints.)
301. David Hilbert (1900) *Mathematische Probleme*, Nachrichten von der königlichen Gesellschaft der Wissenschaften zu Göttingen (mathematisch-physikalische Klasse), pp. 253–296. (English translation [*Mathematical Problems*] by Mary Frances Winston Newson, in Bulletin of the American Mathematical Society 8, 1902, pp. 437–479.)
302. — (1905) *Über die Grundlagen der Logik und Arithmetik*, in Adolf Krazer (ed.) Verhandlungen des dritten Internationalen Mathematiker-Kongresses in Heidelberg vom 8. bis 13. August 1904, B. G. Teubner, Leipzig 1905 (also a Kraus Reprint, Nendeln [Liechtenstein] 1967^R), pp. 174–185. (English translation [*On the foundations of logic and arithmetic*], by George Bruce Halsted, in The Monist 15, 1905, pp. 338–352, and [same title] by Beverly Woodward, in [288], pp. 129–138.)
303. — (1918) *Axiomatisches Denken*, Mathematische Annalen 78, 1918, pp. 405–415. ('Dieser Vortrag ist in der Schweizerischen mathematischen Gesellschaft am 11. September 1917 in Zürich gehalten worden.' Reprinted in [310], pp. 146–156. English translation [*Axiomatic thought*], in [194] 2, pp. 1105–1115. [Anonymous] Romanian translation [*Gîndirea axiomatică*], in [750], pp. 92–104.)
304. — (1922) *Neubegründung der Mathematik: Erste Mitteilung*, Abhandlungen aus dem Seminar der Hamburgischen Universität 1, 1922, pp. 157—177. (English translation [*The new grounding of mathematics, first report*], by William Ewald, in [194], 2, pp. 1115—1134.).
305. — (1923) *Die logischen Grundlagen der Mathematik*, Mathematische Annalen 88, 1923, pp. 151–165. (Reprinted in [310], pp. 178–191. English translation, by William Ewald, in [194], 2, pp. 1134–1148.)
306. — (1926) *Über das Unendliche*, Mathematische Annalen, 95, 1926, pp. 161–190. (French translation [*Sur l'infini*] by André Weil, in Acta Mathematica 48 (1–2), [June] 1926, pp. 91–122. English translation in [288], pp. 367–192.)
307. — (1928) *Die Grundlagen der Mathematik*, Abhandlungen aus dem Seminar der Hamburgischen Universität 6, 1928, pp. 57–85. (English translation, by Stefan Bauer-Mengelberg and Dagfinn Føllesdal, in [288], pp. 464–479.)

308. — (1929) *Probleme der Grundlegung der Mathematik*, in Atti del Congresso Internazionale dei Matematici (Bologna, 3–10 September 1928), 1, Nicola Zanichelli, Bologna 1929, pp. 135–141. (Reprinted, with omissions, as [309].)
309. — (1930) *Probleme der Grundlegung der Mathematik*, Mathematische Annalen 102, 1930, pp. 151–165. (Reprint, with omissions, of [308].)
310. — (1935) Gesammelte Abhandlungen, 3, Julius Springer, Berlin 1935.
311. — (2013) Lectures on the Foundations of Arithmetic and Logic, 1917–1933 (edited by William Ewald, and Wilfried Sieg [with Michael Hallett, Ulrich Majer, and Dirk Schlimm]), Springer Verlag, Berlin & Heidelberg 2013 (= DAVID HILBERT's Lectures on the Foundations of Mathematics and Physics, 1891–1933, volume 3. Hilbert's text in German, editorial notes in English; with an Introduction [in English], by William Ewald and Wilfried Sieg, pp. 1–30).
312. — (2020) Lectures on the Foundations of Arithmetic and Logic, 1894–1917 (edited by William Ewald, Michael Hallett, Ulrich Majer, and Wilfried Sieg), Springer Verlag, Berlin & Heidelberg [2 July] 2020 (= DAVID HILBERT's Lectures on the Foundations of Mathematics and Physics, 1891–1933, volume 2) [forthcoming].
313. David Hilbert, and Paul Bernays (1934–1939, 1969–1970^2) Grundlagen der Mathematik, 1–2, Springer Verlag, Berlin [1] 1934, [2] 1939. (Second edition 1:1969^2, 2:1970^2. Russian translation [Osnovaniĭa matematiki] of the second edition, by Nikolaĭ Makarovič Nagornyĭ, edited by Sergeĭ Ivanovič Adĭan [in 2 volumes], Izdatel'stvo 'Nauka' [Science Publishers], Moskow 1979, 1982^R. French translation [Fondements des mathématiques] of the second edition, by François Gaillard, Eugène Guillaume, and Marcel Guillaume [in 2 volumes], L'Harmattan, Paris 2001. Bilingual [German-English] critical edition, in progress, issued by College Publications, London.)
314. J. Roger Hindley (1969) *The principal type scheme of an object in combinatory logic*, Transactions of the American Mathematical Society 146, 1969, pp. 29–60.
315. — (1989) BCK-*combinators and linear* λ-*terms have type*, Theoretical Computer Science 64, 1989, pp. 97–105.
316. — (1997) Basic Simple Type Theory, Cambridge University Press, Cambridge UK 1997 [*Cambridge Tracts in Theoretical Computer Science 42*].
317. J. Roger Hindley, and David Meredith (1990) *Principal type schemes and condensed detachment*, The Journal of Symbolic Logic 55 (1), 1990, pp. 90–105.
318. J. Roger Hindley, and Jonathan P. Seldin (1986) Introduction to Combinators and λ-Calculus, Cambridge University Press, Cambridge UK 1986 [*London Mathematical Society Student Texts 1*].
319. — (2008) Lambda-Calculus and Combinators, *An Introduction*, Cambridge University Press, Cambridge UK 2008 (Revised from [318].)
320. Henry Hiż (1957) *Complete sentential calculus admitting extensions* (abstract), Summaries of Talks Presented at the Summer Institute for Symbolic Logic, Cornell University, Ithaca NY, 1957, pp. 260–262.
321. — (1958) *Extendible sentential calculus* (abstract), Notices of the American Mathematical Society 5, 1958, p. 34.
322. — (1959) *Extendible sentential calculus*, The Journal of Symbolic Logic 24 (3), 1959, pp. 193–202. (See also [508].)
323. William A. Howard (1980) *The Formulae-as-Types notion of construction*, in [672], pp. 479–490. (A slightly edited version of notes written in 1969.)
324. Yisheng Huang (2006) BCI-*algebras*, Science Press, Beijing 2006. (Distributed by Elsevier Science.)

325. Karlheinz Hülser (ed., tr.) (1987–1988) Die Fragmente zur Dialektik der Stoiker. *Neue Sammlung der Texte mit deutscher Übersetzung und Kommentaren*, 1–4, Frommann-Holzboog, Stuttgart – Bad Cannstatt 1987–1988.
326. Lloyd Humberstone (2011) The Connectives, The MIT Press, Cambridge MA & London 2011.
327. Katerina Ierodiakonou (1990) Analysis in Stoic Logic, PhD Dissertation, University of London [London School of Economics and Political Science, Department of Philosophy, Logic and Scientific Method] 1990 [472 pp.]. (Available via the E-Theses online Service of the British Library.)
328. Yasuyuki Imai, and Kiyoshi Iséki (1966) *On axiom systems of propositional calculi XIV*, in Proceedings of the Japan Academy (Ser. A, Math. Sci.), 42, 1966, pp. 19–22.
329. Andrzej Indrzejczak (1998) *Jaśkowski and Gentzen approaches to natural deduction and related systems*, in Katarzyna Kijania-Placek & Jan Woleński (eds.) The Lvov-Warsaw School and Contemporary Philosophy, Kluwer Academic Publishers, Dordrecht 1998, pp. 253–264.
330. — (2016) *Natural deduction*, Internet Encyclopedia of Philosophy.
331. Afrodita Iorgulescu (2008) Algebras of logic as BCK-algebras, Editura ASE [Academy of Economic Studies Press], Bucharest 2008 [*Colecţia Informatică*].
332. — (2019) *Algebras of logic vs. algebras*, Bucharest, September 2019 [forthcoming].
333. Kiyoshi Iséki (1966) *An algebra related with a propositional calculus*, Proceedings of the Japan Academy 42 (Ser. A, Math. Sci.), 1966, pp. 26–29.
334. — (1980) *On BCI-algebras*, Mathematics Seminar Notes [Kobe University], 8, 1980, pp. 125–130.
335. Kiyoshi Iséki, and Shôtarô Tanaka (1978) *An introduction to the theory of BCK-algebras*, Mathematica Japonica 23 (1), 1978, pp. 1–26.
336. Jacek Juliusz Jadacki (2006) *The Lvov-Warsaw school and its influence on Polish philosophy of the second half of 20th century*, in Jacek [Juliusz] Jadacki, and Jacek Paśniczek (eds.) The Lvov-Warsaw School. *The New Generation*, Rodopi, Amsterdam & New York NY 2006 [*Poznań Studies in the Philosophy of the Sciences and the Humanities 89*], pp. 41–83.
337. — (2016) Stanisław Leśniewski: geniusz logiki [Stanisław Leśniewski: a genius of logic] [Polish], Epigram, Bydgoszcz 2016 [381 pp.]. (A companion to [426], with copious bio-bibliographical material.)
338. Stanisław Jaśkowski (1927) *Teorja dedukcji oparta na dyrektywach załozeniowych* [Deduction theory based on supposition rules] [Polish], Paper presented at the First Congress of the Polish Mathematicians, Lvov, 7–10 September 1927, [4].
339. — (1934) *On the rules of suppositions in formal logic*, Studia Logica 1 (1934), pp. 5–32. (Reprinted in [450], pp. 232–258. Results of 1926, announced in [338].)
340. — (1963) *Über Tautologien in welchen keine Variable mehr als zweimal vorkommt*, Zeitschrift für mathematische Logik und Grundlagen der Mathematik 9, 1963, pp. 219–228.
341. Ingebrigt Johansson (1937) *Der Minimalkalkül, ein reduzierter intutionistischer Formalismus*, Compositio Mathematica 4 (1), 1937, pp. 119–136. (The issue containing the paper has appeared in 1936. See also [487].)
342. John of Salisbury (1955) The Metalogicon, *A Twelfth-Century Defense of the Verbal and Logical Arts of the Trivium* (Translated with an Introduction and Notes by Daniel D. McGarry), University of California Press, Berkeley & Los Angeles CA 1955 [reprinted 1962, newly reprinted by Peter Smith, Gloucester

MA 1972]. (Written in 1159. First printed edition: IOANNES SARESBERIENSIS *Metalogicus*, Apud Hadrianum Beys, Paris 1610.)
343. John Kalman (1974) *An algorithm for Meredith's condensed detachment* (abstract), The Journal of Symbolic Logic 39, 1974, p. 206.
344. — (1983) *Condensed detachment as a rule of inference*, Studia Logica 42, 1983, pp. 443–451.
345. Stig Kanger (1955) *A note on partial postulate sets for propositional logic*, Theoria (Lund) 21, 1955, pp. 99–104, reprinted in [347] 1, pp. 3–7.
346. — (1957) *Provability in Logic*, Almquist & Wiksells Boktryckeri, Uppsala 1957 [*Acta Universitatis Stockholmiensis, Stockholm Studies in Philosophy 1*], reprinted in [347] 1, pp. 8–41. (Originally, PhD Dissertation, University of Stockholm 1957.)
347. — (2001) Collected Papers [= Ghita Holmström-Hintikka, Sten Lindström, and Rysiek Sliwinski (eds.) *Collected Papers of* STIG KANGER *with Essays on his Life and Work*] 1–2, Kluwer Academic Publishers, Dordrecht, Boston & London 2001 [*Synthese Library 303–304*].
348. Hubert C. Kennedy (1973) *What Russell learned from Peano*, Notre Dame Journal of Formal Logic 14 (3), 1973, pp. 367–372. (Reprinted in [352], pp. 28–34.)
349. — (1974) Giuseppe Peano (German translation by Ruth Amsler), Birkhäuser Verlag, Basle etc. 1974.
350. — (1975) *Nine letters from Giuseppe Peano to Bertrand Russell*, Journal of the History of Philosophy 13 (2), 1975, pp. 205–220. (Reprinted in [352], pp. 68–90.)
351. — (1980) Peano – Life and Works of Giuseppe Peano, D. Reidel Publishing Company, Dordrecht etc. 1980. (See also the author's [349].)
352. — (2002) Twelve Articles on Giuseppe Peano, Peremptory Publications, San Francisco CA 2002. ('All but one of these articles originally appeared in various journals in the years 1963–1984.')
353. Anthony Kenny (1995) Gottlob Frege, *An introduction to the Founder of Modern Analytic Philosophy*, Penguin Books, London 1995. (Reprint: Blackwell Publishers, Malden MA 2000^R).
354. Stephen Cole Kleene (1934) *Proof by cases in formal logic*, Annals of Mathematics (2), **35** (3), 1934, pp. 529–544.
355. — (1935) *A theory of positive intergers in formal logic*, American Journal of Mathematics 57 (1), 1935, pp. 153–173, and (2), pp. 219–244. ('Received October 9, 1933. Revised manuscript received June 18, 1934.' See also [354]. Originally, PhD Dissertation, Princeton University 1934, under the supervision of Alonzo Church.)
356. — (1952) Introduction to Metamathematics, Wolters Noordhoff Publishing, Groningen & North-Holland Publishing Company, Amsterdam & Oxford 1952 [*Bibliotheca Mathematica 1*]. (Many reprints, as, e.g., [by American Elsevier] 1971^6, 1974^7, etc.)
357. — (1979) *Origins of recursive function theory*, in Proceedings of the 20th Annual Symposium on Foundations of Computer Science [San Juan, Puerto Rico, 1979], IEEE Press, New York, 1979, pp. 371–382 (See also [358].)
358. — 1981 *Origins of recursive function theory*, Annals of the History of Computing 3 (1), 1981, pp. 52–67.
359. Kevin A. Klement (2003) *Russell's 1903–1905 anticipation of the lambda calculus*, History and Philosophy of Logic 24 (1), 2003, pp. 15–37.
360. Raymond Klibansky (1936) *Standing on the shoulders of giants* [= *Notes and Correspondence: Answer to Query N° 53*, Isis 24, 1935, pp. 107–109], Isis 26

(71, i), [December] 1936, pp. 147–149. (See also: James Dean [*Bibliographies*] *Chartres; Dwarfs/Giants Bibliography*, The ORB: On-line Reference Book for Medieval Studies [ORB Online Encyclopedia], 1996.)

361. William C. Kneale (1957) *Aristotle and the 'consequentia mirabilis'*, Journal of Hellenic Studies 77, 1957, pp. 62–66.
362. William Kneale, and Martha Kneale (1962) The Development of Logic, Clarendon Press, Oxford 1962, 1988^R [with corrections, paperback]. (Romanian translation [Dezvoltarea logicii, 1–2], by Sorin Vieru and Uşer Morgenstern, Editura Dacia, Cluj-Napoca 1975.)
363. Andreĭ Nikolaevič Kolmogorov (1925) *O principe 'tertium non datur'* [On the principle of excluded middle] [Russian], Matematičeskiĭ Sbornik 32 (4), 1924–1925, pp. 646–667. (Paper dated Moskow, September 30, 1925. Reprinted in [368], pp. 45–69. English translation by Jean van Heijenoort, in [288], pp. 416–437, with an Introduction by Hao Wang, pp. 414–416. English version reprinted in [369].)
364. — (1931–1932) *Pis'ma A. N. Kolmogorova k A. Geĭtingu* [Letters of A. N. Kolmogorov to A. Heyting] [Russian], Russian translations (from French and German), with a commentary by Valeriĭ E. Plisko], Uspexi Matematičeskix Nauk 43 (6 [264]), 1988, pp. 75–77. (English translation, by D. L. Johnson, in Russian Mathematical Surveys 43 (6), 1988, pp. 89–93. See also [535].)
365. — (1932) *Zur Deutung der intuitionistischen Logik*, Mathematische Zeitschrift 35, 1932, pp. 58–65. (Paper dated Göttingen, January 15, 1931. Russian translation [*K tolkovaniyu intuicionistskoĭ logiki*], in [368], pp. 142–148. English translation [*On the interpretation of intuitionistic logic*], in [369], pp. 151–158, reprinted in [445], pp. 328–334.)
366. — (1941) *Valeriĭ Ivanovič Glivenko (1897–1940)* [Russian], Uspexi Matematičeskix Nauk 8, 1941, pp. 379–383. (Necrology, including a bibliography: *Published works of V. I. Glivenko*, pp. 382–383.)
367. — (1985) *K rabotam po intuicionistskoĭ logike* [About my papers on intuitionistic logic] [Russian], in [368] p. 393 [Russian] and [369] pp. 451–452 [English].
368. — (1985a) *Izbrannye trudy, 1, Matematika i mexanika* [Selected works 1: Mathematics and mechanics] [Russian], edited by S. M. Nikol'skiĭ, Izdatel'stvo 'Nauka' [Science Publishers], Moscow 1985.
369. — (1991) Selected Works, 1, *Mathematics and Mechanics*, edited by Vladimir M. Tikhomirov [Vladimir Mixaĭlovič Tixomirov], Kluwer Academic Publishers, Dordrecht, etc. 1991.
370. Andreĭ Nikolaevič Kolmogorov, and Al'bert G. Dragalin (1982) Vvedenie v matematičeskuĭu logiku [Introduction to mathematical logic] [Russian], Izdatel'stvo Moskoskogo Universiteta [Moscow University Publishing House], Moscow 1982.
371. — (1984) Matematičeskaĭa logika: dopolnitel'nye glavy [Mathematical logic: supplementary chapters] [Russian], Izdatel'stvo Moskovskogo Universiteta [Moscow University Publishing House], Moscow 1984.
372. Yuichi Komori (1987) BCK-*algebras and lambda-calculus*, Proceedings of the 10th Symposium on Semigroups (Sakado 1986), Josai University, Sakado 1987, pp. 5–11.
373. Tadeusz Kotarbiński 1929 Elementy teorji poznania, logiki formalnej i metodologji nauk [Elements theory of knowledge, formal logic and methodology of science] [Polish], Wydawnictwo Zakładu Narodowogo imienia Ossolińskich [Ossolineum Publishers], Lwów 1929. (Reprint: 1947^R. Second revised edition: Ossolineum, Wrocław 1961^2. Third edition [Jan Woleński (ed.)], *ibid.* 1990^3, etc. English translation in [374].)

374. — 1966 Gnosiology, *The Scientific Approach to the Theory of Knowledge* (English translation by Olgierd Adrian Wojtasiewicz, edited by George Bidwell and C. Pinder), Pergamon Press, Oxford & New York 1966. (English translation of [373].)
375. Jerzy Kotas, and August Pieczkowski (1967) *Scientific works of Stanisław Jaśkowski*, Studia Logica 21 (1), 1967, pp. 7–15.
376. Karst [Christian Peter Jozef] Koymans (1982) *Models of the lambda calculus*, Information and Control [currently: Information and Computation] 52 (3), 1982, pp. 306–332.
377. — (1984) Models of the Lambda Calculus, CWI, Amsterdam 1984 [*CWI Tract 9*] (Originally, PhD Dissertation, University of Utrecht 1984. Reviewed by Giuseppe Longo, in The Journal of Symbolic Logic 52 (1), 1987, pp. 284–285.)
378. Georg Kreisel (1951) *On the interpretation of non-finitist proofs: Part I*, The Journal of Symbolic Logic 16 (4), 1951, pp. 241–267.
379. — (1952) *On the interpretation of non-finitist proofs: Part II. Interpretation of number theory. Applications*, The Journal of Symbolic Logic 17 (1), 1952, pp. 43–58.
380. — (1958) *Mathematical significance of consistency proofs*, The Journal of Symbolic Logic 23 (2), [June] 1958, pp. 155–182.
381. — (1958a) *Hilbert's programme*, Dialectica 12, 1958, pp. 346–372. (Revised in Paul Benacerraf and Hilary Putnam (eds.), Philosophy of Mathematics, *Selected Readings*, Prentice Hall, Englewood Cliffs NJ 1964, pp. 157–180 [= second edition, Cambridge University Press, Cambridge UK 1984^2, pp. 207–238], with a Postscript.)
382. — (1962) *Foundations of intuitionistic logic*, in Ernest Nagel, Patrick Suppes, and Alfred Tarski (eds.) Logic, Methodology and Philosophy of Science 1 (1960), Stanford University Press, Stanford CA 1962, pp. 192–210.
383. — (1965) *Mathematical logic*, in Thomas L. Saaty (ed.) Lectures on Modern Mathematics, 3, John Wiley & Sons, New York etc. 1965 [second printing 1967], pp. 95–195. (Survey paper. 'Additions' in [384], p. 316.)
384. — (1967) *Mathematical logic: what has it done for the philosophy of mathematics?*, in Ralph Schoenman (ed.) Bertrand Russell, Philosopher of the Century, *Essays in His Honour*, Little, Brown and Company & Atlantic Monthly Press, Boston and Toronto 1967 [second printing], pp. 201–272. (Paper dated December 1964. Also: *Addenda to contribution by Georg Kreisel*, in *ibid.*, pp. 315–316, with 'Additions to [383]', p. 316.)
385. — (1968) *A survey of proof theory*, The Journal of Symbolic Logic 33 (3), 1978, pp. 321–388.
386. — (1968a) Elements of Proof Theory, Stanford University, Stanford CA [Lecture Notes for a course given at UCLA, undated typescript, 166 pp., numbered 1–87, 87a, 88–165]. (Cf., e.g., [400], § 4, pp. 398–399, and see also [260] [item in Box 5, Folder 8].)
387. — (1971) *A survey of proof theory II*, in Jens-Erik Fenstad (ed.) Proceedings of the Second Scandinavian Logic Symposium, North-Holland Publishing Company, Amsterdam 1971 [*Studies in Logic and the Foundations of Mathematics 63*] pp. 109–170.
388. — (1973) *Perspectives in the philosophy of pure mathematics*, in Patrick Suppes, Leon Henkin, Athanase Joja, and Grigore C. Moisil (eds.) Logic, Methodology, and Philosophy of Science 4, North-Holland Publishing Company, Amsterdam, etc. 1973 [*Studies in Logic and the Foundations of Mathematics 74*], pp. 255–277.

389. — (1973a) **Fundamental Concepts of Intuitionistic Mathematics**, Lectures given at UCLA, Autumn 1973 [Lecture Notes, 138 pp. consisting of (1) a MS of 119 pp., in the handwriting of the author, and (2) a typescript, of 19 pp., containing a *Review of Course Notes 1973*, pp. 1–5, and a *Summary*, pp. 6–19.] (See also [259] [SC0136, Box 23, Folders 10 and 11].)

390. — (1973b) *A neglected aspect of the formalization of informal mathematics*, Stanford University 1973 [MS, in the author's handwriting, 10 pp.] (cf. the UCLA 'Course on intuitionism 1973', [389], in the Stanford *Nachlass* SC0136).

391. — (1976) *'Der unheilvolle Einbruch der Logik in die Mathematik'*, **Acta Philosophica Fennica 28** (1–3), 1976, pp. 166–187 [= Jaakko Hintikka (ed.) **Essays on Wittgenstein in Honour of G. H. von Wright**, North-Holland Publishing Company, Amsterdam, 1976]. ('The title is a quotation from Wittgenstein's [794].' Revised in [401], 1, [Chapter 15 *The disastruous invasion of logic into mathematics*], pp. 253–270.)

392. — (1977) *On the kind of data needed for a theory of proofs*, in Robin O. Gandy and John Martin Elliott Hyland (eds.) **Logic Colloquium 76** (Proceedings of a conference held in Oxford in July 1976), North-Holland Publishing Company, Amsterdam 1977 [*Studies in Logic and the Foundations of Mathematics 87*], pp. 111–128.

393. — (1979) *Some facts from the theory of proofs and some fictions from general proof theory*, in Jaakko Hintikka, Ilkka Niiniluoto, and Esa Saarinen (eds.) **Essays on Mathematical and Philosophical Logic**, D. Reidel Publishing Company, Dordrecht, etc. 1979, pp. 3–23.

394. — (1979a) *Formal rules and questions of justifying mathematical practice*, in Kuno Lorenz (ed.) **Konstruktionen versus Positionen**, *Beiträge zur Diskussion um die konstruktive Wissenschaftstheorie* [1–2], Walther de Gruyter, Berlin & New York 1979, 1, pp. 99–130.

395. — (1980) *Constructivist approaches to logic*, in Evandro Agazzi (ed.) **Modern Logic, A Survey**, D. Reidel Publishing Company, Dordrecht 1980 [*Synthese Library 149*], pp. 67–91.

396. — (1980a) *Kurt Gödel*, **Biographical Memoirs of Fellows of the Royal Society 26**, The Royal Society, London 1980, pp. 149–224. (*Corrigenda*, in ibid., 27, 1981, p. 697, and in ibid., 28, 1982, p. 718.)

397. — (1980b) *Proof theory and computer science*, Stanford University, Stanford CA, 22 May 1980 [MS, in the handwriting of the author, 10 pp.].

398. — (1984) *Frege's foundations and intuitionistic logic*, **The Monist 67** (1), 1984, pp. 72–91

399. — (1987) *Gödel's excursions into intuitionistic logic*, in Paul Weingartner, and Leopold Schmetterer (eds.) **Gödel Remembered**, *Gödel-Symposium in Salzburg* (10–12 July 1983), Bibliopolis, Naples 1987 [*History of Logic 4*] pp. 65–186.

400. (1987a) *Proof theory: some personal recollections*, in [724] [second edition, i.e. 1987^2] pp. 095–405.

401. Georg Kreisel, and Piergiorgio Odifreddi (2019) **About Logic and Logicians**, *A Palimpsest of Essays by* GEORG KREISEL, Selected and arranged by Piergiorgio Odifreddi, 1. Philosophy, 2. Mathematics, Brasília 2019 [*Lógica no Avião*].

402. Lothar Kreiser (1995) *Freges ausserwissenschaftliche Quellen seines logischen Denkens*, in Ingolf Max, and Werner Stelzner (eds.) **Logik und Mathematik**, *Frege-Kolloquium, Jena 1993*, Walther de Gruyter, Berlin and New York 1995 [*Perspektiven der analytischen Philosophie 5*], pp. 219–225.

403. — (2001) **Gottlob Frege**, *Leben, Werk, Zeit*, Felix Meiner Verlag, Hamburg 2001.

404. Jean-Louis Krivine (1990) Lambda-calcul, types et modèles, Masson, Paris 1990.
405. — (1993) Lambda-calculus, Types and Models, Ellis Horwood Ltd., Upper Saddle River NJ 1993 [Ellis Horwood Series in Computers and Their Applications]. (Augmented English version of [404], English translation by René Cori.)
406. Johannes John Carel Kuiper (2004) Ideas and Explorations, Brouwer's Road to Intuitionism, University of Utrecht, Utrecht 2004 [Quaestiones Infinitae 46, Publications of the Department of Philosophy, University of Utrecht] (PhD Dissertation, University of Utrecht 2004.)
407. Milan Kundera (1993) Les Testaments trahis, Gallimard, Paris 1993. (English version: Testaments Betrayed, translated by Linda Asher, Faber & Faber, London and Boston 1995.)
408. Franz von Kutschera (1962) Zum Deduktionsbegriff der klassischen Prädikatenlogik erster Stufe, in Max Käsbauer, and Franz von Kutschera (eds.) Logik und Logikkalkül, Karl Alber, Freiburg & München 1962, pp. 211–236.
409. — (1989) Gottlob Frege, Eine Einführung in sein Werk, Walter de Gruyter, Berlin & New York 1989.
410. — (1996) Frege and natural deduction, in Matthias Schirn (ed.) Frege, Importance and Legacy, Walter de Gruyter, Berlin 1996 [Perspektiven der analytischen Philosophie 13], pp. 301–304.
411. Joachim Lambek (1980) From lambda-calculus to Cartesian closed categories, in [672], pp. 375–402.
412. Joachim Lambek, and Philip J. Scott (1986) Introduction to Higher Order Categorical Logic, Cambridge University Press, Cambridge UK, etc. 1986 [Cambridge Studies in Advanced Mathematics 7].
413. Edmund Landau (1930) Grundlagen der Analysis (Das Rechnen mit ganzen, rationalen, irrationalen, komplexen Zahlen: Ergänzung zu den Lehrbüchern der Differential- und Integralrechnung), Akademische Verlagsgesellschaft, Leipzig 1930. (Many susbsequent editions: Chelsea Pub. Co, New York 1946^2, 1948^2 reprint, 1960^3, 1965^3 reprint, 1997^4 ['with a complete German-English vocabulary']; Wissenschaftliche Buchgesellschaft, Darmstadt 1970; N. Heldermann, Berlin 2004, etc. English translation [Foundations of Analysis, The arithmetic of whole, rational, irrational, and complex numbers], by F. Steinhardt, Chelsea Publishing Company, New York 1951, 1957^2, 1960^3, 1966^4, etc.)
414. Gregory Landini (1998) Russell's Hidden Substitutional Theory, Oxford University Press, Oxford & New York 1998.
415. — (2012) Frege's Notations: What They Are and How They Mean, Palgrave Macmillan, New York 2012 [History of Analytic Philosophy].
416. Hans Läuchli (1965) Intuitionistic propositional calculus and definably non-empty terms (abstract), The Journal of Symbolic Logic 30 (2), 1965, p. 263, ('Received October 20, 1964'.)
417. — (1970) An abstract notion of realizability for which the intuitionistic logic is complete, in A. Kino et alii (eds.) Intuitionism and Proof Theory, North-Holland Publishing Company, Amsterdam, etc. 1970, pp. 227–234. (An algebraic version of the Curry-Howard Correspondence applying to the full Heyting-Glivenko logic. Reports work done earlier [cca 1964]. Cf., e.g., the abstract [416].)
418. Hugues Leblanc (1979) Existence, Truth, and Provability, State University New York Press, Albany NY 1979. (Collected papers. Reprinted in 1982^R).
419. Javier Legris (2012) Paul Hertz and the origins of structural reasoning, in Jean-Yves Béziau (ed.) Universal Logic, An Anthology – From Paul Hertz to Dov Gabbay, Birkhäuser Verlag, Basel & Springer, New York 2012, pp. 3–10.

420. Stanisław Leśniewski (1916) Podstawy ogólnej teorii mnogości I [Foundations of the general theory of sets I] [Polish], [Prace Polskiego Koła Naukowego w Moskwie, Sekcya matematyczno-pryrodnicza 2], 44 pp., Moskow 1916. (Reprinted in [426], 1, pp. 256–294. English translation, by Dene I. Barnett, in [425], 1, pp. 129–173.)

421. — (1927-1930) *O podstawach matematyki* [On the foundations of mathematics] [Polish], Przegląd Filozoficzny 30, 1927, pp. 164–206, 31, 1928, pp. 261–291, 32, 1929, pp. 60–101, 33, 1930, pp. 77–105, 142–170. (Reprinted in [426], 1, pp. 295–468. English translation, by Dene I. Barnett, in [425], 1, pp. 174–382.)

422. — (1929) *Grundzüge eines neuen Systems der Grundlagen der Mathematik*, Fundamenta Mathematicae 14, 1929, pp. 1–81. (Reprinted in [426], 2, pp. 489–569. English translation [*Fundamentals of a new system of the foundation of mathematics*], by Michael P. O'Neil, in [425], 2, pp. 410–492. This contains §§ 1–11. The sequel, [424] [= § 12], is printed in *ibid.*, pp. 492–605, so that the translation appears in *ibid.*, pp. 410–605]. For the fate of the journal Collectanea Logica, see [697], and [450], p. 88, fn 1.)

423. — (1938) *Einleitende Bemerkungen zur Fortsetzung meiner Mitteilung u.d.T. 'Grundzüge eines neuen Systems der Grundlagen der Mathematik'*, Collectanea Logica (Warsaw) 1, 1938, pp. 1–60. (Reprinted in [426], 2, pp. 570–629. English translation [*Introductory remarks to the continuation of my anticle 'Grundzüge eines neuen Systems der Grundlagen der Mathematik'*], by W. Teichmann and Storrs McCall, in [450] [Paper 6], pp. 116–169, reprinted in [425], 2, pp. 649–710.)

424. — (1938a) *Grundzüge eines neuen Systems der Grundlagen der Mathematik* [Fortsetzung], Collectanea Logica (Warsaw) 1, 1938, pp. 61–144 [separatum, pp. 1–84]. (Reprinted in [426], 2, pp. 630–713. Marked '[*Forsetzung folgt.*]' on p. 84 [reprint p. 713]. English translation – of § 12, continued from [422] – by Michael P. O'Neil, in [425], 2, pp. 492–605.)

425. — (1992) Collected Works (edited by Stanisław J. Surma, Jan T. Srzednicki, and Dene I. Barnett, with an annotated bibliography by V. Frederick Rickey), Kluwer Academic Publishers, Dordrecht & Boston 1992. (In 2 volumes, with continuous page-numbering, 1: pp. i–xvi (Introduction by the editors) and pp. 1–382, 2: pp. 383–794).

426. — (2015) Pisma zebrane / Sobrannye sočineniĭa / Gesammelte Schriften (edited with an Introduction and a Postscript [*Stanisław Leśniewski's Life and Work*], in Polish, by Jacek [Juliusz] Jadacki), Wydawnictwo Naukowe Semper [Semper Scientific Publishers], Warsaw 2015 [Société des Sciences et des Lettres de Varsovie. Bibliothèque de Philosophes]. (In 2 volumes, with continuous page-numbering, 1: pp. 1–468, 2: pp. 469–876. Introduction, 1, pp. 7–12, Postscript, 2, pp. 814–869. The edition contains anastatic reprints of Polish, Russian, and German originals.)

427. Alfred Lindenbaum (1927) *Méthodes mathématiques dans les recherches sur le système de la théorie de déduction* (abstract), Paper presented at the First Congress of the Polish Mathematicians, Lwów, 7–10 September 1927, [4].

428. Edgar George Kenneth López-Escobar (1990) *Remarks on the Church-Rosser property*, The Journal of Symbolic Logic 55 (1), [March] 1990, pp. 106–112.

429. Paul Lorenzen (1955, 1969²) Einführung in die operative Logik und Mathematik, Springer Verlag, Berlin 1955, 1969^2 [*Grundlehren der mathematischen Wissenschaften 78*].

430. Jerzy Łoś, and Roman Suszko (1958) *Remarks on sentential logics*, Koninklijke Nederlandse Akademie van Wetenschappen, Proceedings [Series A. Mathematical

Sciences], 61, 1958, pp. 177–183. (Reviewed by Alonzo Church, in The Journal of Symbolic Logic 40 (4), 1975, pp. 603–604.)
431. Erika Luciano (2011) *Giovanni Vacca's contributions to the historiography of logic*, L&PS – Logic and Philosophy of Science 9 (1), 2011, pp. 275–283.
432. Jan Łukasiewicz (1929, 1958^2) Elementy logiki matematycznej [Elements of mathematical logic] [Polish], Skrypt autoryzowany opracował M. Presburger. Z częściowej subwencji Senatu Akademickiego Uniw. Warsz. Nakładem Komisji Wydawniczej Koła Matematyczno-Fizycznego Słuchaczów Uniwersytetu Warszawskiego, Warszawa 1929. (Wydawnictwo Koła tom XVIII). [Publications Series of the Association of Students of Mathematics and Physics of the Warsaw University. Volume XVIII. Authorised typescript (of lecture notes) prepared by Mojzesz Presburger. Subsidised in part by the Senate of the Warsaw University, Financed by the Publishing Committee of the Association of Students of Mathematics and Physics of the Warsaw University, Warsaw 1929.] (Second edition: Państwowe Wydawnictwo Naukowe [PWN: Polish Scientific Publishers], Warsaw 1958^2. English translation [of the second edition] by Olgierd Adrian Wojtasiewicz, as [441].)
433. — (1930) *Philosophische Bemerkungen zu mehrwertigen Systemen des Aussagenkalküls*, Sprawozdania z posiedzeń Towarzystwa Naukowego Warszawskiego (Wydział III), XXIII, 1930 [= Comptes Rendus des séances de la Société des Sciences et des Lettres de Varsovie (Classe III), vol. XXIII, 1930], pp. 51–77. (Translated by H. Weber, [as *Philosophical remarks on many-valued systems of propositional logic*], in [450], pp. 40–65.)
434. — (1931) *Uwagi o aksjomacie Nicoda i o 'dedukcji uogólniajacej'* [Notes on the axiom of Nicod and on 'generalising deduction'] [Polish], Odbitka z Księgi Pamietkowej polskiego Towarzystwa Filozoficznego we Lwowie. 12.II 1904 – 12.II 1929, Lwów 1931. Skład główny w Księgarni S. A. Książnica-Atlas. [Extract from the Memorial Book of the Polish Philosophical Society in Lwów, 12. ii 1904 – 12. ii 1929, Lwów 1931. Printed by Książnica-Atlas SA.] (Reprinted in [440]. English translation [*Comments on Nicod's axiom on 'generalizing deduction'*], in [442], pp. 179–196.)
435. — (1931a) *Dowód zupełności dwuwartościowego rachunku zdań* [Ein Vollständigkeitsbeweis des zweiwertigen Aussagenkalküls], Sprawozdania z posiedzeń Towarzystwa Naukowego Warszawskiego (Wydział III), XXIV, 1931 [= Comptes Rendus des séances de la Société des Sciences et des Lettres de Varsovie (Classe III), vol. XXIV, 1931; in print: Warsaw 1932].
436. — (1934) *Z historii logiki zdań* [From the history of the logic of propositions] [Polish], Przegląd Filozoficzny 37, 1934, pp. 417–437. (Reprinted in [440]. German translation by the author [*Zur Geschichte der Aussagenlogik*], in Erkenntnis 5, 1935, pp. 111–131. [Anonymous] Romanian translation [*Din istoria logicii propoziţiilor*], from the Polish original, in [750], pp. 119–143. English translation [*On the history of the logic of propositions*], from the German version, by Storrs McCall, in [450], pp. 66–87. English translation [*On the history of the logic of propositions*] from Polish, by Olgierd Adrian Wojtasiewicz, in [442], pp. 197–217.)
437. — (1939) *Der Äquivalenzenkalkül*, Collectanea Logica 1, 1939, pp. 145–169. (English translation [*The equivalential calculus*], by Peter Woodruff, in [450], pp. 88–115.)
438. — (1948) *The shortest axiom of the implicational calculus of propositions*, Proceedings of the Royal Irish Academy (section A) 52 (3), 1948, pp. 25–33.

(Reprinted in [442], pp. 295–305. (Reviewed by Alonzo Church, in The Journal of Symbolic Logic 13 (3), [December] 1948, p. 164. Result of 1936. See also [767].)

439. — (1957^2) Aristotle's Syllogistic, From the Standpont of Modern Formal Logic, Clarendon Press, Oxford 1957^2 [second enlarged edition; first edition 1951^1].

440. — (1961) Z zagadnień logiki i filozofii (*Pisma wybrane*) [Problems of logic and philosophy, Selected writings] [Polish] (edited by Jerzy Słupecki), Państwowe Wydawnictwo Naukowe [PWN: Polish Scientific Publishers], Warsaw 1961.

441. — (1963) Elements of Mathematical Logic (English translation by Olgierd Adrian Wojtasiewicz), Pergamon Press, Oxford etc., & Państwowe Wydawnictwo Naukowe [PWN: Polish Scientific Publishers], Warsaw 1963. (This is a translation of the second edition of [432], published by PWN, Warsaw 1958.)

442. — (1970) Selected Writings (edited by Ludwik Borkowski), North-Holland Publishing Company, Amsterdam & Państwowe Wydawnictwo Naukowe [PWN: Polish Scientific Publishers], Warsaw 1970.

443. Jan Łukasiewicz, and Alfred Tarski (1930) Untersuchungen über den Aussagenkalkül, Comptes Rendus des Séances de la Société des Sciences et des Lettres de Varsovie (Classe II-ème), 1930, pp. 39–50. (English translation [*Investigations into the sentential calculus*] in [734], pp. 38–59, and in [442], pp. 131–152.)

444. Saunders MacLane (1934) Abgekürzte Beweise im Logikkalkul, Huber & Co., Göttingen 1934. (Reprinted in Irving Kaplansky (ed.) SAUNDERS MACLANE Selected Papers, Springer-Verlag, New York 1979, pp. 1–62. Originally, *Inauguraldissertation*, Göttingen 1933, under David Hilbert.)

445. Paolo Mancosu (ed.) (1998) From Brouwer to Hilbert, *The Debate on the Foundations of Mathematics in the 1920s*, Oxford University Press, New York etc. 1998.

446. Paolo Mancosu, Richard Zach, and Calixto Badesa (2009) *The development of mathematical logic from Russell to Tarski: 1900–1935*, in Leila Haaparanta (ed.), The Development of Modern Logic, Oxford University Press, New York, etc. 2009, pp. 318–340.

447. Enrico Martino (2018) Intuitionistic Proof Versus Classical Truth, *The Role of Brouwer's Creative Subject in Intuitionistic Mathematics*, Springer International Publishing 2018 [*Logic, Epistemology, and the Unity of Science 42*].

448. Benson Mates (1948) The Logic of the Old Stoa, PhD Dissertation, University of California, Berkeley CA, June 1948 [594 pp.] [typescript, UCB Library: restricted use; not seen].

449. — (1953) Stoic Logic, University of California Press, Berkeley CA 1953, [*University of California Publications in Philosophy 26*], 1961^2, 1973^R [*California Library Reprint Series*].

450. Storrs McCall (ed.,tr.) (1967) Polish Logic 1920–1939, Clarendon Press, Oxford 1967.

451. William McCune, and Larry Wos (1992) Experiments in automated deduction with Condensed Detachment, in Deepak Kapur (ed.) Proceedings of the Eleventh International Conference on Automated Deduction (CADE-11), Springer-Verlag, New York 1992 [*Lecture Notes in Artificial Intelligence 607*], pp. 209–223, reprinted in [798], [2], pp. 1193–1210.

452. Colin McLarty (2007) *The last mathematician from Hilbert's Göttingen: Saunders MacLane as a philosopher of mathematics*, British Journal of Philosophy of Science 58 (1), 2007, pp. 77–112.

453. Richard L. Mendelsohn (2005) The Philosophy of Gottlob Frege, Cambridge University Press, Cambridge UK etc. 2005.

454. José M. Méndez (1988) *A note on the semantics of minimal intuitionism*, Logique et Analyse 31 (123–124), 1988, pp. 371–378. (See also [637].)
455. Jie Meng (1987) BCI-*algebras and abelian groups*, Mathematica Japonica 32, 1987, pp. 693–696.
456. Jie Meng, and Young Bae Jun (1994) BCK-*algebras*, Kyung Moon Sa Co., Seoul 1994.
457. Carew A. Meredith (1953) *Single axioms for the systems (C,N), (C,0) and (A,N) of the two-valued propositional calculus*, The Journal of Computing Systems 1 (3), 1953, pp. 155–164.
458. — (1953a) *A single axiom of positive logic*, The Journal of Computing Systems 1 (3), 1953, pp. 169–170.
459. — (1966) *Postulates for implicational calculi*, The Journal of Symbolic Logic 31 (1), 1966, pp. 7–9. (Cf. [177], [745], [102], [103].)
460. Carew A. Meredith, and Arthur N. Prior (1963) *Notes on the axiomatics of the propositional calculus*, Notre Dame Journal of Formal Logic 4 (3), 1963, pp. 171–187.
461. — (1968) *Equational logic*, Notre Dame Journal of Formal Logic 9 (3), 1968, pp. 212–226. (Errata in [572].)
462. David Meredith (1974) *Combinatory and propositional logic*, Notre Dame Journal of Formal Logic 15 (1), 1974, pp. 156–160. (Curry-Howard for pure λ-calculus. See also [464], [465], and [468].)
463. — (1977) *In Memoriam: Carew Arthur Meredith (1904–1976)*, Notre Dame Journal of Formal Logic 18 (4), 1977, pp. 513–516.
464. — (1978) *Positive logic and λ-constants*, Studia Logica 37 (3), 1978, pp. 269–295. ('Since Jaśkowski's [340] and C. A. Meredith's results [460] on the system $\mathcal{BCI}$ were published in the same year [i.e., 1963], the author suggests the designation Jaśkowski-Meredith $\mathcal{BCI}$ system [...]' Postscript – March 1978.)
465. — (1979) *Axiomatics for implication*, Notre Dame Journal of Formal Logic 20 (1), 1979, pp. 89–91.
466. — (1979a) Letter to Adrian Rezuş, dated Merrimack NH, 6 October 1979. (On the archeology of 'pure' $\mathcal{BCI}$, the strictly linear implicational logic of C. A. Meredith, [460], and S. Jaśkowski [340]. Cf. Church's [124], Curry's comments in [142], Wajsberg's [790] [= *Metalogische Beiträge 1*, 1937], Appendix to Section 1 [= [450], p. 293], and [464], [465]. See also [751], [752].)
467. — (1979b) Letter to Adrian Rezuş, dated Merrimack NH, 28 December 1979.
468. — (1980) *A positive logic proof procedure*, in [672], pp. 503–510.
469. Robert K. Merton (1965, 1985^{R}, 1993^{R}) On the Shoulders of Giants, Free Press, New York 1965 (See also On the Shoulders of Giants, *A Shandean Postscript*, *The Vicennial Edition*: Harcourt Brace Jovanovich, San Diego CA, New York, London 1985; *ibid.*, *The Post-Italianate Edition*: The University of Chicago Press, Chicago IL 1993, etc.).
470. Robert K. Meyer, and John K. Slaney (1980) Abelian Logic (from A to Z), Research Paper No. 7, Logic Group, Research School of Social Sciences, Australian National University, Canberra ACT 1980.
471. — (1981) *Abelian logic* (abstract), The Journal of Symbolic Logic 46 (2), 1981, pp. 425–426.
472. — (1989) *Abelian logic from A to Z*, in Graham Priest, Richard Sylvan, and Jean Norman (eds.) Paraconsistent Logic, *Essays on the Inconsistent*, Philosophia Verlag, Munich, 1989, pp. 245–288.

473. — (2000) A, *still adorable*, in W. A. Carnielli, M. E. Coniglio, and I. M. Loffreddo D'Ottaviano (eds.), **Paraconsistency**, *The Logical Way to the Inconsistent* (Conference Proceedings of the World Congress on Paraconsistency, dedicated to Newton C. A. da Costa, May 2000, Saõ Paulo, Brazil), Marcel Dekker, New York 2000 [*Lecture Notes in Pure and Applied Mathematics, 228*], pp. 241–260.
474. Mario Mignucci (1993) *The Stoic 'themata'*, in Klaus Döring, and Theodor Ebert (eds.) Dialektiker und Stoiker, *Zur Logik der Stoa und ihrer Vorläufer*, Franz Steiner Verlag, Stuttgart 1993, pp. 217–238.
475. Eugen Gh. Mihăilescu (1966) **Sisteme logice şi forme normale în calculul propoziţional bivalent** [Logic systems and normal forms in two-valued propositional calculus] [Romanian], Editura Academiei Republicii Socialiste România [The Publishing House of the Romanian Academy], Bucharest 1966. (Collected papers.)
476. — (1969) **Logica matematică**, *Elemente de calcul cu propoziţii şi predicate* [Mathematical logic, Elements of propositional and predicate calculus] [Romanian] Editura Academiei Republicii Socialiste România [The Publishing House of the Romanian Academy], Bucharest 1969.
477. Grigoriĭ Efraimovič Minc [Grigori E. Mints] (1975) *Teoriĭa dokazatel'stv: Arifmetika i analiz* [Proof theory: Arithmetic and analysis] [Russian], **Itogi Nauki i Texniki**, *Algebra, Topologiĭa, Geometriĭa* 13, VINITI, Moskow 1975, pp. 5–49. (English translation [*Theory of proofs, arithmetics and analysis*] in Journal of Soviet Mathematics 7 (4), 1977, pp. 501–531. A survey paper.)
478. — (1991) *Proof theory in the USSR 1925–1969*, **The Journal of Symbolic Logic** 56 (2), 1991, pp. 385–424. ('A survey / expository paper.')
479. — (1992) **Selected Papers in Proof Theory**, Bibliopolis, Naples & North-Holland Publishing Company, Amsterdam, etc. 1992 [*Studies in Proof Theory 3*].
480. — (2000) **A Short Introduction to Intuitionistic Logic**, Springer 2000 [*University Series in Mathematics*].
481. Ignacio Miralbell (1987) *La consequentia mirabilis: desarrollo histórico y implicaciones filosóficas*, **Thémata – Revista de filosofia** (Sevilla) 4, 1987, pp. 79–95.
482. Grigore Constantin Moisil (1935) *Recherches sur l'algèbre de la logique*, **Annales Scientifiques de l'Université de Jassy** 22 (1), 1935, pp. 1–118. (The paper is not included in [485], nor in any one of Moisil's paper-collections published in Romania, as, e.g. [484], [486], etc.)
483. — (1936) *Recherches sur le principe d'identité*, **Annales Scientifiques de l'Université de Jassy** 22 (1–4), 1936, pp. 7–56. (Reference from Sergiu Rudeanu [November 2015], corrected from original [*separatum*], by Afrodita Iorgulescu [January 2020]. The paper is not included in [485], nor in any one of Moisil's paper-collections published in Romania.)
484. — (1965) Încercări vechi şi noi de logică neclasică [Old and new essays in non-classical logic] [Romanian] Editura ştiinţifică [Scientific Publishers], Bucharest 1965. (Selected papers on logic; Romanian originals and translations [from French] in Romanian. See also [485])
485. — (1972) **Éssais sur les logiques non-chrysippiennes**, Éditions de l'Académie de la RSR, Bucarest 1972 [821 pp.]. (Collected papers on logic 1938–1972. Mainly reprints of French originals.)
486. — (1976–1992) **Opera matematică**, 1–3 [Mathematical works] [Romanian] Editura Academiei Republicii Socialiste România [The Publishing House of the Romanian Academy], Bucharest 1966–1992 [1:1976, 2:1980, 3:1992] (Collected mathematical papers.)

487. Tim van der Molen (2016) *The Johansson/Heyting letters and the birth of minimal logic*, Preprint, University of Amsterdam, 12 May 2016 [= ILLC Technical Notes Series, X-2016-04].
488. Jules Molk (1885) *Sur une notion qui comprend celle de la divisibilité et sur la théorie générale de l'élimination*, Acta Mathematica 6, 1885 [imprimé le 23 Mai 1884] pp. 1–165 and 166 [*Errata*]. (Originally, PhD Dissertation, Berlin 1885, under Leopold Kronecker.)
489. Jules Molk, and Alfred Pringsheim (1904) *Nombres irrationnels et notion de limite*. (*Exposé, d'après l'article allemand de A. Pringsheim*), in Jules Molk et al. (eds.) Encyclopédie des sciences mathématiques [*Édition française rédigée et publiée d'après l'édition allemande sous la direction de Jules Molk*], tome 1 *Arithmétique et Algèbre*, vol. 1 *Arithmétique*, fasc. 1, Gauthier-Villars, Paris, 1904, pp. 133–160. (Based essentially on Pringsheim's [569]. Molk's additions and comments are marked explicitly in the French text. Reprinted by Éditions Jacques Gabay, Paris 1992 [*Les grands classiques Gauthier-Villars*].)
490. Gregory H. Moore (1988) *The emergence of first-order logic*, in William Aspray, and Philip Kitcher (eds.), History and Philosophy of Modern Mathematics, University of Minnesota Press, Minneapolis MN 1988 [*Minnesota Studies in the Philosophy of Science 11*], pp. 95–135.
491. Ian Mueller (1974) *Greek mathematics and Greek logic*, in John Corcoran (ed.) Ancient Logic and Its Modern Interpretations, D. Reidel Publishing Company, Dordrecht 1974, pp. 35–70.
492. — (1978) *An introduction to Stoic logic*, in John Michael Rist (ed.) The Stoics, University of California Press, Los Angeles / Berkeley CA 1978, pp. 1–26.
493. — (1979) *The completeness of Stoic propositional logic*, Notre Dame Journal of Formal Logic 20 (1), 1979, pp. 201–215.
494. Sara Negri, and Jan von Plato (2001) Structural Proof Theory, Cambridge University Press, Cambridge UK 2001 (digital reprint 2008^R).
495. Marek Nasieniewski (1998) *Is Stoic logic classical?* Logic and Logical Philosophy 6, 1998, pp. 55–61.
496. Rob Nederpelt [Robert Pieter Nederpelt Lazarom] (1977) *Presentation of natural deduction*, Recueil de Travaux de l'Institut Mathématique [Beograd] (n.s.), 2, 1977, pp. 115–126.
497. — (1987) *De taal van de wiskunde, Een verkenning van wiskundig taalgebruik en logische redeneerpatronen* (met een voorwoord van Dirk van Dalen) [The Language of Mathematics, An exploration of the mathematical use of language and the logical proof-patterns (with an Introduction by Dirk van Dalen)] [Dutch], Versluis, Almere [The Netherlands] 1987. (Based on de Bruijn's Eindhoven Lectures [97], 1978, and on [99], 1981, etc. See also [227].)
498. Rob P. Nederpelt, Jan Herman Geuvers, and Roel C. de Vrijer (eds.) (1994) Selected Papers on AUTOMATH, Elsevier Science BV, Amsterdam etc. 1994 [*Studies in Logic and the Foundations of Mathematics 133*]. (See also The AUTOMATH Archive [1].)
499. John von Neumann (1927) *Zur Hilbertschen Beweistheorie*, Mathematische Zeitschrift 26, 1927, pp. 1–46. (Reprinted in JOHN VON NEUMANN Collected Works 1, *Logic, Theory of Sets and Quantum Mechanics* [edited by A. H. Taub], Pergamon Press, New York etc. 1961, pp. 256–300.)
500. Isaac Newton (1675) Letter to Robert Hooke (5 February 1675) = *154. Newton to Hooke*, in H. W. Thurnbull (ed.) The Correspondence of ISAAC NEWTON, 1, 1661–1675, Cambridge University Press, Cambridge UK 1959, pp. 416–417.

501. Jean Nicod (1917–1920) *A reduction in the number of primitive propositions of logic*, Proceedings of the Cambridge Philosophical Society 19, 1917–1920, pp. 32–41.
502. N. H. Niekus, H. van Riemsdijk, and A. S. Troelstra (1981) *Bibliography of A. Heyting*, Nieuw Archief voor Wiskunde (3), 29, 1981, pp. 24–35. (Cf. also the Heyting edition of 1980, [300], pp. iii–xvii.)
503. Pyotr Sergeevič Novikov (1941) *On the consistency of certain logical calculus* [Russian], Matematičeskiĭ Sbornik 12 [54] (2), 1941, pp. 231–261. (See also the reviews by Alonzo Church, in The Journal of Symbolic Logic 11 (4), 1946, pp. 129–131, and Andrzej Mostowski, in The Journal of Symbolic Logic 27 (2), 1962, p. 246.)
504. — 1959 Elementy matematičeskoĭ logiki [Elements of mathematical logic] [Russian], Gosudarstvennoe Izdatel'stvo Fiziko-matematičeskoĭ Literatury, Moscow 1959. (Reprint: Izdatel'stvo 'Nauka', Moskow 1973^R. Translations: French [Dunod, Paris 1964], English [Oliver & Boyd, Edinburgh 1964] [505], Romanian [Editura ştiinţifică, Bucharest 1966], German [VEB Deutscher Verlag der Wissenschaften, Berlin 1973, resp. Friedrich Vieweg & Sohn GmbH, Braunschweig 1973], Italian [Editori Riuniti, Rome 1975], etc.)
505. — 1964 Elements of Mathematical Logic (translation by Leo F. Boron, with a preface and notes by R. L. Goodstein), Oliver and Boyd, Edinburgh 1964. (English translation of [504].)
506. — 1977 Konstruktivnaĭa matematičeskaĭa logika s točki zreniĭa klassičeskoĭ [*Constructive mathematical logic from the classical standpoint*] [Russian] (edited by F. A. Kabakov, B. A. Kušner, and V. V. Dončenko, with a preface by S. I. Adĭan), Izdatel'stvo 'Nauka' [Science Publishers], Moscow 1977 [*Matematičeskaĭa logika i osnovaniĭa matematiki*].
507. — 1979 Izbrannye trudy: *Teoriĭa mnozhestv i funkciĭ, Matematičeskaĭa logika i algebra* [Selected Works: Set Theory and Function Theory, Mathematical Logic and Algebra] [Russian], Izdatel'stvo 'Nauka' [Scientific Publishers], Moskow 1979.
508. Marek Nowak (1992) *On two relatives of classical logic*, Bulletin of the Section of Logic (University of Lódź, Polish Academy of Sciences) 21 (3), 1992, pp. 97–102.
509. Gabriel Nuchelmans (1991) Dilemmatic Arguments. *Towards a History of their Logic and Rhetoric*, North-Holland Publishing Company, Amsterdam, etc. 1991 [*Verhandelingen der Koninklijke Nederlandse Akademie van Wetenschappen, Afdeling Letterkunde* (n.s.), *145*]. (Cf. II.9: Dilemma and consequentia mirabilis, pp. 115–137.)
510. — (1992) *A 17th-century debate on the 'consequentia mirabilis'*, History and Philosophy of Logic 13 (1), 1992, pp. 43–58.
511. John J. O'Connor, and Edmund F. Robertson (1997) *Giovanni Vacca*, December 1997,[5].
512. Hiroakira Ono, and Yuichi Komori (1995) *Logics without the contraction rule*, The Journal of Symbolic Logic 50 (1), 1985, pp. 169–201.
513. Ivan Efimovič Orlov (1928) Isčislenie sovmestnosti predloženiĭ [The calculus of propositional compatibility] [Russian], Matematičeskii Sbornik 35 (3–4), 1928, pp. 263–286. (See also [122], [123], [678], and [14], [170], [171], [49], [50], [703] for further historical cross-references.)
514. Ewa Orłowska (1975) *On the Jaśkowski method of suppositions*, Studia Logica 34 (2), 1975, pp. 187–200.
515. Robert R. O'Toole, and Raymond E. Jennings (2004) *The Megarians and the Stoics*, in Dov M. Gabbay, and John Woods (eds.) Handbook of the History of

Logic 1 (*Greek, Indian and Arabic Logic*), Elsevier / North-Holland Publishing Company, Amsterdam etc. 2004, [Chapter VI], pp. 397–522.

516. Ranganathan Padmanabhan, and Sergiu Rudeanu (2008) **Axioms for Lattices and Boolean Algebras**, World Scientific, Singapore, etc. 2008.

517. Michel Parigot (1997) *Strong normalization for second order classical natural deduction*, The Journal of Symbolic Logic 62 (4), 1997, pp. 1461–1479.

518. Volker Peckhaus (1990) Hilbertprogramm und kritische Philosophie, *Das Göttinger Modell interdisziplinärer Zusammenarbeit zwischen Mathematik und Philosophie*, Vandenhoeck & Ruprecht, Göttingen 1990. (Originally, PhD Dissertation, Universität Erlangen-Nürnberg 1990.)

519. Giuseppe Peano (1894) Notations de logique mathématique [*Introduction au «Formulaire de mathématiques»*], Guadagnini, Turin 1894.

520. — (1895–1903) Formulaire de mathématiques, 1–4, Bocca, Turin 1895 [1], 1898 [2, fasc. 2], 1899 [2, fasc. 3]; Carré & Naud, resp. Gauthier-Villars, Paris 1901 [3], Bocca & Ch. Clausen, Turin 1903 [4].

521. — (1903–1904) *Il latino, quale lingua ausiliare internazionale*, Atti della Reale Accademia delle Scienze di Torino 39, 1903–1904, pp. 273–283.

522. — (1908) Formulario mathematico, 5, Bocca, Turin 1908 (in *latino sine flexione*).

523. Charles Sanders Peirce (1880) [*A Boolian algebra with one constant*], in [526], 4: The Simplest Mathematics, *Book I: Logic and Mathematics (Unpublished Papers)* Paper 1, 4.12–4.20M, and in [527], 4 [1879–1884], Paper 23, pp. 218–221. (A first anticipation of the 'Sheffer stroke' [NAND]. Editors' note, in [526]: 'Untitled paper, c. 1880. Compare H. M. Sheffer's [679], of which this is a striking anticipation. See also [526] [4.]264f. (= [525], here), where the same idea is developed from a different angle.'. Cf. also the *Introduction* to [527], 4 [1879–1884], by Nathan Houser: '[...] in 1880 he [CSP] wrote his short *A Boolian Algebra with One Constant* (item 23), in which he anticipated H. M. Sheffer's paper of 1913 that introduced the stroke function, [679]'.)

524. — (1885) *On the algebra of logic. A contribution to the philosophy of notation*, American Journal of Mathematics (1), 7 (2), 1885, pp. 180–202. (Presented at the National Academy of Sciences, Newport RI, 14–17 October 1884. Reprinted in [526], 3: Exact Logic: *Published Papers*, Paper 13, 3.359–3.403M, and in [527], 5 [1884–1886], Paper 30, pp. 162–190. Among other things, the paper contains a [slightly redundant] complete axiomatization of classical ['Boolian'] propositional logic, in f [*falsum*] and material implication [written $\prec$], with [1] the 'strictly linear' [BCI] axioms [= the first three 'icons': I: $x \prec x$, C: $(x \prec (y \prec z)) \prec (y \prec (x \prec z))$, CB: $(x \prec y) \prec ((y \prec z) \prec (x \prec z))$, [2] 'Peirce's Law' [= the 'fifth icon'], P: $((x \prec y) \prec x) \prec x$, [3] *ex falso quodlibet*: $f \prec x$ [= the 'fourth icon'], and *modus ponens*. The 1885-proposal is independent of Frege's **Begriffsschrift** [Halle 1879] system, and, in a way, superior to the latter, because [1–2] isolate already axiomatically material implication, and classical negation is understood 'inferentially', as $\sim x := x \prec f$. Actually, the 1885-system is also an anticipation of the 'extended propositional logic', i.e., propositionally quantified classical logic, credited to Bertrand Russell and Alfred Tarski, usually [see, e.g. [125], § 28 *Extended propositional calculus and protothetic*, pp. 151–154], because Peirce had also something equivalent to Π and Σ, as *propositional quantifiers*, and was aware of the fact that f [*falsum*] can be defined as $f := \Pi p.p.$ — Cf. also Russell's [647], [649], and [125], *loc. cit.* Notably, Russell uses Peirce's 'fifth icon' in his [647], [649], with his own [colloquial] label, without naming Peirce explicitly, on this very detail, although he was, no doubt, familiar with the 1885-paper before 1903.)

525. — (1902) *The simplest mathematics*, in [526], 4: The Simplest Mathematics, *Book I: Logic and Mathematics (Unpublished Papers)* Paper 7, 4.227–4.323M. (Editors' note: 'Chapter 3 of the *Minute Logic*, dated January–February, 1902 [...]'. See especially [526], 4.264 and the [Editors'] footnote 2, page 215: 'This is another anticipation of the Shefferian stroke-function; cf. 4.12ff.' [i.e., [523], here].)

526. — (1931–1935) Collected Papers [of CHARLES SANDERS PEIRCE], 1–6 (edited by Charles Hartshorne, and Paul Weiss), [The Belknap Press of the] Harvard University Press, Cambridge MA 1931–1935^1, 1960^2.

527. — (1982–2000) The Writings of CHARLES S. PEIRCE [*A Chronological Edition*], 1–6 (edited by Max H. Fisch, Christian J. W. Kloesel, Edward C. Moore, Nathan Houser et alii), Indiana University Press, Bloomington IN, 1988–2000.

528. Judy Pelham (1993) Russell on Propositions and Objects, PhD Dissertation, University of Toronto 1993 [unpublished, not seen].

529. — (1996) *A reconstruction of Russell's substitution theory*, in Mathieu Marion, and Robert S. Cohen (eds.) Quebec Studies in the Philosophy of Science I, *Essays in Honor of Hugues Leblanc*, Kluwer Academic Publishers, Dordrecht, Boston, and London, pp. 123–132. [*Boston Studies in the Philosophy of Science 177*]

530. — (1999) *Russell, Frege, and the nature of implication*, Topoi 18 (2), [September] 1999, pp. 175–184.

531. Francis Jeffry Pelletier (1999) *A brief history of natural deduction*, History and Philosophy of Logic 20 (1), 1999, pp. 1–31.

532. — (2001) *A history of natural deduction and elementary logic textbooks*, in John Woods, and Bryson Brown (eds.) Logical Consequence, *Rival Approaches* 1, Hermes Science Publications, Oxford 2001, pp. 105–138.

533. Dariusz Piętka (2008) *Stanisław Jaśkowski's logical investigations*, Organon 37 (40), 2008, pp. 39–69.

534. Jan von Plato (2017) Saved from the Cellar, *Gerhard Gentzen Shorthand Notes on Logic and the Foundations of Mathematics*, Springer International Publishing [CH] 2017 [*Sources and Studies in the History of Mathematics and Physical Sciences*]. (Reviewed by Adrian Rezuş, in [621].)

535. Valeriĭ Egorovič Plisko (1988) *The Kolmogorov calculus as a part of minimal calculus*, Russian Mathematical Surveys 43 (6), 1988, pp. 95–110. (Russian original in Uspexi Matematičeskix Nauk 43 (6 [264]), 1988, pp. 79–91. See also [364], [536].)

536. — (1989) *A correction [to]* Russian Math[ematical] Surveys *43:6 (1988), 95–110*, Russian Mathematical Surveys 44 (3), 1989, p. 232.

537. Willem Louis van de Poel (1969) *De fundering van de propositionele calculus met behulp van een axioma* [The foundation of propositional calculus with the help of a single axiom] [Dutch], Report ZW 1-1969, pp. 1–6, Mathematisch Centrum [*CWI*], Afdeling Zuivere Wiskunde [Department of Pure Mathematics], Amsterdam 1969. Online @ the CWI [reprints] Repository, Amsterdam. (Conference in the MC-series *Elementaire onderwerpen van hoger standdpunt uit belicht* [Elementary topics from an advanced point of view]. The talk contains an 'elementary' discussion of Jean Nicod's single axiom of 1916, [501], based on the 'Sheffer functor'.)

538. — (1979) *Letter to Adrian Rezuş*, September 1979.

539. — (1988) *Een leven met computers. Afscheidsrede* [A Life With the Computers. A Valediction Speech] [Dutch], Delft University of Technology, October 26, 1988.

540. — (2003) *Curriculum Vitae* [English], Delft University of Technology, April 2003.

541. Willem Louis van de Poel, Chaim E. Schaap, and Gerrit van der Mey (1980) *New arithmetical operators in the theory of combinators*, I–III, Koninklijke

Nederlandse Akademie van Wetenschappen. Proceedings [Series A], 83, 1980, pp. 271–325. (Also in Indagationes Mathematicae 42, 1980.)
542. Witold A. Pogorzelski (1964) *The deduction theorem for Łukasiewicz many-valued propositional calculi*, Studia Logica 15, 1964, pp. 7–19.
543. — (1968) *On the scope of the classical deduction theorem*, The Journal of Symbolic Logic 33 (1), 1968, pp. 77–81. (Reviewed by Mircea Tîrnoveanu, in The Journal of Symbolic Logic 40 (4), 1975, p. 606.)
544. — (1971) *Structural completeness of the propositional calculus*, Bulletin de l'Académie Polonaise des Sciences (Série des sciences mathématiques, astronomique et physiques) 19, pp. 349–351.
545. — (1994) *A minimal implicational logic*, in Jan Woleński (ed.) Philosophical Logic in Poland, Kluwer Academic Publishers, Dordrecht 1994 [*Synthese Library 228*], pp. 213–216. (Reviewed by Kosta Došen, in MRh:03022.)
546. Witold A. Pogorzelski, and Paweł Wojtylak (2001) Cn-*definitions of propositional connectives*, Studia Logica 67, 2001, pp. 1–26
547. — (2008) Completeness Theory for Propositional Logics, Bikhäuser, Basle etc. 2008.
548. Wolfram Pohlers (1989) Proof Theory, *An Introduction*, Springer Verlag, Berlin etc. 1989 [*Lecture Notes in Mathematics 1407*].
549. — (2009) Proof Theory, *The First Step into Impredicativity*, Springer Verlag, Berlin & Heidelberg 2009 [*Universitext*].
550. Henri Poincaré (1894) *Sur la nature du raisonnement mathématique*, Revue de Métaphyisque et de Morale 2, 1894, pp. 371–384. (Reprinted with alterations in [552], pp. 29–44. English translation, by George Bruce Halsted, in [557], pp. 31–42. Reprinted in [194], vol. 2, pp. 972–982.)
551. — (1900) *Du rôle de l'intuition dans la logique et la mathématique*, in Compte rendu du Deuxième Congrès International des Mathématiciens (Paris, 6–12 août 1900), Gauthier-Villars, Paris 1900, pp. 15–30. (Reprinted with alterations in [553], pp. 29–44. English translation, by George Bruce Halsted, in [557], pp. 210–222. Reprinted in [194], vol. 2, pp. 1012–1020.)
552. — (1902) La Science et l'hypothèse, Flammarion, Paris 1902.
553. — (1905) La Valeur de la science, Flammarion, Paris 1905.
554. — (1905–1906) *Les mathématiques et la logique*, Revue de Métaphyisque et de Morale 13, 1905, pp. 815–835, 14, 1906, pp. 14–34, and 294–317. (Reprinted with extensive deletions in [556], Chapters 3–5. Translated by George Bruce Halsted in [557]. English translation [*Mathematics and logic*], by William Ewald, in [194], vol. 2, pp. 1021–1038, 1038–1052, and 1052–1971 resp.)
555. — (1906) *À propos de la logistique*, Revue de Métaphyisque et de Morale 14, 1906, pp. 866–868.
556. — (1908) Science et méthode, Flammarion, Paris 1908.
557. — (1913) The Foundations of Science (translated by George Bruce Halsted), The Science Press, New York 1913.
558. Jean Porte (1982) *Fifty years of deduction theorems*, in J. Stern (ed.) Proceedings of the Herbrand Symposium / Logic Colloquium '81 (Marseilles, July 1981), North-Holland Publishing Company, Amsterdam etc. 1982 [*Studies in Logic and the Foundations of Mathematics 107*], pp. 243–250. (Reviewed by John Corcoran, in MR 757033 [= 85j:03002].)
559. Carl [von] Prantl (1855–1870) Geschichte der Logik im Abendlande, 1–4, Verlag von S. Hirzel, Leipzig, 1855–1870 [1:1855, 2:1861, 3:1867, 4:1870]. Anastatic reprints: Akad. Verlag, Berlin 1955 (in three volumes) resp. Georg Olms, Hildesheim 1997.

560. Dag Prawitz (1965) **Natural Deduction**, *A Proof-Theoretical Study*, Almqvist & Wiksell, Uppsala [Stockholm-Gothenburg-Uppsala] 1965 [*Acta Universitatis Stockholmiensis: Stockholm Studies in Philosophy 3*]. (Reprinted [with corrections] by Dover Publications, Mineola NY 2006 [*Dover Books in Mathematics*]. Originally, PhD Dissertation, University of Stockholm 1965.)
561. — (1971) *Ideas and results in proof theory*, in Jens Erik Fenstad (ed.) **Proceedings of the Second Scandinavian Logic Symposium**, North-Holland Publishing Company, Amsterdam, etc., 1971, pp. 235–307 [*Studies in Logic and the Foundations of Mathematics*].
562. — (1973) *Towards a foundation of general proof theory*, in Patrick Suppes, Leon Henkin, Athanase Joja, and Grigore C. Moisil (eds.) **Logic, Methodology, and Philosophy of Science 4**, North-Holland Publishing Company, Amsterdam, etc. 1973 [*Studies in Logic and the Foundations of Mathematics*], pp. 225–250.
563. — (1974) *On the idea of a general proof theory*, Synthese 27, 1974, pp. 63–77.
564. — (1977) *Meanings and proofs: on the conflict between classical and intuitionistic logic*, Theoria [Lund] 43, 1977, pp. 2–40.
565. — (1979) *Proofs and the meaning and completeness of the logical constants*, in Jaakko Hintikka, Ilkka Niiniluoto, and Esa Saarinen (eds.) **Essays on Mathematical and Philosophical Logic**, D. Reidel Publishing Company, Dordrecht, etc. 1979, pp. 25–40.
566. — (1981) *Philosophical aspects of proof-theory*, in Guttorm Fløistad, and Georg Henrik von Wright (eds.) **Contemporary Philosophy, A New Survey 1**, *Philosophy of Language / Philosophical Logic*, Martinus Nijhoff, The Hague, etc. 1981, pp. 335–277.
567. Dag Prawitz, and Per-Erik Malmnäs (1968) *A survey of some connections between classical, intuitionistic and minimal logic*, in H. Arnold Schmidt, Kurt Schütte, and H.-J. Thiele (eds.) **Contributions to Mathematical Logic**, North-Holland Publishing Company, Amsterdam 1968 [*Studies in Logic and the Foundations of Mathematics 50*], pp. 215–229.
568. Robert Price (1961) *The stroke function in natural deduction*, Zeitschrift für mathematische Logik und Grundlagen der Mathematik 7, 1961, pp. 117–123.
569. Alfred Pringsheim (1898) *Irrazionalzahlen und Konvergenz unendlicher Prozesse* in Wilhelm Franz Meyer (ed.) Encyklopädie der mathematischen Wissenschaften mit Einschluss ihrer Anwendungen, Erster Band, *Arithmetik und Algebra*, Heft 1, I A 3, [7.XI.1898], Druck und Verlag von B. G. Teubner, Leipzig 1898–1904, pp. 47–146. (See also [489].)
570. Arthur N. Prior (1956) *Logicians at play; or Syll, Simp and Hilbert*, Australasian Journal of Philosophy 34 (3), 1956, pp. 182–192.
571. — (1962^2) **Formal Logic**, Clarendon Press, Oxford 1962 (second, revised edition).
572. — (1969) *Corrigendum to C. A. Meredith's and my paper: 'Equational logic'*, Notre Dame Journal of Formal Logic 9 (4) 1969, p. 452.
573. Ruy J. Guerra B. de Queiroz, Anjolina Grisi de Oliveira, and Dov M. Gabbay (2012) **The Functional Interpretation of Logical Deduction**, World Scientific, Singapore, etc. 2012 [*Advances in Logic 5*]. (Reviewed by Adrian Rezuş, in [615].)
574. Willard van Orman Quine (1940) **Mathematical Logic**, Harvard University Press, Cambridge MA 1940. (Revised edition, Harvard University Press, Cambridge MA & London 2003^R.)
575. Roberto Radice (ed.) (1998) **Stoici antichi**. *Tutti i frammenti raccolti da Hans von Arnim* [Testo greco e latino in fronte] (Presentazione di Giovanni Reale), Bompiani, Milan 1998, 2002^R [*Il Pensiero Occidentale*].

576. Frank Plumpton Ramsey (1925) *The Foundations of Mathematics*, Proceedings of the London Mathematical Society (2), 25 (Part 5), 1925, pp. 338–384. (Reprinted in FRANK PLUMPTON RAMSEY The Foundations of Mathematics and Other Logical Essays, edited by Richard Bevan Braithwaite, with a Preface by George Edward Moore, Routledge & Kegan Paul Ltd., London 1931 [*International Library of Psychology, Philosophy and Scientific Method*] [second impression 1950^2], pp. 1–61.)

577. Stephen Read (1999) *Sheffer's stroke: a study in proof-theoretic harmony*, Danish Yearbook of Philosophy 34, 1999, pp. 7–23.

578. Niels Jakob Rehof, and Morten Heine Sørensen (1994) λ_Δ-*calculus*, in TACS '94, *Proceedings of the International Conference on Theoretical Aspects of Computer Software*, Springer Verlag, Berlin etc. 1994 [*Lecture Notes in Computer Science 789*], pp. 516–542. (Also: Preprint Department of Computer Science, University of Copenhagen, 26 October 1993.)

579. Constance Reid (1986) Hilbert-Courant, Springer-Verlag, New York, Berlin etc. 1986 (Previously issued as two separate volumes: Hilbert, Springer Verlag 1970, and Courant in Göttingen and New York, *The Story of an Improbable Mathematician*, Springer Verlag 1976.)

580. Michael D. Resnik (1980) Frege and the Philosophy of Mathematics, Cornell University Press, Ithaca NY 1980.

581. Greg Restall (1994) Logics without Contraction, PhD Dissertation, University of Queensland 1994.

582. Adrian Rezuş (1978) *Restricted solvability* (abstract), Geneva, September 1978. (Based on an idea of Stephen C. Kleene, [355]). For applications, see [584], [587].)

583. — (1980) *Singleton bases for subsets of Λ_K^0 and a result of Alfred Tarski*, Preprint 150, Department of Mathematics, University of Utrecht, April 1980 [ii + 43 pp.]. (Cf. [587], [609].)

584. — (1981) Lambda-conversion and Logic, Elinkwijk BV, Utrecht 1981 [xii + 197 pp.]. (PhD Dissertation, University of Utrecht, [4 June] 1981, under the supervision of Dirk van Dalen and Henk Barendregt.)

585. — (1981a) *Stellingen* [Propositions] [Dutch] to the PhD Dissertation [584], Utrecht, 4 June 1981. (Dutch version by Henk Barendregt.)

586. — (1981b) *Bibliography and Indices* to [39] (first edition 1981), pp. 580–615.

587. — (1982) *On a theorem of Tarski*, Libertas Mathematica (Arlington TX) 2, 1982, pp. 62–95. (Work of 1978–1979. Also: Preprint 227, Department of Mathematics, University of Utrecht, January 1982 [35 pp.]. Cf. [588], and [609].)

588. — (1982a) *On a singleton basis for the set of closed lambda-terms*, Libertas Mathematica (Arlington TX) 2, 1982, p. 94. (Appendix to [587], dated Utrecht, October 1981.)

589. — (1982b) A Bibliography of Lambda-calculi, Combinatory Logics and Related Topics, Mathematisch Centrum [*CWI*], Amsterdam 1982 [i + 86 pp.] [*MC Varia*] [ISBN 90-6196-234-X].

590. — (1983) Abstract AUTOMATH, Mathematisch Centrum [*CWI*], Amsterdam 1983 [vi + 188 pp.] [*Mathematical Centre Tracts 160*] [ISBN 90-6196-256-0].

591. — (1986) Impredicative Type Theories, University of Nijmegen, Faculty of Mathematics and Science, Department of Computer Science, Report KUN–WN–CS TR–85–1986, Nijmegen 1986 [288 pp.]. (Originally, lectures for the Department of Computer Science, University of Nijmegen.)

592. — (1986a) *Semantics of Constructive Type Theory*, Libertas Mathematica [Arlington TX] 6, 1986, pp. 1–82. (Also: Report KUN–WN–CS TR–2–1987 (n.s.), University of Nijmegen, Department of Computer Science.)

336   References

593. — (1986b) AUTOMATH: Syntax and Semantics, Lecture given at the Institute of Advanced Studies, The Australian National University, The *Automated Reasoning Project*, Canberra ACT, September 1986.
594. — (1987) Propositions-as-types Revisited [*1 Higher-order constructive type theory*], University of Nijmegen, Faculty of Mathematics and Science, Department of Computer Science, Report KUN–WN–CS TR–97–1987, Nijmegen [February] 1987 [91 pp.]. (Originally, lectures for the Department of Computer Science, University of Nijmegen.)
595. — (1987a) Varieties of Generalized Functionality, University of Nijmegen, Faculty of Mathematics and Science, Department of Computer Science, Report KUN–WN–CS TR–102–1987, February 1987 [106 pp.] (With an Appendix by Peter J. de Bruin. Originally, lectures for the Department of Computer Science, University of Nijmegen.)
596. — (1987b) Constructions and Propositional Types, University of Nijmegen, Faculty of Mathematics and Science, Department of Computer Science, Internal Report KUN-WN-CS 87–1–1987, April 1987 [64 pp.]. (Originally, lectures for the Department of Computer Science, University of Nijmegen.)
597. — (1987c) *Two 'modal' logics 1*, Paris, June 1987 [revised Nijmegen, August 28, 1990]. (Address for the XII-ème Congrès de L'Académie Roumano-Américaine [ARA], III-ème Section: Mathématiques, Physique, Paris-Sorbonne, France, 22–27 June 1987, @ www.equivalences.org.)
598. — (1987d) *Generalized typed lambda-calculi: recent advances*, Paper contributed to the XII-ème Congrès de L'Académie Roumano-Américaine [ARA] (III-ème Section: Mathématiques, Physique), Paris-Sorbonne, France, 22–27 June 1987. (Summary of a Research Report *NWO-SION* on Constructive Mathematics and Functional Programming Languages, *NWO* [The Dutch National Science Research Foundation], The Hague & *Stichting voor Informatica Onderzoek in Nederland, SION* [The Dutch Foundation for Research in Computer Science], Amsterdam, June 1987.)
599. — (1988) *Paul Hertz und die Ursprung der modernen Beweistheorie*, Nijmegen, June 1988 [unpublished]. (Introduction to a projected, subsequently abandoned, edition PAUL HERTZ Logische Schriften, Nijmegen 1988–1989. Cf. [274], [275], [276], [278], [279], [280], [281], and Gentzen's [223] [on *Satzsysteme*], [282], [283], [284], [285] [on 'the essence of logic'], and possibly, [277] [cognitive science].)
600. — (1988a) *A type-theoretic approach to classical and non-classical logics*, Talk for the International 'Jumelage' Workshop Typed Lambda-Calculi, University of Nijmegen, 14–18 November 1988.
601. — (1989) *What is a 'classical' proof?*, Lecture for a GMD-Kolloquium held at the *GMD*, the German National Research Center, *Abteilung* Karlsruhe [University of Karlsruhe, *GMD Forschungsstelle für Programmstrukturen*, Computing Department], May 1989.
602. — (1989a) *The type theory of classical logic*, Lecture for the Informatica-Colloquium, University of Groningen, Department of Computer Science, June 1989.
603. — (1989b) Lambda, *A λ-calculus interpreter and combinatory code generator*, Nijmegen 1989 [software documentation].
604. — (1990) Classical Proofs, *λ-calculus Methods in Elementary Proof Theory*, Nijmegen [5 August] 1990. (Originally, Lecture Notes for a course given at the Department of Computer Science, University of Nijmegen 1986–1987, reprint @ www.equivalences.org.)

605. — (1991, 1993^{R}) Beyond BHK, Nijmegen, 1 December 1991. (A slightly revised version [dated 20 July 1993, updated bibliographically in 2000] appears online @ www.equivalences.org. Introduction and extended abstract reprinted as [607].)
606. — (1992) *Finitism and proof-operations – On a finitistic concept of classical proof-conversion*, Nijmegen, 27 June – 27 August 1992 [unpublished].
607. — (1993) *Beyond BHK* (extended abstract), in Henk Barendregt, Marc Bezem, and Jan Willem Klop (eds.) Dirk van Dalen Festschrift, University of Utrecht, Utrecht 1993, pp. 114–120 [*Quaestiones Infinitae 5*. Publications of the Department of Philosophy, University of Utrecht].
608. — (2009) *Im Buchstabenparadies – Gottlob Frege and his Regellogik*, Nijmegen, 13 June 2009, revised 16 March 2016; updated version [*Im Buchstabenparadies*] in THIS VOLUME.
609. — (2010) *Tarski's claim, thirty years later*, Nijmegen, 27–28 September – 1 October 2010, in THIS VOLUME.
610. — (2010a) *What is a 'challenging problem'?*, Nijmegen, September-October 2010, revised 12 May 2016, in THIS VOLUME.
611. — (2015) *Classical propositional quantifiers*, Nijmegen, October 2015 – December 2018. Revised as *'Begriffsschrift' as Beweisschrift II*, 9 October 2019 – 15 February 2020 [forthcoming].
612. — (2015a) Review of [43], Studia Logica 103 (6), 2015, pp. 1319–1326.
613. — (2015b) *Old folklore on λ-calculus generators*, Nijmegen, 23 May 2015, revised 3 May 2017, in THIS VOLUME. (Excerpted from [583].)
614. — (2015c) *On seeing out of the fog*, Nijmegen, 24 October 2015, revised 10 November 2017. (Comments on [286]. Absorbed into [630] and revised for THIS VOLUME.)
615. — (2015d) Review of [573], Studia Logica 103 (2), 2015, pp. 447–451.
616. — (2016) An Ancient Logic (*Chrysippus and His Modern Readers I*), LAP – Lambert Academic Publishing, Saarbrücken 2016 [ISBN: 978-3-330-01661-3]. (Revised in THIS VOLUME.)
617. — (2017) *Łukasiewicz, Jaśkowski and Natural Deduction* (*Curry-Howard for Classical Logic*), in Massoud Pourmahdian, and Ali Sadegh Daghighi (eds.) Logic Around the World, AFJ – Andisheh & Farhang-e Javidan Publishing, Tehran 2017 [ISBN: 978-600-6386-99-7], pp. 89–138. (Revised in THIS VOLUME.)
618. — (2017a) *Witness theory for classical logic* (*Inferential systems*), Nijmegen, April 2017. Revised as *Witness theory for inferential systems*, October 2019, in THIS VOLUME.
619. — (2017b) *Cartesian folds and Witness Theory*, Nijmegen, 9–21 June 2017, in THIS VOLUME.
620. — (2017c) *On the Kolmogorov λγ-calculus*, Nijmegen, 26 June 2017, in THIS VOLUME.
621. — (2017d) Review of [534], Studia Logica 107 (3), [June] 2019, pp. 583–589.
622. — (2017e) *Proof-détours in classical logic*, Nijmegen, 16–20 July 2017, revised 31 October 2017, in THIS VOLUME.
623. — (2017f) *Does (η) really matter?* Nijmegen, 15 August 2017, in THIS VOLUME.
624. — (2018) *The scientific activity of Bolesław Sobociński* (*A bio-bibliographical survey*), Nijmegen, 24 June 2018 [forthcoming].
625. — (2018a) *The single-axiom problem*, Nijmegen, [draft] 17 August 2018.
626. — (2019) *Tarski singleton bases: 1925–1932* (*On an allegedly lost 'method of proof' of Alfred Tarski*), Nijmegen, 9 August 2019, revised in THIS VOLUME.
627. — (2019a) *'Begriffsschrift' as Beweisschrift*, Nijmegen 2019, in THIS VOLUME.

628. — (2019b) *The Curry-Howard Correspondence revisited*, Nijmegen 2019, in THIS VOLUME.
629. — (2019c) *Implications and generators*, Nijmegen 2019, in THIS VOLUME.
630. — (2019d) *...en avançant dans le brouillard...*, Nijmegen 2019, in THIS VOLUME.
631. — (2019e) *The witness theory of Abelian groups*, Nijmegen 2019 [forthcoming].
632. — (2020) Review of [130] [forthcoming].
633. V. Frederick Rickey (2018) *Professor Bolesław Sobociński and logic at Notre Dame*, Cornwall NY, 30 May 2018 (to apper in the Proceedings of the Canadian Society for History and Philosophy of Mathematics [CSHPM], Montréal CA).
634. — (2019) *Sobociński's Nachlass*, Cornwall NY, October 2019 [forthcoming].
635. John Alan Robinson (1965) *A machine-oriented logic based on the resolution principle*, Journal of the Association for Computing Machinery 12 (1) [January] 1965, pp. 23–41.
636. — (1979) Logic, Form and Function, *The Mechanization of Deductive Reasoning*, Edinburgh University Press, Edinburgh 1979.
637. Gemma Robles, and José M. Méndez (2005) *Two versions of minimal intuitionism with the CAP. A note*, Theoria: Revista de Teoría, Historia y Fundamentos de la Ciencia 20 (2), 2005, pp. 183–190.
638. Paul C. Rosenbloom (1950) Elements of Mathematical Logic, Dover Publications, Inc., New York 1950 [*The Dover Series in Mathematics and Physics*]. (Reviewed by I. L. Novak, in Bulletin of the American Mathematical Society 58 (2), 1952, pp. 266–268. Reprint: Dover Publications, Mineola NY 2005^R.)
639. J. Barkley Rosser (1935) *A mathematical logic without free variables*, Annals of Mathematics (2), 36, 1935, pp. 127–150, and Duke Mathematical Journal 1, 1935, pp. 329–355. (Originally, PhD Dissertation, Princeton University 1935, under the supervision of Alonzo Church.)
640. — (1942) *New sets of postulates for combinatory logics*, The Journal of Symbolic Logic 7 (1), 1942, pp. 18–27.
641. — (1953) *Logique combinatoire et λ-conversion*, in J. BARKLEY ROSSER Deux esquisses de logique, Gauthier-Villars, Paris & E. Nauwelaerts, Louvain 1953, pp. 3–31.
642. — (1971) Letter to Henk Barendregt, 20 December 1971.
643. — (1984) *Highlights of the history of lambda-calculus*, Annals of the History of Computing 6 (4), 1984, pp. 337–349.
644. Gian-Carlo Rota (1997) *The phenomenology of mathematical proof*, Synthese 111 (2), 1997, pp. 183–196. (Also in Fabrizio Palombi (ed.) GIAN-CARLO ROTA Indiscrete Thoughts, Birkhäuser, Basel, Boston etc. 1997 [reprinted 2008^R], Chapter XI, pp. 134–150.)
645. Sergiu Rudeanu (1963) *Axiomele laticilor şi ale algebrelor booleene* [Axioms for lattices and Boolean algebras] [Romanian], Editura Academiei R. P. Romîne [The Publishing House of the Romanian Academy], Bucharest 1963. (See also the revision [516].)
646. Wim Ruitenburg (1991) *The unintended interpretations of intuitionistic logic*, in Thomas Drucker (ed.) Perspectives on the History of Mathematical Logic, Birkhäuser, Boston, Basel & Berlin 1991 (reprinted 2008^R) [*Modern Birkhäuser Classics*], pp. 134–160.
647. Bertrand Russell (1903) The Principles of Mathematics, Cambridge University Press, Cambridge UK 1903. (Second edition 1937^2. Reprint ['Third Imprint'] by George Allen & Unwin, London, etc. 1979^R.)

648. — (1903a) *Functions* [MS circa 1903], in **Foundations of Logic 1903–05** = Alasdair Urquhart, and Albert C. Lewis (eds.) BERTRAND RUSSELL The Collected Papers 4, Routledge, London & New York 1994, pp. 49–73. ('The use of a vertical bar for function application indicates that the manuscript was probably written in the later part of 1903.' [editorial note at **3a**, p. 49].)
649. — (1906) *The theory of implication*, **American Journal of Mathematics 28** (2), 1906, pp. 159–202. (Cf. Chapter 1, in Gregory H. Moore (ed.) [The McMaster University Edition of] BERTRAND RUSSELL The Collected Papers 5, *Toward 'Principia Mathematica'*, *1905–08, 1st Edition*, Routledge, London and New York NY 2014. For Russell's use of 'Peirce's Law', see, e.g., [18].)
650. Girolamo Saccheri SJ (1697) **Logica Demonstrativa** *quam una cum thesibus ex tota philosophia decerptis, dependendam proponit Joannes Franciscus Caselette...*, Typis Ioannis Baptistae Zappatae, Augustae Taurinorum [Turin] 1697^1, in 8°. (Reprint *Mit einer Einleitung von Wilhelm Risse. Nachdruck der Ausgabe 1697* [Domenico Paolini] Turin, by Olms Verlag, Hildesheim & New York 1980, 1983^R. There are two Italian translations: **Logica dimostrativa**, a bilingual Latin-Italian edition, *testo latino al fronte*, edited by Paolo Pagli & Corrado Mangione, Bompiani, Milan 2011 [*Il pensiero occidentale*], containing the Latin text with *face à face* Italian translation of the edition issued by Heinrich Noethen, Augustae Ubiorum MDCCXXXV [Köln 1735], and **Logica dimostrativa, 1–2**, edited by Massimo Mugnai & Massimo Girondino, Edizioni della Normale [Scuola Normale Superiore di Pisa, Centro di Ricerca Matematica], Pisa 2012, whose first volume contains an anastatic reprint of the edition issued in Ticino Regia MDCCI [Pavia 1701], with the Italian translation in the second volume, pp. 1–213.)
651. — (1733) **Euclides ab omni naevo vindicatus**; *sive conatus geometricus quo stabiliuntur prima ipsa universae geometriae principia, auctore Hieronymo Saccherio Societatis Jesu; in Ticinensi Universitate Matheseos Professore Opusculum Exmo Senatui Mediolanensi ab auctore dicatum...*, Paolo Antonio Montano [typographer], Mediolani [i.e., Milan] 1733, xvi + 142 pp., in 4°. (Part I, Latin-English text: **Girolamo Saccheri's Euclides Vindicatus**, edited and translated by George Bruce Halsted, Open Court, Chicago & London 1920, reprint: Chelsea Publishing Company, New York 1986^R. There are *partial* translations of Part I: in *German*, by Paul Stäckel and Friedrich Engel, in **Die Theorie der Parallellinien von Euklid bis auf Gauss**, B. G. Teubner Verlag, Leipzig 1895, pp. 31–135, reprinted by the Johnson Reprint Corporation, New York & London 1968^R, in *Italian*, by Giuseppe Boccardini, in *L'Euclide emendato del P[adre] Gerolamo Saccheri...*, Ulrico Hoepli, Milan 1904, and in *Interlingue*, by C. E. Sjöstedt, in C. E. Sjöstedt (ed.) **Le axiome de paralleles de Euclides a Hilbert**, *un problema cardinal in le evolution del geometrie. Excerptes in faccimile... e traduction in le lingue international auxiliari Interlingue, Interlingue-Fundation*, Uppsala and *Bokvörlaget Natur och Kultur*, Stockholm 1986, pp. 96–175. Part II has been translated into English by Linda Allegri in her thesis: **The Mathematical Works of Girolamo Saccheri, S.J. (1667–1733)** [PhD Dissertation, Columbia University 1960, 272 pp.], University Microfilms, Ann Arbor MI 1960. A new edition of the Latin text with an Italian translation: **Euclide liberato da ogni macchia** (edited, with translation and notes by Pierangelo Frigerio, and an introduction by Imre Tóth and Elisabetta Cattanei) has been published by Bompiani, Milan 2001 [*Il pensiero occidentale*]. Yet another complete Italian translation: **Euclide vendicato da ogni nero, 1–2**, edited by Vincenzo De Risi, has been issued by the Edizione della Normale [Scuola

Normale Superiore di Pisa], Pisa 2011, with [1] an edition of the text and [2] an anastatic reproduction of the Latin original. The new [complete] English edition: **Euclid Vindicated from Every Blemish** (edited and annotated by Vincenzo De Risi, translated by George Bruce Halsted, and Linda Allegri), Birkhäuser [Springer International Publishing], Basle 2014 [*Classic Texts in the Sciences*] is based on the Italian edition of 2011 and on the English translations mentioned before. For Saccheri's use of *consequentia mirabilis*, see [20].)

652. Kenneth M. Sayre (2014) **Adventures in Philosophy at Notre Dame**, University of Notre Dame Press, Notre Dame IN 2014.
653. Michael Scanlan (2000) *The known and unknown H. M. Sheffer*, **Transactions of the Charles S. Peirce Society 36** (2), 2000, pp. 193–224.
654. Matthias Schirn (ed.) (1996) **Frege, Importance and Legacy**, Walter de Gruyter, Berlin & New York 1996 [*Perspektiven der analytischen Philosophie 13*].
655. Moses Schönfinkel (1924) *Über die Bausteine der mathematischen Logik*, **Mathematische Annalen 92** (34), 1924, pp. 305–316. (English translation [*On the building blocks of mathematical logic*], by Stefan Bauer-Mengelberg, in [288], pp. 357–367, with an introduction by Willard van Orman Quine [pp. 355–357]. French translation [*Sur les éléments de construction de la logique mathématique*], by Geneviève Vandevelde, in **Mathématiques et sciences humaines 112** (28e année), 1990, pp. 5–26, with 'Analyse et notes' by Jean-Pierre Ginsti, pp. 5–11. The *first translation* of this paper was in *Romanian*, viz. [*Despre elementele constitutive ale logicii matematice*], in [750], pp. 105–118. [The translator, not mentioned in print, was Uşer Morgenstern, of Bucharest.)
656. Kurt Schütte (1960) **Beweistheorie**, Springer Verlag, Berlin 1960 [*Grundlehren der mathematischen Wissenschaften 103*].
657. — (1977) **Proof Theory** (translated by John N. Crossley), Springer Verlag, Berlin etc. 1977 [*Grundlehren der mathematischen Wissenschaften 225*]. (A translation of [656].)
658. Peter Schroeder-Heister (1997) *Frege and the resolution calculus*, **History and Philosophy of Logic 18** (2), 1997, pp. 95–108.
659. — (1999) *Gentzen style features in Frege*, **Abstracts of the 11th International Congress of Logic, Methodology and Philosophy of Science**, Cracow, Poland (August 1999), Cracow, 1999, p. 449.
660. — (2002) *Resolution and the origins of structural reasoning: early proof-theoretic ideas of Hertz and Gentzen*, The Bulletin of Symbolic Logic 8 (2) [June] 2002, pp. 246–265.
661. Karl Schröter (1952) *Deduktiv abgeschlossene Mengen ohne Basis*, **Mathematische Nachrichten 7** (5), 1952, pp. 293–304. (Results of 1937.)
662. Helmut Schwichtenberg (1994) **Proof Theory**, Lecture Notes, *Mathematisches Institut der Universität München*, summer semester 1994, Munich, July 1994. [iii + 103 pp.] (See also [663].)
663. Helmut Schwichtenberg, and Anne S. Troelstra (1996, 2000[2]) **Basic Proof Theory**, Cambridge University Press, Cambridge UK, 1996, 2000[2] (second edition, 'revised and expanded') [*Cambridge Tracts in Theoretical Computer Science 43*]. (First edition reviewed by Harold Schellinx, in **Journal of Logic, Language, and Information 7**, 1998, pp. 221–223, and by Roy Dyckhoff, in **The Journal of Symbolic Logic 63**, (4), 1998, pp. 1605–1606. Second edition reviewed by Jeremy Avigad, in **ACM SIGACT News 32** (2), [June] 2001, pp. 15–19.)
664. Dana S. Scott (1963) **A System of Functional Abstraction**, Lectures given at the University of California, Berkeley, during 1962–1963. Lecture Notes [MS, 146 pp.],

dated Stanford University, September 1963. (From the Preface: 'The preliminary version for the exposition is written for the Seminar in Foundations, Stanford University, Fall 1963.')

665. — (1970) *Constructive validity*, in M. Laudet, D. Lacombe, L. Nolin, and M. Schützenberger (eds.) Symposium on Automatic Demonstration (held at Versailles / France, December 1968), Springer Verlag, Berlin etc. 1970 [*Lecture Notes in Mathematics 125*], pp. 237–275. (On the Curry-Howard Correspondence.)

666. — (1973) *Models for various type-free lambda-calculi*, in Patrick Suppes, Leon Henkin, Athanase Joja, and Grigore C. Moisil (eds.) Logic, Methodology, and Philosophy of Science 4, North-Holland Publishing Company, Amsterdam, etc. 1973 [*Studies in Logic and the Foundations of Mathematics 74*], pp. 157–187.

667. — (1976) *Data types as lattices*, SIAM Journal of Computing 5 (3), [September] 1976, pp. 522–587.

668. — (1980) *Lambda calculus: some models, some philosophy*, in Jon Barwise, H. Jerome Keisler, and Kenneth Kunen (eds.) The Kleene Symposium, North-Holland Publishing Company, Amsterdam etc. 1980 [*Studies in Logic and the Foundations of Mathematics 101*], pp. 223–265.

669. — (1993) *A type-theoretical alternative to* ISWIM, CUCH, OWHY, Theoretical Computer Science 121 (1–2), [6 December] 1993, 411–440. (Paper 'first written in [Oxford, October] 1969 and circulated privately.')

670. Jonathan P. Seldin (1989) *Normalization and excluded middle I*, Studia Logica 48 (2), 1989, pp. 193–217.

671. — (2009) *The logic of Curry and Church*, in Dov M. Gabbay, and John Woods (eds.) Handbook of the History of Logic 5, *Logic from Rusell to Church*, North-Holland Publishing Company, Amsterdam, etc. 2009, pp. 819–873. (Also: Preprint University of Lethbridge, Department of Mathematics and Computer Science, Lethbridge, Alberta (Canada), 9 February 2006 [75 pp.].)

672. Jonathan P. Seldin, and J. Roger Hindley (eds.) (1980) To H. B. Curry, *Essays on Combinatory Logic, Lambda Calculus and Formalism*, Academic Press, London, etc. 1980.

673. Peter Selinger (2002) *The lambda calculus is algebraic*, Journal of Functional Programming 12 (6), 2002, pp. 549–566.

674. Sextus Empiricus [Gentien Hervet] (1718) *Sexti Empirici* Opera Graece et Latine Pyrrhoniarum Institutionum Libri III. *Cum Henr. Stephani Versione Et Notis.* Contra Mathematicos, Sive Disciplinarum Professores, Libri VI. Contra Philosophos Libri V. *Cum Versione Gentiani Herveti &c.*, [*Lipsiae ... Anno M DCC XVIII*], Johann Albert Fabricius, Leipzig 1718.

675. Sextus Empiricus [Robert Gregg Bury] (1933) *Sextus Empiricus* [Opera 1–4] (Translated by Robert Gregg Bury), Harvard University Press, Cambridge MA & William Heinemann, London 1933 [many reprints].

676. Sextus Empiricus [Richard Bett] (2005) Against the Logicians [= Adv. Math. vii–viii] (translated and edited by Richard Bett), Cambridge University Press, Cambridge UK 2005 [*Cambridge Texts in the History of Philosophy*].

677. Sextus Empiricus [Antonio Russo] (1975) [*Sesto Empirico*] Contro i logici [= Adv. Math. vii–viii] (Introduzione, traduzione e note di Antonio Russo), Editori Laterza, Bari & Roma 1975 [*Piccola biblioteca filosofica Laterza 94*].

678. Moh Shaw-Kwei (1950) *The deduction theorems and two new logical systems*, Methodos 2 (5), 1950, pp. 56–75. (Reviewed by Nicholas Rescher, in The Journal of Symbolic Logic 17 (2), 1952, pp. 153–154.)

679. Henry M. Sheffer (1913) *A set of five postulates for Boolean algebras, with application to logical constants*, Transactions of the American Mathematical Society 14, 1913, pp. 481–488.
680. — (1913a) *A set of postulates for Boolean algebra* (abstract), Bulletin of the American Mathematical Society 19, 1913, p. 283.
681. Wilfried Sieg (1988) *Hilbert's Program, sixty years later*, The Journal of Symbolic Logic 53 (2), 1988, pp. 338–348.
682. — (1994) *Eine neue Perspektive für das Hilbertsche Programm*, Dialektik, 1994, pp. 163–180.
683. — (1997) *Step by recursive step: Church's analysis of effective calculability*, Bulletin of Symbolic Logic 3 (2), 1997, pp. 154–180.
684. — (1999) *Hilbert's programs: 1917–1922*, Bulletin of Symbolic Logic 5 (1), 1999, pp. 1–44.
685. — (2000) *Towards finitist proof theory*, in [269], pp. 95–114.
686. — (2009) *Hilbert's proof theory*, in Dov Gabbay and John Woods (eds.) Handbook of the History of Logic 5, *Logic from Russell to Church*, Elsevier, Amsterdam 2009, pp. 127–190.
687. — (2012) *In the shadow of incompleteness: Hilbert and Gentzen*, in Peter Dybjer, Sten Lindström, Erik Palmgren, and Göran Sundholm (eds.) Epistemology versus Ontology, *Essays on the Philosophy and Foundations of Mathematics in Honour of Per Martin-Löf*, Springer, Berlin etc. 2012, pp. 87–127. (Reprinted in [688], pp. 155–192.)
688. — (2013) Hilbert's Programs and Beyond, Oxford University Press, New York etc. 2013.
689. Peter Simons (1996) *The Horizontal*, in Matthias Schirn (ed.) Frege, Importance and Legacy, Walter de Gruyter, Berlin & New York 1996 [*Perspektiven der analytischen Philosophie 13*], pp. 280–300.
690. — (2014) *Jan Łukasiewicz*, The Stanford Encyclopedia of Philosophy, revised 6 June 2014.
691. Jerzy Słupecki (1953) *Über die Regeln des Aussagenkalküls*, Studia Logica 1, 1953, pp. 19–40.
692. Timothy J. Smiley (1955) Natural Systems of Logic, Doctoral Dissertation, University of Cambridge UK [184 pp.]
693. Raymond M. Smullyan (1968) First-Order Logic, Springer Verlag, Berlin etc. 1968 [*Ergebnisse der Mathematik und ihrer Grenzgebiete 68*].
694. Bolesław Sobociński (1932) *Z badań nad teorią dedukcji* [Investigations on the theory of deduction] [Polish], Przegląd Filozoficzny 35, 1932, pp. 171–193 [= Księga Pamiątkowa Koła Filozoficznego Słuchaczy Uniwersytetu Warszawskiego, Warsaw 1932, pp. 3–25]. (Master's Thesis, Warsaw 1932. English translation [by Adrian Rezuş] of the Introduction and § 1 in THIS VOLUME.)
695. — (1935) *Aksjomatyzacja implikacyjno-koniunkcyjnej teorii dedukcji* [Axiomatization of the implicative-conjunctive theory of deduction] [Polish], Przegląd Filozoficzny 38, 1935, pp. 85–95.
696. — (1939) *Z badań nad prototetyką* [Investigations on protothetic] [Polish], Collectanea Logica (Warsaw) 1, 1939, pp. 171–176. (See [697], for an English translation by the author, and [699], for a new translation by Zbigniew Jordan, published in [450], pp. 201–206.) [Reviewed by Alonzo Church, in The Journal of Symbolic Logic 15 (1), 1950, p. 64.]
697. — (1949) *An investigation of protothetic*, Cahiers de l'Institut d'Études Polonaises en Belgique (Brussels) 5 [v + 44 pp.]. (This is a translation of [696] by the author.

See also the new translation by Zbigniew Jordan, without the Introduction [pp. 7–27, giving an account of the ill-fated periodical Collectanea Logica], in [450], pp. 201–206. Also reprinted, with some omissions, in [702], pp. 69–88. Reviewed by Alonzo Church, in The Journal of Symbolic Logic 15, 1960, p. 64.)

698. — (1955–1956) *On well-constructed axiom systems*, Rocznik Polskiego Towarzystwa Naukowego na Obczyinie [Yearbook of the Polish Society of Arts and Sciences Abroad] (London), 6, 1955–1956, pp. 54–65. (Polish translation [*W sprawie dobrze skonstruowanej aksjomatyki*], by Józef Andrzej Stuchliński, in Filozofia Nauki (Warsaw) 12 (1), 2004, pp. 123–136. Reviewed by Hugues Leblanc, in The Journal of Symbolic Logic 22 (4), 1957, pp. 358–359.)

699. — (1967) *An investigation of protothetic*, in [450], pp. 201–206. (A new translation of [696], by Zbigniew Jordan.)

700. Morten Heine Sørensen, and Paweł Urzyczyn (2006) Lectures on the Curry-Howard Isomorphism, Elsevier Science BV, Amsterdam etc., 2006. (Originally, Lecture Notes for a one-semenster graduate / PhD course held at DIKU, Copenhagen, 1998–1999. DIKU [= University of Copenhagen, Department of Computer Science] Report 98/14, May 1998.)

701. Jan T. J. Srzednicki, V. Frederick Rickey, and J. Czelakowski (eds.) (1984) Leśniewski's Systems: Ontology and Mereology, Martinus Nijhoff Publishers [Kluwer Academic Publishers], The Hague, etc. & Ossolineum [The Publishing House of the Polish Academy of Sciences], Wrocław – Warsaw 1984. [*Nijhoff International Philosophy Series 13*]

702. Jan T. J. Srzednicki, and Zbigniew Stachniak (eds.) (1998) Leśniewski's Systems: Protothetic, Kluwer Academic Publishers, Dordrecht 1998. [Reviewed by Arianna Betti, Studia Logica 68 (3), 2001, pp. 401–404.]

703. Werner Stelzner (2002) *Compatibility and relevance: Bolzano and Orlov*, Logic and Logical Philosophy 10, 2002, pp. 137–171.

704. Sören Stenlund (1972) Combinators, λ-Terms, and Proof Theory, D. Reidel Publishing Company, Dordrecht 1972.

705. Walter P. van Stigt (1979) *The rejected parts of Brouwer's dissertation on the Foundations of Mathematics*, Historia Mathematica 6 (4), 1979, pp. 385–404.

706. — (1990) Brouwer's Intuitionism, North-Holland Publishing Company, Amsterdam 1990 [*Studies in the History and Philosophy of Mathematics 2*]. (Reviewed by Craig Smoryński, in The American Mathematical Monthly 101 (8), 1994, pp. 799–802.)

707. Jon R. Stone (2006) The Routledge Book of World Proverbs, Routledge, London 2006.

708. Kristian Støvring (2006) *Extending the extensional lambda calculus with surjective pairing is conservative*, Logical Methods in Computer Science 2 (1), 2006, pp. 1–14. (Also: Research Report BRICS RS-05-35, DAIMI, Department of Computer Science, University of Aarhus, Aarhus [Denmark], November 2005.)

709. Göran Sundholm (1983) *Constructions, proofs and the meanings of logical constants*, Journal of Philosophical Logic 12 (2), 1983, pp. 151–172.

710. — (2000) *When, and why, did Frege read Bolzano?* in Timothy Childers (ed.) LOGICA Yearbook 1999, Filosofia, Prague 2000, pp. 164–174.

711. Alexandru Surdu (1976) Elemente de logică intuiționistă [Elements of intuitionistic logic] [Romanian], Editura Academiei Republicii Socialiste România [The Publishing House of the Romanian Academy], Bucharest 1976. (Historical background. See also [286].)

712. — (1977) Neointuiționismul [The neointuitionism] [Romanian], Editura Academiei Republicii Socialiste România [The Publishing House of the Romanian Academy], Bucharest 1977. (Cf., mainly, Part I, *IV.2 Intuiționismul logic* [Logical intuitionism], pp. 63–92.)

713. Jonathan Swift (1726) Gulliver's Travels, or Travels into Several Remote Nations of the World. In *Four Parts. By Lemuel Gulliver, First a Surgeon, and then a Captain of Several Ships*, Printed for Benj[amin] Motte, London [28 October 1726]. (Cf. Herbert Davis et al. (eds.) The Prose Works of JONATHAN SWIFT [in 14 vols., with an *Index* in vol. 14], Blackwell, Oxford 1939–1974. The Travels are printed in vol. 11: 1941, reprint 1965 etc.)

714. William W. Tait (1967) *Intensional interpretations of functionals of finite type I*, The Journal of Symbolic Logic 32, 1967, pp. 198–212.

715. — (1981) *Finitism*, Journal of Philosophy 78, 1981, pp. 524–556. (See also [717], [718].)

716. — (2001) *Gödel's unpublished papers on foundations of mathematics*, Philosophia Mathematica (3) 9 (1), 2001, pp. 87–126. (A review of [248], 3, *Unpublished Essays and Letters* [1995].)

717. — (2002) *Remarks on finitism*, in Siegfried Sieg, Richard Sommer, and Carolyn Talcott (eds.) Foundations of Mathematics, *Essays in Honor of Solomon Feferman*, A. K. Peters / CRC Press, Wellesley MA & ASL, Urbana IL 2002 [*Lecture Notes in Logic 15*] pp. 410–419. (See also [715], [718].)

718. — (2005) The Provenance of Pure Reason, *Essays in the Philosophy of Mathematics and Its History*, Oxford University Press, Oxford 2005 [*Logic and Computation in Philosophy*].

719. — (2006) *Gödel's interpretation of intuitionism*, Philosophia Mathematica (3) 14 (2), 2006, pp. 208–228.

720. — (2006a) *Gödel's correspondence on proof theory and constructive mathematics*, Philosophia Mathematica (3) 14 (1), 2006, pp. 76–111. (A review of [248], 4–5, *Selected Correspondence* [2002].)

721. — (2010) *Gödel on intuition and on Hilbert's finitism*, in Solomon Feferman, Charles Parsons, and Stephen Simpson (eds.) Kurt Gödel, *Essays for His Centenial*, Cambridge University Press, Cambridge MA & ASL, Urbana IL 2010 [*Lecture Notes in Logic 33*], pp. 88–108.

722. Masako Takahashi (1989) *Parallel reductions in lambda-calculus*, Journal of Symbolic Computation 7 (2), [February] 1989, pp. 113–123.

723. — (1995) *Parallel reductions in lambda-calculus*, Information and Computation 118 (1), [April] 1995, pp. 120–127.

724. Gaisi Takeuti (1975, 1987^2, 2013^R) Proof Theory, North-Holland Publishing Company, Amsterdam etc. 1987 [*Studies in Logic and the Foundations of Mathematics 81*]. (Second, revised edition 1987^2. The second edition has been reprinted by Dover Publications Inc., Mineola NY 2010^R.)

725. — (1982) Work of Paul Bernays and Kurt Gödel, in L. Jonathan Cohen, Jerzy Łoś, Helmut Pfeiffer, and Klaus-Peter Podewski (eds.) Logic, Methodology and Philosophy of Science VI (Hannover 1979), North-Holland Publishing House, Amsterdam etc. 1982 [*Studies in Logic and the Foundations of Mathematic 104*], pp. 77–85.

726. Alfred Tarski (1923) O wyrazie pierwotnym logistyki [*On the primitive term of logistic*] [Polish], PhD Dissertation, University of Warsaw, Warsaw 1923. (Published, partially, in Polish: [727], in French: [728], and [729], as well as in English: [734], Paper I, pp. 1–23.)

727. — (1923a) *O wyrazie pierwotnym logistyki* [*On the primitive term of logistic*] [Polish], Przegląd Filozoficzny 26, 1923, pp. 68–89.
728. — (1923b) *Sur le terme primitif de la logistique*, Fundamenta Mathematicae 4, 1923, pp. 196–200.
729. — (1924) *Sur les 'truth functions' au sens de MM. Russell et Whitehead*, Fundamenta Mathematicae 5, 1924, pp. 59–74.
730. — (1930) *O niektórych podstawowych pojęciach metamatematyki* [*Über einige fundamentalen Begriffe der Metamathematik*], Odbitka ze sprawozdań z posiedzcen Towarzystwa Naukowego Warszawskiego XXIII. 1930. Wydział III. [*Separatum* from Comptes Rendus des séances de la Société des Sciences et des Lettres de Varsovie 23, 1930, Classe III, Varsovie 1930, pp. 22–29]. (Reprinted in [737], 1, pp. 313–320. English translation [*On some fundamental concepts of metamathematics*], by J. H. Woodger, in [734], pp. 30–37.)
731. — (1936) *O ogruntowaniu naukowej semantyki* [On the foundations of scientific semantics] [Polish], Przegląd Filozoficzny 39, 1936, pp. 50–57. (German translation [*Grundlegung der wissenschaftlichen Semantik*], by the author, in Actes du Congrès International de Philosophie Scientifique (Sorbonne, Paris 1935), 3, *Langage et pseudo-problèmes*, Hermann & Cie, Paris 1936 [*Actualités scientifiques et industrielles 390*] pp. 1–8. English translation [*On the establishment of scientific semantics*, by J. H. Woodger, in [734], pp. 401–408.)
732. — (1936a) *O pojęciu wynikania logicznego* [On the concept of logical consequence] [Polish], Przegląd Filozoficzny 39, 1936, pp. 58–68. (German translation [*Über den Begriff der logischen Folgerung*], by the author, in Actes du Congrès International de Philosophie Scientifique (Sorbonne, Paris 1935), 7 *Logique*, Hermann & Cie, Paris 1936 [*Actualités scientifiques et industrielles 394*], pp. 1–11. English translation [*On the concept of logical consequence*], by J. H. Woodger, in [734], pp. 409–420. [Anonymous] Romanian translation [*Cu privire la noțiunea de consecință logică*] (from Polish, German and English), in [750], pp. 283–294. For a new English translation, see [739].)
733. — (1938) *Ein Beitrag zur Axiomatik der Abelschen Gruppen*, Fundamenta Mathematicae 30, 1938, pp. 253–256.
734. — (1956) Logic, Semantics, Metamathematics, *Papers from 1923 to 1938* (translated by J. H. Woodger), Clarendon Press / Oxford University Press, Oxford 1956. (A second revised edition has been issued by John Corcoran (ed.) at the Hackett Publishing Company, Indianapolis IN, in 1983, with an Introduction and Index by the editor.)
735. — (1968) *Equational logic and equational theories of algebras*, in Kurt Schütte (ed.) Contributions to Mathematical Logic, North-Holland Publishing Company, Amsterdam 1968, pp. 275–288.
736. — (1969) *Truth and proof*, Scientific American 220 (6), [June] 1969, pp. 63–70, 75–77.
737. — (1986) Collected Papers, 1–4 (edited by Steven R. Givant, and Ralph N. McKenzie), Birkhäuser Verlag, Basle, Boston & Stuttgart 1986 [*Contemporary Mathematicians*]. (1: 1921–1934, 2: 1935–1944, 3: 1945–1957, 4: 1958–1979.)
738. — (1986a) *What are logical notions?* (edited by John Corcoran), History and Philosophy of Logic 7 (2), 1986, pp. 143–154. (Originally, a lecture delivered, on 16 May 1966, at Bedford College, University of London. Edited from a revised typescript of 1982.)

739. — (2002) *On the concept of following logically* (translated [from Polish and German] by Magda Stroińska and David Hitchcock), History and Philosophy of Logic 23 (3), 2002, pp. 155–196. (A revised translation of [732].)
740. TeReSe (2003) = Marc Bezem, Jan Willem Klop, and Roel de Vrijer (eds.) Term Rewriting Systems, Cambridge University Press, Cambridge UK 2003 [*Cambridge Tracts in Theoretical Computer Science 55*].
741. Christian Thiel (1965) Sinn und Bedetung in der Logik Gottlob Freges, Anton Hain, Meisenheim am Glan 1965. (Originally, PhD Dissertation, Erlangen-Nürnberg 1965. English translation by T. J. Blakeley [Sense and Reference in Frege's Logic], D. Reidel Publishing Company, Dordrecht 1968.)
742. — (1982) *From Leibniz to Frege: mathematical logic between 1679 and 1879*, in L. Jonathan Cohen, Jerzy Łoś, Helmut Pfeiffer, and Klaus-Peter Podewski (eds.) Logic, Methodology and Philosophy of Science VI (Hannover 1979), North-Holland Publishing Company, Amsterdam etc. 1982 [*Studies in Logic and the Foundations of Mathematics 104*], pp. 755–770. (On Saccheri's use of *consequentia mirabilis*, pp. 760–762.)
743. Paul B. Thistlewaite, Michael A. McRobbie, and Robert K. Meyer (1985) *Advanced relevant proving techniques for relevant logics*, Logic Group Research Paper 19, Research School of Social Sciences, Australian National University, Canberra ACT 1985.
744. — (1987) Automated Theorem-Proving in Non-Classical Logics, Pitman Publishing Ltd., London & Wiley, New York 1987.
745. Ivo Thomas (1962) *Finite limitations on Dummett's LC*, Notre Dame Journal of Formal Logic 3 (3), 1962, pp. 170–174. (Cf. [177], [102], [103], [459].)
746. — (1974) *On Meredith's sole positive axiom*, Notre Dame Journal of Formal Logic 15 (3), 1974, p. 477.
747. Pavel Tichý (1988) The Foundations of Frege's Logic, Walter de Gruyter, Berlin & New York 1988 [*Grundlagen der Kommunikation – Foundations of Communication*]. (Cf. Chapter XIII *Inference*, pp. 234–253 (§§ 45–47), esp. § 47. *Sequents*, pp. 248–253. On Gentzen sequents vs Frege's GGA, pp. 252–253.)
748. Richard Tieszen (1992) *What is a proof?*, in [168], pp. 57–76.
749. Mircea Tîrnoveanu (1964) Elemente de logică matematică, 1, *Logica propoziţiilor bivalente* [Elements of mathematical logic 1: Logic of two-valued propositions] [Romanian], Editura didactică şi pedagogică [Educational Publishing House], Bucharest 1964 [517 pp.]. (No other volume in print.)
750. Mircea Tîrnoveanu, and Gheorghe Enescu (eds.) (1966) Logică şi filozofie. *Orientări în logica modernă şi fundamentele matematicii* [Logic and Philosophy: Trends in Modern Logic and the Foundations of Mathematics] [Romanian] (wth a Preface by Grigore C. Moisil), Editura politică [Political Publishing House], Bucharest 1966 [*Materialismul dialectic şi ştiinţele moderne 11*].
751. Peter Trigg (1989) Abstraction in Combinatory Systems Containing B, B', M Phil thesis, University College of Swansea, Wales UK 1989.
752. Peter Trigg, J. Roger Hindley, and Martin W. Bunder (1994) *Combinatory abstraction using B, B' and friends*, Theoretical Computer Science 135 (2), 1994, pp. 405–422. (An expanded version of Peter Trigg's M. Phil. thesis, Swansea 1989 [751].)
753. Anne S. Troelstra (1968) *The scientific work of A. Heyting*, Compositio Mathematica 20, 1968, pp. 3–12. (With 'A Bibliography of A. Heyting', pp. 9–11.)
754. — (1969) Principles of Intuitionism, Springer Verlag, Berlin, etc. [*Lecture Notes in Mathematics 95*].

References 347

755. — (1978) *A. Heyting on the formalization of intuitionistic mathematics*, in E. M. J. Bertin, H. J. M. Bos, and A. W. Grootendorst (eds.) Two Decades of Mathematics in the Netherlands 1920–1940, *A Retrospection on the Occasion of the Bicentennial of the Wiskundig Genootschap*, Mathematisch Centrum [*CWI*], Amsterdam, 1978, pp. 153–175. (Excerpts from [289], pp. 153–162, with a *Commentary* by A. S. Troelstra, pp. 163–175.)

756. — (1980) *The interplay between logic and mathematics: intuitionism*, in Evandro Agazzi (ed.) Modern Logic, *A Survey*, D. Reidel Publishing Company, Dordrecht, etc. 1980, pp. 197–221.

757. — (1981) *Arend Heyting and his contribution to intuitionism*, Nieuw Archief voor Wiskunde (3) 29, 1981, pp. 1–23.

758. — (1983) *Logic in the writings of Brouwer and Heyting*, in Vittorio Michele Abrusci, Ettore Casari, and Massimo Mugnai (eds.) Atti del Convegno Internazionale di Storia della Logica (San Gimignano, 4–8 December 1982), Cooperativa Libreria Universitaria Editrice, Bologna 1983, pp. 193–210.

759. — (1989) *Index of the Heyting Nachlass*, ITLI Prepublication Series X–89–03, University of Amsterdam 1989.

760. — (1990) *On the early history of intuitionistic logic*, in Petio P. Petkov (ed.) Mathematical Logic, *Proceedings of the Heyting '88 Summer School and conference on Mathematical Logic* (13–23 September 1988, Chaika near Varna, Bulgaria), Plenum Press, New York, and London 1990, pp. 3–17.

761. — (1990a) *Remarks on intuitionism and the philosophy of mathematics*, in Giovanni Corsi, and Giovanni Sambin (eds.) Atti del Congresso 'Nuovi Problemi della Logica e della Filosofia della Scienza' (Viareggio, 8–13 January 1990), 2 *Logica*, Cooperativa Libreria Universitaria Editrice, Bologna 1990, pp. 211–228.

762. — (2011) *History of constructivism in the twentieth century*, in Juliette Kennedy, and Roman Kossak (eds.) Set Theory, Arithmetic, and Foundations of Mathematics, *Theorems, Philosophies*, Cambridge University Press, Cambridge UK 2011 [*Lecture Notes in Logic 36*], pp. 150–179. (Also: ITLI Prepublication Series ML-91-05, University of Amsterdam 1991.)

763. Anne S. Troelstra (ed.) (1973) Metamathematical Investigation of Intuitionistic Arithmetic and Analysis, Springer Verlag, Berlin, New York etc. 1973 [*Lecture Notes in Mathematics 344*]. (See also *Corrections and additions to 'Metamathematical investigation of intuitionistic arithmetic and analysis'*, Report Department of Mathematics, University of Amsterdam 1974, and *Corrections to some publications*, Report X-2018-2, Department of Mathematics, University of Amsterdam, 10 December 2018, containing 'Corrections [as of] 22 June 2009'.)

764. Anne S. Troelstra, and Dirk van Dalen (1988) Constructivism in Mathematics, 1–2, North-Holland Publishing Company, Amsterdam [*Studies in Logic and the Foundations of Mathematics 121, 123*].

765. Kazimierz Trzęsicki (2007) *Polish logicians' contribution to the world's informatics*, Studies in Logic, Grammar and Rhetorics 11 (24), 2007, pp. 15–34.

766. Alan N. Turing (1938) *Systems of logic based on ordinals*, Proceedings of the London Mathematical Society (2), 45, 1938, pp. 161–228. (Reprinted in [159], pp. 155–222.)

767. Richard Tursman (1968) *The shortest axioms for the implicational calculus*, Notre Dame Journal of Formal Logic 9 (4), 1968, pp. 351–358.

768. Dolph Ulrich (1999) *New axioms for positive implication*, Bulletin of the Section of Logic (University of Łódź, Polish Academy of Sciences) 28 (1), 1999, pp. 39–42.

769. — (2001) *A legacy recalled and a tradition continued*, Journal of Automated Reasoning 21 (2), [August] 2001, pp. 97–122.
770. — (2007) *Single axioms for C-pure Intuitionism*, @ [7].
771. — (2007a) *A single axiom for C-pure $\mathcal{R}$*, @ [7]. (See [582], [587], [202], [801], [802], [803], etc.)
772. — (2007b) *Single axioms for Meredith's $\mathcal{BCK}$*, @ [7]. (See also [800].)
773. — (2007c) *Single axioms for Meredith's system $\mathcal{BCI}$*, @ [7]. (See also [800].)
774. Paul van Ulsen (2000) E. W. Beth als logicus [E. W. Beth as a logician] [Dutch], PhD Dissertation, University of Amsterdam [UvA] 2000 [*ILLC Dissertation Series 2000-04*]. (Cf. Chapter 9 *Implicatieve systemen* [Implicational systems], and, specifically, § 9.1.2 *Toevoeging van andere operatoren* [Addition of other operators]: *Minimaal-calculus, negatie* [Minimal calculus, negation], pp. 244–246, and *Intuïtionistische propositionele fragmenten* [Intuitionistic propositional fragments], pp. 247–129 [on Kanger's Rule].)
775. — (2001) *Index of the Troelstra Archive* [*Rijksarchief*, Noord-Holland], Amsterdam 2001 [= E-print ILLC X-2003-01].
776. Alasdair Urquhart (1972) *A general theory of implications* (abstract), The Journal of Symbolic Logic 37 (2), 1972, page 443.
777. — (1973) *The Semantics of Entailment*, PhD Dissertation, University of Pittsburgh, 1973 [66 pp.].
778. — (2012) *Henry M. Sheffer and notational relativity*, History and Philosophy of Logic 33 (1), 2012, pp. 33–47.
779. Alasdair Urquhart, and Judy Pelham (1994) *Russellian propositions*, in Dag Prawitz, Brian Skyrms, and Dag Westerståhl (eds.) Logic, Methodology, and Philosophy of science IX: *Proceedings of the Ninth International Congress of Logic, Methodology, and Philosophy of Science* (Uppsala, Sweden, August 7–14, 1991), Elsevier, Amsterdam & New York 1994 [*Studies in Logic and the Foundations of Mathematics 134*] pp. 307–326.
780. Wim Veldman (1981) Investigations in Intuitionistic Hierarchy Theory, Krips Repro, Meppel 1981 [x + 234 pp.]. (PhD Dissertation [mathematics], University of Nijmegen, [20 May] 1981, under the supervision of Johan J. de Jongh.)
781. — (2000) *In memoriam J. J. de Jongh (1915–1999)*, Nieuw Archief voor Wiskunde (5), 1, (1), [March] 2000, pp. 16–17.
782. — (2019) *Treading in Brouwer's footsteps*, Nijmegen, September 2019 [forthcoming]. (On the 'Nijmegen school of intuitionism'.)
783. Sorin Vieru (1968) *Însemnări despre ontologia lui Frege* [Notes on Frege's ontology] [Romanian], Revista de filosofie (Bucharest) 2, 1968. (Reprinted in [786], pp. 191–213.)
784. — (1977) *Studiu introductiv* [Introduction] [Romanian], to: GOTTLOB FREGE Scrieri logico-filosofice [Logisch-philosophische Schriften] [Romanian] 1 (Romanian translations by Sorin Vieru), Editura științifică și enciclopedică [Scientific and Encyclopedic Publishers] Bucharest 1977 [*Clasicii filosofiei universale*], pp. V–LIV (No other volume in print. The Introduction reprinted as: *O privire de ansamblu asupra operei lui Frege* [A global view on the work of Frege], in [786], pp. 5–64.)
785. — (1991) *Constituirea paradigmei moderne a logicii: câteva contribuțiii fregeene* [The birth of the modern paradigm of logic: some Fregean contributions] [Romanian], in Crizantema Joja, Sorin Vieru, Călin Candiescu & Alexandru Surdu (eds.) Orientări contemporane în filosofia logicii [Contemporary trends in the philosophy of logic] [Romanian], Editura științifică [Scientific Publishers], Bucharest

1991. (Reprinted in [786], pp. 191–213. Cf. also 'Frege and the new paradigm for logic', in [228], pp. 272–277.)
786. — (2000) Încercări de logică 2, *Studii fregeene* [Logical essays 2: Frege studies] [Romanian] Editura Paideia, Bucharest 2000.
787. Roel de Vrijer (1987) **Surjective Pairing and Strong Normalization,** *Two Themes in Lambda-calculus*, Eburon, Delft 1987 (PhD Dissertation, University of Amsterdam 1987).
788. Mordchaj Wajsberg (1931) *Aksjomatyzacja trójwartościowego rachunku zdań* [Axiomatissation of the trivalent propositional calculus] [Polish], **Comptes Rendus des séances de la Sociéte des Sciences et des Lettres de Varsovie** (Classe III) **24**, 1931, pp. 126–145. (Originally, PhD Dissertation, University of Warsaw 1931. English translation [*Axiomatization of the three-valued propositional calculus*], by B. Gruchman and Storrs McCall, in [450], pp. 264–284, reprinted in [791], pp. 12–29.)
789. — (1932) *Ein neues Axiom des Aussagenkalküls in der Symbolik von Sheffer*, **Monatsheften für Mathematik und Physik 39** (2), pp. 259–262. (English translation [*A new axiom of propositional calculus in Sheffer's symbols*, in [791], pp. 37–39.)
790. — (1937–1939) *Metalogische Beiträge I, II*, **Wiadomości Matematyczne 43** 1937, pp. 1–38, and **47**, 1939, 119–139. (English translation, by Storrs McCall and P. Woodruff, in [450], pp. 285–334, reprinted in [791], pp. 172–214.)
791. — (1977) **Logical Works** (edited by Stanisław J. Surma), Zakład Narodowy imenia Ossolińskich,Wydawnictwo Polskiej Akademii Nauk [Ossolineum Publishers & The Publishing House of the Polish Academy of Sciences], Wrocław 1977. (Reviewed by Storrs McCall, in **The Journal of Symbolic Logic 48** (3), 1983, pp. 873–874.)
792. Alfred North Whitehead, and Bertrand Russell (1910–1913) **Principia Mathematica, 1–3**, Cambridge University Press, Cambridge UK 1:1910, 2:1912, 3:1913. (Reprint: Merchant Books 2009. Second edition: Cambridge University Press, Cambridge UK 1:1925; 2–3:1927. Russian translation [**Osnovaniĭa matematiki, 1–3**] of the second edition, by Ĭu. N. Radaev, and I. S. Frolov [edited by G. P. Ĭarovoĭ, and Ĭu. N. Radaev], Samarskiĭ Gosudarstvennyĭ Universitet [The Publishing House of the Samara State University], Samara [formerly (1935–1991) Kuĭbyšev] 2005.)
793. — (1962) **Principia Mathematica to *56**, Cambridge University Press, Cambridge UK 1962 [*Cambridge Mathematical Library*]. (Many reprints.)
794. Ludwig Wittgenstein (1956) **Bemerkungen über die Grundlagen der Mathematik / Remarks on the Foundations of Mathematics** (edited by G. H. von Wright, R. Rhees, and G. E. M. Anscombe; translated [into English] by G. E. M. Anscombe), Basil Blackwell, Oxford 1956, 1964^R, 1967^R, 1978^R [revised and reset] (reprints 1989^R, 1994^R, 1998^R, etc.), and MIT Press, Cambridge MA & London 1967^R [bilingual, pbk]. (Also in LUDWIG WITTGENSTEIN **Werkausgabe 6** [German text only], Suhrkamp Verlag, Frankfurt am Main 1984. Lectures of 1937–1944. See also [391].)
795. — (1976) **Lectures on the Foundations of Mathematics**, Cambridge [UK] 1939, (edited by Cora Diamond from the notes of R. G. Bosanquet, Norman Malcolm, Rush Rhees, and Yorick Smythies), The Harvester Press, Ltd., Hassocks, Sussex UK & Cornell University Press, Ithaca NY 1976. (Reviewed by Georg Kreisel, in **Bulletin of the American Mathematical Society 84** (1), 1978, pp. 79–90.)
796. Jan Woleński (1985) *Filozoficzna Szkoła Lwowsko-Warszawska* [The Lvov-Warsaw Philosophical School] [Polish], PWN, Warsaw 1985. (English version: **Logic and**

Philosophy in the Lvov-Warsaw School, Kluwer Academic Publishers, Dordrecht 1989.)

797. — (2015) *Lvov-Warsaw School*, The Standford Encyclopedia of Philosophy, revised 18 October 2015.

798. Larry Wos (2000) Collected Works, 1–2 [1 *Exploring the Power of Automated Reasoning*, 2 *Applying Automated Reasoning to Puzzles, Problems, and Open Questions*] (edited by Gail W. Pieper), World Scientific, Singapore, etc. 2000.

799. — (2001) *Conquering the Meredith single axiom*, Journal of Automated Reasoning 27 (2), 2001, pp. 175–199.

800. — (2006) *The joy of solving a decades-old mystery: success with the BCK and BCI logics*, Mathematics and Computer Science Division, Argonne National Laboratory, Argonne IL, 6 March 2006, revised 6 November 2009 [51 pp.].

801. — (2007) *A startling result, some challenges met, but some still remain: more coping with complex expressions*, Mathematics and Computer Science Division, Argonne National Laboratory, Argonne IL, 2 February 2007, revised 6 December 2009 [45 pp.].

802. — (2007a) *The subformula strategy: coping with complex expressions*, Mathematics and Computer Science Division, Argonne National Laboratory, Argonne IL, 18 December 2007, revised 18 May 2009 [30 pp.].

803. Larry Wos, and Robert [Bob] Veroff (2001) *Challenge problems with condensed detachment*, Computer Science Department, School of Engineering, The University of New Mexico, Albuquerque NM, 6 June 2001 ['last updated on May 8, 2011 by Bob Veroff'].

804. Crispin Wright (1983) Frege's Conception of Numbers as Objects, Aberdeen University Press, Aberdeen 1983 [*Scots Philosophical Monographs 2*]. (See also [252].)

805. Urszula Wybraniec-Skardowska (2009) *Polish logic – some lines from a personal perspective*, Publications of the Institute for Logic, Language and Computation, Amsterdam [July] 2009.

806. Silvia Yu, and Laura Hildago (2004) Alonzo Church Papers (C0948) 1924–1995: *Alonzo Church, 1903-1995, A Finding Aid*, Manuscripts Division, Department of Rare Books and Special Collections, Princeton University Library, Princeton NJ 2004.

807. Richard Zach (2015) *Natural deduction for the Sheffer stroke and Peirce's arrow (and any other truth-functional connective)*, Preprint: University of Calgary, Department of Philosophy, 22 April 2015. Published in Journal of Philosophical Logic 46 (2), [April] 2016, pp. 183–197.

808. Eduard Zeller (1879, 1923^5, 1990^R) Die Philosophie der Griechen in ihrer geschichtlichen Entwicklung 3.1. *Die nacharistotelische Philosophie*, Fues, Leipzig 1879^1, [...], O. R. Reisland, Leipzig 1923^5. (Reprint: Georg Olms, Hildesheim 1990^R.)

809. — (1892, 1892^R, 1962^R) The Stoics, Epicurean and Sceptics (translation by Oswald J. Reichel), Longmans, Green and Co, London 1880, 1892^R. (Revised reprint: Russell and Russell, New York 1962^R. This is a partial translation of [808], i.e., 3.1., *Erster Abschnitt* only.)

Internal cross-references

1978-, 38, 253, 255, 335
1980-, 38, 217–220, 227, 228, 235, 241–243, 335
1981-, viii, 35, 37, 38, 53, 73–75, 79, 102, 140, 229, 230, 246, 251, 253, 288, 303, 304, 335
1981a, 79, 335
1981b, 54, 335
1982-, ix, 38, 55, 120, 121, 137, 140, 190, 192, 206, 222, 227, 228, 230, 235, 237, 241, 243, 246, 248, 251, 253–256, 335
1982a, 54, 243, 248, 254, 256, 335
1982b, 335
1983-, 9, 53, 102, 197, 212, 233, 288, 290, 335
1983- (with H. Barendregt), 102, 212, 298
1986-, 42, 53, 102, 198, 199, 201, 211, 290, 304, 335
1986a, 53, 102, 335
1986b, 336
1987-, 53, 211, 251, 290, 304, 336
1987a, 211, 290, 304, 336
1987b, 212, 290, 304, 336
1987c, 252, 336
1987d, 290, 336
1988-, 141, 288, 336
1988a, 336
1989-, 288, 336
1989a, 288, 336
1989b, 336
1990-, 53, 55, 60, 63, 65, 68, 71, 72, 101, 102, 147–150, 156, 170–172, 177, 194–198, 206, 207, 213, 227, 229, 233, 253, 254, 272, 288, 291, 336
1991-, ix, 53–55, 60, 61, 63, 65–67, 97, 101, 102, 147–150, 156, 160, 170–172, 176, 177, 193–198, 201, 207, 213, 253, 288, 289, 337
1992-, 337
1993-, 53, 55, 60, 63, 65, 97, 101, 102, 147–150, 156, 171, 253, 288, 291, 337
2009-, 48, 54, 82, 129, 133, 137, 140, 147, 172, 184, 288, 291, 337
= Part III, Chapter 9, 103
2010-, 29, 38, 54, 55, 137, 190, 192, 227, 228, 235, 236, 241, 253–256, 288, 337
= Part IV, Chapter 13, 217
2010a, 337
= Part I, Chapter 2, 19
2012-, 179, 256
2015-, 48, 102, 191, 229, 337
2015a, 37, 102, 169, 220, 272, 337
2015b, 254, 337
= Part IV, Chapter 15, 241
2015c, 254, 256, 337
2015d, 334, 337
2016-, 49, 50, 65, 71, 102, 129, 139, 142, 162, 183, 184, 198, 200, 211, 214, 272, 288, 291, 337
= Part III, Chapter 8, 79
2017-, 15, 39, 41, 46–48, 54–56, 58–60, 63, 65, 69, 73, 76, 82, 128, 138, 144, 146, 147, 156, 214, 227, 229, 233, 288, 290, 291, 337

= Part III, Chapter 11, 181
2017a, 39, 41, 43, 46, 47, 54, 65–69, 76,
    147–149, 153, 156, 162, 170, 171,
    179, 214, 227, 229, 231, 254, 272,
    279, 288, 337
= Part II, Chapter 4, 55
2017b, 39, 65, 67–70, 72, 147–149, 153,
    156, 159, 161, 162, 164–166, 168,
    170, 171, 176, 177, 179, 195, 214,
    227, 229, 231, 254, 272, 279, 288,
    291, 337
= Part II, Chapter 3, 33
2017c, 72, 148, 149, 156, 229, 235, 288,
    291, 337
= Part I, Chapter 5, 65
2017d, 72, 212, 214, 227, 229, 272, 288,
    291, 310, 337
2017e, 15, 170, 212, 214, 229, 272, 288,
    291, 337
= Part II, Chapter 5, 69
2017f, 337
= Part I, Chapter 7, 73
2018-, 291, 337
2018a, 245, 248, 253–256, 290, 291, 337
2019-, 38, 54, 237, 253–255, 288, 290, 337
= Part IV, Chapter 14, 227
2019a, 15, 48, 49, 52, 54, 209, 212–214,
    234, 248, 253, 291, 293, 337
= Part III, Chapter 10, 127
2019b, 49, 54, 176, 253, 338
= Part III, Chapter 12, 209

2019c, 290, 338
= Part IV, Chapter 16, 245
2019d, 253, 338
= Part I, Chapter 1, 3
2019e, 338
2020-, 338

Part I, 3
Part I, Chapter 1 = 2019d, 3
Part I, Chapter 2 = 2010a, 19
Part II, 33
Part II, Chapter 3 = 2017b, 33
Part II, Chapter 4 = 2017a, 55
Part II, Chapter 5 = 2017c, 65
Part II, Chapter 6 = 2017e, 69
Part II, Chapter 7 = 2017f, 73
Part III, 79
Part III, Chapter 8 = 2016-, 79
Part III, Chapter 9 = 2009-, 103
Part III, Chapter 10 = 2019a, 127
Part III, Chapter 11 = 2017-, 181
Part III, Chapter 12 = 2019b, 209
Part IV, 217
Part IV, Chapter 13 = 2010-, 217
Part IV, Chapter 14 = 2019-, 227
Part IV, Chapter 15 = 2015b, 241
Part IV, Chapter 16 = 2019c, 245
Part IV, Chapter 17 = Appendix, 257
Part V, 271
Part V, Chapter 18 = Synopsis, 271
Part V, Chapter 19 = Vademecum, 287

Index of subjects

abstraction operator, 17, 34, 36, 150, 187
  dyadic, 35
  see abstractor, 187
abstractor, 34, 35
  $\int$, 40
  Σ, 200, 201
  basic, 36
  definable, 35
  dyadic, 35
    $\int$, 70, 165, 188
    Σ, 70, 201
    split, 70, 188
  linear, 34
  mixed, 34, 188
    $\mathcal{E}$, 176
    χ, 188, 196, 197
    χ (Jaśkowski rule), 196, 197
  monadic, 34–37
    $\Delta \equiv \gamma$ (Rehof-Sørensen), 63
    ϵ, 62, 171
    γ, 59, 171
    $\gamma \equiv \gamma^{\mathrm{G}}$ ('Glivenko γ' in $\boldsymbol{\lambda}_{\beta\eta}$), 148
    $\hat{\gamma}$, 67
    $\hat{\gamma} \equiv \gamma^{\mathrm{K}}$ (Kolmogorov), 66
    λ, 74, 137, 188
    ∂, 56, 171
    ε, 171
    degenerated ϖ, 171
  monotony, 35
    monadic, 74
  n-adic, 188
  polyadic, 34, 188
  pure, 34
  scalar, 34, 35
  strict, 34, 37
  strictly linear, 34
  see also w-operator (abstractor), 275
algebra, v, vi, 186
  abstract, 5, 133, 187
  Boolean, 94, 99, 211
  combinatory, 35
  cylindric, 187
  free, 190
  polyadic, 187
  universal, 150
algebraic, 11, 14, 17, 35, 36
  equation, 11
  expression, 10
  logic, 5, 182, 292
  number, 11
  operator, 34, 150
  structure, 5, 36
    lattice, 5
  variety, 35, 36, 150, 161
α-conversion, 34, 288
analysis
  classical, 6, 44, 53, 129
AUTOMATH, 9, 26, 53, 90, 102, 197, 207, 233, 290, 295, 298, 301, 302, 306, 329, 335, 336

basis
  axiomatic, 245
  non-organic, 229
  normal, 253
  singleton, 38, 228, 229, 235, 241, 253–257

354   Index of subjects

  Łukasiewicz (for material implication), 254
  Łukasiewicz Ł$_{29,1932}$ (for CN-calculus), 237
  axiomatic, 245
  Meredith (for λ), 253
  Meredith (for $\lambda[\to]$), 254
  non-organic, 231, 241, 242, 254
  normal, 245, 253, 255–257
  organic, 254
  Sobociński, 231
  Sobociński S$_{36,1927}$ (for CN-calculus), 232
  Tarski (for CN-calculus, 1925), 227
  Tarski T$_{53,1925}$ (for CN-calculus), 234
Beweisumweg
  see proof-détour, 212
BHK-interpretation, 8, 50, 94, 157, 210, 211, 287
  bibliography, ix

C-monoid, 36, 45, 46, 58, 59, 63, 163, 206, 229
  ~ extended λ-calculus, 59
categorical λ-model
  see λ-model, 63
category theory, 58, 76, 144, 187, 229
  λ-model
    categorical, 59
  λ-model
    see also categorical λ-model, 63
  cartesian pair, 58, 75
  category
    as a monoid, 59
    cartesian closed, 59, 163
  commuting diagram, 134, 144
  folklore, 56, 206
  internal language, 206
  monoid
    cartesian, 58
    cartesian closed, 59, 206
  morphism, 134
characteristica universalis, 4
Church numerals, 230, 241
Church-Rosser
  property, 324
  theorem, 24, 195
Clavius' Law, 61, 62, 147, 238

Clavius' Rule, 61, 65, 67, 122, 193
closure
  operator, 16
  under a rule, 16
combinator, 34
  basic, 36
  non-organic, 229
  organic, 229
  see also w-combinator, 276
completeness, 123
  for classical logic, 94, 95
  for propositional logics, 146, 333
  Post, 17
  Stoic logic, 83
  structural, 333
  see incompleteness theorem, 5
computability
  Turing, viii
condensed detachment, 121, 137, 186, 190, 222, 230, 232, 237, 248, 252, 255–257
  see detachment, 121
  see modus ponens, 121
confluence, 24, 71, 75, 76
  confluence theorem, 195
  see Church-Rosser theorem, 24
consequence relation, 5, 16, 209, 333
  structural, 17, 325, 333
consequentia mirabilis, 6, 51, 61, 62, 65, 67, 122, 140, 147, 155, 156, 189, 193, 231, 238, 296, 298, 303, 320, 328, 330, 340, 346
consistency, 44, 76
  for first-order logics, 5
  for propositional logics, 146
  in Frege's Grundgesetze, 293
continuation, 66, 188
contradiction, 82
  in Ancient Greece, 82
  Stoic
    see Stoic polar opposition, 83
contraposition, 6, 117, 121, 122, 132, 133, 152, 160, 191, 234
CuCh, 242, 300
Curry-Howard Correspondence, vii, ix, 49, 53, 54, 68, 75, 80, 90, 101, 102, 162, 167, 176, 181, 195, 209–212, 218, 219, 225, 287, 290, 304, 313, 323

Index of subjects   355

cut (substitution)
  cut-elimination, 18, 144, 272, 290
  cut-free logic, 15
  cut-rule, 15
cut operator, 35, 36
  application, 38
  associated to a monadic abstractor, 36
  definability, 38
  explicit definability, 38
  see also w-operator (binary), 274

De Morgan 'laws', 6, 87
  ≡ Ockham / De Morgan 'laws', 87
decidability, 3, 156
deduction theorem, 15, 50, 83, 100, 121,
    130, 132, 134, 135, 141, 143,
    178, 184, 191, 194, 204, 219, 314,
    333, 341
  λ, 131, 135, 136, 142, 151, 167, 214
  conditionalisation, 115, 119
  for $\mathcal{R}[\to]$ (Church), 251
  for many-valued logics, 333
definability
  λ-, viii
  explicit, 16, 36
detachment
  see modus ponens, 184
disjunction
  'disjunction property', 167, 176
  classical, 86
  classical (inclusive), 162, 167
  intuitionistic, 76, 167, 176, 188, 210
double negation, 6, 17, 85, 87, 100, 101,
    122, 132, 198, 200
  'laws', 6, 51, 52, 84
  elimination, 52, 131, 132, 142, 143,
    166, 191
  introduction, 131, 132, 142, 143, 166,
    191
  isomorphism, 52
  primitive(s), 212
  proof-normalisation, 132, 159–162, 168,
    169, 198, 206, 212
  Stoic
    built-in at atomic level, 84

entailment
  as a connective, 82
  first-degree, 82
  Stoic
    see Stoic entailment, 82
equational
  class, 35, 36
  theory, 35
ex contradictione quodlibet, 154, 189,
    195
ex falso quodlibet, 51, 62, 65, 67, 146,
    154–156, 159, 167, 189, 231, 248
ex falso rule
  see ex falso quodlibet, 67
explosion
  see ex contradictione quodlibet, 189

Fermat-Wiles theorem, 19, 20, 27
fibration, 50, 96–98, 100, 200
  meaning, 96
  semantic, 96, 98, 200
  syntactic, 98, 100
fold, 35, 36
  ∂, 36
  λ, 36
  as sub-cartesian theory, 39, 40
  cartesian, 39
    β-only, 40
    $\partial\pi$, 40
    $\partial\pi\Pi$, 40
    as cyclic theory, 44
    as n-fold, 45
    has infinitely many inversions, 45
    has infinitely many mono-folds, 45
  combinatory, 35
  cyclic, 44
    as n-fold, 45
    has infinitely many inversions, 45
    has infinitely many mono-folds, 45
  di-fold, 36
    $\rho\partial$, 44
    as cyclic theory, 44
  extensional, 35
  Kolmogorov
    β-only, 39, 65, 66
    $\lambda^K \hat{\gamma}$, 66
    λ^K_β, 39, 65, 66
  mono-fold, 36, 161
    λ, 36
    ∂, 44
    ρ, 44
    basic properties, 36
    cyclic, 45

cyclic $\partial\Delta$, 45
cyclic $\rho\Delta$, 45
folklore, 36
strict, 37
n-fold, 36
see also λ-calculus, 35
formalisation, 8, 9, 14, 53, 137, 287
  AUTOMATH, 9, 53
  Glivenko-Heyting, 8
  proof formalisation, 9, 53, 137, 287
formularian, 3–9, 12–18, 128–130, 132, 140, 143, 144, 146, 155, 156, 158, 163, 167, 174, 176, 182, 209–212, 214, 245, 247, 251, 255, 256, 289
Frege', 293
Frege-Church
  pairs, 37, 40, 60, 65, 220–224, 230, 234–237, 241
  sequences, 37, 40, 65, 222, 230, 241
functionality theory, 271

Galois theory, 20
generator, 245
  non-organic, 242
    Halleck (shortest known), 242
    Rosser, 242
  organic
    Meredith (shortest known), 243
    normal, 257
  see singleton basis, 38
generic arity (gen-arity), 188
Gentzen's Hauptsatz, 5, 18
Gentzenisation, 144, 272

Heraclitus' metaphysics of conflict, 82
Heyting Arithmetic **HA**, 210
Hilbert's epsilon, 41, 49, 147, 172, 177–179
homomorphism
  of formulas, 17
HoTT, 76
  book, 290
Howard-Hindley-Scott λ-calculus
  see weak λ-calculus $\lambda_\mathbf{o}$, 35

identity
  'law', 6
    Stoic, 82
incompleteness theorem, 5
interlingua, 4

inversion, 44–46, 52, 58, 59, 101, 163
  $\bar{\tau}$, 161
  Δ, 161
  curry-ing/un-curry-ing, 45
  qua degenerated mono-fold, 161
iteration, 58, 205
  infinite iteration effect, 59

jump (non-local cotrol), 188

Kanger's Rule, 250, 274, 299, 348
Kolmogorov
  $\lambda_\beta^K\hat{\gamma}$-calculus $\lambda_\beta\gamma_\mathbf{K}$, 66
  $\lambda_\beta^K\hat{\gamma}$-calculus $\lambda_\beta\gamma_\mathbf{K} \equiv \lambda^K\hat{\gamma}$, 61, 66
  translation, 61, 66

λ-abstractor
  see monadic abstractor (λ), 121
λ-calculus (historical)
  $\lambda_\beta\delta$I (Church, strict), 53
  pure λ_βI (Church, strict), 75
  pure $\lambda_{\beta\eta}$ (Russell), 220
  pure λ_β (Frege), 220
λ-calculus, 288
  λ, 56
  $\lambda \equiv \lambda_{\beta\eta}$ (pure), 36
  $\lambda\epsilon \preceq \lambda\partial$, 62
  $\lambda\epsilon\varpi \sim \lambda\partial$, 62
  $\lambda\mu$, 63, 101, 102
  $\lambda\pi$ (extended), 36, 40, 47, 56
  $\lambda\partial$, 47, 56, 195
  $\lambda\partial\mathbf{p}$, 56
  $\lambda\partial\mathbf{p}_{\beta\eta} \equiv \lambda\partial\mathbf{p}$, 47
  $\lambda\partial\mathbf{p}_\beta$, 47, 56
  $\lambda\partial\mathbf{p}_\zeta$, 47
  λ_Δ (Rehof-Sørensen), 63, 335
  $\partial\int$, 47
  $\partial\int_\zeta$, 47
  $\partial\lambda^*$, 47
  $\partial\lambda^*_\zeta$, 47
  pure λ_β, 73
  pure $\lambda_{\beta\eta}$, 74
  pure $\lambda_\mathbf{o}$, 74
  scalar
    $\lambda\partial\Lambda$, 199
    $\partial\lambda^*\Lambda^*$, 201
    $\partial\lambda^*\Sigma$, 200
  type-free, 219
  typed, 53

inferential $\lambda\gamma$-calculus, 52, 55
   $\lambda^K\hat{\gamma} \preceq \lambda$, 66
   $\lambda^K\hat{\gamma}$ (Kolmogorov), 66
   $\lambda\gamma$, 43, 55, 59
   $\lambda\gamma \preceq \lambda\partial\mathbf{p}$, 55
   $\lambda\gamma\tau$, 43, 59
   $\lambda\gamma_\mathbf{G} \preceq \lambda$, 60
   $\lambda\gamma_\mathbf{G}$ (Glivenko), 60
   $\lambda\bar{\gamma} \sim \lambda\mu$ (Parigot), 43
   $\lambda_\beta\gamma_\mathbf{K} \equiv \lambda^K\hat{\gamma}$, 60, 61
   $\lambda_\beta\gamma_\mathbf{K}$ [τ^K-extension] $\preceq \lambda$, 61
   scalar (first-order), 53
inferential $\lambda\varepsilon$-calculus
   $\lambda\varepsilon$ (Curry), 43
inferential $\lambda\varepsilon\varpi$-calculus
   $\lambda\varepsilon\varpi$, 43
   $\lambda\varepsilon\varpi \sim \lambda\gamma$, 61
   $\lambda^K\hat{\varepsilon}\hat{\varpi}$ (Kolmogorov), 67
inferential CBV λ-calculus
   $\lambda_\mathbf{v}$-CS (Felleisen), 288
weak λ-calculus
   $\lambda_\mathbf{o}$ (Howard-Hindley-Scott), 74
λ-model
   categorical, 59, 206
   type-free, 59
   see also type-free λ-calculus, 59
lingua characteristica
   see characteristica universalis, 129
[LNC]
   see non-contradiction ('law'), 6
logic
   $\mathcal{A}$, 252, 256
   $\mathcal{BCI}$, 140, 249, 252, 255, 317, 348, 350
   $\mathcal{BCI}\Delta^\gamma$, 252, 253
   $\mathcal{BCK}$, 140, 255, 348, 350
   $\mathcal{H}$, 174, 175
   $\mathcal{H}_3$ (Hiż), 16, 17
   $\mathcal{LL}$ (Girard), 252
   $\mathcal{M}$, 174, 175
   $\mathcal{R}$, 250, 251, 255
   'Lattice $\mathcal{R}$', 88, 251
   abelian, 252, 327, 328
   affine (Girard) $\sim$ linear, 53
   algebraic, 5, 182, 292
   Aristotelian, 13
   Brouwerian, 8, 156, 176, 287
   Chrysippean, 12, 81, 89, 102
   Chrysippean $\equiv$ classical, 13, 94
   classical, 8, 10, 13, 15–18, 28, 33, 48–50, 52, 53, 68, 69, 71, 72, 80, 81, 84, 87, 88, 94, 97, 99–102, 127, 131, 132, 140, 141, 146, 153, 156–158, 161, 162, 167, 180, 181, 183, 185, 188, 193–195, 197, 200, 207, 209–214, 245, 247, 252, 255
      first-order, 48, 70, 128, 130, 180, 181, 219
      Principle of Omniscience, 213
      propositional, 55, 61, 66–68, 94, 99, 122, 144, 148, 149, 162, 163, 165, 166, 194, 195, 204, 211, 227, 229, 231, 234, 236–238, 248–250, 255, 257
      propositionally quantified, 146, 181, 198, 248
      second-order, 44, 48, 181
      subsystems, 253, 256
   classical (extended), 188
      $\equiv$ propositionally quantified, 146
   classical (pure implicational), 252, 254
   combinatory, 9, 191, 224
      Rosser, 74
      Schönfinkel-Curry, 74
   comparative, 252
   cut-free, 15
   Frege-logic
      higher-order, 288
   Fregean
      see classical logic, 155
   Hiż see $\mathcal{H}_3$ (Hiż), 182
   illative, 271
   inferential, 51
   intuitionistic, 5, 8–11, 15, 50, 53, 55, 61, 65, 66, 75, 94, 97, 140, 141, 145, 155–157, 159, 160, 167, 173, 174, 176, 188, 209–214, 239, 247, 287, 289, 290, 292, 293
      Glivenko-Heyting, 8, 97, 155, 160, 173, 174, 210, 212
      Kolmogorov, 11
   linear, 53
   linear (Girard), 84, 88, 251, 287
   linear (Girard) $\sim$ strictly linear, 53
   logic of strict negation $\mathcal{D}$ (Curry)
      see inferential $\lambda\varepsilon$-calculus, 67
   many-valued, 251

358 Index of subjects

mathematical, 4, 9, 12, 14, 27, 80, 111, 130, 183
Minimalkalkül, 8, 52, 65, 66, 133, 140, 145, 146, 151, 154, 155, 157, 160, 167, 173, 174, 209, 210, 212, 213, 289, 290
modal, 157
  (Lewis), 53, 287
  (Lewis) $S4$, 251
multi-valued, 16
ordinal, 37, 53, 207
paraconsistent, 82
Post-complete, 17
post-Fregean, 80, 83, 210
quantifier-free, 17
Regellogik, 15, 82, 102, 115, 123, 128, 129, 141, 181, 183, 221, 291, 342
relevance, 53, 88, 207, 251, 287
  non-distributive, 88, 247, 251
  purely implicational, 251
Satzlogik, 14, 82, 115, 128, 130, 181, 183, 221
Stoic, 12, 80, 81, 83, 84, 91, 102, 183, 288, 293, 299, 308, 312, 318, 326, 329
  as entailment logic, 82
strict negation $\mathcal{D}$ (Curry), 156
strictly linear, 53
substructural, 88, 140, 251
traditional, 12
logical form, 81, 83, 84
logistique, 4, 6

model theory, 17, 292
  Plotkin-Scott Graph Model, 74
  Scott D_∞-model, 74, 75, 271
  Tarskian, 5, 18, 27, 108, 292, 311
modus
  Aristotelian syllogistic, 6
  BARBARA, 119
  ponendo ponens, 128, 129, 131-133, 135-137, 142, 143, 150, 151, 167, 211, 250
  tollendo tollens, 128, 131, 132, 142, 144, 152-154, 157, 158
modus ponens, 14, 26, 28, 29, 121, 178, 184-187, 189-194, 201, 204, 217, 227, 228, 234, 243, 245, 252, 253, 257

  as valid entailment, 187
  see modus ponendo ponens, 128
modus tollens
  see modus tollendo tollens, 128
monoid, 10, 37, 161
  λ-calculus, 58
  composition, 58
  identity, 58
  see also C-monoid, 58
monotony, 34, 35, 38-43, 45, 46, 56, 59, 66, 73, 148, 150, 151, 161, 163, 164, 166, 168, 173, 177-179, 195, 199-202, 205
  failure, 35
  quasi-monotony, 60

natural deduction, 15, 49, 55, 80, 97, 101, 102, 115, 131, 138, 143, 144, 149, 150, 181-183, 186, 189, 193, 194, 197, 203, 206, 209, 229, 233, 257, 290, 291, 332
negation, 211
  'surface negation', 140, 197
  Chrysippean, 83
  classical, 11, 48, 68, 70, 87, 98, 130, 146, 183, 188, 196, 211, 214, 234, 250, 257, 258, 266, 288
  ≡ self-incompatibility, 48
  ≡ Stoic polar opposition, 87
  inferential, 48, 146, 196, 252, 287
  intuitionistic, 6, 8, 210
  Stoic, 83, 85
  three-valued (Hiż), 17
  see also double negation, 6
non-contradiction
  'law', 6, 17, 50, 82, 146, 150-160, 162, 204, 209, 278
non-local control, 188
normalisation, 71, 72, 76, 272
notational relativity, 255

opposition
  square (Aristotle), 82
  see Stoic logical conflict, 82
  see Stoic polar opposition, 82
OTTER, 28, 252, 257, 309

pairing, 39, 46, 49, 205
  combinator (pure), 27, 217
  extensional, 39, 40, 163

Index of subjects 359

≡ surjective, 39
non-extensional, 46, 70, 166, 168, 204
non-surjective ≡ non-extensional, 70
scalar, 41, 177, 179
  ≡ mixed, 41
  see Hilbert's epsilon, 41
surjective, 39, 40, 56, 60, 74, 101, 163, 164, 168, 229
undefinable in λ, 40
Peano Arithmetic, 5, 11, 25, 53, 145
Peirce's Law, 62, 68, 147, 191, 211, 238, 248–250
Peirce's Rule, 62, 250
problem
  interpretation, 10, 11
  mathematical, 11
proof-complexity, 212, 255
  and notational relativity, 255
  formularian, 255
proof-conversion, 101, 196, 201
proof-détour, 69, 71, 150, 160, 196, 212, 213
  elimination, 71, 132, 196
    double negation, 132
  intuitionistic pseudo-détours, 213
proof-isomorphism, 101, 154, 157, 186, 201, 212, 213
  double negation, 128
proof-reduction, 150, 196, 197
  rules, 197
  see proof-détour elimination, 196
protothetic, 188, 218, 342
provability, 3, 8, 15, 128, 132, 141–144, 151, 153–155, 158, 160–163, 168, 174, 182, 185, 203, 209, 213, 248, 250

quantifier, 33, 188
  ∃ (intuitionistic), 76, 201
  ∃ (classical), 174
  ∃ (minimal), 176
  ∀ (classical)
    no-counterexample interpretation, 41
  ∀ (classical first-order), 48
  'existence property', 176
  'truth-value quantifier', 146, 189
  classical, 15, 179, 181, 182, 189, 198, 249
  first-order, 34, 63, 174, 179, 181, 182, 187, 198

Fregean, 115, 172
higher-order, 174
propositional, 14, 34, 145, 181, 182, 188, 189, 198, 248, 249
pseudo-existence, 174
  Kolmogorov, 174, 176
rules, 15, 146, 201
second-order, 34, 63, 181, 182
Stoic, 81, 102

recursive function, 319
  general, viii
redex, 71, 212
reductio ad absurdum, 6, 49, 52, 71, 89, 99, 155, 156, 159, 171, 178, 194–196, 204, 232
  in Eleats, 82
  inferential, 50, 62, 67, 68, 148, 150, 155, 196
  restricted, 51
reduction, 74, 212
  see also proof-reduction, 74
refutation, 15, 17, 18, 82, 89, 211
  defective (Aristotle), 82
  Socrates' maieutic, 82
  sophistic, 82
  Stoic, 83
rejection, 18, 82, 89
  Stoic
    see Stoic refutation, 83
  see refutation, 82
replacement, 17
rule
  admissibility, 16, 17, 141, 182
  admissibility (cut), 140, 141
  case-analysis $\mathcal{E}$, 188
  derivability, 15–17, 132, 141–143, 151, 154, 155, 158, 160, 162, 168, 182, 250
  see also witness operator, 273
  see w-operator, 273

scalar, 33, 69
  equality $\doteq$, 33
  operator, 33
    see also w-operator (scalar), 275
  pairing, 177
  substitution, 33
  theory, 177
set theory

360   Index of subjects

ordered pair (Kuratowski), 150
ordinals, 150
Zermelo-Fraenkel **ZFC**, 25
Zermelo-Fraenkel **ZF**, 150
Sheffer's functor, 72, 85, 102, 166, 203
   see Stoic **nand**, 85
shuffling, 116, 137
solvability
   local, 255
Sophists, 82
   use of contradictions, 82
Stoic
   connector, 84, 85, 87, 94
      ≡ binary connective (sundesmos), 84
      additive (disjunction-like), 86
      improper polar pair ≡ sub-polar, 86
      improper polars **iff** vs **xor**, 85
      multiary links, 85
      multiplicative (conjunction-like), 86
      polar pair, 85
      proper polar pair, 86
      proper polars **if** vs **more**, 85
      proper polars **nand** vs **and**, 85
      proper polars **or** vs **nor**, 85
      proper polars **since** vs **less**, 85
   dilution, 91
   double negation
      atomic, 84
      complex, 87
   entailment
      valid, 88
   fallacies of relevance, 91
   indemonstrable
      T1–T2, 97
      T1–T3, 86
      T3, 85
      T4–T5, 86, 97
   logic, 87
   meaning postulates, 87
   negation
      atomic, 84
      complex, 87
      meaning postulates, 87
      nand, 85
      iff, 86
      xor, 86
Stoic basic concepts, 81
   entailment (argument, syllogism), 81
   polar opposition (logical conflict), 81

proposition (axiōma), 81
rule of inference (thema), 81
Stoic contradiction
   see Stoic polar opposition, 83
Stoic entailment, 82
   ≡ argument, logos, sullogismos, 82
   assumptions, 82
   assumptions ≡ lēmmata, 82
   conclusion, 82
      ≡ sumperasma ≡ epiphora, 82
   dyadic, 82
   invalid, 83
   monadic, 82
      ≡ monolemmatic argument, 82
   projection, 82
   valid, 83
      primitive ≡ Stoic indemonstrable, 83
   validity criterion, 83
Stoic logical conflict
   see Stoic polar opposition, 83
Stoic polar opposition, 82, 85, 87, 94
   atomic, 82, 84
   complex, 85
Stoic proposition
   as abstract entity, 81
   atomic, 83
      popositional constant, 84
      proper atoms, 84
   complex, 81, 84, 86
   complex: additive, 86
   complex: multiplicative, 86
   simple
      ≡ atomic, 81
Stoics, 13, 80, 87, 293, 299, 329
structure
   algebraic, 5
   applicative, 35
   combinatory, 36
subformula property, 5
substitution, 17, 33, 35, 113, 120, 139, 146, 172, 173, 190, 224, 237, 288
   as endomorphism, 190
   atomic, 139
   cut-rule (Hertz-Gentzen), 141
   global, 140, 141
   instance, 6, 17, 26, 120, 137, 147, 150, 154, 155, 159, 252
   local, 141
   most general, 190, 237, 257

Index of subjects   361

rule, 15, 26, 27, 29, 115, 121, 131, 189, 192, 217, 227, 228, 234, 245, 253
   scalar, 33, 44
   simultaneous, 33, 113
   uniform, 120, 137, 190, 245
subterm, 34, 288
syllogistic
   Aristotelian, 6, 13, 147, 309, 326
   Stoic, 309
syntactic identity ≡, 34
system
   'System F' ≡ $\lambda\Lambda$ (Girard), 199
   axiomatic, 14, 128, 189
   L-system, 18, 49, 87, 197, 290
   N-system
      see natural deduction, 290
   sequent system, 87, 102, 115, 144, 197, 209, 290

Tarski's Claim, 27, 29, 38, 217–220, 227, 228, 236
tautology
   non-organic, 229
   organic, 229
term, 33
   closed, 34, 288
   individual, 34
   linear ($\mathcal{BCK}$), 53
   open, 34, 288
   strictly linear ($\mathcal{BCI}$), 53
tertium non datur, 6, 65, 68, 156, 320
theory
   combinatory, 36
   equational, 36
topology
   set-theoretic, 5, 16
transcendental
   number, 11
translation
   Gödel, 8
   Gentzen, 8
   Kolmogorov
      see Kolmogorov translation, 8
truth-table, 87, 211, 262
   semantics, 146

unification, 190, 221
   theorem, 237
unreliable
   principle of logic, 6, 50

rule of inference, 50

valuation, 211, 262
variable
   bound, 34, 288
   free, 34, 288

w-combinator
   (Ω), 278
   (Δ^γ), 278
   (Δ^∂), 278
   (∇^γ), 278
   (∇^∂), 278
   (τ), 278
   ($\dot{\varpi}$), 278
   ($\bar{\tau}$), 278
   ($Ł^\gamma$), 277
   ($Ł^\partial$), 277
   (A), 276
   (BI), 276
   (B), 276
   (CB$^\lambda$), 276
   (CB$^\partial$), 277
   (CB$^\rho$), 277
   (CB^M), 277
   (CB), 276
   (CC), 277
   (CI), 277
   (CO$^\gamma$), 277
   (CO$^\partial$), 277
   (CS), 277
   (C$^\lambda$), 276
   (C$^\partial$), 276
   (C$^\rho$), 276
   (C^M), 276
   (C), 276
   (D), 277
   (E$^\gamma$), 277
   (E$^\partial$), 277
   (I), 277
   (KI), 277
   (K), 277
   (M$_A$), 277
   (M$'_A$), 277
   (O$^\gamma$), 278
   (O$^\partial$), 278
   (P$^\gamma$), 278
   (P$^\partial$), 278
   (S$_M$), 278
   (S), 278

362   Index of subjects

(T), 278
(W), 278
w-operator (abstractor)
  $(\bar{\gamma})$, 275
  $(\chi)_1$, 275
  $(\chi)_2$, 275
  (ϵ), 275
  (γ), 275
  $(\int)$, 275
  $(\lambda^{\triangleright})$, 275
  $(\lambda^{\triangleleft})$, 275
  $(\wp^{\gamma})$, 275
  $(\wp^{\partial})$, 275
  (∂), 275
  (ρ), 275
  (ε), 275
  see also abstractor, 275
w-operator (singulary)
  (Δ), 273
  (∇), 273
  $(\pi_1)_1$, 273
  $(\pi_1)_2$, 273
  $(\pi_2)_1$, 273
  $(\pi_2)_2$, 273
  (τ), 273
  (ϖ), 273
  $(\bar{\tau})$, 274
  $(\mathsf{Cl}_{\vdash})$, 273
w-operator (binary)
  $(\cdot)$, 274
  $(\circ)$, 275
  $(\triangleright)$, 274
  $(\triangleright^{+})$, 274
  $(\star)$, 274
  $(\ltimes)$, 274
  $(\pi)_1$, 274
  $(\pi)_2$, 275
  $(\square)$, 275
  $(\breve{\triangleright})$, 274
  $(\triangleleft)$, 274
  $(\mathsf{D}_{\vdash})$, 274
  see also cut operator, 274
w-operator (scalar)
  $(\blacktriangleright)$, 275
  (Λ), 276
  (Π_1), 276
  (Π_2), 276
  (Π), 276
  (Σ), 276
witness
  atom, 33
  combinator
    see w-combinator, 276
  operator, 33
    see w-operator, 273
  term, 33, 69
  variable, 33
witness theory, vi–ix, 9, 38, 47–49, 51–55, 67, 71, 90, 100, 101, 128, 144, 148, 162, 165, 166, 178, 186, 195, 197, 199, 200, 207, 210, 253, 257, 278, 288

Index of names

Abrusci, V. M., 141, 295
  (ed.), 347
Ackermann, W., 177, 292, 296
Adĭan, S. I., 296, 330
  (ed.), 317
Adachi, T., 63, 206
Adler, M.
  (ed.), 296
Agazzi, E.
  (ed.), 322, 347
D'Agostino, M., 102, 296
Alexander of Aphrodisias, 97, 101, 296
Algra, K. A.
  (ed.), 298, 299
Allegri, L., 339
  (tr.), 340
Alves, E. H., 296, 330
Amsler, R.
  (ed.), 319
Anderson, A. R., 75, 82, 88, 189, 197, 207, 247, 250, 251, 290, 296, 304, 313
Anellis, I. H., 146, 184, 296, 339
Angelelli, I., 293, 296, 340
Angstl, H.
  (ed.), 304
Anscombe, G. E. M
  (ed.), 349
  (tr.), 349
Apollonius Discolus, 86
Arana, A., 307
Aristophanes, 110
Aristotle, 12, 13, 20, 23, 24, 82, 97, 122, 147, 153, 175, 184, 198

von Arnim, H. F.
  (ed.), 80, 101, 296, 334
Artemov, S. N., 292, 297
Asher, L., 14
  (tr.), 323
Asimov, I., 114
Aspray, W.
  (ed.), 329
Atherton, C., 81, 297
van Atten, M., ix, 7, 13, 156, 213, 287, 289, 297, 306, 315
  (ed.), 289, 297, 300
  (tr.), 289, 297, 300
van Atten, V.
  (ed.), 297
  (tr.), 297
Auerbach, D. A., 306
Ausiello, G., 242, 297
Austin, J. L.
  (ed.), 309
Avigad, J., 13, 315, 340

Baader, F., 69, 70, 212, 297
Bacon, F., 21
Badesa, C., 289, 326
Baldassari, M., 101, 297
Barendregt, H., viii, ix, 9, 25, 36–38, 40, 53, 54, 69, 70, 73, 79, 102, 169, 191, 207, 212, 219–221, 223, 224, 229, 230, 241–243, 288, 297, 298, 304, 311, 335, 337
  (ed.), 337
Barnes, J., 91, 97, 102, 298
  (ed.), 299, 301

Barnes, Jennifer
  (tr.), 301
Barnett, D. I.
  (ed.), 324
Barwise, J.
  (ed.), 305, 341
Bauer-Mengelberg, S.
  (tr.), 309, 312, 316, 340
Bäuerle, R.
  (ed.), 308
Bažanov [Bazhanov], V. A., 298, 330
Becker, O., 80, 297, 298
Behmann, H., 224
Bejan, C.
  (tr.), 296
Bellissima, F., 67, 102, 122, 147, 193, 298
Belnap Jr., N., ix, 26, 54, 75, 82, 88, 146, 189, 197, 207, 247, 250, 251, 290, 296, 304, 313
Benacerraf, P.
  (ed.), 301, 312, 321
van Bethem Jutting, L. S., 9, 298
Bernard de Chartres, 12
Bernays, P., 90, 141, 180, 183, 250, 292, 299, 310, 314, 317, 344
Bernini, S.
  (ed.), 303
Bertin, E. M. J.
  (ed.), 347
Beth, E. W., 93, 102, 211, 250, 291, 299, 348
  (ed.), 301
Bett, R.
  (ed.), 341
  (tr.), 341
Betti, A., 343
Bezem, M., 69, 70, 212
  (ed.), 337, 346
Béziau, J.-Y.
  (ed.), 314, 323
Bidwell, G.
  (ed.), 321
Bimbó, K., 290, 299
Birkhoff, G., vii, 35, 150, 161
Blakeley, T. J.
  (tr.), 346
Blanchette, P. A., 293, 299
Bobzien, S., 81, 83, 86, 91, 102, 211, 291, 293, 299

Boccardini, G.
  (tr.), 339
Bocheński, J. M., 299
Bogoiu, F.
  (tr.), 296
Böhm, C., ix, 54, 220, 241–243, 300
Boldini, P.
  (ed.), 297
Bolzano, B., 293, 343
Bonotto, C.
  (ed.), 303
Boole, G., 5, 80
Boolos, G., 293, 299
Borel, E., 6
Borkowski, L., 182, 300
  (ed.), 326
Boron, L. F.
  (tr.), 330
Bos, H. M. J.
  (ed.), 347
Bosanquet, R. G.
  (ed.), 349
Bourbaki, N., 23
Bourdeau, M.
  (ed.), 297
  (tr.), 297, 300
Bouveresse, J.
  (tr.), 300
Boyle, R., 13
Braithwaite, R. B.
  (ed.), 335
van Briemen, J. W., 220, 222, 242
Britzelmayr, W., 304
Brouwer, L. E. J., v, 5–9, 13, 18, 50, 97, 147, 156, 157, 210, 211, 213, 287–290, 297, 300, 301, 305, 306, 309, 311, 315, 326, 343, 347, 348
Brown, B.
  (ed.), 332
de Bruin, P. J., 336
de Bruijn, N. G., ix, 9, 15, 26, 53–55, 102, 187, 191, 197, 210, 212, 271, 290, 298, 301, 302, 306, 329
Brunschwig, J., 102, 301, 307
Bułecka, A., 258
Bull, R. A., 247, 251, 302, 327, 346
Bunder, M., 251, 302, 305, 327
Burge, T., ix
  (ed.), 304, 338

Burgess, J. P., 299
Burnyeat, M. F.
  (ed.), 301
Bury, R. G.
  (tr.), 92, 341
Bynum, T. W.
  (ed.), 309

Caianiello, E. R.
  (ed.), 300
Candiescu, C., 293, 302
  (ed.), 348
Cardano, G., 6, 61, 65, 67, 102, 122, 155, 189, 193, 231, 302, 303
Cardone, F., 220, 288, 303
Carnielli, W. A.
  (ed.), 328
Casari, E., 252, 303
  (ed.), 347
Cattanei, E., 339
Charpentier, É
  (ed.), 305
Childers, T.
  (ed.), 343
Chrysippus, viii, 13, 17, 18, 49, 50, 71, 79–83, 85, 87, 91, 92, 94, 96, 99, 102, 122, 128, 179, 183, 184, 198, 200, 201, 211, 288, 291, 293, 307
Church, A., ix, 24, 27, 37, 38, 40, 48, 53, 60, 65, 70, 73, 75, 99, 128, 191, 195, 207, 220–224, 227, 230, 234–237, 241, 251, 271, 281, 288, 303, 304, 325–327, 330, 331, 338, 341–343, 350
Clavius, C., 122, 147, 193, 304
Cohen, L. J.
  (ed.), 344, 346
Cohen, R. S.
  (ed.), 313, 315, 332
Coniglio, M. E.
  (ed.), 328
Copeland, B. J., 248
  (ed.), 302, 305
Coquand, T., 66, 289, 305
Corcoran, J., 102, 305, 333
  (ed.), 305, 312, 329, 345
Cori, R.
  (ed.), 308
Corsi, G.
  (ed.), 303, 347
da Costa, N. C. A., 328
Creţia, P., 81
Crossley, J. N.
  (tr.), 340
Curien, P.-L., 63, 206, 305
Curry, H. B., vii, ix, 16, 33, 37, 49, 53–55, 61, 67–69, 74, 75, 80, 90, 101, 102, 140, 156, 162, 167, 181, 182, 192, 195, 209–211, 218–220, 224, 225, 229–231, 242, 247, 271, 287, 288, 290, 304–306, 308, 313, 323, 327, 341
Czelakowski, J.
  (ed.), 343

van Daalen, D. T., 290, 306
van Dalen, D., ix, 7, 53, 54, 157, 207, 219, 289, 290, 297, 306, 329, 335, 337, 347
  (ed.), 300
Davis, H.
  (ed.), 344
Davis, M.
  (ed.), 306, 311
Dawson, J. W., 304
Dean, J.
  (ed.), 320
Dekkers, W., 298
Democritus, 13
Demopoulos, W., 293
  (ed.), 306
Descartes, R., 12, 24
Detlefsen, M., 5, 290, 292, 306, 307
  (ed.), 306, 307, 346
Dezani, M., 242, 300
Diamond, C.
  (ed.), 349
Dodgson, C. L., 80, 307
Dončenko [Donchenko], V. V.
  (ed.), 330
Döring, K.
  (ed.), 308, 328
Došen, K., 307, 330, 333
  (ed.), 307
Dragalin, A. G., 289, 292, 297, 307, 320
Dreben, B., 314
Dresden, A.
  (tr.), 300

Duba, B., 66, 308
Dubikajtis, L., 182, 307
Dufour, R.
  (ed.), 101, 307
  (tr.), 101, 307
Dummett, M., 247, 289, 293, 302, 307, 327, 346
Dunn, M., 75, 82, 189, 197, 207, 251, 290, 296, 304, 313
Duns Scotus, 193
Durfee, G., 75, 307
Dybjer P.
  (ed.), 342
Dyckhoff, R., 340

Ebert, P. A.
  (ed.), 310
  (tr.), 310
Ebert, T.
  (ed.), 308, 328
Eco, U., 12, 307
Edwards, P., 299
Egli, U., 102, 179, 307, 308
  (ed.), 308
Einstein, A., 7, 12, 141
Eltz, R.
  (ed.), 304
Eminescu, M., 104, 308
Enderton, H.
  (ed.), 304, 338
Enescu, G.
  (ed.), 316, 325, 340, 345, 346
Engel, F.
  (tr.), 339
Engeler, E., 75, 210, 308
Etchemendy, J., 16, 308
Euclid, 67, 80, 193, 305, 339
Ewald, W., 292, 308, 312, 317
  (ed.), 308, 316, 317, 333
  (tr.), 308, 316

Feferman, S., 9, 177, 308, 344
  (ed.), 312, 344
Felleisen, M., 63, 66, 288, 308
Fenstad, J.-E.
  (ed.), 311, 321
Fermat, P., 19, 27
Ferro, R.
  (ed.), 303
Feys, R., 192, 305, 310

Ficken, F. A., 303
Fisch, M. H.
  (ed.), 332
Fischer, M. J., 66
Fitch, F. B., 53, 97, 102, 197, 251, 290, 308
Fitelson, B., ix, 28, 252, 255, 256, 309, 348
Fløistad, G.
  (ed.), 334
Føllesdal, D.
  (tr.), 316
Forder, H. G., 252, 309
Fortenbaugh, W. W.
  (ed.), 298
Fraenkel, A., 306
Franchella, M., 289, 290, 309
Frede, M., 102, 309
Frege, G., viii, 3–5, 13–15, 17, 36, 37, 40, 52, 60, 65, 80, 82, 83, 87, 89–91, 99, 102, 104–121, 126–131, 133, 136–147, 152, 153, 155, 158, 172–174, 181–184, 191, 196, 210, 211, 213, 214, 220–224, 230, 234–237, 241, 246, 288, 291, 293, 296, 299, 302, 306, 307, 309, 313, 315, 319, 322, 323, 326, 331, 332, 340, 342, 343, 346, 348–350
Frege, K. A., 127
Freud, S., 109
Freudenthal, H., 289, 310, 315
Friedman, D. P., 308
Frigerio, P.
  (ed.), 339
Frolov, I. S.
  (tr.), 349
Furth, M.
  (tr.), 310

Gödel, K., 5, 8, 11–13, 18
Gabbay, D. M., 290, 314, 323, 334, 337
  (ed.), 303, 306, 313, 330, 341, 342
Gagnon, L. S., 102, 310
Gaillard, F.
  (tr.), 317
Galenus, 86, 101, 310
Galileo, 12, 24
Galois, É., 20
Gandon, S.

(tr.), 306
Gandy, R. O.
  (ed.), 322
Garrido, A.
  (ed.), 182, 291, 310
Gauss, C. F., 339
Gentzen, G., viii, 5, 8, 15, 18, 49, 55,
    75, 79, 80, 87, 90–92, 97, 102,
    108, 115, 119, 132, 140, 141, 160,
    182–184, 196–198, 211, 212, 229,
    272, 288–290, 292, 310, 313, 332,
    340, 342
Georgacarakos, G. N., 289, 310
George, A.
  (ed.), 313
Geuvers, H., 302, 311, 329
  (ed.), 302, 329
Ghilezan, S., 102, 298
Gillies, D., 293, 311, 349
  (ed.), 311
Ginsti, J.-P., 340
Girard, J.-Y., 16, 53, 84, 87, 102, 199,
    251, 289, 290, 304, 311
Givant, S. R.
  (ed.), 345
Glivenko, V. I., 5, 6, 8, 18, 55, 60, 65, 68,
    150, 156, 160, 210, 213, 287–289,
    291, 293, 311, 320, 323
Glymour, C.
  (ed.), 297
Goble, L.
  (ed.), 306
Gödel, K., 184, 289, 292, 306, 311, 312,
    315, 322, 344
Goldfarb, W.
  (ed.), 314
  (tr.), 314
Gonseth, F., 299
Gonzalez, W. J., 289, 312
Goodstein, R. L., 330
Gould, J. B., 102, 312
Grassmann, H., 127, 293
Grassmann, R., 127
Grattan-Guinness, I., 4, 5, 7, 219, 289,
    293, 312, 314
Griffin, T. G., 66, 288
Griss, G. F. C., 309
Grootendorst, A. W.
  (ed.), 347

Gross, W., 242, 300
Gruchman, B.
  (tr.), 349
Guenthner, F.
  (ed.), 306
Guillaume, E.
  (tr.), 317
Guillaume, M.
  (tr.), 317

Haaparanta, L.
  (ed.), 326
Hale, B., 293, 313
Halleck, J., ix, 28, 29, 224, 239, 242, 252,
    313
Haller, R., 313
Hallett, M., 292, 313
  (ed.), 317
Halmos, P., 187
Halsted, G. B.
  (ed.), 339
  (tr.), 316, 333, 339, 340
Hartshorne, C.
  (ed.), 332
Hartwig, D., 313, 321
Hazen, A. P., 102, 144, 182, 289, 290, 313
Heck Jr., R. G., 293, 313
Hegel, G. W. F., 12, 14, 313
Heidegger, M., 12, 21, 22
van Heijenoort, J., 129, 293, 313–315
  (ed.), 309, 311, 314–316, 320, 340
  (tr.), 312, 320
Heinzmann, G.
  (ed.), 297
Helman, G., 75, 304, 313
Hendricks, V. F.
  (ed.), 306, 313, 342
Henkin, L.
  (ed.), 321, 334, 341
Heraclitus, 82, 153
Herbrand, J., 184, 219, 314
Hertz, P., 90, 92, 102, 115, 119, 141, 183,
    295, 299, 314, 315, 323
Hertz, R., 141
Hervet, G.
  (tr.), 92, 341
Hesseling, D. E., 8, 12–14, 18, 315
von Heusinger, K.
  (ed.), 308

Index of names

Heyting, A., 8, 18, 97, 156, 157, 160, 210, 213, 287, 289, 290, 292, 297, 309, 310, 315, 316, 320, 323, 329, 330, 346, 347
  (ed.), 301
  (tr.), 300
Hidalgo, L., 303, 350
Hieb, R., 308
Hilbert, D., viii, 3–5, 7, 13, 41, 49, 90, 130, 141, 155, 172, 174, 177, 179, 180, 183, 191, 246, 247, 291, 292, 299, 306, 308, 313, 316, 317, 321, 326, 339, 342
Hindley, J. R., ix, 35, 53, 54, 74, 102, 121, 137, 140, 190, 191, 207, 220, 229, 230, 248, 288, 298, 303–306, 317, 327
  (ed.), 210, 302, 317, 323, 327, 341
Hintikka, J., 93, 102, 211, 291
  (ed.), 322, 334
Hitchcock, D.
  (tr.), 346
Hiż, H., 15–17, 182, 317
Hogemann, F.
  (ed.), 313
Hollack, J.
  (ed.), 301
Holmström-Hintikka, G.
  (ed.), 319
Hooke, R., 12, 329
van der Horst, P. W.
  (ed.), 298
Houser, N., 331
  (ed.), 332
Howard, W., vii, ix, 9, 35, 49, 53–55, 68, 74, 75, 80, 90, 101, 102, 162, 167, 181, 195, 209–211, 218, 219, 225, 287, 290, 304, 313, 317, 323
Huang, Y., 251, 317
Hülser, K., 101
  (ed.), 318
  (tr.), 318
Humberstone, L., ix, 248, 290, 318
Husserl, E., 9, 12
Hyland, J. M. E.
  (ed.), 322

Idel'son, A. V.
  (ed.), 310

  (tr.), 310
Ierodiakonou, K., 83, 92, 102, 318
  (ed.), 298
Imai, Y., 251, 318
Indrzejczak, A., 102, 182, 318
Inwood, B.
  (ed.), 299
Iorgulescu, A., ix, 251, 318, 328
Iséki, K., 251, 318

Jadacki, J. J., 182, 188, 318, 324
Jaeschke, W.
  (ed.), 313
Jaquette, D.
  (ed.), 297
Jaśkowski, S., viii, 15, 55, 80, 97, 99, 102, 115, 119, 138, 143, 144, 181–185, 188, 189, 194–199, 206, 207, 221, 225, 229, 251, 253, 288, 290, 291, 307, 313, 318, 321, 327, 330, 332, 337
Jeffrey, R.
  (ed.), 299
Jennings, R. E., 102, 330
Johansson, I., 8, 52, 65, 66, 210, 213, 247, 289, 329
John of Salisbury, 12, 318
Johnson, D. L.
  (tr.), 320
Joja, A.
  (ed.), 321, 334, 341
Joja, C.
  (ed.), 348
de Jongh, J. J., 348
Jordan, Z.
  (tr.), 342, 343
Jørgensen, K. F. (ed.), 306, 313, 342
Jun, Y. B., 327

Kabakov, F. A.
  (ed.), 330
Kalman, J., 121, 137, 190, 230, 248, 319
Kamareddine, F.
  (ed.), 305
  (tr.), 305
Kanger, S., 250, 290, 299, 319
Kant, I., 12, 14, 24, 308
Kaplansky, I.
  (ed.), 326
Kapur, D.

(ed.), 326
Käsbauer, M.
  (ed.), 323
Keisler, H. J.
  (ed.), 305, 341
Kennedy, H. C., 5, 319
Kennedy, J.
  (ed.), 347
Kenny, A., 293, 319
Kieffer, J. S., 86
  (tr.), 310
Kijania-Placek, K.
  (ed.), 318
Kino, A.
  (ed.), 323
Kitcher, P.
  (ed.), 329
Klau, C., 122, 304
  see Clavius, C., 67, 193
Kleene, S. C., 9, 38, 73, 255, 288–290, 305, 319, 341
Klein, F.
  (ed.), 6
Klement, K. A., 220, 319
Klibansky, R., 12, 319
Kloesel, J. W.
  (ed.), 332
Klop, J. W., 69, 70, 212
  (ed.), 337, 346
Kneale, M., 122, 320
Kneale, W. C., 122, 147, 320
Knuth, D. E., 111
Kohlbecker, E., 308
Kolmogorov, A. N., 8, 10, 11, 18, 39, 55, 60, 61, 65–68, 72, 76, 149, 155–157, 174, 209, 210, 213, 287–289, 291, 293, 305, 306, 310, 315, 320, 332, 337
Komori, Y., 53, 251, 320, 330
Kossak, R.
  (ed.), 347
Kotarbiński, T., 266, 320, 321
Kotas, J., 182, 321
Kowalski, E.
  (tr.), 305
Koymans, K., 63, 75, 206, 321
Kratzer, A.
  (ed.), 316

Kreisel, G., ix, 4, 9, 41, 177, 180, 289, 292, 298, 301, 308, 310, 313, 321, 322, 349
Kreiser, L., 4, 293, 322
Krivine, J.-L., 288, 323
Kronecker, L., 6, 7
Kueker, D. W.
  (ed.), 302
Kuiper, J. J. C., 289, 290, 323
Kundera, M., 13, 14, 323
Kunen, K.
  (ed.), 305, 341
Kuratowski, K., 150
Kušner [Kushner], B. A., 297
  (ed.), 330
von Kutschera, F., 102, 115, 293, 323
  (ed.), 323
Kuzičev [Kuzichev], A. S., 298

Lacombe, D.
  (ed.), 302
Ladrière, J., 310
Lafont, Y.
  (tr.), 311
Lambek, J., viii, 36, 45, 46, 58, 59, 63, 161, 163, 206, 229, 323
Landau, E., 9, 298, 323
Landau, H. G., 303
Landini, G., 128, 323
Largeault, J.
  (ed.), 300
  (tr.), 300
Läuchli, H., 55, 323
Laudet, M.
  (ed.), 302
Leblanc, H., 250, 290, 299, 323, 332, 343
Legris, J., 102, 141, 314, 323
  (tr.), 314, 323
Leibniz, G. W., 4, 12, 17, 21, 22, 35, 127, 129, 150, 346
Lesne, A.
  (ed.), 305
Leśniewski, S., ix, 5, 126, 130, 146, 184, 188, 196, 217–219, 224, 225, 229, 253, 266, 267, 291, 318, 324, 343
Lessing, D., 114
Leucippus, 13
Lewis, A, C.
  (ed.), 339

Lewis, C. I., 287
Lindenbaum, A., 146, 324
Lindström, S.
 (ed.), 319, 342
Loffredo D'Ottaviano, I. M.
 (ed.), 328
Longo, G., 321
López-Escobar, E. G. K., 71, 324
Lorenz, K.
 (ed.), 322
Lorenzen, P., 16, 140, 324
Łoś, J., 17, 325
 (ed.), 344, 346
Luciano, E., 4, 325
Łukasiewicz, J., viii, 5, 12, 13, 16, 29,
 53, 55, 61, 80, 82, 83, 97, 99,
 102, 121, 123, 126, 130, 143, 144,
 146, 181–185, 188–191, 193, 194,
 196, 198–200, 203, 206, 207, 217,
 218, 223–225, 227–232, 234, 237,
 238, 243, 248, 250–252, 254, 255,
 257–259, 261, 262, 266, 267, 291,
 292, 325, 326, 333, 337, 342
 (tr.), 325

MacLane, S., 55, 210, 308, 326
Majer, U.
 (ed.), 317
Mal'cev [Maltsev], A. I., vii, 35, 150, 161,
 293
Malcolm, N.
 (ed.), 349
Malmnäs, P.-E., 334
Mancosu, P., 289, 326
 (ed.), 301, 311, 315, 320, 326
Mangione, C.
 (ed.), 303
Mannoury, G., 301
Mansfeld, J., 298
 (ed.), 299
Marion, M.
 (ed.), 313
Markov, A. A., 293, 316
Martin-Löf, P., 53, 76, 290, 342
Martino, E., 290, 326
Mates, B., 80, 102, 326
Mathieu, M.
 (ed.), 332
Max, I.
 (ed.), 322
Mazzucchelli, L.
 (ed.), 300
McCall, S., 349
 (ed.), 291, 318, 324–327, 342, 343, 349
 (tr.), 324–326, 349
McCune, W., 137, 326
McEwan, A.
 (tr.), 307
McGarry, D. G.
 (tr.), 318
McKenzie, R. N.
 (ed.), 345
McLarty, C., 210, 326
McRobbie, M. A., 346
Mendelsohn, R. L., 293, 326
Mendelson, E.
 (tr.), 307, 311
Méndez, J. M., 289, 327
Meng, J., 251, 327
Menne, A.
 (ed.), 304
Meredith, C. A., 53, 55, 120, 121,
 137, 186, 190, 210, 218–220, 222,
 228, 230, 237, 239, 243, 246–248,
 251–257, 292, 302, 327, 334, 346,
 348, 350
Meredith, D., ix, 121, 137, 186, 190, 191,
 218, 228, 248, 304, 305, 317, 327
Merton, R. K., 12, 327
Meyer, R. K., ix, 54, 252, 327, 328, 346
Meyer, W. F.
 (ed.), 334
Mignucci, M., 102, 328
Mihăilescu, E. G., 291, 328
Minc [Mints], G. E., 289, 290, 292, 297,
 298, 328
 (ed.), 310
Miralbell, I., 122, 147, 328
Moerdijk, I., 307
Moisil, G. C., ix, 5, 290–292, 328, 346
 (ed.), 321, 334, 341
van der Molen, T., 289, 318, 329
Molk, J., 6, 7, 329, 334
 (ed.), 329
Moore, Edward C.
 (ed.), 332
Moore, George Edward, 335
Moore, Gregory H., 289, 329

(ed.), 339
Moraux, P., 298
Morgenstern, U.
  (tr.), 320, 340
Mostowski, A., 330
Mueller, I., 102, 329
Mugnai, M.
  (ed.), 347
Müller, T.
  (ed.), 106, 310
Murthy, C. R., 63, 66, 288

Nagel, E.
  (ed.), 316, 321
Nagornyĭ, N. M.
  (tr.), 317
Nasieniewski, M., 91, 102, 329
Nederpelt, R., 290, 302, 311, 329
  (ed.), 302, 329
Negri, S., 75, 290, 329
von Neumann, J., 224, 292, 329
Newton, I., 12, 329
Nicod, J., 203, 259, 267, 325, 330, 332
Niekus, J.
  (ed.), 316
Niekus, N. H., 290, 330
Niiniluoto, I.
  (ed.), 322, 334
Nikol'skiĭ [Nikolski], K.
  (ed.), 305
Nikol'skiĭ, S. M.
  (ed.), 320
Nipkow, T., 69, 70, 212, 297
Nogina, E., 297
Nolin, L.
  (ed.), 302
Norman, J.
  (ed.), 327
Novak, I. L., 338
Novikov, P. S., 84, 102, 293, 296, 330
Nowak, M., 17, 182, 317, 330
Nuchelmans, G., 122, 147, 330

O'Connor, J. J., 4, 330
O'Neil, M. P.
  (tr.), 324
O'Toole, R. R., 102, 330
Obtułowicz, A., 63, 206
Odifreddi, P.
  (ed.), 298, 308, 322

de Oliveira, A. G., 290, 334, 337
Ono, H., 251, 330
Orlov, I. E., 298, 330, 343
Orłowska, E., 182, 330

Padmanabhan, R., 292, 331
Pagli, P., 67, 102, 122, 147, 193, 298
Palmgren, E.
  (ed.), 342
Palombi, F.
  (ed.), 338
Panaitescu-Perpessicius, D.
  (ed.), 308
Parigot, M., 44, 63, 101, 102, 272, 288, 331
Parmenides, 153
Parsons, C., 312
  (ed.), 299, 344
Peano, G., 3–5, 9, 11, 13, 14, 25, 53, 80, 128, 130, 145, 182, 319, 331
Peckhaus, V., 292, 331
  (ed.), 297
Pedersen, S. A.
  (ed.), 306, 313, 342
Peirce, C. S., 5, 14, 55, 62, 68, 80, 82, 85, 87, 98, 100–102, 115, 130, 145–148, 150, 155, 166, 181, 183, 184, 188–191, 196, 203, 211, 238, 248–250, 296, 331, 332, 350
Pelham, J., 17, 128, 332, 348
Pelletier, F. J., 102, 144, 182, 290, 313, 332
  (ed.), 313
Peremans, W., 306
Petkov, P. P.
  (ed.), 347
Pieczkowski, A., 182, 321
Piętka, D., 182, 332
Pieper, G. W.
  (ed.), 350
Pier, J.-P.
  (ed.), 311
Pinder, C.
  (ed.), 321
Plato, 12, 23, 81, 110
von Plato, J., 55, 75, 102, 212, 289, 290, 310, 329, 332, 337
  (ed.), 289, 332
Plisko, V. E., 289, 320, 332

372   Index of names

(ed.), 320
(tr.), 320
Plotkin, G., 40, 66, 74, 75
Podewski, K.-P.
 (ed.), 344, 346
van der Poel, W. L., ix, 220, 222, 242,
 243, 254, 332
Pogorzelski, W. A., 17, 146, 184, 333
Pohle, H., 108
Pohlers, W., 289, 333
Poincaré, H., 5–8, 306, 333
Porte, J., 17, 184, 219, 314, 333
Pos, H., 301
 (ed.), 301
Post, E., 87
Pourmahdian, M.
 (ed.), 181
von Prantl, C., 80, 97, 102, 333
Prawitz, D., 55, 65, 68, 72, 75, 80, 87, 97,
 101, 102, 150, 170, 196, 201, 210,
 288, 290, 334
 (ed.), 348
Presburger, M., 267
Price, R., 102, 334
Priest, G.
 (ed.), 327
Pringsheim, A., 7, 329, 334
Prior, A. N., 137, 218, 220, 222, 243, 248,
 252, 254–256, 290, 292, 305, 327,
 334
Putnam, H.
 (ed.), 301, 312, 321

de Queiroz, R. G. B., 290, 334, 337
Quine, W. V. O., 108, 111, 126, 334

Radaev, I. N.
 (ed.), 349
 (tr.), 349
Radice, R., 101
 (ed.), 334
Ramsey, F. P., 7, 335
Razborov, A.
 (ed.), 308
Read, S., 102, 335
Rehof, N. J., 63, 101, 102, 288, 335
Reichel, O. J.
 (tr.), 350
Reid, C., 7, 335
Rescher, N., 341

Resnik, M. D., 130, 293, 335
Restall, G., 251, 335
Retoré, C., 311
Révész, Géza, 301
Révész, György E., 75
Reynolds, J., 53, 199
Rhees, R.
 (ed.), 349
Rickey, V. F., ix, 228, 239, 291, 324, 338
 (ed.), 343
van Riemsdijk, H., 290, 330
 (ed.), 316
De Risi, V.
 (ed.), 339
Rist, J. M.
 (ed.)., 329
Rivenc, F.
 (tr.), 300
Robertson, E. F., 4, 330
Robinson, J. A., 102, 190, 221, 237
Robles, G., 289
Rocha, A.
 (tr.), 311
Roddenberry, G., 114
Rosenbloom, P. C., 338
Rossberg, M.
 (ed.), 310
 (tr.), 310
Rosser, J. B., 24, 73, 74, 192, 195, 220,
 222, 241, 242, 251, 288, 304
Rota, G.-C., 9, 338
de Rouilhan, P.
 (tr.), 300
Rudeanu, S., 292, 328, 331
Rudner, R. S.
 (ed.), 304
Ruitenburg, W., 289
Ruja, H., 303
Runia, D. T.
 (ed.), 298
Russell, B., 4, 5, 9, 13, 80, 82, 83, 87,
 107, 109, 128, 130, 141, 145–147,
 155, 172, 181, 184, 188, 196, 220,
 288, 289, 296, 304, 306, 312, 319,
 326, 331, 332, 338, 339, 341, 342,
 348, 349
Russo, A., 341
 (tr.), 341

Saarinen, E.

(ed.), 322, 334
Saaty T. L.
   (ed.), 321
Saccheri, G., 296, 339, 340, 346
Sadegh Daghighi, A.
   (ed.), 181
Sambin, G.
   (ed.), 303, 347
Sayre, K. M., 291, 340
Scanlan, M., 102, 340
Schellinx, H., 340
Schilpp, P. A.
   (ed.), 312
Schirn, M., 293
   (ed.), 323, 342
Schlimm, D.
   (ed.), 317
Schmetterer, L.
   (ed.), 322
Schmidt, A.
   (ed.), 334
Schneider, H. J., 302
Schoenman, R.
   (ed.), 321
Schönfinkel, M., 74, 128, 220, 224, 242, 288, 340
Schröder, B.
   (ed.), 106, 310
Schroeder-Heister, P., 115, 141, 340
Schröter, K., 245, 340
Schurn, M.
   (ed.), 340
Schütte, K., 84, 87, 102, 340
   (ed.), 334, 345
Schützenberger, M.
   (ed.), 302
Schwichtenberg, H., 75, 157, 289, 290, 340
Scott, D. S., viii, ix, 16, 35, 40, 73–75, 163, 220, 271, 289, 307, 340, 341
Scott, P. J., viii, 36, 45, 58, 59, 63, 161, 163, 206, 229, 323
Seiller, T., 311
Seldin, J. P., ix, 61, 75, 102, 191, 229, 288, 306, 317, 341
   (ed.), 210, 302, 305, 317, 323, 327, 341
   (tr.), 305
Selinger, P., 36, 161, 341
Serebrianikov, O. F.

(tr.), 300
Sextus Empiricus, 92, 101, 122, 341
Shakespeare, W., 124
Shaw-Kwei, M., 330, 341
Sheffer, H. M., 85, 102, 166, 203, 268, 331, 332, 342, 348–350
Sheffler, I.
   (ed.), 304
Shofield, M.
   (ed.), 299, 301
Sieckmann, J.
   (ed.), 302
Sieg, W., 180, 292, 304, 317, 342
   (ed.), 317, 344
Simons, P., 17, 342
Simpson, S.
   (ed.), 344
Singleton, R. R., 303
Sjösted, C. E.
   (ed.), 339
Skyrms, B.
   (ed.), 348
Slaney, J. K., 252, 327, 328
Sliwinski, R.
   (ed.), 319
Słupecki, J., 182, 300, 342
Smiley, T. J., 251, 342
Smoryński, C., 343
Smullyan, R. M., 93, 342
Smythies, Y.
   (ed.), 349
Sobociński, B., 5, 188, 217, 218, 228–232, 234, 237, 241, 243, 253, 254, 256–258, 261–266, 291, 292, 324, 337, 338, 342, 343
Socrates, 82, 107, 109–111, 114
Sommer, R.
   (ed.), 344
Sørensen, M. H., 44, 49, 55, 63, 101, 102, 195, 219, 272, 288, 290, 335, 343
Specker, E. P., 141
Spon, K.
   (ed.), 303
Spurr, J., ix
Srzednicki, J. T., 184, 188
   (ed.), 324, 343
Stöckler, M.
   (ed.), 298
Stachniak, Z., 184, 188

374   Index of names

(ed.), 343
Stäckel, P.
  (tr.), 339
Statman, R., 298
von Stechow, A.
  (ed.), 308
Steele, T. B.
  (ed.), 300
Steenrod, N. E., 303
Stelzner, W., 330, 343
  (ed.), 298, 322
Stenlund, S., 219, 290, 343
Stern, J.
  (ed.), 333
van Stigt, W. P., 289, 290, 300, 343
  (tr.), 301
Stone, J. R., 104, 343
Støvring, K., viii, 46, 74, 75, 163, 343
Stroińska, M.
  (tr.), 346
Stuchliński, J. A.
  (tr.), 343
Stuhlmann-Laeiz, R.
  (ed.), 106, 310
Sundholm, G., 7, 290, 293, 297, 343
  (ed.), 297, 300, 342
  (tr.), 297, 300
Suppes, P.
  (ed.), 316, 321, 334, 341
Surdu, A., 8, 343, 344
  (ed.), 348
Surma, S.
  (ed.), 324, 349
Suszko, R., 17, 325
Sweer, J. H., 303
Swift, J., 108, 344
Sylvan, R.
  (ed.), 327
Szabo, M.
  (ed.), 310
  (tr.), 310

Tait, W., 180, 289, 292, 344
Takahashi, M., 344
Takeuti, G., 289, 344
Talcott, C.
  (ed.), 344
Tanaka, S., 251, 318
Tarski, A., ix, 5, 13, 15, 16, 18, 27–29, 38, 55, 83, 108, 126, 130, 131, 146, 183, 184, 187, 188, 191, 192, 196, 206, 207, 217–221, 224, 225, 227–229, 231, 232, 234, 236–238, 241, 243, 248–250, 252–255, 257, 259, 261, 262, 267, 289, 291, 292, 326, 331, 344–346
  (ed.), 316, 321
  (tr.), 345
Taub, A. H.
  (ed.), 329
Taylor, P.
  (tr.), 311
Teichmann, W.
  (tr.), 324
Terence [Publius Terentius Afer], 104
Theodosius of Bithynia, 193, 305
Theophrastus of Eresus, 298
Thiel, C., 293, 346
Thiele, H.-J.
  (ed.), 334
Thierry de Chartres, 12
Thistlewaite, P. B., 346
Thomas, I., 220, 247, 327, 346
Thurnbull, H. W.
  (ed.), 329
Tichý, P., 115, 130, 293, 346
Tieszen, R., 290, 346
Tixomirov [Tikhomirov], V. M.
  (ed.), 320
Tinbergen, J., 301
Tîrnoveanu, M., 291, 333, 346
  (ed.), 316, 325, 340, 345, 346
Todorčević, S. S.
  (ed.), 308
Tóth, I., 339
Trigg, P., 305, 327
Troelstra, A. S., ix, 7, 9, 54, 75, 157, 289, 290, 297, 312, 315, 330, 340, 346–348
  (ed.), 289, 290, 316, 317
Trzęsicki, K., 347
Turing, A., viii, 20, 288, 303, 347
Tursman, R., 326, 347
Tymieniecka, A.-T.
  (ed.), 299

Ulrich, D., 29, 137, 223, 243, 248, 254–256, 295, 347, 348
van Ulsen, P., 102, 211, 250, 290, 348

Urquhart, A., 102, 128, 251, 348
  (ed.), 339
Urzyczyn, P., 44, 49, 55, 63, 101, 102, 195, 219, 272, 288, 290, 343

Vacca, G., 4, 325, 330
Valentini, S.
  (ed.), 303
Vandervelde, G.
  (tr.), 340
Veldman, W., 290, 348
Veroff, R., 230, 248, 255, 256, 348, 350
Vieru, S., 293, 348, 349
  (ed.), 348
  (tr.), 309, 320, 348
de Vrijer, R., 69, 70, 74, 75, 212, 349
  (ed.), 302, 329, 346

van der Waals Jr., J., 301
Wajsberg, M., 5, 189, 221, 229, 243, 250, 253, 254, 257, 268, 290, 291, 327, 349
Wallies, M.
  (ed.), 296
Wang, H., 320
Wartofsky, M. W.
  (ed.), 315
Weber, H.
  (tr.), 325
Wei, W.
  (ed.), 297
Weil, A.
  (tr.), 316
Weingartner, P.
  (ed.), 322
Weiss, P.
  (ed.), 332
Westerståhl, D.
  (ed.), 297, 348
Weyl, H., 7
Weyl, J., 303
Whitehead, A. N., 172, 181, 349
Wiesner, J.
  (ed.), 298
Wiles, A., 19, 20, 27
Wilhelmy, A.
  (ed.), 304
Winston Newson, M. F.
  (tr.), 316
Wittgenstein, L., 4, 7, 87, 289, 290, 322, 349
Wojtasiewicz, O. A.
  (tr.), 300, 321, 325, 326
Wojtylak, P., 146, 333
Woleński, J., 182, 349, 350
  (ed.), 318, 320, 333
Wood, C.
  (ed.), 308
Woodger, J. H.
  (tr.), 345
Woodruff, P.
  (tr.), 325, 349
Woods, J.
  (ed.), 303, 313, 330, 332, 341, 342
Woodward, B.
  (tr.), 316
Woś, L., 28, 137, 230, 248, 255, 256, 326, 348, 350
Wright, C., 293, 313, 350
  (ed.), 310
  (tr.), 310
von Wright, G. H., 322
  (ed.), 334, 349
Wrightson, G.
  (ed.), 302
Wybraniec-Skardowska, U., 350
  (ed.), 182, 291, 310

Xenophon, 110

Yankov, V. A.
  (tr.), 316
Yarovoy [Ĭarovoĭ], G. P.
  (ed.), 349
Yokouchi, H., 63
Yu, S., 303, 350

Zach, R., 102, 289, 326, 350
Zanardo, A.
  (ed.), 303
Zeller, E., 80, 97, 102, 350

www.ingramcontent.com/pod-product-compliance
Lightning Source LLC
Chambersburg PA
CBHW050120170426
43197CB00011B/1661